Sea Surface Temperature Retrievals from Remote Sensing

Sea Surface Temperature Retrievals from Remote Sensing

Special Issue Editors

Jorge Vazquez-Cuervo
Xiaofeng Li

MDPI • Basel • Beijing • Wuhan • Barcelona • Belgrade

MDPI

Special Issue Editors
Jorge Vazquez-Cuervo
Jet Propulsion Laboratory/
California Institute of Technology
USA

Xiaofeng Li
National Environmental Satellite,
Data, and Information Service
National Oceanic and Atmospheric Administration
USA

Editorial Office
MDPI
St. Alban-Anlage 66
4052 Basel, Switzerland

This is a reprint of articles from the Special Issue published online in the open access journal *Remote Sensing* (ISSN 2072-4292) from 2017 to 2018 (available at: https://www.mdpi.com/journal/remotesensing/special_issues/SST_RS)

For citation purposes, cite each article independently as indicated on the article page online and as indicated below:

LastName, A.A.; LastName, B.B.; LastName, C.C. Article Title. *Journal Name* **Year**, *Article Number, Page Range.*

ISBN 978-3-03897-479-6 (Pbk)
ISBN 978-3-03897-480-2 (PDF)

Contents

About the Special Issue Editors

Jorge Vazquez-Cuervo received his B.S. in Physics, with honors, from the University of Miami in 1980. He earned a Master of Science in Physical Oceanography from the Graduate School of Oceanography at the University of Rhode Island in 1984. He earned his PhD in Geological Sciences from the University of Southern California in 1991. In 1993, Dr. Vazquez-Cuervo served as a visiting scientist at the Institute of Marine Science in Barcelona, Spain. He has worked as a Research Scientist at the Jet Propulsion Laboratory/California Institute of Technology since 1984 and currently serves as the Project Scientist supporting sea surface temperature and sea surface salinity for NASA's Physical Oceanography Distributed Active Archive PO.DAAC. Dr. Vazquez-Cuervo serves as the NASA representative to the Group for High Resolution Sea Surface Temperature (GHRSST). He is a member of the American Geophysical Union and The Oceanography Society and was adjunct faculty at Azusa Pacific University. He is funded under contract with the National Aeronautics and Space Administration.

Xiaofeng Li received a B.S. degree in optical engineering from Zhejiang University, Hangzhou, China, in 1985; an M.S. degree in physical oceanography from the First Institute of Oceanography, State Oceanic Administration, Qingdao, China, in 1992; and a Ph.D. degree in physical oceanography from North Carolina State University, Raleigh, NC, USA, in 1997. Since 1997, he has been with the National Environmental Satellite, Data, and Information Service (NESDIS), NOAA, College Park, MD, USA, where he is involved in developing many operational satellite ocean remote sensing products. He has authored more than 140 peer-reviewed publications and edited four books. Dr. Li is an Associate Editor of *IEEE Transactions on Geoscience and Remote Sensing and International Journal of Remote Sensing*, the Ocean Section Editor-in-Chief of Remote Sensing, and an Editorial Board Member of *International Journal of Digital Earth* and *Big Earth Data*.

Preface to "Sea Surface Temperature Retrievals from Remote Sensing"

Since the launch of NOAA-7 on 23 June 1981, sea surface temperatures (SSTs) have been derived from Earth-orbiting satellites. Thus, SSTs provide the longest time series of satellite-derived ocean parameters, now spanning nearly four decades. Sensors such as the Moderate Resolution Imaging Spectroradiometer (MODIS) and the Visible Infrared Imaging Radiometer Suite (VIIRS) have provided increased spatial resolution, while Geostationary Platforms provide information on the diurnal cycle and increased cloud-free detection of infrared-derived SSTs. Microwave sensors continue to provide all-weather capability. The dual view capability of the Along-Track Scanning Radiometer (ATSR) allows for the accurate determination of atmospheric corrections, critical for climate data records.

This collection covers topics ranging from the detailed error analysis of SSTs to new applications employed, for example, in the study of the El Niño–La Niña Southern Oscillation, lake temperatures, and coral bleaching. New techniques for interpolation and algorithm development are presented, including improvements for cloud detection. Analysis of the pixel-to-pixel uncertainties provides insight into applications for high spatial resolutions. New approaches for the estimation and evaluation of SSTs are presented. The collection provides an excellent overview of the current technology, while also highlighting new technologies and their applications to new missions.

Overall, 17 papers have been published in this Special Collection. Below is a very brief highlight of each paper.

The authors of "Sea Surface Temperature (SST) Variability of the Eastern Coastal Zone of the Gulf of California" use sea surface from the Pathfinder Advanced Very High Resolution Sea Surface Temperature Data Set as well as Moderate Resolution Imaging Spectroradiometer to examine the space-time variability of SST in the eastern part of the Gulf of California. The authors found a seasonal cycle as well as decreasing values from south to north. In "Quality Assessment of Sea Surface Temperature from ATSR of the Climate Change Initiative (Phase 1)", direct comparisons are made between SSTs from Along-Track Scanning Radiometer (ATSR) instruments and drifting buoys. Robust Standard Deviations ranging from 0.2 K to 0.3 K were derived.

In "Confirmation of ENSO-Southern Ocean Teleconnections Using Satellite-Derived SST", the authors examined teleconnections between SSTs in the Antarctic and Southern Oceans. The authors' primary conclusion is that significant correlations exists between the oceanic basins, which influence cooling and warming trends. The authors of "Spatio-Temporal Interpolation of Cloudy SST Fields Using Conditional Analog Data Assimilation" apply an analog data assimilation technique to create cloud-free level 4 SST products. A specific example using ocean current data is highlighted.

The article "Optimal Estimation of Sea Surface Temperature from AMSR-E" uses an optimal estimation (OE) technique developed by the European Space Agency. The OE technique is applied to a set of drifting buoys to estimate SSTs from the Aqua AMSR-E microwave instrument.

"Exploring Machine Learning to Correct Satellite-Derived Sea Surface Temperatures" applies four statistical models to a set of co-located satellite/in situ data to determine which model best predicts the errors. The advantage of the machine learning approach is that it considers the non-linearity of the errors.

The authors of "The Accuracies of Himawari-8 and MTSAT-2 Sea-Surface Temperatures in the Tropical Western Pacific Ocean" compare the quality of two geostationary Japanese satellites,

MTSAT-2 and Himwari-8. Overall, comparisons with the TOGA-TAO measurements in the Western Tropical Pacific showed the Himawari-8 SSTs had greater accuracies by approximately 0.15 K.

In the "Role of El Niño Southern Oscillation (ENSO) Events on Temperature and Salinity Variability in the Agulhas Leakage Region", the authors specifically examine the relationship between the variability in the Agulhas retroflection and ENSO. Overall, using SST, sea surface salinity, and a model, they determine that a strong relationship exists between ENSO and the Agulhas retroflection.

The authors of "Stability Assessment of the (A)ATSR Sea Surface Temperature Climate Dataset from the European Space Agency Climate Change Initiative" show that the reprocessing of the (A)ATSR data, under ESA's Climate Change Initiative (CCI), leads to stability estimates of 0.01 degrees Kelvin/decade.

In the article "Bayesian Cloud Detection for 37 Years of Advanced Very High Resolution Radiometer (AVHRR) Global Area Coverage (GAC) Data", the authors apply the method of statistical probability (Bayesian) to remove clouds from the AVHRR SST record. Comparisons with in situ data indicate a reduction in bias and STD by 10%, critical for application to climate studies and the longest satellite-derived measurement of SST from the AVHRR satellite.

The authors of "The Role of Advanced Microwave Scanning Radiometer 2 Channels within an Optimal Estimation Scheme for Sea Surface Temperature" found that a 0.37-K accuracy could be achieved using an optimal estimation approach for the derivation of SSTs. Critical to achieving this accuracy was prior knowledge of errors, inclusive of instrument errors.

In "Remote Sensing of Coral Bleaching Using Temperature and Light: Progress towards an Operational Algorithm", the authors use SST from the Advanced Very High Resolution Radiometer (AVHRR) to assess the impacts of temperature on coral bleaching. Their overall conclusion is that the prediction of coral bleaching is dependent on both SST and light.

"Reconstruction of Daily Sea Surface Temperature Based on Radial Basis Function Networks" shows that the Radial Basis Function Network improves on traditional OI approaches for gap filling, especially under cloudy conditions. Comparisons with in situ data show that errors are reduced to 0.48 K for the specific example given during the passage of a hurricane.

The article "Submesoscale Sea Surface Temperature Variability from UAV and Satellite Measurements" shows that in a comparison between MODIS-derived SSTs and experimental infrared-derived in situ SSTs in the Arctic, significant subpixel variability exists. The highest variability occurred in areas of high density and SST fronts.

In the article "Environmental Variability and Oceanographic Dynamics of the Central and Southern Coastal Zone of Sonora in the Gulf of California", 1-km SST data are used to examine the variability in the Gulf of California and the correlation with climate indices, including ENSO. High resolution data were used for both SST and Chlorophyll-a in the Sonora region off the Gulf of California.

The paper "Determining the Pixel-to-Pixel Uncertainty in Satellite-Derived SST Fields" examines the spatial variability of the data and applicability to submeso-scale and frontal studies. Two methods are applied that use wavenumber spectra and variograms to determine the uncertainty.

In the paper "Evaluation of the Multi-Scale Ultra-High Resolution (MUR) Analysis of Lake Surface Temperature", the MUR SST data is validated in three different lakes: a small, a medium, and a large sized lake. Larger biases were directly linked to cloud issues, as well as the application of the ice mask. Overall trends from the buoys and the MUR dataset were well correlated.

Jorge Vazquez-Cuervo, Xiaofeng Li
Special Issue Editors

remote sensing

MDPI

Article

Sea Surface Temperature (SST) Variability of the Eastern Coastal Zone of the Gulf of California

Carlos Manuel Robles-Tamayo [1], José Eduardo Valdez-Holguín [2,*], Ricardo García-Morales [3], Gudelia Figueroa-Preciado [4], Hugo Herrera-Cervantes [5], Juana López-Martínez [6] and Luis Fernando Enríquez-Ocaña [2]

[1] Postgraduate Master's Program in Biosciences, Division of Biological and Health Sciences of the University of Sonora, Luis Donaldo Colosio s/n, Colonia Centro, Hermosillo 83000, Sonora, Mexico; ctamayo93@gmail.com

[2] Department of Scientific and Technological Research of the University of Sonora, Luis Donaldo Colosio s/n, Colonia Centro, Hermosillo 83000, Sonora, Mexico; fernando.enriquez@unison.mx

[3] CONACYT, Center for Biological Research of the Northwest S.C. Nayarit Unit (UNCIBNOR+), Calle Dos No. 23. Cd del Conocimiento., Tepic 63173, Nayarit, Mexico; rgarcia@cibnor.mx

[4] Department of Mathematics of the University of Sonora, Luis Encinas y Rosales s/n, Colonia Centro, Hermosillo 83000, Sonora, Mexico; gfiguero@mat.uson.mx

[5] Center for Scientific Research and Higher Education of Ensenada, La Paz Unit, Calle Miraflores No. 334, La Paz 23050, B.C.S., Mexico; hherrera@cicese.mx

[6] Center for Biological Research of the Northwest S.C. Sonora Unit, Guaymas Campus Km. 2.35 Camino al Tular del Estero de Bacochibampo, Heroica Guaymas 85400, Sonora, Mexico; jlopez04@cibnor.mx

* Correspondence: eduardo.valdez@unison.mx

Received: 13 July 2018; Accepted: 17 August 2018; Published: 8 September 2018

Abstract: The coastal zones are areas with a high flow of energy and materials where diverse ecosystems are developed. The study of coastal oceanography is important to understand the variability of these ecosystems and determine their role in biogeochemical cycles and climate change. Sea surface temperature (SST) analysis is indispensable for the characterization of physical and biological processes, and it is affected by processes at diverse timescales. The purpose of this work is to analyze the oceanographic variability of the Eastern Coastal Zone of the Gulf of California through the study of the SST from time series analysis of monthly data obtained from remote sensors (AVHRR-Pathfinder Version 5.1 and Version 5 resolution of 4 km, MODIS-Aqua, resolution of 4 km) for the period 1981 to 2016. The descriptive analysis of SST series showed that the values decrease from south to north, as well as the amplitude of the warm period decrease from south to north (cold period increase from south to north). The minimum values occurred during January and February, and ranged between 18 and 20 °C; and maximum values, of about 32 °C, arose in August and September. Cluster analysis allowed to group the data in four regions (south, center, midriff islands and north), the spectral analysis in each region showed frequencies of variation in scales: Annual (the main), seasonal, semiannual, and interannual. The latter is associated with the El Niño and La Niña climatological phenomena.

Keywords: sea surface temperature (SST); oceanographic variability; Eastern Coastal Zone; Gulf of California

---

1. Introduction

The coastal zones are wide geographical areas with intense physical, chemical and biological interactions with strong exchange of energy and materials between the terrestrial environment, aquatic environment and the atmosphere [1]. The coastal zones have high primary productivity levels and play important role in the biogeochemical cycles [2]. These areas support high abundance of natural resources, providing refuge, feeding and spawning areas for diverse organisms, and enhancing activities like tourism, aquaculture and fisheries [3].

The Gulf of California is a marginal sea located in the Eastern Pacific Ocean, surrounded by arid environment that encompasses the Baja California Peninsula, and the states of Sonora, Sinaloa, and Nayarit, with an average length and width of 1400 and 150–200 km respectively and basins (deepening to the south) separated by sills [4,5]. Tidal currents, the transfer of wind moment, upwelling and high solar heating determine a strong physical dynamics [5]. Northwest Winds from December to May produce intense upwelling processes off the Eastern Coast. During these "winter conditions", nutrients supply enhance the growth of phytoplankton communities. While, from July to October, the prevailing winds from the southeast, are weak and do not have enough energy to break the strong thermal stratification of the water column in summer, these "summer conditions" do not have an effect on the levels of phytoplankton biomass on the western coast [5,6]. The months of June and November are considered periods of transition between both conditions [7]. On the other hand, the Baja California Peninsula has mountain ranges that prevent the low-level clouds from the Pacific Ocean influencing in the gulf generating cloud-free most of the time with exceptions in summer when the tropical air from the south moves into the gulf [8].

The Eastern Coast of the Gulf of California is characterized by diverse water bodies, that play an important role in fisheries. This coastal zone has a coastal plain with an extensive deltaic river formed by the Colorado, Sonora, Yaqui and Mayo Rivers [9]. The seasonal coastal winds develop a sea surface circulation along the coast, to the south during October–March, and to the north during June–September [8,10].

Sea surface temperature (SST) is considered the most important variable in oceanography, it is considered an essential climate variable (ECV) and essential ocean variable (EOV) [11], because influences many physical, chemical and biological properties of oceans, and is an effective indicator of changes in marine ecosystems [12]. SST is considered the most important variable in oceanography. The SST temporal spatial distribution is useful in the location of thermal fronts, current systems in the oceans and the exchange of thermal energy between the ocean and the atmosphere [13]. In the recent years, the anthropogenic activities have increased contributing to climate change through modifications of physical and chemical aspects in the marine ecosystems [14,15]. Global warming associated with anthropogenic climate change impacts both mean SST, and as well as the thermal and atmospheric processes that affect ocean circulations. It also has influence on the physiology, behavior and demographic aspects of organism, altering size, structure, range of distribution, and abundance of populations consequently generating changes the trophic routes and the community and functions of the ecosystem [15]. The changes in the SST have as a consequence, alterations in the marine biological processes, from individuals to ecosystems, in local to global scales, impacting the ecosystem services [16]. The SST as an ECV is of great importance for the study, monitoring and management of the marine environment since it allows to quantitatively estimate recent changes and their effect on ecosystem services [12].

When analyzing SST variability, it is important to study its trends of change associated with environmental modifications that can occur of a natural or anthropogenic form [17]. From this perspective, remote sensing is a technique that can provide a high temporal resolution with a wide coverage of environmental variables; besides, their cost is lower than in situ measurements from boats that are not considered optimal when looking for long and large regions. The use of satellite images, through remote sensing, has allowed having accurate information on a global scale describing the physical and biological aspects of the oceans [18]. One of the disadvantages of remote sensors are uncertainty of SST retrievals in the presence of clouds, atmospheric gases and aerosols [19]. The analysis of SST in marine environments can be a challenge due to the variety of available remote sensors characteristics and resolution (spatial, spectral, radiometric and temporal). The coastal areas are sites with large spatial and temporal fluctuations, showing a high complexity, their study requires the analysis of high-resolution processes [20,21]. Remote sensing provides adequate resolution of SST, a very important variable in oceanography that contributes to their long-term studies [20].

Several time series analyses have shown that sea surface temperature in the Gulf of California varies on seasonal to interannual scales [22,23]. Soto-Mardones et al. [22] found a decrease in the sea surface temperature from south to north, and also that the annual scale is responsible for most of the SST variability oscillating in phase with minor north-south variations. Escalante et al. [24] also reported this decrease of the SST average from south to north along the Gulf of California, with clear differences between the warm (summer and autumn) and cold (winter and spring) conditions for the entire gulf. Robles and Marinone [25], as well as Ripa and Marinone [26] found a clear seasonal SST variability across the Guaymas Basin in the Central Gulf of California: Winter conditions extend from December to April and summer conditions from June to October with transition periods in May and November. Valenzuela-Sánchez [23] worked in the Central Region of the Gulf of California focusing on SST off Guaymas Bay, concluding that the annual cycle dominates following the semiannual signal. Interannual variability was associated with the cyclonic north equatorial circulation composed of the North Equatorial Countercurrent, the North Equatorial Current and the Costa Rica Current [27], El Niño-La Niña, and also with the Interdecadal Variability of the Pacific Ocean [28]. García-Morales et al. [29] analyzed meso-scale phenomena, and their influence on SST in the southern and central region of the coastal area of the State of Sonora. They showed that environmental conditions in the Gulf of California have a great influence, seasonal and interannual, of the meteorological and oceanographic processes variability of the Pacific Ocean. The studies in the eastern side of the Gulf of California have done in coastal lagoons and bays, but there are lack of studies along this coastal zone to analyze its oceanographic variability. These areas are vulnerable to natural and anthropogenic changes, and it is required to provide more environmental and oceanographic information along this area through the SST analysis considered important in the marine ecosystems. According to Heras-Sánchez [30], there are clear differences in the Western and Eastern Coast of the Gulf of California, with clear variability in the SST values. Lon-term observations in coastal zones are important for the analysis and prediction of changes in marine ecosystems allowing developing an adequate management of the marine and coastal resources. The SST analysis allows us to establish ecological characterizations, change trends associated with environmental and oceanographic factors, how it influences the ecosystem and its possible effect on the distribution and abundance of marine resources, promoting knowledge of oceanography of the Eastern Coastal Zone by remote sensing. The Gulf of California is often cloud free, which makes it ideal for observations using satellite remote sensing techniques [8]. Therefore, the objective of this work is to describe the spatial and temporal variability (regional and monthly) of the SST in the Eastern Coastal Zone of the Gulf of California using a 415-month (September 1981–March 2016) series of remote sensor databases.

2. Materials and Methods

2.1. Study Area

This study comprised the Eastern Coastal Zone of the Gulf of California (from the north of Sinaloa State to the delta of the Colorado River) (Figure 1). This coastal region has a great diversity of costal ecosystems with several important fishing ports like Yavaros, Guaymas and Puerto Peñasco, and highly productive irrigated farmlands are located in the coastal plain; it also has a high fertility induced by coastal upwellings [31]. The central region has a seasonal pattern of biological production, pigment concentration, pigment-rich water is found at northeast side (continental side) during early winter (November to December), and sometime later this water expands to southwestern region (peninsula side). Summer conditions contracts the rich water in the opposite direction, remaining a small rich area in the northeastern central coastal zone [32–34]. In the Midriff Islands area, there is a vertical distribution of nutrients constantly associated to tidal mixing [35]. Hidalgo-González et al. [36] indicated near from the Tiburón Island and Ángel de la Guarda Island there are sills and deep basins forming tidal streams developing a particular behavior in the distribution of the temperature with different levels in the south of Ángel de la Guarda and Tiburón Island [37]. We retrieved SST data from 17 georeferenced sites (sampling points), from the Eastern Coast of the Gulf of California. These sampling points were 60 km separated from each other and 30 km from the coastline, that ranges from the northern Sinaloa to northern Sonora (Table 1).

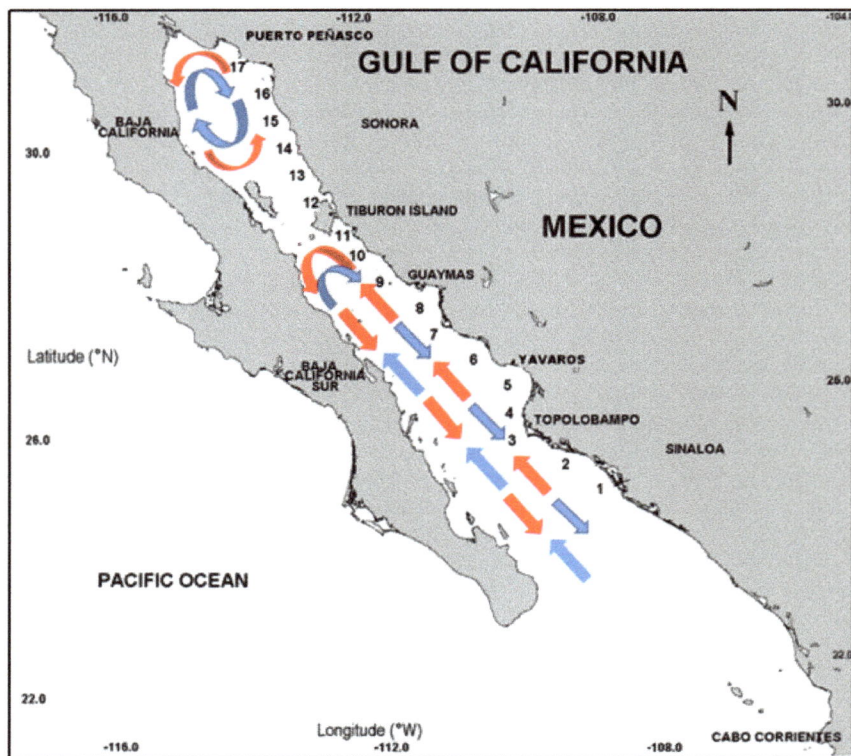

Figure 1. Map of the Eastern Coastal Zone of the Gulf of California showing the location of the sampling points where monthly sea surface temperature (SST) was measured for environmental and oceanographic analyses. Ocean surface circulation: Warm period (red) and cold period (blue) [38].

Table 1. List of geographic coordinates of the sampling points of the Eastern Coastal Zone of the Gulf of California.

Sampling Point	Geographic Coordinates	Sampling Point	Geographic Coordinates	Sampling Point	Geographic Coordinates
1	24°48′N, 108°24′W	8	27°36′N, 111°W	15	30°24′N, 113°18′N
2	25°12′N, 108°54′W	9	28°N, 111°36′W	16	30°48′N, 113°24′W
3	25°36′N, 109°42′W	10	28°24′N, 112°W	17	31°12′N, 113°48′W
4	25°60′N, 109°42′W	11	28°42′N, 112°12′W		
5	26°24′N, 109°42′W	12	29°12′N, 112°42′W		
6	26°48′N, 110°12′W	13	29°36′N,112°54′W		
7	27°12′N, 110°48′W	14	30°N, 113°6′W		

2.2. Oceanographic Characterization and Obtaining Data of the Sea Surface Temperature (SST)

The SST data were obtained of the Physical Oceanography Distributed Active Archive Center (PODAAC) of the Jet Propulsion Laboratory California Institute of Technology (https://podaac.jpl.nasa.gov/) and OceanColor Web (https://oceancolor.gsfc.nasa.gov/) from NASA, processed at level 3 and downloaded in HDF (Hierarchical Data Format). The data processing from PODAAC at level 3 consist on variables mapped on uniform space-time. The all-pixel SST files contain values for each pixel location, including those contaminated with clouds or other sources of error. The Overall Quality Flag values may be used to filter out these unwanted values. The SST value in each pixel location is an average of the highest quality AVHRR Global Area Coverage (GAC) observations available in each roughly 4 km bin. First-guess SST, the Pathfinder algorithm uses a first guess SST based on the Reynolds Optimally Interpolated SST (OISST), Version 2 product. The OISST V2 is also used in the quality control procedures. That is this parameter indicates the number of AVHRR GAC observations falling in each approximately 4 km bin, as well as standard deviation of the observations in each 4 km bin. Overall Quality Flag, the overall quality flag is a relative assignment of SST quality based on a hierarchical suite of tests. The Quality Flag varies from 0 to 7, with 0 being the lowest quality and 7 the highest. For most applications, using SST observations with quality levels of 4 to 7 is typical, this being the quality indicator that was used to validate the AVHRR GAC images, since the of the pixels of poor values (level 0) are discarded in the first tests of condition, quality or error. For applications requiring only the best-available observations (at the expense of the number of observations), use quality levels of 7 only [39].

The data processing from OceanColor are of geophysical variables that have been projected onto a well-defined spatial grid over a well-defined time period, each file contains an equirectangular projection and a registration of structure square cells grids grid of floating-point values for a single geophysical parameter. NASA standard processing and distribution of the SST products from the MODIS sensors is now performed using software developed by the Ocean Biology Processing Group (OBPG). The OBPG generates Level-2 SST products using the Multi-Sensor Level-1 to Level-2 software (l2gen), which is the same software used to generate MODIS ocean color products. The SST algorithm and quality assessment logic are the responsibility of the MODIS Science Team Leads for SST (currently P. Minnett and R. Evans of the Rosenstiel School of Marine and Atmospheric Science at the University of Miami). The description is valid for both the standard products distributed by the OBPG through the ocean color web and the products delivered to the Physical Oceanography DAAC, where the latter are subsequently repackaged for GHRSST distribution. MODIS Aqua Global Level 3 Mapped Thermal SST products consists of sea surface temperature (SST) data derived from the 11 and 12 um thermal IR infrared (IR) bands (MODIS channels 31 and 32). Daily, weekly (8 day), monthly and annual MODIS SST products are available at both 4.63 and 9.26 km spatial resolution and for both daytime and nighttime passes. This particular dataset is the MODIS Aqua, thermal-IR SST level 3, 4 km, daily, daytime product (ftp://podaac-ftp.jpl.nasa.gov/allData/modis/L3/docs/modis_sst.html).

The sensors used of the PODAAC portal to obtain the SST data were Advanced Very High Resolution Radiometer (AVHRR-Pathfinder Version 5.1) for the period from September 1981 to January 1985 and Advanced Very High Resolution Radiometer (AVHRR-Pathfinder Version 5) for the period from February 1985 to June 2002. These sensors correspond to day period in Celsius degrees (°C) with pixel size 4×4 km^2. However, due the AVHRR-Pathfinder sensor only presents SST data until 2009, the other sensor used was Aqua MODIS Sea Surface Temperature (11μ daytime) in scale of degree centigrade (°C) with a resolution of 4 km from the OceanColor Web for the period from July 2002 to March 2016 with the objective to construct and continue a database of 415 months. This last sensor is of the new generations of radiometers that combine a wider range of spectral measurement with improvements in technology, the values measured with both sensors are similar [40]. The satellite images were processed with the Windows Image Manager (WIM), WimSoft version 9.06 Software (Copyright Mati Kahru 1995–2015) to obtain the SST monthly mean data of the 17 sampling points transect of the coastal zone.

2.3. Analysis and Processing Monthly Data of SST

Before performing the statistical analysis that will be presented here, SST data were previously inter-calibrated to adjust the values from the different sensors used, by means of lineal regression in each one of the data pair where the SST were obtained. This process was developed using a five years' period where both sensors coincide (July 2002–July 2007), a strong correlation between both data and an intersection that is not significantly different from zero indicated that the values are practically similar. Monthly SST comparisons of different sensors have done to validate the combination of these sensors, obtaining a similar distribution pattern and a high correlation coefficient justifying the analysis of long SST time series from different sensors [40]. Each regression equation (AVHRR-Pathfinder VS Aqua MODIS) was used to adjust the SST data of the Aqua MODIS.

Once the sensor data were inter-calibrated, a cluster analysis was carried out, with the purpose of grouping the sampling points with greater homogeneity among them, and greater difference (statistical distance) between the groups. A nonparametric test (Kruskal-Wallis, $p < 0.05$) was used to show the statistical differences between the groups obtained from the cluster analysis. A descriptive analysis that includes tables, box plots and Receiver Operating Characteristic (ROC) curves were done to analyze the climatology and time series by regions and during the different months. The ROC curves, is a statistical tool, which are widely used for assessing the performance of classification algorithms. The class assignment is made comparing a score or measurement, with a threshold: If the score is above the threshold, it is assigned to one class, otherwise, is assigned to the other class [41]. In this way, ROC curves allows us to identify if we could differentiate between two regions, based on their mean SST values. On the other hand, a simple way to see if the classifier has the ability to discriminate between two populations or in this case, two regions, is checking if the Area Under the Curve (AUC) is significantly greater than 0.5, the chance diagonal. From an averaged time series of each group, a Fast Fourier Transform was performed to obtain the spectral density. The software Statistical Software version 7.5 and R Software version 3.3.2 performed all the analyzes.

3. Results

3.1. Distribution and General Variability of SST

The overall SST distribution (Figure 2) evince a latitudinal gradient. The warmest SST values were located at the entrance of the gulf, with average values higher than 26 °C, and the coldest SST values from the Midriff Islands to the north, with average values lower than 24 °C.

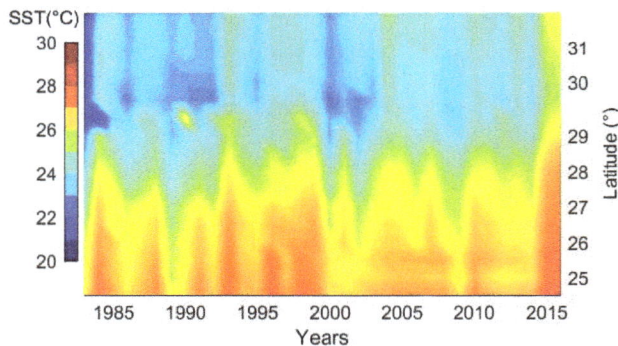

Figure 2. Hovmoller diagram of the Sea Surface Temperature (SST) of Eastern Coastal Zone of the Gulf of California.

The coastal zone SST rapidly increases from minimum values of 18 to 20 °C (January and February) to maximum temperatures of 30 to 32 °C (August and September). Two well defined periods were observed in the annual SST pattern for the Eastern Coast of the Gulf of California: Warm period with SST higher than 25 °C, and cold period with SST lower than 25 °C, May–June and November are transitions periods with temperatures around the 25 °C (Figure 3).

6

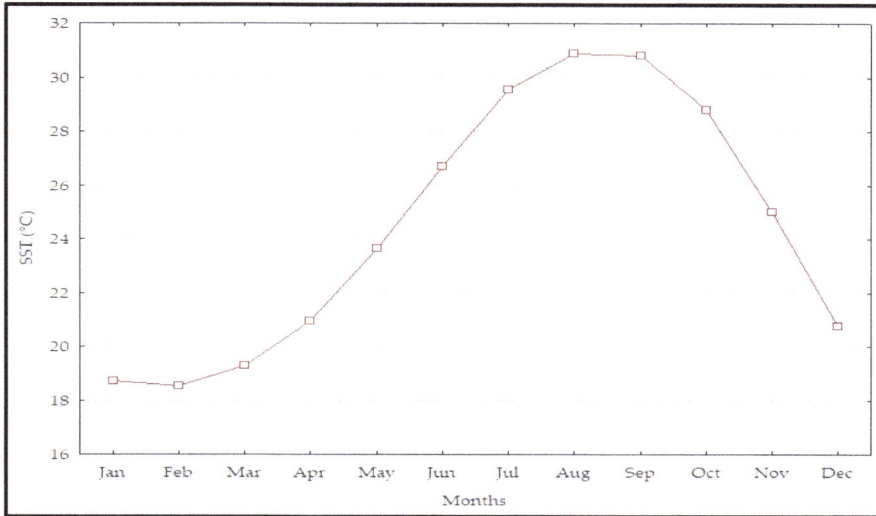

Figure 3. Climatology of the SST of the Eastern Coastal Zone of the Gulf of California.

3.2. Characterization of Regions

Cluster analysis clearly grouped the SST values in four clusters from 20 linkage distance units (Figure 4). The first group includes from the first to the third sampling point, South Region. The second group, Central Region, comprises from the fourth to the tenth sampling point. The third group is the Midriff Islands Region that covers only the eleventh sampling point, and the fourth group is the North Region that spans from the twelfth to the seventeenth sampling point. These regions will be analyzed from now on (Table 2).

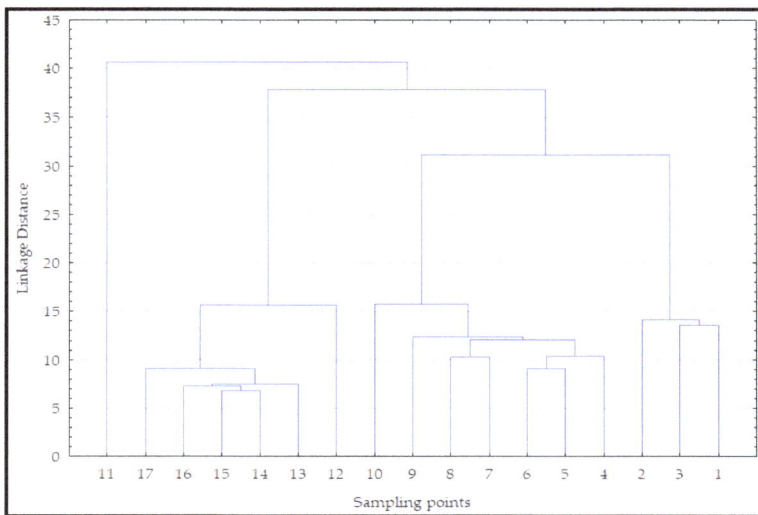

Figure 4. Cluster analysis of the sampling points of the Eastern Coastal Zone of the Gulf of California.

Table 2. Classification of the Regions of the sampling points of the Eastern Coastal Zone of the Gulf of California.

Sampling Point	Region	Sampling Point	Region	Sampling Point	Region
1	South Region	8	Central Region	15	North Region
2	South Region	9	Central Region	16	North Region
3	South Region	10	Central Region	17	North Region
4	Central Region	11	Midriff Islands Region		
5	Central Region	12	North Region		
6	Central Region	13	North Region		
7	Central Region	14	North Region		

3.3. Time Series and Climatology

The averaged time series for SST of each group showed values between 12 and 36 °C. The South Region presented the narrowest range, between 17 and 32 °C (Figure 5), while Central Region has a range of values between 16 and 32 °C (Figure 6). On the other hand, Midriff Islands Region (Figure 7) presented the widest range of values (12 to 36 °C) and the North Region (Figure 8) has the similar range as the Central Region.

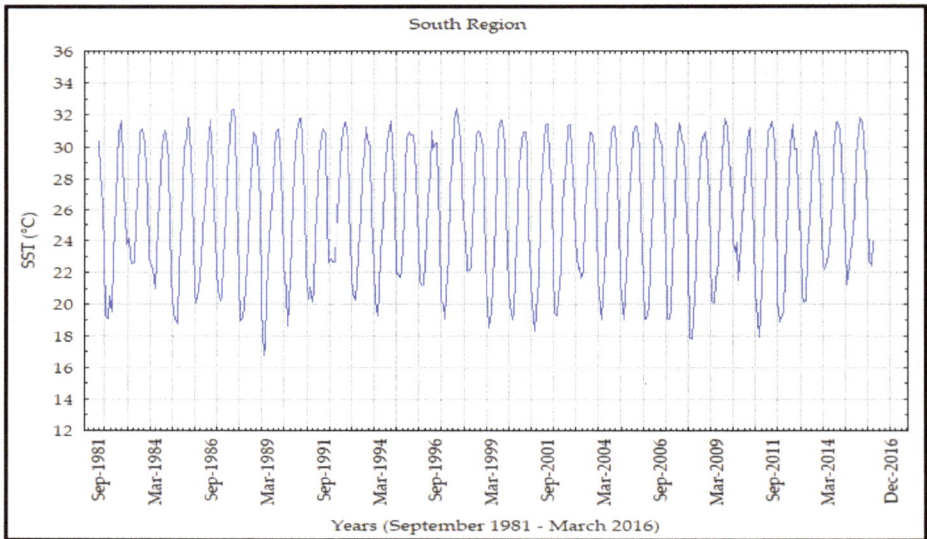

Figure 5. Time series of the SST of the South Region of Eastern Coastal Zone of the Gulf of California.

The SST anomalies varied from −5.5 to 4.3 °C. The South Region has the narrowest range between −2.9 and 2.8 °C (Figure 9) and the Midriff Islands Region showed the widest range of values between −5.5 and 4.3 °C (Figure 10). The Central and North Region anomalies varied from −4.1 to 2.9 °C (Figure 11) and −4.8 and 2.4 °C (Figure 12). The SST anomalies were compared with a database of Southern Oscillation Index (SOI) from the Equatorial Pacific Ocean downloaded from ClimateDateGuide (https://climatedataguide.ucar.edu/climate-data/nino-sst-indices-nino-12-3-34-4-oni-and-tni) with a range between −3.4 and 4.2 indicating that the positive anomalies are associated to El Niño event and a negative anomalies are associated to La Niña event.

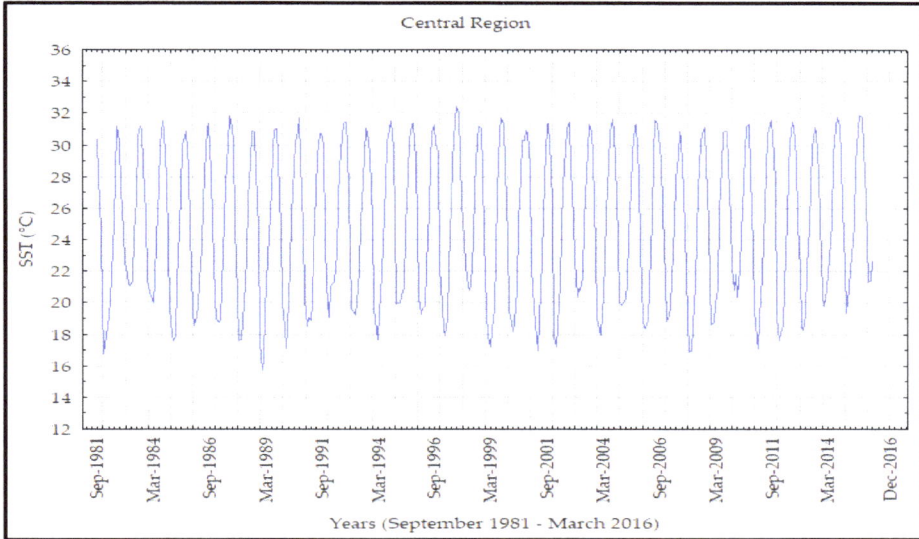

Figure 6. Time series of the SST of the Central Region of the Eastern Coastal Zone of the Gulf of California.

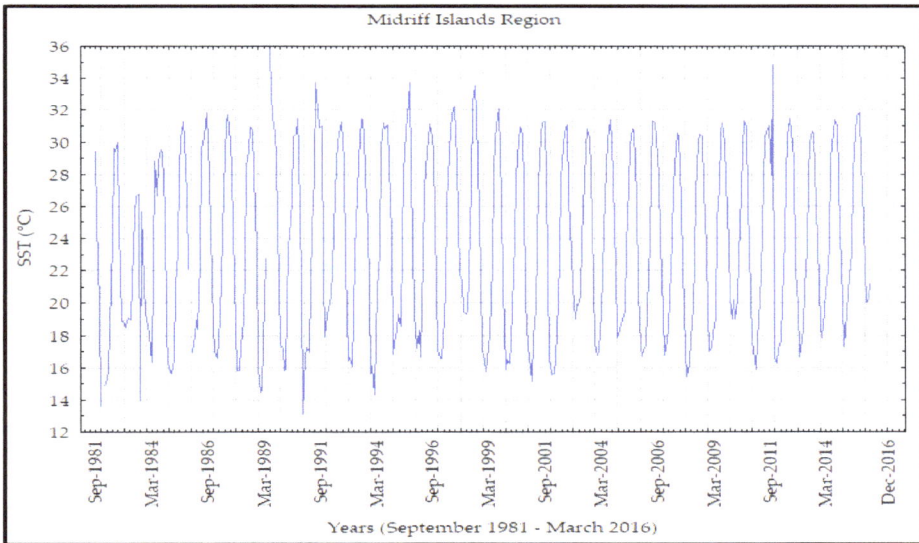

Figure 7. Time series of the SST of the Midriff Islands Region of Eastern Coastal Zone of the Gulf of California.

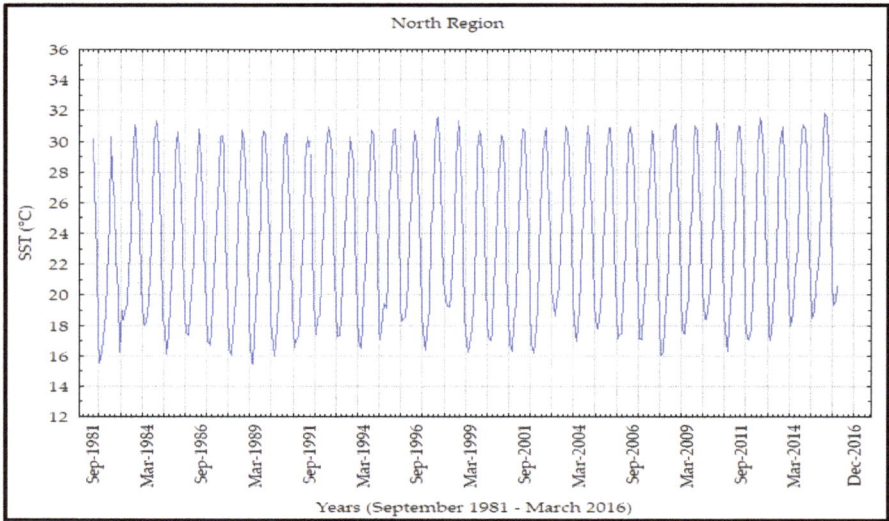

Figure 8. Time series of the SST of the North Region of the Eastern Coastal Zone of the Gulf of California.

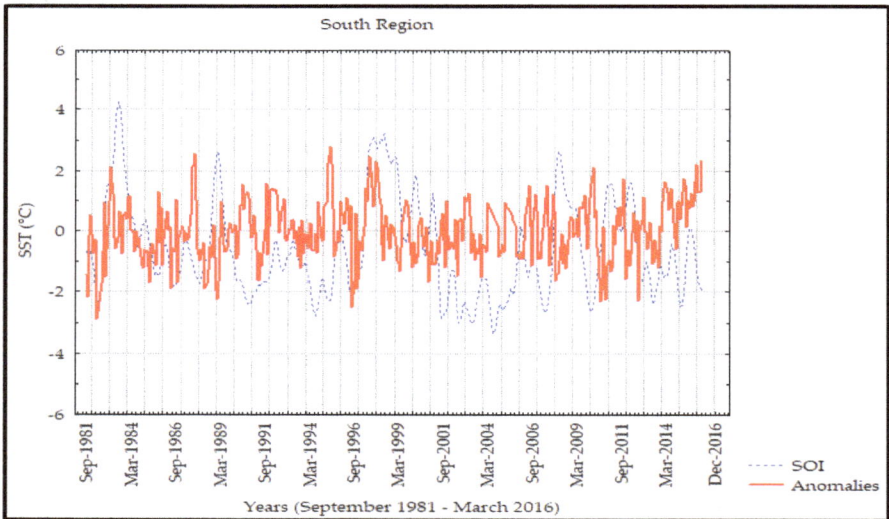

Figure 9. Time series of the SST anomalies of the South Region of the Eastern Coastal Zone of the Gulf of California.

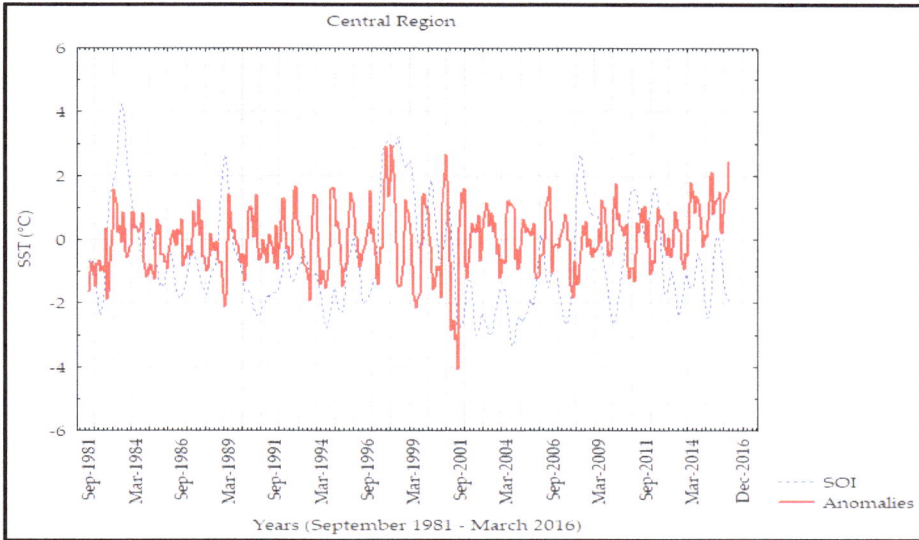

Figure 10. Time series of the SST anomalies of the Central Region of the Eastern Coastal Zone of the Gulf of California.

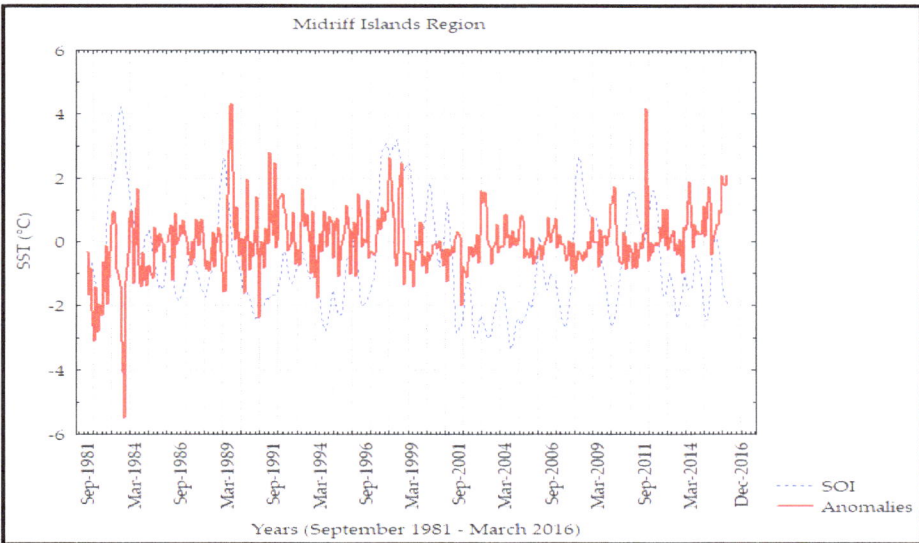

Figure 11. Time series of the SST anomalies of the Midriff Islands Region of the Eastern Coastal Zone of the Gulf of California.

Some overall statistics for SST are described in Table 3. This table contains the mean, standard deviation, as well as the maximum and minimum of SST, for each region, and clearly, we can observe a decrease in the SST mean value from south to north region.

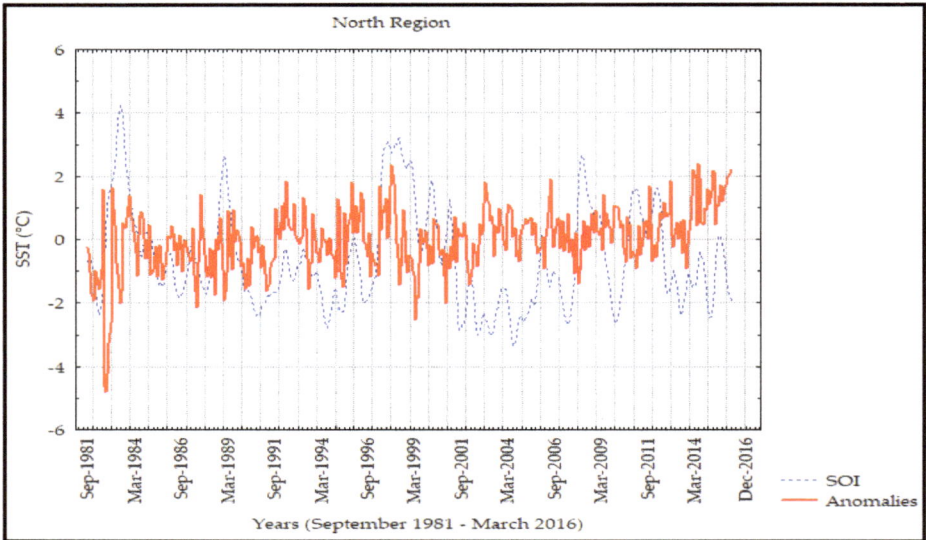

Figure 12. Time series of the SST anomalies of the North Region of the Eastern Coastal Zone of the Gulf of California.

Table 3. Descriptive statistics of SST for each region of the Eastern Coastal Zone of the Gulf of California.

Region	Mean ± SD (°C)	Maximum (°C)	Minimum (°C)
South	25.81 ± 4.28	32.43	16.76
Central	24.80 ± 4.77	32.41	15.71
Midriff Islands	23.67 ± 5.68	36.30	11.90
North	23.49 ± 5.01	31.83	15.48

For each of these regions a similar variability was observed, with different time intervals of the transitions periods around 25 °C (Figure 13). The South Region present the transition to summer between April and May reaching maximum values in August and September for seven months of warm period with a mean of 28.53 °C and a maximum of 32.43 °C, while the transition to winter starts between November and December reaching the minimum values in February for five months of cold period with a mean of 22.10 °C and a minimum of 16.76 °C. The Central Region begins its transition to summer between May and June reaching the maximum values in August and September for seven months of warm period with a mean of 28.55 °C and maximum of 32.41 °C, while the transition to winter starts again between November and December obtaining minimum values in January and February for six months of cold period with a mean of 21.12 °C and a minimum of 15.71 °C. The Midriff Islands Region begins the transition to summer between May and June with maximum values in August for five months of warm period with a mean of 28.77 °C and maximum of 36.30 °C, while the transition to winter that starts between October and November with minimum values in January for seven months of cold period with a mean of 20.08 °C and a minimum of 11.90 °C. The North Region has a summer transition that begins in June with maximum values in August for four months of warm period with a mean of 29.01 °C and a maximum of 31.83 °C, while the transition to winter begins between October and November having minimum values in February for eight months with a mean of 20.76 °C and a minimum of 15.48 °C.

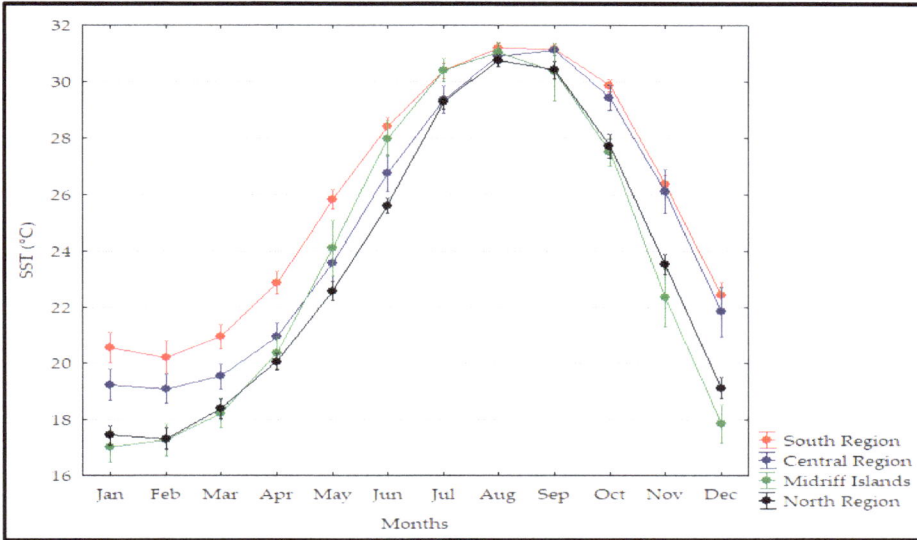

Figure 13. Climatology of the SST of each region of the Eastern Coastal Zone of the Gulf of California.

According to the statistics for SST during each one of the periods (warm and cold) that are described in Tables 4 and 5, the SST gradients are reversed in each period. During the warm period, the months of summer decreased in time from south to north, as well as SST mean values increased from south to north. During the cold period, the months of winter increased in duration and the SST mean values decreased from south to north.

Table 4. Descriptive statistics of SST for each region of the Eastern Coastal Zone of the Gulf of California (warm period).

Region	Time	Mean $\pm$ SD (°C)	Maximum (°C)	Minimum (°C)
South	7 Months	28.53 $\pm$ 3.00	32.43	19.49
Central	6 Months	28.55 $\pm$ 2.96	32.41	18.94
Midriff Islands	5 Months	28.77 $\pm$ 3.32	36.30	13.92
North	4 Months	29.01 $\pm$ 2.18	31.83	24.03

Table 5. Descriptive statistics of SST for each region of the Eastern Coastal Zone of the Gulf of California (cold period).

Region	Time	Mean $\pm$ SD (°C)	Maximum (°C)	Minimum (°C)
South	5 Months	22.10 $\pm$ 2.68	28.03	16.76
Central	6 Months	21.12 $\pm$ 3.06	30.79	15.71
Midriff Islands	7 Months	20.08 $\pm$ 3.99	34.85	11.90
North	8 Months	20.76 $\pm$ 3.55	29.47	15.48

For the comparison of the monthly SST values in the warm and cold period between the four regions, a Kruskal-Wallis test was performed. The monthly SST values of the warm period between these regions were not statistically different ($p = 0.7763$); however, the monthly SST winter values were statistically different ($p < 0.0001$). Differences of monthly SST mean of the cold period values were confirmed by a Bonferroni multiple comparison post hoc test. Significant p values (2-Tailed) are denoted with an asterisk in Table 6, where can be observed that only North and Central Regions are not statistically significant.

Table 6. Multiple comparison p values (2-Tailed) of the monthly values of the cold periods with a Bonferroni adjustment of each one of the regions of the Eastern Coastal Zone of the Gulf of California. Asterisk (*) denotes probability values of statistical differences.

	South	Central	Midriff Islands	North
South		0.003880 *	0.000000 *	0.000000 *
Central	0.003880 *		0.000094 *	0.339130
Midriff Islands	0.000000 *	0.000094 *		0.048219 *
North	0.000000 *	0.339130	0.048219 *	

A Box-Whisker plot for winter SST mean values can be observed in Figure 14, where a clear difference in variability can be seen. Graphically the differences between the regions obtained in Table 5 were presented using ROC Curves based on the mean SST values. It is clear that during the warm period the South and Midriff Islands Regions are practically similar with a low sensitivity and specificity levels (0.617, 0.462) and a confidence interval for the Area Under the Curve (AUC) of (0.47, 0.584), where 0.5 is into this interval (Figure 15), being not possible to discriminate between these regions. However, during the cold period the results were different obtaining a high sensitivity but low specificity levels (0.943, 0.473) and a confidence interval of (0.656, 0.755) for AUC, obtaining clear differences between these regions (Figure 16).

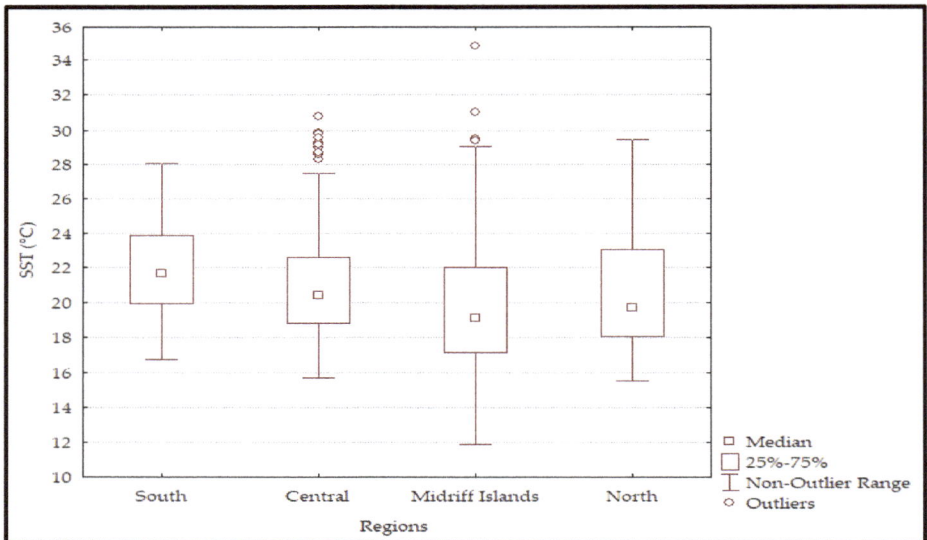

Figure 14. Variance distribution of the monthly SST values during cold period of each one of the regions of the Eastern Coastal Zone of the Gulf of California.

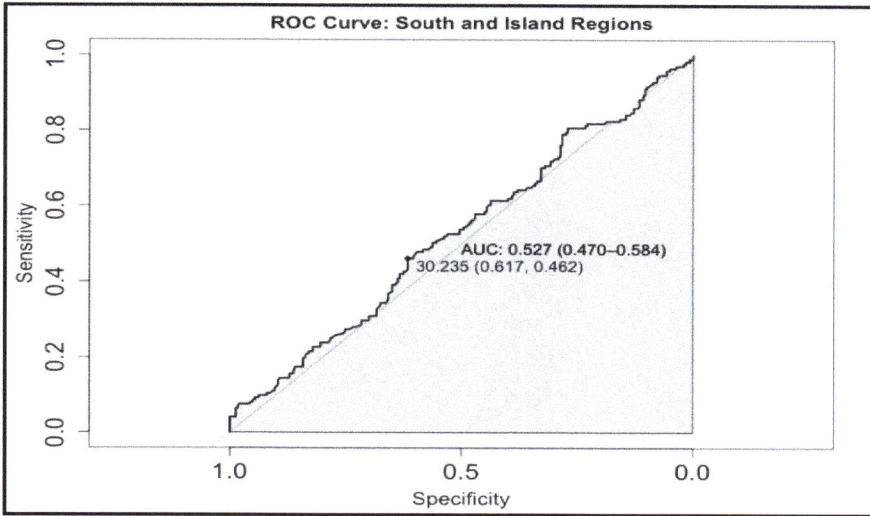

Figure 15. ROC curve analysis of the South and Midriff Islands Regions of the Eastern Coastal Zone of the Gulf of California (warm period).

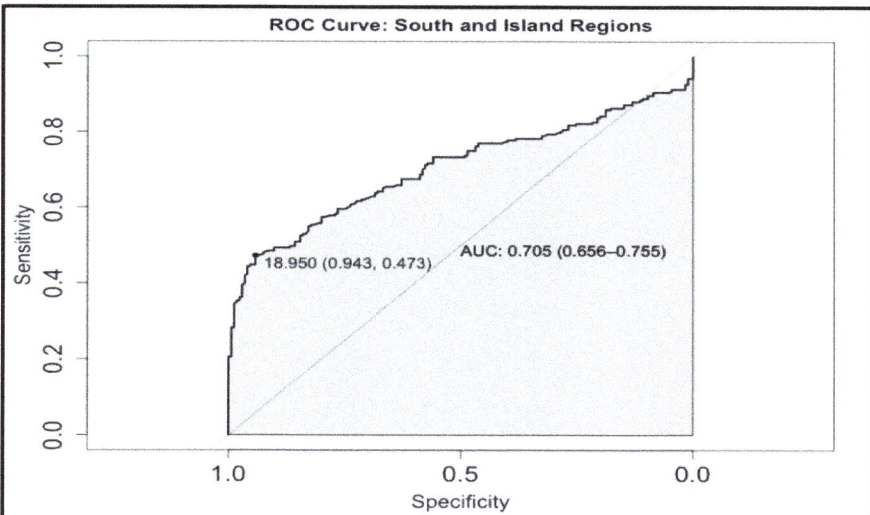

Figure 16. ROC curve analysis of the South and Midriff Islands Regions of the Eastern Coastal Zone of the Gulf of California (cold period).

3.4. Fourier Analyses

The SST spectral analysis of the regions (Figures 17–20) showed that the main frequencies are annual associated with maximum levels in summer and minimum levels in winter, semiannual (six months), and also interannual (periods of three to five years), whose frequencies of variation are associated to global climatological-oceanographic phenomena such as El Niño and La Niña.

In addition to all these frequencies, also a seasonal variability can be observed in some regions (South, Central and Midriff Islands), but not at the North Region. The spectral analysis showed an increase in the semiannual frequency from the South to the North Region as well an increase in the seasonal frequency in the Midriff Islands Region reaching almost the same semiannual spectral density level. The interannual frequency rise in the Midriff Islands Region, as well, a possible rise of the decadal frequency.

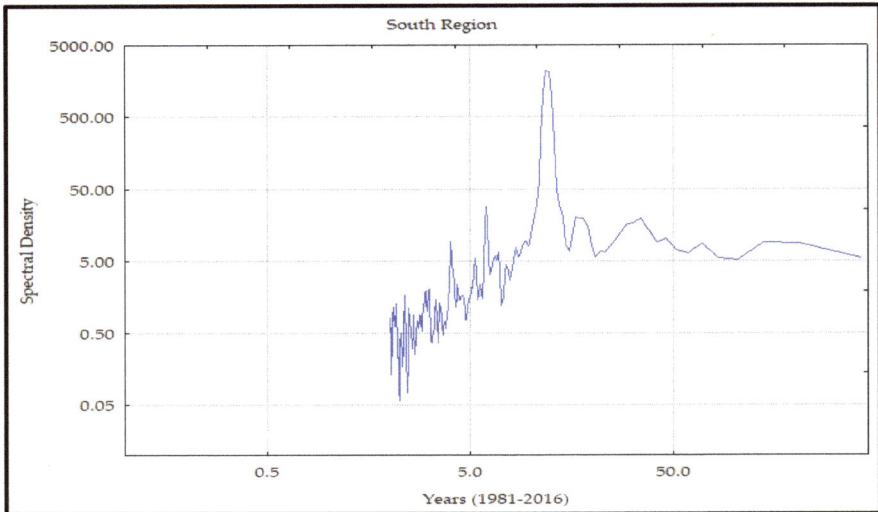

Figure 17. Spectral Analysis for the SST of the South Region of the Eastern Coastal Zone of the Gulf of California.

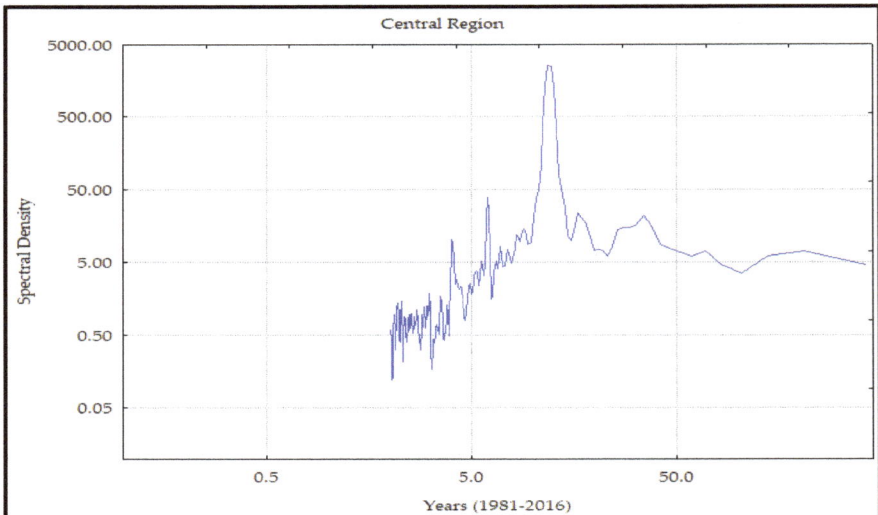

Figure 18. Spectral Analysis for the SST of the Central Region of the Eastern Coastal Zone of the Gulf of California.

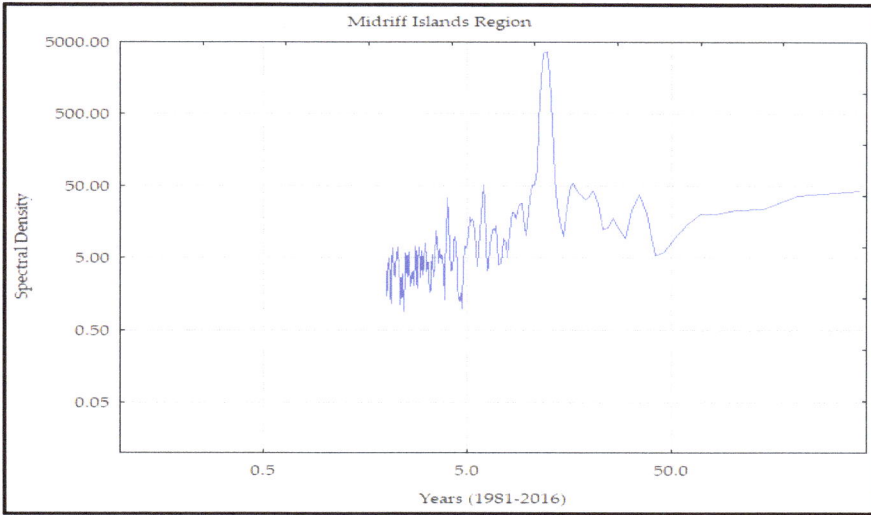

Figure 19. Spectral Analysis for the SST of the Midriff Islands Region of the Eastern Coastal Zone of the Gulf of California.

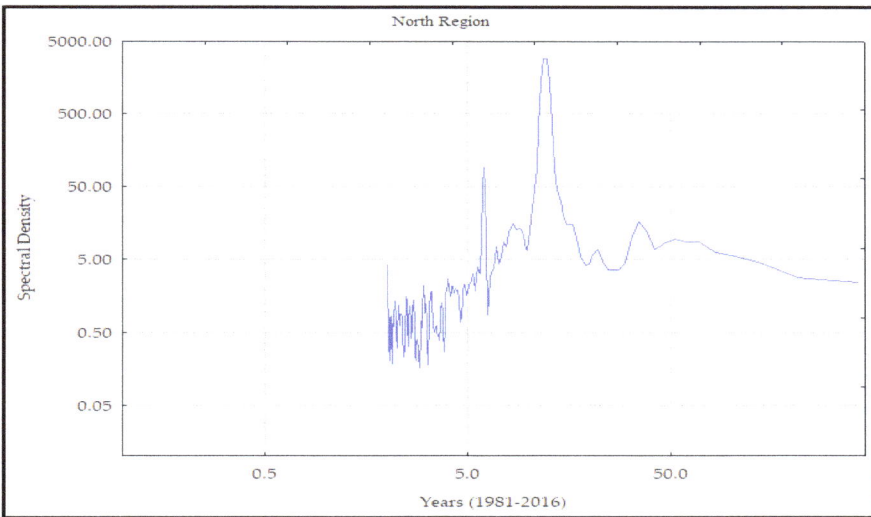

Figure 20. Spectral Analysis for the SST of the North Region of the Eastern Coastal Zone of the Gulf of California.

4. Discussion

The SST during the year is similar to the values reported by Robles and Marinone [25] and Valenzuela-Sánchez [23] allowing establishing a general climatology, highlighting the occurrence of two clearly defined periods. A warm and a cold period, with very short transition periods. The climatology of the Gulf of California has been described by Roden [42], showing clear latitudinal differences and from the east and west coasts, related to atmospheric circulation and mountain ranges that influences its circulation. The SST mean values of the study area decreased from south to north during the warm period, and reversed during the cold period. This gradient from south to north is similar to the results for whole Gulf of California described by Soto-Mardones et al. [22], Escalante et al. [24] and Hidalgo-González and Álvarez-Borrego [43]. This SST distribution, during the warm period, can be explained by the fact that the direct communication with the Tropical Pacific Ocean allows the entry of Equatorial Surface Water [29,44], and inside the Gulf of California it is modified by greater solar irradiance and evaporation effects [42], generating this gradient. During the cold period, SST decrease inside the Gulf by physical dynamics such tidal and wins mixing [25], mainly in the Midriff Islands and North Regions. In addition, the effect of the continental meteorological conditions in the northern Gulf of California are important in the decrease of the SST [42,45]. Low temperatures are found all around the year in the Midriff Islands, where winds and tides [46], hydraulic jump [8] and strong currents through the narrow channels mix the water column. López et al. [47] reported a decrease on SST in this region, coldest SST were associated to tidal pumping and mean flow in the sills in the southern part of this region, developing a particular circulation pattern that generates persistent upwelling of deep waters causing low SST values, high productivity and well mixed conditions throughout the water column. A more constant SST at the entrance of the Gulf allows these gradients to exist, as well as, the varying duration of warm or cold periods. The further south the duration of the warm period is longer, and the cold period is shorter.

Based on the Cluster analysis, four regions were obtained: South, Central, Midriff Islands and North. Different regionalizations have made for the whole Gulf based on different aspects: Round [48] on the phytoplankton remains in sediments. Santamaría-del-Angel et al. [6] used a time series of Chl *a* in a number of stations throughout the gulf and for a short period (eight years). Hidalgo-González and Álvarez-Borrego [43] analyzed Chl *a* data during the cold season along the Gulf of California and Heras-Sánchez [30] used a combination of SST and Chl *a* for a period of 18 years included in their regionalizations the coastal zone of the Gulf. Our results agree with Hidalgo-González and Álvarez-Borrego [43], and also are very similar with Heras-Sánchez [30], who identified four regions of the Gulf of California with three trophic levels with a clear seasonality. The results of this work are different from Santamaría-del-Angel et al. [6], because they obtained 14 regions using eight years chlorophyll weekly composites of the Coastal Zone Color Scanner data. The high chlorophyll variability associated to physical phenomena such as the movement of water masses and the upwelling systems generated a greater number of regions along the Gulf of California.

The South Region is directly influenced by the tropical Pacific, and has a more homogenous temperature distribution throughout the year. During the summer with the northward displacement of the Inter-Tropical Convergence Zone (ITCZ) [27], there is advection of water of tropical origin. The SST climatology of this region, maximum values during August and September and minimum during January and February, agreed with the observed by Castro et al. [49] at the entrance of the Gulf of California, showing a seasonal increase of the thermocline, due to the exchanges that take place when alternating the inflow and outflow of water masses. These conditions reported by Castro et al. [49] can extend to the Central Region, and agree with the one reported by Robles and Marinone [25] in the Guaymas Basin. They found winter conditions from December to April and summer conditions from June to October, with an increase of SST from 16° in February–March to 31 °C in August. The SST values were similar to Valenzuela-Sánchez [23], Soto-Mardones et al. [22] and García-Morales et al. [29], as well as in this study. Bernal et al. [50] report that the variability in this region is caused by the effects of seasonal winds and the oceanographic influence from the Pacific Ocean, the seasonal coastal winds develop a sea surface circulation along the Eastern Coast, to the south during October–March and to the north during June–September [8]. Robles and Marinone [25] suggested an advection process of the California Current Water and indicated that Subtropical Subsurface Water may occur around the year in this region, but vertical mixing during winter attenuated their characteristics. Midriff Islands and North Regions presented the lowest minimum in comparison with the ones observed in

Central and South Regions. This minimum value is due to the mixing processes in the Midriff Islands Region, system of fronts influenced by seasonal winds and the development of coastal upwellings in the Eastern Coast [8]. Strong currents take place at the channels between the islands, particularly at the Sonora Coast, El Infiernillo Channel is a narrow and shallow channel between Tiburon Island and mainland that allows a well-mixed water column. The North Region is considered a shallow area [4,22], and subjected to a strong tidal mixing in a vertical form as well by the effect of the gravity currents [51] that modulate the distribution pattern of the temperature, the lowest averaged SST observed. However, the maximum temperature values associated with the summer months cause the climatology between each of the regions to cover the same range of values, indicating little difference between them, probably due to the stratification effect, causing the surface temperature of the sea to be constant in each of the regions. In this region, the influence of the continental climate, surrounded by desert with high solar irradiance may be part important of higher temperatures during warm periods.

The range of average annual values was very wide, and almost of the same magnitude, in all the regions of the Eastern Coast of the Gulf. In such a way, the main frequency of variation in each of the regions was the annual, same frequency that was observed by Garcia-Morales et al. [29], Lavín et al. [28] and Herrera-Cervantes et al. [52]. Soto-Mardones et al. [22] showed that the annual scale determines most of the SST variability with small variations from south to north and a warming and cooling process in the entire gulf. The annual variability in the gulf is determined by the influence of the Tropical Pacific Ocean, through the displacement of the Inter-Tropical Converge Zone (ITCZ) that develops latitudinal displacements of all the current systems and consequently modulating the SST values at a seasonal scale [38,53] as well by the influence of the continental climate [42,45]. In addition, the semiannual variability was observed in the four regions. Soto-Mardones [22] showed an increase of the semianual and annual amplitude to the north, with a maximum in the Islands Region. The winds in the gulf are strong and dominant from the northwest (NW) in winter-spring, and weak in summer–autumn with a main component of the southeast (SE) [42,45], upwelling is present in the Eastern Coast during the cold period, while in the warm period the Eastern Coast presents a strong stratification [22,45], with a great semiannual variability in SST. Furthermore, associated to this, winds, surface circulation changes from warm (coastal circulation to the north) to cold (coastal circulation to the south) periods [8]. In the region of the islands, this effect is more important because in this region other physical processes increase semiannual variability: Tide mixing and water and heat flows between the north and central gulf [4,38]. The seasonal cycle is present in these regions, except in the Northern Region, where it is clear that they dominate the annual and semiannual signal, possibly due to a greater influence of the continental desert climate. Ripa and Marinone [26] indicated a significant seasonal cycle associated to the interaction with the atmosphere through turbulent diffusion of heat, as well horizontal physical processes that contribute in the heat balance in the upper layers. A five-year signal was detected in each of the regions of the coastal zone, this interannual variability is associated with periods such as the El Niño Southern Oscillation. This signal increases from south to Midriff Islands, and then decreased to northern region, the influence of these processes (El Niño–La Niña) in the Gulf have been described by Baumgartner and Christensen [27], Robles and Marinone [25] and Lavín et al. [28]. Lavín et al. [28] explained that the positive anomalies are related to warm water advection of El Niño and negative anomalies associated to La Niña events, with positive anomalies larges in the Central Gulf and the negatives did not have a pattern. Robles and Marinone [25] mention that this signal is attenuated in the central region by mixing processes in this place; in this work, this signal is greater in the region of the Midriff Islands and present in the northern region, according to Herrera-Cervantes et al. [52] concludes that it can be observed in the northern region. Although the spectral analysis requires repeating 10 times the observations of a particular period, it is evident in the region of the islands a variability associated with decadal changes; in the other regions, it is not observed. This frequency of variability, as well as a greater variability in interannual scales, El Niño-La Niña, suggests that the region of the islands is potentially more susceptible to long period events.

The SST variability analysis indicated that this variable is determined mainly by physical and climatological processes in different timescales, with an influence in the coastal ecosystems either a positive or negative effect and consequently to the distribution of the organism. García-Morales et al. [29] found indicating that the SST and chlorophyll *a* (Chl *a*) in the central coastal zone of Sonora can influence on the pelagic ecosystems providing productive and biologically rich habitats of diverse species, some of them of commercial interest reported this effect. Nevárez-Martínez et al. [54] obtained similar results when analyzing the distribution and abundance of the Pacific sardine (*Sardinops sagax*) in the Gulf of California, and its relationship with the environment, determining that the distribution is influenced by the SST and winds that causes upwelling effects. On the other hand, García-Morales et al. [55] determined the influence of environmental variability in the distribution of whales in the Gulf of California, based on the SST and Chl *a*, they obtained that the largest number of whales were during the cold season of the Gulf of California and the lowest number during the warm season, concluding that the SST influence the relative abundance of the whales while the concentration of Chl *a* influences its distribution.

5. Conclusions

There is an increase (decrease) of SST mean value from south to north region during the warm (cold) period, with different duration in time and effects in the SST mean values.

A clear semiannual variability in the SST climatology was observed with maximum values in August and September while the minimum values were during January and February. Ranges of the transition periods during summer and winter were different.

Statistical analysis showed that the Eastern Coastal Zone of the Gulf of California can be grouped in four regions. During cold period mean monthly SST values were significant different, something that is not observed during warm period due to homogenization process in the water column.

Based on the results in this research, the four regions of the coastal zone of Sonora have different climatology and transitions periods.

In each one of the regions, the annual variability was the main frequency of variability followed by the variability associated with semiannual, seasonal and interannual events.

The SST analysis showed that the variability is determined by physical and climatological and processes that present different timescales, developing an influence in oceanographic processes as well as in environmental conditions of the coastal zone of Sonora.

Author Contributions: C.M.R.-T. and J.E.V.-H. wrote and edited this article, processed, analyzed and interpreted the sensor of SST data. R.G.-M., H.H.-C., J.L.-M. and L.F.E.-O. reviewed and edited this article. J.L.-M. is responsible of the project of Consejo Nacional de Ciencia y Tecnología (CONACYT). G.F.-P. did the statistical analysis of the data obtained from the satellite images.

Funding: National Council of Science and Technology (CONACYT) project: Respuestas poblacionales de algunas especies marinas del Golfo de California al Cambio Climático Global. Clave CB-2015-256477.

Acknowledgments: This research was financed by the project of Consejo Nacional de Ciencia y Tecnología (CONACYT) Respuestas poblacionales de algunas especies marinas del Golfo de California al Cambio Climático Global. Clave CB-2015-256477. C.M.R.-T. is a CONACYT fellow.

Conflicts of Interest: The authors declare no conflict of interest.

References

1. Yañez-Arancibia, A.; Day, J.W. La Zona Costera frente al cambio climático: Vulnerabilidad de un sistema biocomplejo e implicaciones en el manejo costero. In *Impactos del Cambio Climático Sobre la Zona Costera*, 1st ed.; Yañez-Arancibia, A., Ed.; Instituto Nacional de Ecología (INE-SEMARNAT): México D.F.; Instituto de Ecología A.C. (INECOL): Xalapa, Veracruz, México; Texas Sea Grant College Program: College Station, TX, USA, 2010; pp. 12–35, ISBN 978 607-7579-17-5.

2. Hernández-Ayón, J.M.; Camacho-Ibar, V.F.; Mejía-Trejo, A.; Cabello-Pasini, A. Variabilidad del CO_2 total durante eventos de surgencia en Bahía San Quintín, Baja California, México. In *Carbono en Ecosistemas Acuáticos*, 1st ed.; Hernández-de la Torre, B., Gaxiola-Castro, G., Eds.; Instituto Nacional de Ecología, SEMARNAT: México D.F.; Centro de Investigación Científica y de Educación Superior de Ensenada, CICESE: Ensenada, Baja California, México, 2007; pp. 187–200, ISBN 978-968-817-855-3.

3. Martínez-López, A.; Cervantes-Duarte, R.; Reyes-Salinas, A.; Valdez-Holguín, J.E. Cambio estacional de clorofila *a* en la Bahía de La Paz, Baja California Sur. *Hidrobiológica* **2001**, *11*, 45–52.

4. Lavín, M.F.; Marinone, S.G. An overview of the physical oceanography of the Gulf of California. In *Nonlinear Processes in Geophysical Fluid Dynamics*, 1st ed.; Velasco-Fuentes, O.U., Sheinbaum, J., Ochoa-de la Torre, J.L., Eds.; Kluwer Academic Publishers: Dordrecht, The Netherlands, 2003; pp. 173–204, ISBN 978-94-010-0074-1.

5. Álvarez-Borrego, S. Flujos de carbono en los Golfos de California y México. In *Carbono en Ecosistemas Acuáticos*, 1st ed.; Hernández-de la Torre, B., Gaxiola-Castro, G., Eds.; Instituto Nacional de Ecología, SEMARNAT: México D.F.; Centro de Investigación Científica y de Educación Superior de Ensenada, CICESE: Ensenada, Baja California, México, 2007; pp. 337–353, ISBN 978-968-817-855-3.

6. Santamaría-del-Angel, E.; Álvarez-Borrego, S.; Muller-Karger, F.E. Gulf of California biogeographic regions based on coastal zone color scanner imagery. *J. Geophys. Res.* **1994**, *99*, 7411–7421. [CrossRef]

7. Álvarez-Borrego, S. Physical, chemical and biological oceanography of the Gulf of California. In *The Gulf of California: Biodiversity and Conservation*, 2nd ed.; Brusca, R., Ed.; The University of Arizona Press: Tucson, AZ, USA, 2010; pp. 24–48, ISBN 978-0816500109.

8. Badan-Dangon, A.; Koblisnksy, C.J.; Baumgartner, T. Spring and summer in the Gulf of California: Observations of surface thermal patterns. *Oceanol. Acta* **1985**, *8*, 13–22.

9. Álvarez-Arellano, A.D.; Gaitán-Moran, J. Lagunas costeras y el Litoral Mexicano: Geología. In *Lagunas Costeras y el Litoral Mexicano*, 1st ed.; de la Lanza-Espino, G., Cáceres-Martínez, C., Eds.; Universidad Autónoma de Baja California Sur: La Paz, Baja California Sur, México, 1994; pp. 13–74, ISBN 968-896-048-9.

10. Reyes, A.C.; Lavín, M.F. Effects of the autumn–winter meteorology upon the surface heat lost in the Northern Gulf of California. *Atmósfera* **1997**, *10*, 101–123.

11. Ocean Observations Panel for Climate (OOPC). EOV Spec Sheet: Sea Surface Temperature, Global Ocean Observations System (GOOS) Panel for Physics, EOV-SeaSurfaceTemperature v5.2. Available online: http://goosocean.org/index.php?option=com_oe&task=viewDocumentRecord&docID=17466 (accessed on 8 August 2018).

12. Filipponi, F.; Valentini, E.; Taramelli, A. Sea Surface Temperature changes analysis, an Essential Climate Variable for Ecosystems Service Provisioning. In Proceedings of the 2017 9th International Workshop of Multitemporal Remote Sensing Images (MultiTemp2017), Brugge, Belgium, 27–29 June 2017; Institute of Electrical and Electronic Engineers (IEEE): New York, NY, USA, 2017; pp. 244–251.

13. Rojas-Acuña, J.; Eche-Llenque, J.C. La temperatura de la superficie del mar peruano a partir de imágenes AVHRR/NOOA (2000–2003). *Rev. Investig. Fis.* **2006**, *9*, 24–30.

14. Harley, C.D.G.; Hughes, A.R.; Hultgren, K.M.; Miner, B.G.; Sorte, C.J.B.; Thornber, C.S.; Rodríguez, L.F.; Tomanek, L.; Williams, S.L. The impacts of climate change in coastal marine ecosystems. *Ecol. Lett.* **2006**, *9*, 228–241. [CrossRef] [PubMed]

15. Doney, S.C.; Ruckelshaus, M.; Duffy, J.E.; Barry, J.P.; Chan, F.; English, C.A.; Galindo, H.M.; Grebmeier, J.M.; Hollowed, A.W.; Knowlton, N.; et al. Climate change impacts on marine ecosystems. *Annu. Rev. Mar. Sci.* **2012**, *4*, 11–37. [CrossRef] [PubMed]

16. Brierley, A.S.; Kingsford, M.J. Impacts of climate change on marine organisms and ecosystems. *Curr. Biol.* **2009**, *19*, 602–614. [CrossRef] [PubMed]

17. Morales-Hernández, J.C.; Carrillo-González, F.M.; Farfán-Molina, L.M.; Cornejo-López, V.M. Cambio de cobertura vegetal en la región de Bahía de Banderas, México. *Rev. Colomb. Biotecnol.* **2016**, *18*, 17–29. [CrossRef]

18. Longhurst, A.R. *Ecological Geography of the Sea*, 2nd ed.; Elsevier Academic Press: San Diego, CA, USA, 2007; p. 5, ISBN 978-0-12-455521-1.

19. Gentemann, C. Three way validation of MODIS and AMRS-E sea surface temperatures. *J. Geophys. Res.* **2014**, *119*, 2583–2598. [CrossRef]

20. Klemas, V. Remote Sensing Techniques for studying Coastal Ecosystems: An Overview. *J. Coast. Res.* **2011**, *27*, 2–17. [CrossRef]

21. Rajeesh, R.; Dwarakish, G.S. Satellite oceanography-A review. *Aquat. Procedia* **2015**, *4*, 165–172. [CrossRef]
22. Soto-Mardones, L.; Marinone, S.G.; Parés-Sierra, A. Variabilidad espaciotemporal de la temperatura superficial del mar en el Golfo de California. *Cienc. Mar.* **1999**, *25*, 1–30. [CrossRef]
23. Valenzuela-Sánchez, C.G. Tendencia de la Temperatura Superficial del Mar, Nivel Medio del Mar e Incidencia de Vientos en la Región Central del Golfo de California. Master's Thesis, Maestría en Biociencias de la Universidad de Sonora, Hermosillo, Sonora, México, 25 August 2016.
24. Escalante, F.; Valdez-Holguín, J.E.; Álvarez-Borrego, S.; Lara-Lara, J.R. Variación temporal y espacial de la temperatura superficial del mar, clorofila *a* y productividad primaria en el Golfo de California. *Cienc. Mar.* **2013**, *39*, 203–215. [CrossRef]
25. Robles, J.M.; Marinone, S.G. Seasonal and interannual Thermohaline variability in the Guaymas Basing in the Gulf of California. *Cont. Shelf Res.* **1987**, *7*, 715–733. [CrossRef]
26. Ripa, P.; Marinone, S.G. Seasonal variability of temperature, salinity, velocity and sea level in the central Gulf of California, as inferred from historical data. *Q. J. R. Meteorol. Soc.* **1989**, *115*, 887–913. [CrossRef]
27. Baumgartner, T.R.; Christensen, N. Coupling of the Gulf of California to large-scale interannual climatic variability. *J. Mar. Res.* **1985**, *43*, 825–848. [CrossRef]
28. Lavín, M.F.; Palacios-Hernández, E.; Cabrera, C. Sea Surface Temperature Anomalies in the Gulf of California. *Geofís. Int.* **2003**, *42*, 363–375.
29. García-Morales, R.; López-Martínez, J.; Valdez-Holguín, J.E.; Herrera-Cervantes, H.; Espinosa-Chaurand, L.D. Environmental Variability and Oceanographic Dynamics of the Central and Southern Coastal Zone of Sonora in the Gulf of California. *Remote Sens.* **2017**, *9*, 925. [CrossRef]
30. Heras-Sánchez, M.C. Biosimulación de la Producción Primaria en el Golfo de California. Ph.D. Thesis, Doctorado en Ciencias en Especialidad en Biotecnología del Instituto Tecnológico de Sonora, Ciudad Obregón, Sonora, México, 22 June 2018.
31. Merino-Ibarra, M. The coastal zone of Mexico. *Coast. Manag.* **1987**, *15*, 27–42. [CrossRef]
32. Álvarez-Borrego, S.; Lara-Lara, J.R. The physical environment and primary production in the Gulf of California. In *The Gulf and Peninsular Province of the Californias*, 1st ed.; Dauphin, J.P., Simoneit, B.R.T., Eds.; American Association of Petroleum Geologist: Tulsa, OK, USA, 1991; pp. 555–567, ISBN 978162981130.
33. Lluch-Cota, S.E. Coastal upwelling in the Eastern Gulf of California. *Oceanol. Acta* **2000**, *23*, 731–740. [CrossRef]
34. Espinosa-Carreón, L.; Valdez-Holguín, E. Variabilidad interanual de clorofila en Golfo de California. *Ecol. Apl.* **2007**, *6*, 83–92. [CrossRef]
35. Hernández-Ayón, J.M.; Chapa-Balcorta, C.; Delgadillo-Hinojosa, F.; Camacho-Ibar, V.F.; Huerta-Díaz, M.A.; Santamaría-del-Angel, E.; Galindo-Bect, S.; Segovia-Zavala, J.A. Dinámica del carbono inorgánico disuelto en la Región de las Grandes Islas del Golfo de California. *Cienc. Mar.* **2013**, *39*, 183–201. [CrossRef]
36. Hidalgo-González, R.M.; Álvarez-Borrego, S.; Zirino, A. Mezcla en la Región de las Grandes Islas del Golfo de California: Efecto en la pCO_2 superficial. *Cienc. Mar.* **1997**, *23*, 317–327.
37. Álvarez-Borrego, S. Oceanografía de la Región de las Grandes Islas. In *Bahía de Los Ángeles: Recursos Naturales y Comunidad. Línea Base 2007*, 1st ed.; Danemann, G.D., Ezcurra, E., Eds.; Secretaría del Medio Ambiente y Recursos Naturales, Instituto Nacional de Ecología: México D.F.; Pronatura Noroeste A.C.: Ensenada, Baja California, México; San Diego Natural History Museum: San Diego, CA, USA, 2008; pp. 45–65, ISBN 978-968-817-891-1.
38. Lavín, M.F.; Beier, E.; Badan, A. Estructura hidrográfica y circulación del Golfo de California: Escalas estacional e interanual. In *Contribuciones a la Oceanografía Física en México*, 1st ed.; Monografía No. 3; Lavín, M.F., Ed.; Unión Geofísica Mexicana: Ensenada, Baja California, México, 1997; pp. 141–172, ISBN 968-7829-00-1.
39. Kilpatrick, K.A.; Podesta, G.P.; Evans, R. Overview of the NOAA/NASA Advanced Very High Resolution Radiometer Pathfinder algorithm for sea surface temperature and associated matchup database. *J. Geophys. Res.* **2001**, *106*, 9179–9197. [CrossRef]
40. Allega, L.; Cozzolino, E.; Pisoni, J.P.; Piccolo, M.C. Comparison of SST L3 products generated from the AVHRR and MODIS sensors in front of the San Jorge Gulf, Argentina. *Rev. Teledetec.* **2017**, *50*, 17–26. [CrossRef]
41. Krzanowski, W.J.; Hand, D.J. *ROC Curves for Continues Data*, 1st ed.; Chapman and Hall/CRC Monographs on Statistics and Applied Probability: Boca Raton, FL, USA, 2009; p. 37, ISBN 978-1-4398-0021-8.

42. Roden, G.I. Oceanographic and meteorological aspects of the Gulf of California. *Pac. Sci.* **1958**, *12*, 21–45.
43. Hidalgo-González, R.; Álvarez-Borrego, S. Chlorophyll profiles and the water column structure in the Gulf of California. *Oceanol. Acta* **2001**, *24*, 19–28. [CrossRef]
44. Torres-Orozco, E. Análisis Volumétrico de las Masas de Agua del Golfo de California. Master's Thesis, Maestría en Oceanografía Física del Centro de Investigación Científica y Educación de Ensenada, Ensenada, Baja California, México, 1993.
45. Álvarez-Borrego, S. Gulf of California. In *Estuaries and Enclosed Areas*, 1st ed.; Ketchum, B.H., Ed.; Elsevier Scientific Publishing Company: Amsterdam, The Netherlands, 1983; Volume 26, pp. 427–449, ISBN 0-444-41921-7.
46. Fu, L.L.; Holt, B. Internal waves in the Gulf of California: Observations from a spaceborne radar. *J. Geophys. Res.* **1984**, *89*, 2053–2060. [CrossRef]
47. López, M.; Candela, J.; Argote, M.L. Why does the Ballenas Channel have the coldest SST in the Gulf of California? *Geophys. Res. Lett.* **2006**, *33*. [CrossRef]
48. Round, F.E. The phytoplankton of the Gulf of California-Part I. Its composition, distribution and contribution to the sediments. *J. Exp. Mar. Biol. Ecol.* **1967**, *1*, 76–97. [CrossRef]
49. Castro, R.; Mascarenhas, A.S.; Durazo, R.; Collins, C.A. Variación estacional de la temperatura y salinidad en la entrada del Golfo de California, México. *Cienc. Mar.* **2000**, *26*, 561–583. [CrossRef]
50. Bernal, G.; Ripa, P.; Herguera, J.C. Variabilidad oceanográfica y climática en el Bajo Golfo de California: Influencias de trópico y pacifico norte. *Cienc. Mar.* **2001**, *27*, 595–617. [CrossRef]
51. Lavín, M.F.; Godínez, V.; Álvarez, L.G. Inverse-estuarine features of the Upper Gulf of California. *Estuar. Coast. Shelf Sci.* **1998**, *47*, 769–795. [CrossRef]
52. Herrera-Cervantes, H.; Lluch-Cota, D.B.; Gutiérrez-de-Velasco, G.; Lluch-Cota, S.E. The ENSO signature in sea-surface temperature in the Gulf of California. *J. Mar. Res.* **2007**, *65*, 589–605. [CrossRef]
53. Flores-Morales, A.L.; Páres-Sierra, A.; Marinone, S.G. Seasonal variability of the sea surface temperature in the Eastern Tropical Pacific Ocean. *Geof. Int.* **2009**, *48*, 337–349.
54. Nevárez-Martínez, M.O.; Lluch-Belda, D.; Cisneros-Mata, M.A.; Santos-Molina, J.P.; Martínez-Zavala, M.A.; Lluch-Cota, S.E. Distribution and abundance of the Pacific sardine (*Sardinops sagax*) in the Gulf of California and their relation with the environment. *Prog. Oceanogr.* **2001**, *49*, 565–580. [CrossRef]
55. García-Morales, R.; Pérez-Lezama, E.L.; Shirasago-Germán, B. Influence of environmental variability of baleen whales (suborden Mysticeti) in the Gulf of California. *Mar. Ecol.* **2017**, *38*. [CrossRef]

remote sensing

Article

Quality Assessment of Sea Surface Temperature from ATSRs of the Climate Change Initiative (Phase 1)

Christoforos Tsamalis * and Roger Saunders

Met Office, Fitzroy Road, Exeter EX1 3PB, UK; roger.saunders@metoffice.gov.uk
* Correspondence: christoforos.tsamalis@metoffice.gov.uk

Received: 30 January 2018; Accepted: 16 March 2018; Published: 21 March 2018

Abstract: Sea Surface Temperature (SST) observations from space have been made by the Along Track Scanning Radiometers (ATSRs) providing 20 years (August 1991–April 2012) of high quality data. As part of the ESA Climate Change Initiative (CCI) project, SSTs have been retrieved from the ATSRs. Here, the quality of CCI SST (Phase 1) from ATSRs is validated against drifting buoys. Only CCI ATSR SSTs (Version 1.1) are considered, to facilitate the comparison with the precursor dataset ATSR Reprocessing for Climate (ARC). The CCI retrievals compared with drifting buoys have a median difference slightly larger than 0.1 K. The median SST difference is larger in the tropics (~0.3 K) during the day, with the night time showing a spatially homogeneous pattern. ATSR-2 and AATSR show similar performance in terms of Robust Standard Deviation (RSD) being 0.2–0.3 K during night and about 0.1 K higher during day. On the other hand, ATSR-1 shows increasing RSD with time from 0.3 K to over 0.6 K. Triple collocation analysis has been applied for the first time on TMI/ATSR-2 observations and for daytime conditions when the wind speed is greater than 10 m/s. Both day and night results indicate that since 2004, the random uncertainty of drifting buoys and CCI AATSR is rather stable at about 0.22 K. Before 2004, drifting buoys have larger values (~0.3 K), while ATSR-2 shows slightly lower values (~0.2 K). The random uncertainty for AMSR-E is about 0.47 K, also rather stable with time, while as expected, the TMI has higher values of ~0.55 K. It is shown for the first time that the AMSR-E random uncertainty changes with latitude, being ~0.3 K in the tropics and about double this value at mid-latitudes. The SST uncertainties provided with the CCI data are slightly overestimated above 0.45 K and underestimated below 0.3 K during the day. The uncertainty model does not capture correctly the periods with instrument problems after the ATSR-1 3.7 μm channel failed and the gyro failure of ERS-2. During the night, the uncertainties are slightly underestimated. The CCI SSTs (Phase 1) do not yet match the quality of the ARC dataset when comparing to drifting buoys. The value of the ARC median bias is closer to zero than for CCI, while the RSD is about 0.05 K lower for ARC. ARC also shows a more homogeneous geographical distribution of median bias and RSD, although the differences between the two datasets are small. The observed discrepancies between CCI and ARC during the period of ATSR-1 are unexplained given that both datasets use the same retrieval method.

Keywords: ATSRs; sea surface temperature; CCI; ARC; validation; drifting buoys; AMSR-E; triple collocation

1. Introduction

Sea Surface Temperature (SST) is an Essential Climate Variable (ECV) for which there are available observations continuously since 1850 [1,2] made mainly by ships, but also from drifting and moored buoys during the recent decades. SST is directly related to and often dictates the exchanges of heat, momentum and gases between the ocean and the atmosphere [3], making it an important geophysical parameter for climate variability monitoring and prediction, operational weather and ocean forecasting, ecosystem assessment and military operations [4]. Because of its significance, SST observations have been made operationally from space using the AVHRR instruments since 1981, as they offer the advantage of global coverage in contrast to in situ measurements [5,6]. Other satellite instruments designed for SST retrievals are the ATSRs, with ATSR-1 on board ERS-1 from August 1991 and its successor instruments ATSR-2, and AATSR, on board ERS-2 and ENVISAT, respectively, providing 20 years of high quality global SST observations [7]. Indeed, the three ATSRs SSTs outperform not only in comparison to other satellite instruments like AVHRRs, but also, their SST retrievals are of about equal or even better quality to in situ instruments [8,9]. Thus, the measurements from ATSRs have been used to assess the quality of in situ SST observations [10], to bias correct other satellite SST retrievals either directly [11,12] or through an SST analysis system [13], and to estimate the background error covariance parameters in SST analysis systems [14].

The ATSRs produce accurate SST retrievals thanks to their well-calibrated blackbodies and the low-noise infrared detectors cooled by a pair of Stirling-cycle coolers [15], while their dual view capability allows for better cloud detection and correction of atmospheric absorption and aerosols [16–18]. All three ATSRs observed Earth with four channels at 1.6, 3.7, 10.8 and 12 μm, while ATSR-2 and AATSR had three additional visible/NIR channels. The pre-launch calibration of the ATSRs demonstrated that the radiometric noise was below 0.05 K (at a reference of 270 K) in all thermal IR channels and was stable throughout the lifetimes of ATSR-2 and AATSR, and although variable for ATSR-1, the total drift of the mission was only 0.1 K [19]. The temporal stability of AATSR and its good radiometric calibration, especially for the 10.8 μm channel, have been verified against the high spectral resolution interferometer IASI on MetOp-A, which is the reference used by GSICS [20,21]. Similarly, ATSR-2's good radiometric calibration consistency to the level of 0.1 K has been verified by comparing with the high spectral resolution spectrometer AIRS [21].

The CCI SST [22], part of the ESA's Climate Change Initiative (CCI) [23], is an effort to produce a complete and homogeneous dataset of SST designed specifically with the climate quality criteria in mind, i.e., high accuracy and stability, while at the same time providing uncertainties per pixel. Phase 1 of CCI SST covers the period 1991–2012 when data from both ATSRs and AVHRRs are available. The project attempts to increase the global coverage on a daily scale by the use of AVHRRs, which have a larger swath width than ATSRs (2900 vs. 500 km), by combining their observations with the high quality SST retrievals based on ATSRs. The purpose of this study is to assess the quality of the SST retrievals from Phase 1 of the CCI project using comparisons with drifting buoys. The results will be compared with respective match-ups from the ATSR Reprocessing for Climate (ARC) dataset, which was the precursor of the CCI SST project [7,24,25]. As ARC involved only measurements from the ATSRs, in order to facilitate the comparison between CCI and ARC, hereafter, the focus is on the validation of CCI SST from ATSRs only. In Section 2, the datasets are described. Section 3 presents the assessment of CCI SST based on match-ups with drifting buoys, while in Section 4, similar results for ARC are given followed by the comparison with CCI SST. Finally, the conclusions are given in Section 5.

2. Datasets

2.1. Satellite SST Retrievals

The description of Phase 1 CCI SST project (CCI) is given by Merchant et al. [22] and the references therein, while for ARC, see Merchant et al. [7]. A short description is provided here focusing mainly on the differences and the similarities between CCI and ARC. In CCI, the retrieval of SST from ATSR-2 and AATSR brightness temperature observations is based on optimal estimation theory following the work of Merchant et al. [26]. In contrast, in ARC, the inversion of SST from ATSRs observations was based on retrieval coefficients estimated from radiative transfer simulations [27,28]. However, the SST from ATSR-1 in CCI also used the retrieval method of coefficients, in order to account for the impacts of the stratospheric aerosols from the Pinatubo eruption and the less stable performance of ATSR-1. Both CCI and ARC SST retrievals are independent of in situ measurements in contrast to the majority of other satellite SST datasets, which are calibrated against them [29–32]. In both datasets, the cloud mask is based on a probabilistic (Bayesian) approach [33] (with minor differences concerning the versions of the radiative code and the NWP model), while dust aerosols are treated as clouds (i.e., masked) using the infrared dust index [34,35]. The ARC dataset provides more flexibility with SST retrievals calculated from both nadir and dual views either for two or three (during night time) channels, while CCI retrievals are based on dual view only, because of the better atmospheric correction [7,25], with two channels during daytime and three channels during night time. An exception to this is the period for ATSR-1 after the failure of 3.7 μm in May 1992, when the two 10.8 and 12 μm channels are used for both day and night retrievals. In addition to the skin temperature, which is the SST retrieved by the IR radiometers, both ARC and CCI provide the SST at a depth of 20 cm (ARC also provides depths at 1 and 1.5 m). The depth SST is based on a parameterization of the ocean skin effect [36] in order to facilitate the comparison with drifting buoys and the creation of climate datasets merging satellite and in situ observations [37]. In CCI, the depth SST is given at a standardized local time of 10:30 a.m./p.m. in contrast to the actual time of observation in ARC. The Equator crossing time for ATSR-1/2 was 10:30 a.m./p.m. and 30 min earlier for AATSR, stable (deviations less than 2 min) for all three satellite instruments during their lifetime [38]. Finally, the spatial resolution of the two datasets is different with ARC being 0.1° and CCI 0.05°, for the L3U product, which is the assessed product here (L3U stands for SST data from a single orbit file remapped and/or averaged onto a regular grid).

For the application of the triple collocation (or three-way error) analysis [39,40], in order to estimate the random uncertainty of CCI SSTs, use was made of SST retrievals from microwave (MW) observations obtained from TRMM Microwave Imager (TMI) [41] and Advanced Microwave Scanning Radiometer for EOS (AMSR-E) [42]. TMI was on board the TRMM satellite with a low-inclination orbit covering only the tropics and subtropics 40° S–40° N and providing good quality SST observations from 1998–2014 with variable Equator crossing times. AMSR-E on board the AQUA platform measured SST globally from 2002–2011 with an Equator crossing time of 1:30 am/pm. For both instruments, the SST retrieval developed by Remote Sensing Systems (Version 7) was used, with a spatial resolution of 0.25°. The SST retrieval algorithm for both TMI and AMSR-E is a physically-based two-stage regression that expresses SST in terms of the brightness temperatures for all the available channels of each instrument [42,43]. Here, TMI data are used from January 1998–July 2002 and AMSR-E data from August 2002–September 2011.

2.2. In Situ Observations and Quality Control

The assessment of the CCI dataset is based on comparisons with SST observations from drifting buoys available through International Comprehensive Ocean-Atmosphere Data Set (ICOADS) Version 2.5 [44] for the period 1991–1996 and the Global Telecommunication System (GTS) for the period 1997–2012. Drifting buoys had a conservative quality control developed at the Met Office in order to eliminate buoys with gross errors [45]. The spatial collocation criterion between satellites and drifting buoys is 0.05° for CCI and 0.1° for ARC. In both cases, the maximum time window is 3 h, but when several observations from drifting buoys are available, the closest in time to the satellite overpass is chosen. It should be noted that the quality assessment of the ARC dataset provided by Lean and Saunders [25] was based mainly on the comparison of drifting buoy SST with ARC at a depth of 1 m. As the CCI depth is calculated only at 20 cm, the respective ARC depth (at 20 cm) will be used for consistency. However, according to the results of Lean and Saunders [25], the comparison statistics are similar for different SST depths, especially at night.

In both datasets, a common threshold has been applied to eliminate match-ups with a difference between buoys and ATSRs greater or equal to ±3 K. In this way, problematic drifting buoy observations that have passed the above-mentioned quality control and/or cloud affected satellite SST retrievals were eliminated. Although, drifting buoys datasets exist with more elaborate quality control [44,46,47], here the same dataset as in Lean and Saunders [25] is used in order to provide a direct comparison between CCI and ARC. The constant threshold has been chosen because the number of match-ups increases with time following the number of available drifting buoys. This can be seen in Figure 1a by the solid lines for day (red) and night (blue) when the number of match-ups before filtering increases from ~500 in 1991 to more than 30,000 in 2011. Note that the first data from ATSR-1 became available in August 1991, while AATSR ended in April 2012. The choice of the threshold is based on the annual standard deviation of the ATSRs-buoys unfiltered differences which is relatively close to 1 K for both datasets (Figure 1b). Thus, the selected threshold of 3 K is roughly a three sigma elimination of outliers, being mostly a conservative option that does not remove too many pairs.

The annual percentage of match-ups filtered out by the 3 K threshold is in general close to 1% for both datasets either for daytime or night time retrievals (Figure 1a, dashed lines). Nevertheless, there are years when the percentage of filtered match-ups is greater than 2%. For CCI these years are 1998–2001 and 2005, while for ARC (not shown) are 1998 and 2005. The years 1998 and 2005 appear in both datasets, with the percentage of collocations filtered out in 1998 during CCI daytime retrievals approaching 6%. The differences in the actual number of percentages between CCI and ARC indicate that the outliers are not only due to errors in drifting buoys. Furthermore, the difference in quality of drifting buoys from ICOADS and GTS can be clearly seen in Figure 1, with ICOADS buoys (1991–1996) having both smaller rejection percentage and standard deviation than the GTS data (1997–2012). In order to understand better the reason for the outliers, Figure 2a presents the bias (CCI minus drifting buoy SST) for daytime retrievals in 2005, the year with the larger number of outliers ($n = 568$). It can be observed that the outliers generally follow tracks rather than found in isolated locations or specific regions. This suggests drifting buoys with gross errors passed the conservative quality control. In an effort to locate possible regions that are prone to outliers, Figure 2b shows the geographical distribution of the daytime rejected match-ups for the whole period of CCI. For convenience, the number is provided only for 5° grid boxes with more than four rejected match-ups. Two regions that have systematic outliers (more than 30) for 2005 are the Mediterranean Sea and the Japan/East China Sea. In the next section, the location of the outliers will be reassessed in light of the CCI evaluation results.

Throughout this paper, the term bias is used, which by definition implies that drifting buoys provide unbiased observations. Certainly this is not true, at least not for all individual drifting buoys. Another study will focus on the quality of drifting buoys SST observations.

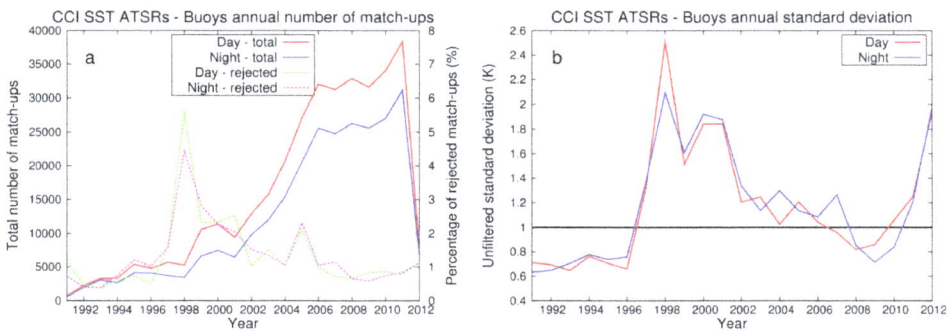

Figure 1. (**a**) Time series of the total number of match-ups (solid lines, left axis) before the application of ±3 K filter during day (red) and night (blue) and of the percentage of eliminated match-ups due to the ±3 K filter (dashed lines, right axis) during day (green) and night (magenta). (**b**) Annual time series of the standard deviation between CCI and drifting buoys before the application of ±3 K filter during day (red) and night (blue).

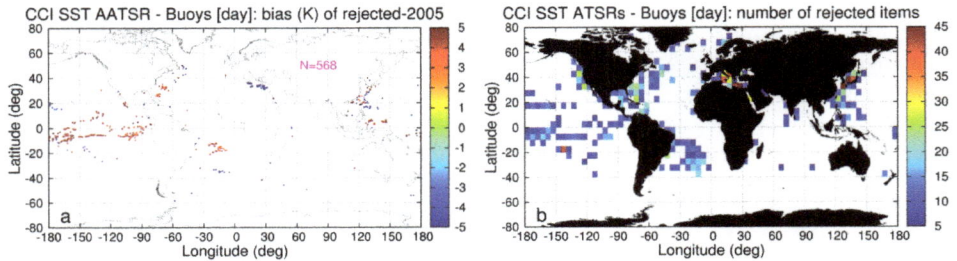

Figure 2. (a) SST difference of CCI minus drifting buoys for filtered match-ups in 2005 during daytime. (b) Number of CCI-buoys filtered match-ups for the period 1991–2012 in 5° boxes during daytime.

3. CCI SST Validation

3.1. Individual Performance

Firstly, the evaluation of each of the three ATSRs will be examined independently against drifting buoys. Table 1 presents the results for whole period with data for each instrument: ATSR-1 (2 August 1991–29 May 1996), ATSR-2 (1 June 1995–31 May 2003) and AATSR (24 July 2002–8 April 2012). Note that for ATSR-2, there are no data for the first six months of 1996 due to the scan mirror failure, while the quality of the data (when available) is reduced during the period mid-January to mid-June 2001 due to the gyro failure [7]. The mean and the median bias (CCI minus drifting buoys) is 0.10–0.13 K for all three instruments both for daytime and night time retrievals indicating a slight overestimation of depth SST by CCI. On the other hand, there are differences among the three instruments regarding the standard deviation or the Robust Standard Deviation (RSD) and between daytime and night time retrievals. During day, the RSD (standard deviation) is 0.55 (0.67) K for ATSR-1, 0.39 (0.53) K for ATSR-2 and 0.28 (0.43) K for AATSR. The respective values during night are 0.55 (0.69) K for ATSR-1, 0.28 (0.46) K for ATSR-2 and 0.22 (0.39) K for AATSR. The standard deviation is larger than RSD as it is more affected by pairs with bigger differences, while RSD better describes the distribution of the SST differences (RSD is 1.4826-times the median absolute deviation). Daytime retrievals have larger values than night time, unsurprising given the use of the 3.7 μm channel during night, which offers a better correction for atmospheric absorption, especially regarding water vapour in the Tropics. The 3.7 μm channel in ATSR-1 worked only for the first 10 months, meaning that the two channels retrieval was used for both day and night for the rest of its lifetime. This explains the absence of a difference at the standard deviation or RSD between day and night for ATSR-1. The performance of the instruments improves with time with ATSR-1 having the worst results and AATSR the best in terms of RSD and standard deviation. However, the apparent improvement with time from Table 1 can also reflect an improvement in terms of performance of the drifting buoys. The results of Table 1 are similar to those reported in Merchant et al. [22], although both the median bias and the RSD are slightly larger here. This probably reflects the less stringent quality control applied here for the drifting buoys, given that the number of match-ups is lower in the study of Merchant et al. [22].

In order to eliminate the variable performance of drifting buoys with time in the evaluation of ATSRs, Table 2 presents the statistics for the periods when the nominal operations of two instruments overlap. During the last seven months of 1995, both ATSR-1 and ATSR-2 were taking observations in tandem. The daytime results of the common period (Table 2) are similar to above-mentioned results for the whole lifetime of ATSR-1 and ATSR-2, with the exception of the mean bias for ATSR-2, which is reduced to 0.05 K. This confirms that ATSR-2 was performing better than ATSR-1. The common period between ATSR-2 and AATSR lasted for about 10 months. The daytime retrievals of ATSR-2 between August 2002 and May 2003 have a similar quality to the whole period (Table 1), but for AATSR both the standard deviation and the RSD are increased approaching the values of ATSR-2. The opposite can be observed for the night time retrievals with AATSR showing the same performance to the lifetime results and the ATSR-2 standard deviation/RSD improving and matching the respective values of AATSR (Table 2). This fact indicates that ATSR-2 and AATSR CCI SSTs are essentially of similar quality, with AATSR being only very slightly better.

Table 1. Statistics (mean bias, standard deviation, median bias, Robust Standard Deviation (RSD) and match-ups number) of the comparisons between CCI ATSRs and drifting buoys. The period with available data for each instrument is indicated in parentheses, while daytime and night time retrievals are treated separately. [†] No data for ATSR-2 during the first 6 months of 1996.

Algorithm	Mean Bias (K)	St. Deviation (K)	Median Bias (K)	RSD (K)	Number
		ATSR-1 (08/1991–05/1996)			
Day	0.13	0.67	0.13	0.55	16,241
Night	0.10	0.69	0.12	0.55	13,832
		ATSR-2 (06/1995–05/2003) [†]			
Day	0.10	0.53	0.10	0.39	66,029
Night	0.11	0.46	0.12	0.28	45,861
		AATSR (07/2002–04/2012)			
Day	0.11	0.43	0.12	0.28	274,795
Night	0.10	0.39	0.11	0.22	217,622

Table 2. As in Table 1 but for the common periods (indicated in parentheses) when observations from two ATSRs are available. For ATSR-1/2, only daytime comparisons are shown, because the night time retrieval algorithms are different due to the early failure of ATSR-1 3.7 µm channel.

Algorithm	Mean Bias (K)	St. Deviation (K)	Median Bias (K)	RSD (K)	Number
		ATSR-1 (06/1995–12/1995)			
Day	0.11	0.71	0.12	0.56	3609
		ATSR-2 (06/1995–12/1995)			
Day	0.05	0.57	0.07	0.39	3791
		ATSR-2 (08/2002–05/2003)			
Day	0.09	0.52	0.11	0.39	10,794
Night	0.14	0.40	0.15	0.25	8203
		AATSR (08/2002–05/2003)			
Day	0.08	0.50	0.10	0.34	11,429
Night	0.12	0.40	0.12	0.25	8719

3.2. Geographical Distribution

Figure 3 presents the maps of median bias, RSD and number of collocations for the 20 years period separately for day and night. The median bias and RSD are calculated for every 5° grid box with at least 20 match-ups inside it for the whole period of ATSRs. Thus, the white boxes indicate lack of sufficient number of collocations. Regarding the overlapping periods between satellites (Table 2), data from the best performing satellite have been used i.e., ATSR-2 and AATSR (instead of ATSR-1 and ATSR-2). It can be seen that during day the CCI depth SST is significantly warmer than the buoys SST in the tropics ∼0.3 K, while in general for the other regions, the median difference is inside the range [−0.1, 0.1] K with CCI being mainly slightly warmer than the buoys. Median differences less than −0.1 K can be observed in the Black Sea and off the coast of Arabian Peninsula. On the other hand, during night time, the CCI is found again warmer than buoys around 0.1–0.2 K, but now, the difference is more homogeneous spatially. Higher differences are found over the northern Indian Ocean and the Western Pacific. Especially, in the case of the north Indian Ocean, there is a sharp transition with the bias being cold off the Arabian Peninsula becoming warm close to the Indian Peninsula. A less significant gradient can be also observed over the eastern part of tropical Atlantic. The cold bias off western Africa and Arabian Peninsula may be related to dust aerosols, which are abundant in these two areas [48,49]. The location of the daytime warm bias in the tropics seems to be related to the absorption of IR radiation by water vapour, which is retrieved less accurately with the two-channel algorithm, used during the day. A very similar geographical pattern for the median difference between the CCI SST analysis and drifting buoys has been found in Merchant et al. [22].

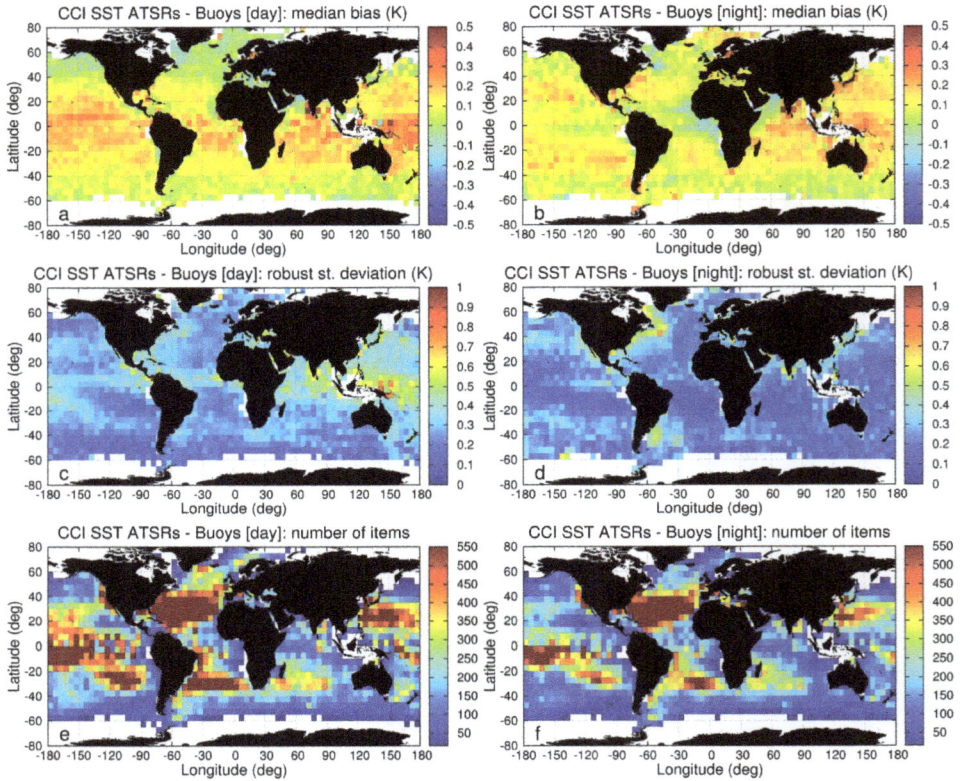

Figure 3. The median bias (**a,b**), robust standard deviation (**c,d**) and number of match-ups (**e,f**) of ATSRs' CCI SST against drifting buoys averaged in 5° × 5° boxes for the period: August 1991–April 2012. The results for daytime (night time) observations are presented in left (right) column. Gridboxes with less than 20 match-ups appear white.

Regarding the robust standard deviation of the difference between CCI and buoys, during the day, larger values ∼0.6 K are found over the Maritime Continent (SE Asia) and Northwest Pacific (Figure 3c). During the night, the spatial distribution is again more homogeneous than daytime and only off the eastern coast of Canada, and the south-western Atlantic Ocean RSD reaches 0.5 K (Figure 3d). The geographical distribution of match-ups is not spatially homogeneous (Figure 3e,f), but this fact does not impact the median difference or the robust standard deviation.

It should be kept in mind that the number of drifting buoys increased considerably with time (Figure 1a and Table 1), especially after 2003, meaning that the maps reflect mostly the performance of

AATSR. The same analysis has been repeated for each one of the three ATSRs (not shown); the results are similar to Figure 3a,c for ATSR-2, and the geographical patterns are very similar, although the values for the bias and the RSD are slightly larger. For ATSR-1, it is difficult to assess as there are far fewer collocations to give statistically-significant results about the geographical patterns of median bias and the RSD.

It can be useful to investigate if there is a specific geographical distribution for the collocations with the largest differences between CCI and drifting buoys. Figure 4 presents the mean bias and number of match-ups for the collocations with absolute difference larger than 1 K. The threshold of 1 K is chosen arbitrarily taking into account that collocations with difference greater than ±3 K are discarded (Section 2), while an important number of match-ups should be left in order to arrive to some conclusions. The comparison of Figure 4a with Figure 3a indicates that the pairs with $|\Delta SST| > 1$ K during daytime contribute to the warm bias observed in the tropics and the western Pacific. For the other regions, the large differences from these collocations cancel out more or less giving a mean value close to 0 K. It is important to note that the number of the collocations per grid box with large SST differences is about 5–15 (Figure 4b) and much lower, at least an order of magnitude, than the total number of collocations (Figure 3e). For the night time retrieval (not shown), there are even fewer pairs with absolute SST differences larger than 1 K. The only geographic patterns emerging during night are the warm biases over the north-western Pacific and around the Indian Peninsula (similarly to Figure 3b). The identified regions with systematic large positive SST differences probably indicate deficiencies in the SST retrieval rather than issues with observations from drifting buoys.

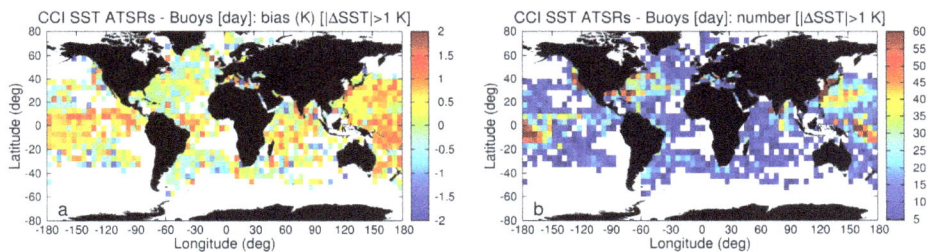

Figure 4. Maps of mean bias (**a**) and number of match-ups (**b**) for the collocations with $|\Delta SST| > 1$ K during daytime. Gridboxes with less than 5 match-ups appear white. Note the different scales to Figure 3.

3.3. Time Series

In order to have a clear indication of how the statistics are evolving with time given that Figure 3 reflects mostly the results of AATSR, Figure 5 presents the time series of CCI for the median bias and the RSD. The median bias is almost always positive with the exception of the last months of 1991 and the years 1997 and 1998 for the daytime retrieval only, when it reaches −0.1 K. From 2002 onwards, the daytime and night time retrievals have the same monthly global median bias of about 0.12 K, a fact clearly reflecting the stability of AATSR. Similar values of the median bias for the daytime and the night time retrievals are also found for the ATSR-1 period, but for ATSR-2 from 1995–2001, there are differences between day and night, although these are in general smaller than 0.1 K. The similar values between day and night for the ATSR-1 after 27 May 1992 (when the 3.7 μm channel failure occurred) are not surprising as the retrieval algorithms are the same, although the daytime cloud mask also uses the 1.6 μm channel. When switching between ATSR-1 and ATSR-2 (the ATSR-2 scan mirror failed in December 1995 and restarted in July 1996), an abrupt change of the median bias can be observed during night (for daytime observations a change also exists, but it is less significant). This is as expected as ATSR-2 uses the 3.7 μm channel in the night time algorithm. The median bias of ATSR-1 is not stable temporally increasing from 1991–1993 reaching 0.3 K and then mainly decreasing towards 0 K. For ATSR-2, the changes are more of stepwise nature with the median bias being mostly stable between them, especially for the night time retrieval. These step changes occur around January 1997 (decrease from 0.15 K–0.05 K during night) and then around October 1999 (increase from 0.05 K–0.2 K during night), with the changes being slightly larger in magnitude for the daytime retrieval. The median bias for the whole 20 years of ATSRs CCI is almost always within ±0.2 K with the exception of the periods 1993–1994 and 2000.

The temporal evolution of the robust standard deviation is simpler than the median bias (Figure 5b). Always, the night time RSD is lower than the daytime by about 0.1 K (for ATSR-1 after May 1992, there are no differences as the day and night retrieval algorithms are the same, using only the 10.8 and 12 μm channels). For ATSR-1, the RSD increases with time from 0.3–0.4 K to about 0.6–0.7 K with the most rapid change occurring during 1992. On the other hand, for ATSR-2 and AATSR, the RSD decreases slightly with time being lower for AATSR and the lowest values of ∼0.21 K (0.28 K) for night time (daytime) obtained during 2008–2010. The switches from ATSR-1 to ATSR-2 are obvious with the RSD reducing from ∼0.6 K to 0.3–0.4 K. However, it is important to note that the RSD of the first months of ATSR-1 (before the failure of 3.7 μm channel) is very similar to the values found during the first months of ATSR-2 operations. The switch from ATSR-2 to AATSR is smooth, especially for the night time retrieval. There is an apparent jump for the daytime RSD from 0.35 K to 0.6 K during February–May 2001 due to the gyro failure on ERS-2.

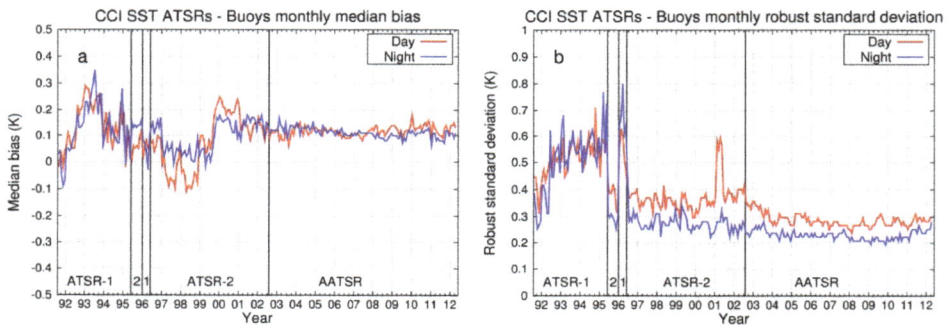

Figure 5. Global monthly time series of median bias (**a**) and robust standard deviation (**b**) of CCI against drifting buoys. The red (blue) line is for daytime (night time) retrievals. The black vertical lines indicate the switch of the time series from one instrument to another as indicated at the bottom of each panel.

Another way to investigate the temporal performance of the CCI against drifting buoys is by presenting how the percentages of match-ups for fixed SST threshold differences evolve. Figure 6 shows the evolution for differences less than 0.1 K or 0.2 K and larger than 1 K or 2 K. The thresholds are ad hoc, but the lowest values represent mostly the optimal cases will the largest thresholds are indicative for possible significant (i.e. in terms of magnitude) events. Similarly with the RSD (Figure 5b), there is a clear separation between night and day (again except for the period after the failure of 3.7 μm channel on ATSR-1), with the night retrievals showing better performance. This stratification between day and night retrievals breaks down for the $|\Delta SST| > 2$ K and the AATSR period for the differences larger than 1 K. Likewise for RSD, the performance improves with time as the percentage within 0.1 or 0.2 K increases, and the percentage larger than 1 K decreases. About 23% (30%) of day (night) retrievals are within 0.1 K from drifting buoys SST. The percentage of collocations within 0.2 K is almost always larger than 40% (50%) for day (night) retrievals. However, the percentages for the ATSR-1 period after the 3.7 μm channel failure are lower being around 14% (27%) within 0.1 K (0.2 K). Poor performance for the same period of ATSR-1 can also be observed in the percentage of match-ups which are larger than 1 K (Figure 6b), reaching values larger than 20% for some months towards the end of the mission. Despite this ATSR-1 period, the 1 K percentage is around 5%. Interestingly, the percentage of collocations that are larger than 2 K does not fluctuate a lot with time being 0.5–1%, although the ATSR-1 period shows more variability from one month to the next. Regarding the period of the ERS-2 gyro failure (beginning of 2001), there is an obvious decrease of the percentages within 0.1 K or 0.2 K and an increase of the percentage larger than 1 K for the daytime retrievals. Nevertheless, the night time retrievals do not seem to be impacted significantly by the gyro failure. In addition, the gyro failure does not affect the percentage of collocations larger than 2 K, which means that any degradation of the CCI due to the gyro failure is within 2 K. In Figure 6b, a spike can be observed for both 1 K and 2 K percentages in August/September 2009, which does not coincide with any known instrumental issue for AATSR, thus indicating an issue with the drifting buoys or at least their quality control.

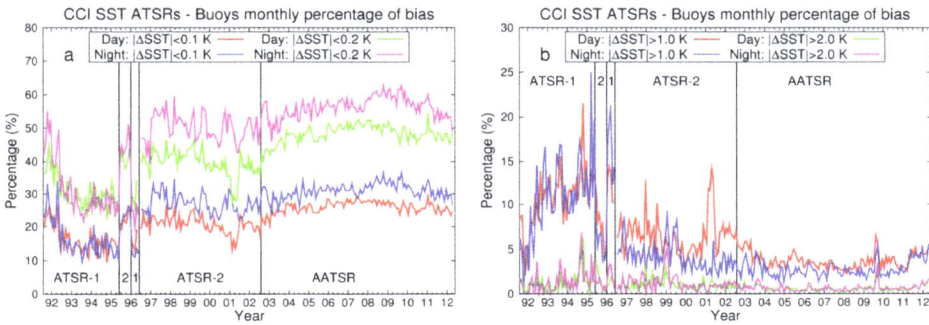

Figure 6. Temporal evolution of monthly percentages of CCI and drifting buoys SST absolute differences for various thresholds separately for daytime and night time retrievals. (**a**) Within 0.1 K and 0.2 K. (**b**) Larger than 1 K and 2 K.

3.4. Triple Collocation Analysis

The triple collocation or three way error analysis uses three independent datasets observing the same quantity (here SST) in order to calculate the standard deviation of the error for every one given the fact that the observations are uncorrelated [39,40]. If the assumptions behind the triple collocation hold, the standard deviation of the error is the random error for observations of each instrument (or more correctly the random uncertainty). Here, the latest version of AMSR-E SST is used (Version 7, which differs from the versions used by Lean and Saunders [25] and O'Carroll et al. [39]). For the first time to the best of our knowledge, the results are extended backwards in time by using the SST observations of TMI (Version 7.1) before August 2002. Thus, ATSR-2/TMI are used for the years 1998–2001 and AATSR/AMSR-E for the years 2003–2011, while for 2002, both sets are used. The match-ups of ATSRs with buoys are collocated with the closest observation of MW SST (provided with resolution of 0.25°) in a time window of 180 min for all three observations. As previously (Section 2), only collocations with absolute differences less than 3 K are used for any combination of the three datasets. In the past, the triple collocation approach has been used only for night time SST observations in order to avoid the impact of the diurnal thermocline [50–52]. However, here, for the first time to the best of our knowledge, the triple collocation is applied also during day, only when the wind speed observed from the MW imagers is larger than 10 m/s. This wind threshold is based on the results of previous studies [50,51,53,54], as it is expected that when the wind speed is above 10 m/s, the diurnal SST cycle should be absent especially for the ATSR overpass time of 10:30 LT.

Figure 7 presents the annual results of the triple collocation analysis. It can be noted that after 2003 both AATSR and drifting buoys have a stable standard deviation of the error ∼0.22 K both during day and night. For the period before 2004, the daytime results indicate that ATSR-2 and AATSR are at the same level, while the night time values suggest that ATSR-2 is slightly better. The drifting buoys before 2004 show a worse quality with values for the standard deviation of the error of 0.3 K or above for the daytime results. Regarding AMSR-E, the night time results indicate a rather stable value of ~0.47 K, but for the daytime conditions this value increases to about 0.6 K. The TMI has higher values than AMSR-E for the standard deviation of the error being 0.55 K during night and even larger during day. The poorer performance of TMI in comparison to AMSR-E is not surprising given that TMI lacks the 7 GHz channel, which is more sensitive to SST [42,43]. The daytime results before 2004 show a lot of annual variability for all three types of instruments. This variability coincides with a number of collocations less than 300 (cyan line) and even less than 100 for 1999 to 2001. Note that for 1998, there are not even 20 collocations during daytime to produce statistically meaningful results. The variability is expected as the number of collocations reduces to lower than about 500 [40], so the daytime results before 2004 should be interpreted with caution.

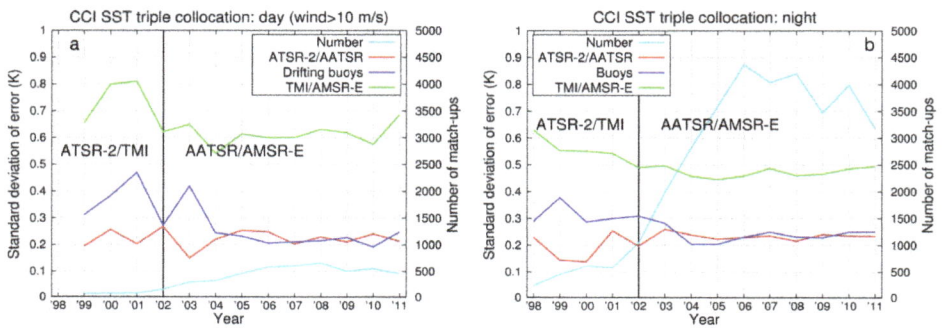

Figure 7. Triple collocation annual results. For the period January 1998–July 2002 observations from ATSR-2/TMI are used, while from August 2002–September 2011, AATSR/AMSR-E are used. The cyan line indicates the number of collocations (right axis), while the red, blue and green lines the standard deviation of the error (left axis) for ATSR-2/AATSR, drifting buoys and TMI/AMSR-E, respectively. (a) During daytime when the TMI/AMSR-E wind speed is larger than 10 m/s and (b) during night time.

It is important to know if the results of the triple collocation vary geographically. Figure 8 presents the Hovmoller diagrams for the standard deviation of the error of the three instrument types (IR imagers, in situ and MW imagers) together with the number of collocations. It can be seen that for the ATSRs (Figure 8a) and the drifting buoys (Figure 8b), the results are rather stable both in time and latitude with the majority of the grid boxes having values between 0.1 K and 0.3 K. The observed variability around the value of 0.22 K and mainly few values outside the range of [0.1, 0.3] K are the result of relatively low number of collocations, especially when the number is less than ~100 (Figure 8d). On the other hand, AMSR-E does show a variable behaviour with latitude with observations north of 35°S having a standard deviation of error ~0.3 K, while south of 35°S, the value doubles to 0.5–0.6 K. This characteristic performance of AMSR-E is stable with time. To the best of our knowledge, this is the first time that the latitudinal variability of AMSR-E SST is reported. While such a latitudinal dependence is expected for an MW SST retrieval using the 11-GHz channel [43], this should not be the case for the 7-GHz channel, which is available on AMSR-E. Regarding TMI, the standard deviation of error is more stable with values 0.5–0.6 K, similar to what is found in Figure 7b. There is also an indication that TMI performance degrades south of 35°S, but as the number of collocations is small (about 60) for this latitude, it is hard to arrive at firm conclusions. The small number of collocations can sometimes prevent the application of the triple collocation approach, e.g., the white grid boxes in Figure 8 for 2010 and 2011 in northern tropics. It is important to note that the triple collocation during night samples mainly the Southern Hemisphere mid-latitudes (Figure 8d). This fact and the variable performance of AMSR-E with latitude could explain the differences seen in Figure 7 between day and night for this instrument.

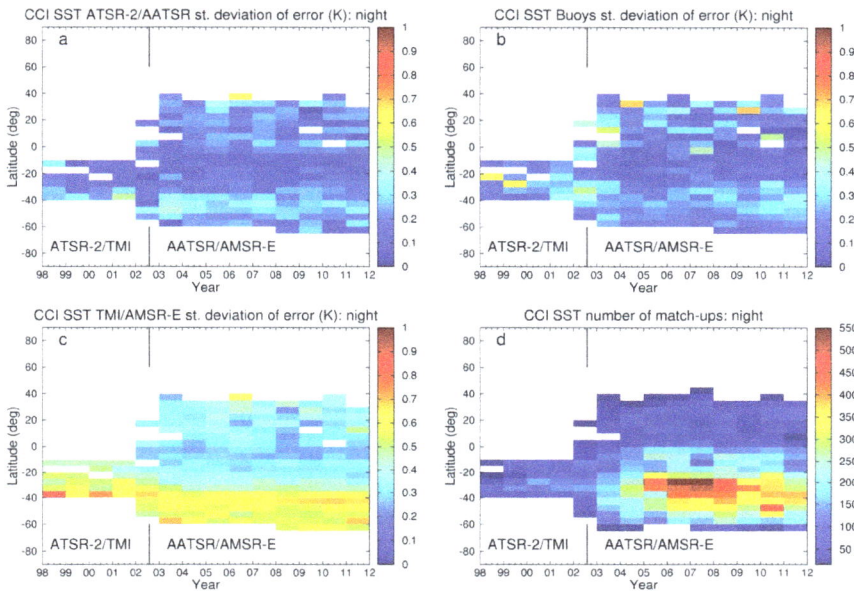

Figure 8. Hovmoller diagrams of triple collocation during night. Standard deviation of the error for (**a**) ATSR-2/AATSR, (**b**) drifting buoys and (**c**) TMI/AMSR-E, while (**d**) shows the number of collocations. Until end of July 2002, the observations of ATSR-2/TMI have been used and afterwards those of AATSR/AMSR-E. Only grid boxes with at least 20 match-ups are plotted.

Previous studies using the triple collocation approach have found for the drifting buoys standard deviation of the error values of 0.23 K [39], 0.26 K (median value) [29], 0.15–0.19 K [25], 0.20 K [55] and 0.21–0.22 K [56], which are very close to the value of ~0.22 K reported here. The respective values found for AATSR are 0.16 K [39], 0.14 K [25] and 0.15–0.30 [56], which are somehow smaller than the value of ~0.22 K of this study, but in agreement with Xu and Ignatov [56], who have also applied the triple collocation to the CCI. The results for AMSR-E standard deviation of the error are 0.42 K [39], ~0.48 K [25] and 0.28 K [55]. The first two results compare favourably with the value of ~0.47 K found here. The much lower value for AMSR-E reported by Gentemann [55] despite using the same dataset version to this study could be a result of averaging the SST from the IR imager (MODIS in Gentemann [55]) to the spatial resolution of the AMSR-E (i.e., 0.25°).

3.5. Validation of Uncertainty

One significant advantage of the CCI dataset is that each SST value is provided with its uncertainty [57,58]. Thus, it is important to assess not only the quality of the SST, but also its associated uncertainty. Figure 9a,b presents how the mean bias and the standard deviation vary against the binned CCI uncertainty for day and night, respectively. Ideally, the bias (red crosses) should be 0 K and the standard deviation (blue squares) should lie on the green line given by the equation $y = \sqrt{0.2^2 + x^2}$, where 0.2 K is the uncertainty of the drifting buoys and x the uncertainty of CCI. Here, the assumption is made that the uncertainty of the drifting buoys is 0.2, which is in line with the above-mentioned results and studies and the review paper by Kennedy [59]. For CCI daytime retrievals, the mean bias increases slightly with increasing CCI uncertainty. Above 0.9 K, the mean bias is oscillating around 0 K, but there, the number of match-ups is very small (right axis, cyan bars). The number distribution indicates that the majority of daytime CCI has uncertainty values between 0.4 K and 0.6 K. The standard deviation lies rather close, although below to the theoretical line for the uncertainty interval [0.45, 0.9] K, an indication of a slight overestimation of CCI daytime uncertainty. Below the uncertainty level of 0.3 K, the standard deviation is almost constant at 0.4 K, which is higher than the expected value of 0.2 K, thus indicating an underestimation of CCI uncertainty. In total, about 78% of the match-ups are within the combined uncertainty (CCI and drifting buoys), while 94% of them are within twice the combined uncertainty.

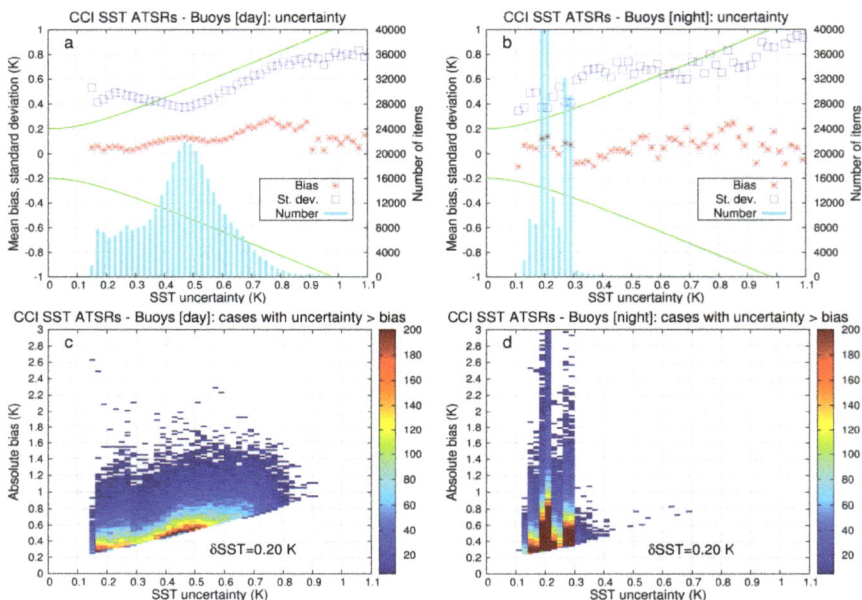

Figure 9. (Top row) Binned mean bias (red crosses) and standard deviation (blue squares) of the difference CCI minus drifting buoys against the CCI uncertainty for (**a**) day and (**b**) night. The size of the bin is 0.02 K. The cyan bars indicates the number of match-ups for each bin (right axis). The green line is the theoretical value of the standard deviation for the match-ups by assuming that the standard deviation of the error for the drifting buoys is 0.2 K. (Bottom row) 2D histograms of the absolute bias versus CCI uncertainty for the number of match-ups with SST difference larger than the combined uncertainty for (**c**) day and (**d**) night. Again, it is assumed that the standard deviation of the error (δSST) for the drifting buoys is 0.2 K, and the size of the bin is 0.02 K for both axis. Only grid boxes with at least five pairs having absolute SST difference larger the combined uncertainty are plotted.

Regarding the CCI night time uncertainty, the situation is more complicated as in general the measurements have an uncertainty of either ~0.2 K or ~0.28 K (Figure 9b). Only a small number of CCI match-ups have uncertainties in the interval [0.13, 0.3] K, and a very limited number has uncertainty larger than 0.3 K. The bias for the majority of match-ups (having uncertainties of about 0.2 or 0.28 K) is 0.1 K, and the standard deviation is close to the theoretical value, but now mostly indicating an underestimation of the CCI night time uncertainty. The underestimation is more evident for the rest of the match-ups, although not too far away from the theoretical value, while now, the mean bias is closer to 0 K. The poorer performance of the night time uncertainties in comparison to daytime is also reflected in the percentages of the match-ups within the combined uncertainty being 57% and 84% for them within twice the value of the combined uncertainty. Given that uncertainty of the drifting buoys is the same between day night, this suggests that the CCI night time uncertainty model needs to be improved.

It is useful to further understand to what degree the uncertainty validation of the CCI depends on the uncertainty value of the drifting buoys and the goodness of the CCI uncertainty model. Figure 9c,d presents the two-dimensional histograms of the absolute bias versus the CCI uncertainty for day and night. In these two figures, only pairs with SST difference outside the expected combined CCI and drifting buoys uncertainties are considered. For the daytime conditions, the majority of the pairs is very close to the border, lying almost parallel to it. This means that with a slight increase of either the CCI or the drifting buoys' uncertainty, these pairs would be successful in terms of uncertainty validation. More specifically regarding the CCI, any uncertainty increase should occur only for the uncertainties in the region of 0.2 K. The increased number of match-ups outside the combined uncertainty in the interval 0.4–0.6 K coincides with the bulk of the uncertainty assignment (Figure 9a). It is worth noting that only a limited number of match-ups (less than 40) shows absolute SST differences larger than 0.8 K not captured by the uncertainty model. On the other hand, the underestimation of CCI night time uncertainty is evident from Figure 9d. It is worth considering that the number of match-ups indicating an underestimation of uncertainty is not huge (compare Figure 9b,d), but still, the absolute SST difference reaches up to 0.8 K when the estimated CCI uncertainty is only 0.2 K.

Figure 10 presents the monthly evolution of the percentage for the match-ups that have differences within the combined uncertainty. In order to verify the impact that the uncertainty of drifting buoys can have on these percentages, three values (0.15 K, 0.20 K and 0.25 K) are chosen, which are compatible with literature and this study. During the day, the percentage of match-ups within the uncertainty improves with time from 70–85%, and this is in line with the previously mentioned percentage of 78% for the whole dataset. However, both the periods after the failure of the ATSR-1 3.7 µm channel (May 1992) and during the ERS-2 gyro failure (beginning of 2001) have lower percentages down to 60% indicating that the uncertainty model does not capture these events well. Although it could be argued that the apparent temporal improvement of the percentage is due to drifting buoys, the difference during day among the three percentages (one for each δSST) is too small to justify this (at least for the δSST considered here). For night time retrievals the percentage within the uncertainty shows obvious monthly fluctuations before 2006, while there is some temporal improvement from 60–75%. Surprisingly, it is the first period of ATSR-2 (till the end of 1996) that shows the poorest performance, about 10% lower than ATSR-1. Now, the choice of the drifting buoys uncertainty has a considerable impact on the percentage by increasing it about 10% when δSST increases from 0.15 K–0.25 K. This is expected given that the night time CCI has a similar uncertainty to drifting buoys (Figure 9b).

To summarize, the daytime uncertainty of CCI is better than the respective one during night, but still in both cases the underlying uncertainty model of CCI should be further improved in order to eliminate the temporal evolution as far as possible and to take into account known periods with degraded instrument performance.

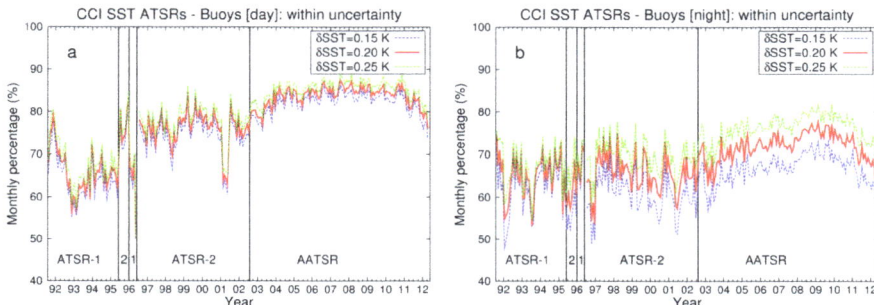

Figure 10. Monthly percentages within combined uncertainty (CCI and drifting buoys) for three different values of uncertainty for the drifting buoys (δSST) during (**a**) day and (**b**) night.

4. Comparison with ARC

The ARC SST was the precursor dataset of the CCI, so it is useful to make a comparison between the two. For this purpose, the match-up dataset of Lean and Saunders [25] is used here, which covers the period August 1991–December 2009 (thus no AATSR/drifting buoys collocations for 2010–2012). Both the observations and the collocation approach used by Lean and Saunders [25] are the same as described in Section 2. Similarly to the quality control applied for CCI (Section 2), a ±3 K threshold is used to eliminate the outliers not removed by the conservative quality control. It is reminded that a three-sigma annual filter has been used in Lean and Saunders [25], while the validation has been performed mainly for 1 m depth SST, which could explain some minor differences between the two studies.

Figure 11 is similar to Figure 3, but presents the comparison between ARC and drifting buoys for the period 1991–2009. The significant warm biases seen in the CCI dataset are absent in ARC both during day and night or at least much less prominent, with ARC being in general in the range [−0.1, 0.1] K from drifting buoys, particularly for night time. The regions appearing consistently with warm SST biases in respect to buoys for both CCI and ARC are off the coast of Indian Peninsula, the Maritime Continent (but not for night time ARC) and the Gulf of Mexico, with the region around the Indian Peninsula being the more challenging (median bias about 0.5 K). During the day, the median bias of CCI is lower than ARC in the Gulf of Mexico and the region of Kuroshio current. There are a few regions with cold biases in ARC, although these are less obvious than CCI with values close to −0.1 K. In terms of robust standard deviation, again the ARC performs better than CCI, and especially during night, the RSD is almost everywhere less than 0.2 K.

However, the direct comparison between Figures 3 and 11 is not really valid as they do not cover exactly the same period, despite the fact that AATSR CCI was pretty stable for its whole lifetime (Section 3). For this reason, Table 3 presents the statistics of the comparison between both CCI and ARC with drifting buoys. The results for all statistical measures of CCI are very similar (maximum difference ±0.03 K) between Tables 1 and 3, despite that the period examined in the Tables is different for all three ATSRs. This shows the robustness of the overall statistics to small changes of the time period considered for CCI. Returning to the comparison between CCI and ARC, it can be seen that the number of match-ups is slightly different between the two (Table 3) with ARC having in general more match-ups. This is the outcome of different spatial resolutions 0.05° for CCI vs. 0.1° for ARC. All statistical parameters show better agreement of ARC with drifting buoys than CCI. Nevertheless, the differences are not large and in general are within ±0.05 K. However, the difference in the statistics between CCI and ARC for ATSR-1 is puzzling, especially regarding the biases. The SST retrieval method is the same between the two datasets, and the only difference is the spatial resolution.

In order to investigate the temporal evolution of quality between CCI and ARC, Figure 12 presents the annual median bias and robust standard deviation. The better performance of ARC than CCI in terms of median bias (being closer to 0 K) can be seen for the whole period from 1991–2009. However, CCI is more consistent regarding the differences between day and night for the periods of ATSR-1 and AATSR. During the ATSR-2 period (1997–2000), the performance of CCI is almost identical to ARC for the daytime retrieval. The sign of the median bias is not the same before 2000 during night, with CCI presenting a warm bias and ARC having a cold bias. The difference in performance between CCI and ARC during night reaches about 0.25 K in 1993–1994 (Figure 12a). As mentioned previously, this big difference is surprising as both CCI and ARC use practically the same cloud mask and SST retrieval method in this period. Concerning the time evolution of the RSD, this is very similar between CCI and ARC, with CCI having larger values by ∼0.05 K from ARC. It is interesting to note that ARC daytime is not impacted by the switch from ATSR-1 to ATSR-2 in 1996, while the CCI night time is not impacted by the gyro failure on ERS-2 in 2001 (in accordance with Figure 5). Furthermore, the daytime ARC RSD is very similar to the night time RSD of CCI.

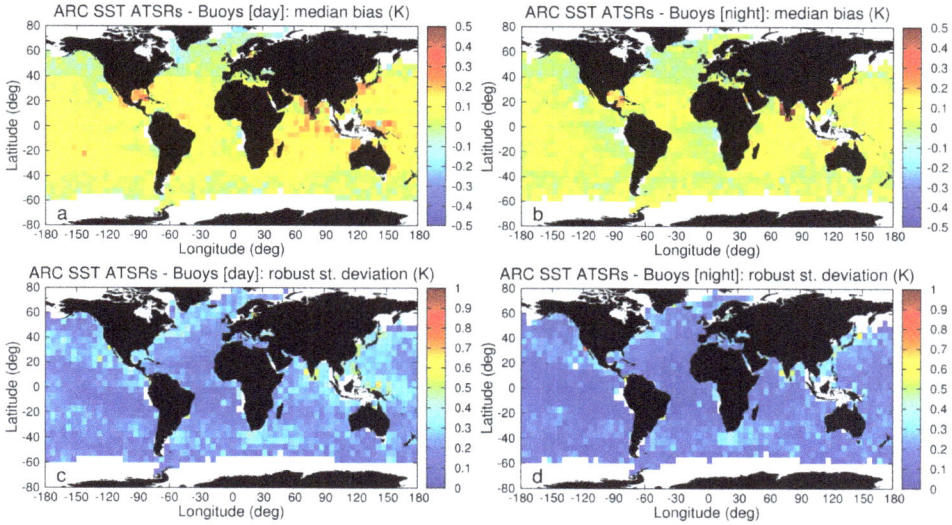

Figure 11. The median bias (**a,b**) and robust standard deviation (**c,d**) of ARC SST minus drifting buoys averaged in 5° × 5° boxes for the period: August 1991–December 2009. The results for daytime (night time) observations are presented in left (right) column. Gridboxes with fewer than 20 match-ups appear white.

Table 3. Same as Table 1. Statistics are provided both for CCI and ARC against drifting buoys for the period indicated in the parentheses.

Algorithm	Mean Bias (K)	St. Deviation (K)	Median Bias (K)	RSD (K)	Number
CCI					
ATSR-1 (08/1991–05/1995)					
Day	0.15	0.65	0.14	0.53	11,060
Night	0.13	0.67	0.14	0.53	9373
ATSR-2 (07/1996–06/2002)					
Day	0.10	0.53	0.10	0.39	50,125
Night	0.10	0.46	0.11	0.28	33,605
AATSR (08/2002–12/2009)					
Day	0.11	0.43	0.11	0.28	194,786
Night	0.11	0.38	0.11	0.22	152,962
ARC					
ATSR-1 (08/1991–05/1995)					
Day	0.01	0.61	0.02	0.46	13,963
Night	−0.09	0.58	−0.07	0.42	10,779
ATSR-2 (07/1996–06/2002)					
Day	0.08	0.48	0.08	0.31	50,639
Night	0.04	0.43	0.04	0.24	37,333
AATSR (08/2002–12/2009)					
Day	0.08	0.37	0.09	0.21	185,114
Night	0.07	0.34	0.06	0.18	156,662

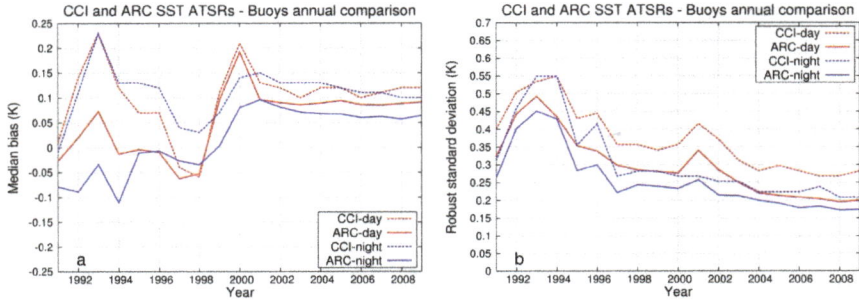

Figure 12. Annual (**a**) median bias and (**b**) robust standard deviation of CCI (dotted lines) and ARC (solid lines) against drifting buoys. The results are given for day (red lines) and night (blue lines) during 1991 to 2009.

Figure 13 shows the monthly evolution of percentages for four ad hoc thresholds (the same as Figure 6), but for ARC. In this way, performance changes at sub-annual time scale are more easily discerned, thus providing complementary information to Figure 12. The superiority of ARC is confirmed with percentages of collocations inside ±0.1 K or ±0.2 K being about 10% higher than CCI. It is worth noting that for night time ARC the percentage within 0.2 K reaches 70% after 2005. Regarding the percentages with absolute differences larger than 2 K, there is no difference between CCI and ARC, although ARC shows smoother evolution. For the percentage of $|\Delta SST|$ larger than 1 K, ARC displays lower values than CCI, especially during the ATSR-1 period by ~5%. The failure of ATSR-1 3.7 μm channel (May 1992) and the ERS-2 gyro (beginning of 2001) are clearly seen in the time series, together with the switch from ATSR-1 to ATSR-2. Although, in ARC, the first months of ATSR-2 have higher percentages (Figure 13a) than the respective period of ATSR-1 (before the 3.7 μm channel failure), as CCI the two instruments seem to have a similar quality.

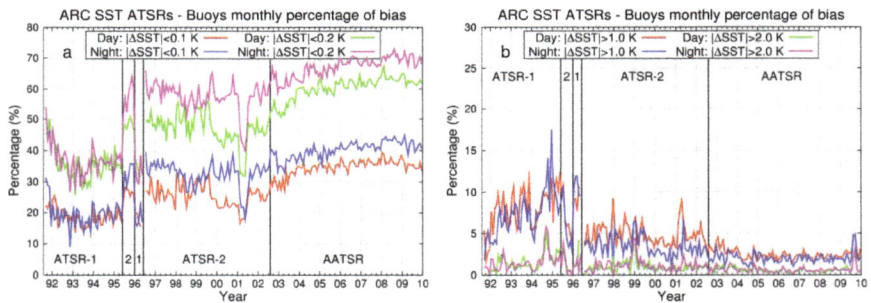

Figure 13. Same as Figure 6, but for ARC during August 1991 to December 2009.

5. Summary and Conclusions

The CCI SSTs (Phase 1, Version 1.1) from the ATSRs covering the period 1991–2012 have been assessed against collocated observations from drifting buoys and also compared with the corresponding comparisons for the ARC (the precursor dataset of CCI SST). Only the performance of CCI SSTs from the ATSRs has been assessed here. For the comparison of the CCI and ARC datasets, it should be noted that apart from ATSR-1, the retrieval methods for the skin SSTs are different. Furthermore, the ARC dataset is provided at a spatial resolution of 0.1°, while CCI has a finer resolution of 0.05°. Because of this, the CCI standard deviation is expected to be worse than ARC due to higher random and sampling uncertainties [57,58,60].

The median bias of CCI is slightly larger than 0.1 K and the robust standard deviation (RSD) in the range 0.22–0.55 K lower for night time than daytime conditions, and this improves with instrument generation (AATSR has the smallest and ATSR-1 the largest) (Table 1). However, considering overlapping periods for the instruments indicates that the performance of ATSR-2 is similar to AATSR (Table 2). The median bias is larger in the tropics ~0.3 K during the day, with lower values (0.1–0.2 K) and more homogeneous distribution during the night. Some regions show negative biases, e.g., due to the dust outflow from the Sahara and Arabian deserts for night time conditions. The Maritime Continent has a peak for daytime RSD at about 0.6 K. The median bias is fairly stable (0.1–0.15 K) since 2000–2001, but variable before then. The comparable performance of ATSR-2 and AATSR is demonstrated in the time series of the RSD being 0.2–0.3 K during night and about 0.1 K higher during the day. On the other hand, ATSR-1 shows increasing RSD with time from 0.3 K to over 0.6 K. Forty to sixty percent of the match-ups have absolute SST differences less than 0.2 K (higher percentages during night), except for the periods after the failure of the 3.7 μm channel on ATSR-1 and the gyro failure on ERS-2. Similarly, less than 10% of match-ups show absolute SST differences larger than 1 K (again except for the above mentioned periods), decreasing to 5% for the AATSR period.

Triple collocation analysis has been used for the first time, to the best of our knowledge, for the period 1998–2002 using TMI and ATSR-2 observations, extending backwards in time the estimation of the random uncertainty of SST from drifting buoys, IR imagers and MW imagers. The application of triple collocation under daytime conditions when the wind speed is larger than 10 m/s provides meaningful results for the period of AATSR/AMSR-E. Both day and night results indicate that, since 2004, drifting buoys and CCI AATSR have a random uncertainty of about 0.22 K, stable with time. Before 2004, drifting buoys have larger values (~0.3 K), while ATSR-2 shows slightly lower values (~0.2 K). The random uncertainty for AMSR-E is about 0.47 K, stable with time, while TMI has as expected higher values of ~0.55 K. The Hovmoller diagrams of the triple collocation during night indicate that the analysis samples mainly the Southern Hemisphere mid-latitudes, producing mostly latitudinally homogeneous results (in addition to time) for the drifting buoys and the ATSR-2/AATSR CCI SST. Surprisingly, the performance of AMSR-E is much better in the tropics (~0.3 K), almost half of the value found in the mid-latitudes, which is comparable to TMI. This is the first time to the best of our knowledge that the AMSR-E SST random uncertainty has been reported to change with latitude.

The CCI provides uncertainty for each SST retrieval, and its validation showed that the model used during the daytime slightly overestimates uncertainty in the interval [0.45, 0.9] K and underestimates below 0.3 K. Small underestimations of the uncertainty are also seen for the night time conditions, but now for some of the match-ups, the uncertainties provided are too low. The time evolution of the percentage of match-ups within combined uncertainty (drifting buoys and CCI) is 70–80% for daytime, though the model does not capture the instrument failure periods of the ATSR-1 3.7 μm channel and the ERS-2 gyro. The corresponding percentages for night are lower, being 60–70%. However, the night percentages depend on the assigned value of uncertainty for the drifting buoys being at the same level, which is not the case during the day.

Results show that CCI (Phase 1) does not yet match the quality of ARC SST retrievals when comparing to drifting buoys. The value of the ARC median bias is closer to 0 K than CCI, and the RSD is about 0.05 K lower for ARC. This means that the night time CCI RSD more or less matches the ARC RSD during the day. Although the results of the comparison against drifting buoys are not very different, the superiority of ARC can also be seen in the more homogeneous geographical distribution of median bias and RSD than CCI. However, there are regions (e.g., Gulf of Mexico, Kuroshio) where CCI is better than ARC, while the region around the Indian Peninsula has proven difficult for both datasets. The same picture emerges for the temporal evolution of the percentages of $|\Delta SST|$ within 0.1/0.2 K or larger than 1 K with ARC manifesting better performance than CCI. This fact probably indicates that the SST retrieval approach based on coefficients estimated from radiative transfer simulations performs better than the optimal estimation for dual view observations; even if for single view observations, the optimal estimation retrieval provides better results [26]. An alternative possibility is that the optimal estimation retrieval has not been implemented correctly in CCI, e.g., by not removing all the subtle biases of the radiances. Nevertheless, the observed discrepancies between CCI and ARC during the period of ATSR-1 are unexplained given that both datasets use the same retrieval method; especially concerning bias, as the standard deviation of CCI is expected to be higher than ARC due to the finer spatial resolution.

It should be noted that although the CCI (Phase 1) SSTs from the ATSRs are not as good as ARC, they are in line or even better than other SST datasets based on thermal infrared (TIR) observations [30–32,61]. The CCI also includes SSTs from the Advanced Very High Resolution Radiometers (AVHRRs) and an SST analysis using both ATSR and AVHRR [22]. The combination of the high quality SSTs provided by the ATSRs and the global coverage provided by AVHRRs offers an attractive dataset for climate studies. Indeed, the CCI SST dataset is already being used successfully [56,62] for climate research. Some of the shortcomings in the CCI SSTs identified here are being addressed in Phase 2 of the CCI, while other enhancements could be incorporated in future versions [63].

Acknowledgments: TMI and AMSR-E data are produced by Remote Sensing Systems and sponsored by the NASA Earth Science MEaSUREs DISCOVER Project. Data are available at the www.remss.com. Both CCI and ARC SST datasets can be downloaded from CEDA http://catalogue.ceda.ac.uk. Christoforos Tsamalis was supported by the Joint UK BEIS/Defra Met Office Hadley Centre Climate Programme (GA01101). Roger Saunders has been partially funded by the ESA CCI CMUG project (RFQ/3-12943/09/I-OL).

Author Contributions: C.T. designed and performed the analysis and wrote the paper. R.S. contributed in the analysis and the writing of the paper.

Conflicts of Interest: The authors declare no conflict of interest.

Abbreviations

The following abbreviations are used in this manuscript:

AATSR	Advanced Along Track Scanning Radiometer
AIRS	Atmospheric InfraRed Sounder
AMSR-E	Advanced Microwave Scanning Radiometer for EOS
ARC	ATSR Reprocessing for Climate
ATSR	Along Track Scanning Radiometer
AVHRR	Advanced Very High Resolution Radiometer
CCI	Climate Change Initiative
ΔSST	Difference of SST
δSST	Uncertainty of SST
ECV	Essential Climate Variable
ENVISAT	ENVIronmental SATellite
EOS	Earth Observing System
ERS	European Remote Sensing
ESA	European Space Agency
GSICS	Global Space-based Inter-Calibration System
GTS	Global Telecommunication System
IASI	Infrared Atmospheric Sounding Interferometer
ICOADS	International Comprehensive Ocean-Atmosphere Data Set
IR	InfraRed
MetOp	Meteorological Operational
MODIS	MODerate resolution Imaging Spectroradiometer
NIR	Near InfraRed
NWP	Numerical Weather Prediction
RSD	Robust Standard Deviation
SST	Sea Surface Temperature
TMI	TRMM Microwave Imager
TRMM	Tropical Rainfall Measuring Mission
MW	MicroWave

References

1. Kennedy, J.J.; Rayner, N.A.; Smith, R.O.; Parker, D.E.; Saunby, M. Reassessing biases and other uncertainties in sea surface temperature observations measured in situ since 1850: 1. Measurement and sampling uncertainties. *J. Geophys. Res.* **2011**, *116*, D14103.
2. Kennedy, J.J.; Rayner, N.A.; Smith, R.O.; Parker, D.E.; Saunby, M. Reassessing biases and other uncertainties in sea surface temperature observations measured in situ since 1850: 2. Biases and homogenization. *J. Geophys. Res.* **2011**, *116*, D14104.
3. Emery, W.J.; Castro, S.; Wick, G.A.; Schluessel, P.; Donlon, C. Estimating sea surface temperature from infrared satellite and in situ temperature data. *Bull. Am. Meteorol. Soc.* **2001**, *82*, 2773–2785.
4. Donlon, C.; Rayner, N.; Robinson, I.; Poulter, D.J.S.; Casey, K.S.; Vazquez-Cuervo, J.; Armstrong, E.; Bingham, A.; Arino, O.; Gentemann, C.; et al. The Global Ocean Data Assimilation Experiment High-resolution Sea Surface Temperature Pilot Project. *Bull. Am. Meteorol. Soc.* **2007**, *88*, 1197–1213.
5. Deschamps, P.Y.; Phulpin, T. Atmospheric correction of infrared measurements of sea surface temperature using channels at 3.7, 11 and 12 μm. *Bound. Lay. Meteorol.* **1980**, *18*, 131–143.
6. Casey, K.S.; Brandon, T.B.; Cornillon, P.; Evans, R. The past, present and future of the AVHRR Pathfinder SST program. In *Oceanography from Space: Revisited*; Barale, V., Gower, J.F.R., Alberotanza, L., Eds.; Springer: Dordrecht, The Netherlands, 2010; pp. 273–287.
7. Merchant, C.J.; Embury, O.; Rayner, N.A.; Berry, D.I.; Corlett, G.K.; Lean, K.; Veal, K.L.; Kent, E.C.; Llewellyn-Jones, D.T.; Remedios, J.J.; et al. A 20 year independent record of sea surface temperature for climate from Along-Track Scanning Radiometers. *J. Geophys. Res.* **2012**, *117*, C12013.
8. Corlett, G.K.; Barton, I.J.; Donlon, C.J.; Edwards, M.C.; Good, S.A.; Horrocks, L.A.; Llewellyn-Jones, D.T.; Merchant, C.J.; Minnett, P.J.; Nightingale, T.J.; et al. The accuracy of SST retrievals from AATSR: An initial assessment through geophysical validation against in situ radiometers, buoys and other SST data sets. *Adv. Space Res.* **2006**, *37*, 764–769.
9. Kennedy, J.J.; Smith, R.O.; Rayner, N.A. Using AATSR data to assess the quality of in situ sea-surface temperature observations for climate studies. *Remote Sens. Environ.* **2012**, *116*, 79–92.
10. Atkinson, C.P.; Rayner, N.A.; Roberts-Jones, J.; Smith, R.O. Assessing the quality of sea surface temperature observations from drifting buoys and ships on a platform-by-platform basis. *J. Geophys. Res.* **2013**, *118*, 3507–3529.
11. Blackmore, T.; O'Carroll, A.; Fennig, K.; Saunders, R. Correction of AVHRR Pathfinder SST data for volcanic aerosol effects using ATSR SSTs and TOMS aerosol optical depth. *Remote Sens. Environ.* **2012**, *116*, 107–117.
12. Le Borgne, P.; Marsouin, A.; Orain, F.; Roquet, H. Operational sea surface temperature bias adjustment using AATSR data. *Remote Sens. Environ.* **2012**, *116*, 93–106.
13. Roberts-Jones, J.; Fiedler, E.K.; Martin, M.J. Daily, global, high-resolution SST and sea ice reanalysis for 1985–2007 using the OSTIA system. *J. Clim.* **2012**, *25*, 6215–6232.
14. Roberts-Jones, J.; Bovis, K.; Martin, M.J.; McLaren, A. Estimating background error covariance parameters and assessing their impact in the OSTIA system. *Remote Sens. Environ.* **2016**, *176*, 117–138.
15. Llewellyn-Jones, D.; Edwards, M.C.; Mutlow, C.T.; Birks, A.R.; Barton, I.J.; Tait, H. AATSR: Global-change and surface-temperature measurements from Envisat. *ESA Bull.* **2001**, *105*, 11–21.
16. Noyes, E.J.; Minnett, P.J.; Remedios, J.J.; Corlett, G.K.; Good, S.A.; Llewellyn-Jones, D.T. The accuracy of the AATSR sea surface temperatures in the Caribbean. *Remote Sens. Environ.* **2006**, *101*, 38–51.
17. O'Carroll, A.G.; Saunders, R.W.; Watts, J.G. The measurement of the sea surface temperature by satellites from 1991 to 2005. *J. Atmos. Ocean. Technol.* **2006**, *23*, 1573–1582.
18. Reynolds, R.W.; Gentemann, C.L.; Corlett, G.K. Evaluation of AATSR and TMI satellite SST data. *J. Clim.* **2010**, *23*, 152–165.
19. Smith, D.; Mutlow, C.; Delderfield, J.; Watkins, B.; Mason, G. ATSR infrared radiometric calibration and in-orbit performance. *Remote Sens. Environ.* **2012**, *116*, 4–16.
20. Illingworth, S.M.; Remedios, J.J.; Parker, R.J. Intercomparison of integrated IASI and AATSR calibrated radiances at 11 and 12 μm. *Atmos. Chem. Phys.* **2009**, *9*, 6677–6683.
21. Bali, M.; Mittaz, J.P.; Maturi, E.; Goldberg, M.D. Comparisons of IASI-A and AATSR measurements of top-of-atmosphere radiance over an extended period. *Atmos. Meas. Tech.* **2016**, *9*, 3325–3336.

22. Merchant, C.J.; Embury, O.; Roberts-Jones, J.; Fiedler, E.; Bulgin, C.E.; Corlett, G.K.; Good, S.; McLaren, A.; Rayner, N.; Morak-Bozzo, S.; et al. Sea surface temperature datasets for climate applications from Phase 1 of the European Space Agency Climate Change Initiative (SST CCI). *Geosci. Data J.* **2014**, *1*, 179–191.

23. Hollmann, R.; Merchant, C.J.; Saunders, R.; Downy, C.; Buchwitz, M.; Cazenave, A.; Chuvieco, E.; Defourny, P.; de Leeuw, G.; Forsberg, R.; et al. The ESA Climate Change Initiative: Satellite data records for essential climate variables. *Bull. Am. Meteorol. Soc.* **2013**, *94*, 1541–1552.

24. Merchant, C.J.; Llewellyn-Jones, D.; Saunders, R.W.; Rayner, N.A.; Kent, E.C.; Old, C.P.; Berry, D.; Birks, A.R.; Blackmore, T.; Corlett, G.K.; et al. Deriving a sea surface temperature record suitable for climate change research from the along-track scanning radiometers. *Adv. Space Res.* **2008**, *41*, 1–11.

25. Lean, K.; Saunders, R.W. Validation of the ATSR Reprocessing for Climate (ARC) dataset using data from drifting buoys and a three-way error analysis. *J. Clim.* **2013**, *26*, 4758–4772.

26. Merchant, C.J.; Le Borgne, P.; Marsouin, A.; Roquet, H. Optimal estimation of sea surface temperature from split-window observations. *Remote Sens. Environ.* **2008**, *112*, 2469–2484.

27. Embury, O.; Merchant, C.J. A reprocessing for climate of sea surface temperature from the along-track scanning radiometers: A new retrieval scheme. *Remote Sens. Environ.* **2012**, *116*, 47–61.

28. Embury, O.; Merchant, C.J.; Filipiak, M.J. A reprocessing for climate of sea surface temperature from the along-track scanning radiometers: Basis in radiative transfer. *Remote Sens. Environ.* **2012**, *116*, 32–46.

29. Xu, F.; Ignatov, A. Evaluation of in situ sea surface temperatures for use in the calibration and validation of satellite retrievals. *J. Geophys. Res.* **2010**, *115*, C09022.

30. Petrenko, B.; Ignatov, A.; Kihai, Y.; Stroup, J.; Dash, P. Evaluation and selection of SST regression algorithms for JPSS VIIRS. *J. Geophys. Res.* **2014**, *119*, 4580–4599.

31. Kilpatrick, K.A.; Podesta, G.; Walsh, S.; Williams, E.; Halliwell, V.; Szczodrak, M.; Brown, O.B.; Minnett, P.J.; Evans, R. A decade of sea surface temperature from MODIS. *Remote Sens. Environ.* **2015**, *165*, 27–41.

32. Marsouin, A.; Le Borgne, P.; Legendre, G.; Pere, S.; Roquet, H. Six years of OSI-SAF METOP-A AVHRR sea surface temperature. *Remote Sens. Environ.* **2015**, *159*, 288–306.

33. Merchant, C.J.; Harris, A.R.; Maturi, E.; MacCallum, S. Probabilistic physically based cloud screening of satellite infrared imagery for operational sea surface temperature retrieval. *Q. J. R. Meteorol. Soc.* **2005**, *131*, 2765–2755.

34. Merchant, C.J.; Embury, O.; Le Borgne, P.; Bellec, B. Saharan dust in nighttime thermal imagery: Detection and reduction of related biases in retrieved sea surface temperature. *Remote Sens. Environ.* **2006**, *104*, 15–30.

35. Good, E.J.; Kong, X.; Embury, O.; Merchant, C.J.; Remedios, J. An infrared desert dust index for the Along-Track Scanning Radiometers. *Remote Sens. Environ.* **2012**, *116*, 159–176.

36. Horrocks, L.A.; Candy, B.; Nightingale, T.J.; Saunders, R.W.; O'Carroll, A.; Harris, A.R. Parameterizations of the ocean skin effect and implications for satellite-based measurement of sea-surface temperature. *J. Geophys. Res.* **2003**, *108*, 3096.

37. Donlon, C.J.; Minnett, P.J.; Gentemann, C.; Nightingale, T.J.; Barton, I.J.; Ward, B.; Murray, M.J. Toward improved validation of satellite sea surface skin temperature measurements for climate research. *J. Clim.* **2002**, *15*, 353–369.

38. Veal, K.L.; Corlett, G.K.; Ghent, D.; Llewellyn-Jones, D.T.; Remedios, J.J. A time series of mean global skin SST anomaly using data from ATSR-2 and AATSR. *Remote Sens. Environ.* **2013**, *135*, 64–76.

39. O'Carroll, A.G.; Eyre, J.R.; Saunders, R.W. Three-way error analysis between AATSR, AMSR-E, and in situ sea surface temperature observations. *J. Atmos. Ocean. Technol.* **2008**, *25*, 1197–1207.

40. Zwieback, S.; Scipal, K.; Dorigo, W.; Wagner, W. Structural and statistical properties of the collocation technique for error characterization. *Nonlinear Process. Geophys.* **2012**, *19*, 69–80.

41. Wentz, F.J.; Gentemann, C.; Smith, D.; Chelton, D. Satellite measurements of sea surface temperature through clouds. *Science* **2000**, *288*, 847–850.

42. Chelton, D.B.; Wentz, F. Global microwave satellite observations of sea surface temperature for numerical weather prediction and climate research. *Bull. Am. Meteorol. Soc.* **2005**, *86*, 1097–1115.

43. Gentemann, C.L.; Meissner, T.; Wentz, F.J. Accuracy of satellite sea surface temperatures at 7 and 11 GHz. *IEEE Trans. Geosci. Remote Sens.* **2010**, *48*, 1009–1017.

44. Woodruff, S.D.; Worley, S.J.; Lubker, S.J.; Ji, Z.; Freeman, J.E.; Berry, D.I.; Brohan, P.; Kent, E.C.; Reynolds, R.W.; Smith, S.R.; et al. ICOADS Release 2.5: extensions and enhancements to the surface marine meteorological archive. *Int. J. Climatol.* **2011**, *31*, 951–967.

45. O'Carroll, A.G.; Watts, J.G.; Horrocks, L.A.; Saunders, R.W.; Rayner, N.A. Validation of the AATSR meteo product sea surface temperature. *J. Atmos. Ocean. Technol.* **2006**, *23*, 711–726.

46. Atkinson, C.P.; Rayner, N.A.; Kennedy, J.J.; Good, S.A. An integrated database of ocean temperature and salinity observations. *J. Geophys. Res.* **2014**, *119*, 7139–7163.

47. Xu, F.; Ignatov, A. In situ SST Quality Monitor (iQuam). *J. Atmos. Ocean. Technol.* **2014**, *31*, 164–180.

48. Peyridieu, S.; Chédin, A.; Capelle, V.; Tsamalis, C.; Pierangelo, C.; Armante, R.; Crevoisier, C.; Crépeau, L.; Siméon, M.; Ducos, F.; et al. Characterisation of dust aerosols in the infrared from IASI and comparison with PARASOL, MODIS, MISR, CALIOP, and AERONET observations. *Atmos. Chem. Phys.* **2013**, *13*, 6065–6082.

49. Tsamalis, C.; Chédin, A.; Pelon, J.; Capelle, V. The seasonal vertical distribution of the Saharan Air Layer and its modulation by the wind. *Atmos. Chem. Phys.* **2013**, *13*, 11235–11257.

50. Gentemann, C.L.; Donlon, C.J.; Stuart-Menteth, A.; Wentz, F.J. Diurnal signals in satellite sea surface temperature measurements. *Geophys. Res. Lett.* **2003**, *30*, 1140.

51. Gentemann, C.L.; Minnett, P.J.; Le Borgne, P.; Merchant, C.J. Multi-satellite measurements of large diurnal warming events. *Geophys. Res. Lett.* **2008**, *35*, L22602.

52. Merchant, C.J.; Filipiak, M.J.; Borgne, P.L.; Roquet, H.; Autret, E.; Piolle, J.; Lavender, S. Diurnal warm-layer events in the western Mediterranean and European shelf seas. *Geophys. Res. Lett.* **2008**, *35*, L04601.

53. Gentemann, C.L.; Wentz, F.J.; Mears, C.A.; Smith, D.K. In situ validation of Tropical Rainfall Measuring Mission microwave sea surface temperatures. *J. Geophys. Res.* **2004**, *104*, C04021.

54. Morak-Bozzo, S.; Merchant, C.J.; Kent, E.C.; Berry, D.I.; Carella, G. Climatological diurnal variability in sea surface temperature characterized from drifting buoy data. *Geosci. Data J.* **2016**, *3*, 20–28.

55. Gentemann, C.L. Three way validation of MODIS and AMSR-E sea surface temperatures. *J. Geophys. Res.* **2014**, *119*, 2583–2598.

56. Xu, F.; Ignatov, A. Error characterization in iQuam SSTs using triple collocations with satellite measurements. *Geophys. Res. Lett.* **2016**, *43*, 1–9.

57. Bulgin, C.E.; Embury, O.; Corlett, G.; Merchant, C.J. Independent uncertainty estimates for coefficient based sea surface temperature retrieval from the Along-Track Scanning Radiometer instruments. *Remote Sens. Environ.* **2016**, *178*, 213–222.

58. Bulgin, C.E.; Embury, O.; Merchant, C.J. Sampling uncertainty in gridded sea surface temperature products and Advanced Very High Resolution Radiometer (AVHRR) Global Area Coverage (GAC) data. *Remote Sens. Environ.* **2016**, *177*, 287–294.

59. Kennedy, J.J. A review of uncertainty in in situ measurements and data sets of sea surface temperature. *Rev. Geophys.* **2014**, *52*, 1–32.

60. Liu, Y.; Minnett, P.J. Sampling errors in satellite-derived infrared sea-surface temperatures. Part I: Global and regional MODIS fields. *Remote Sens. Environ.* **2016**, *177*, 48–64.

61. O'Carroll, A.G.; August, T.; Le Borgne, P.; Marsouin, A. The accuracy of SST retrievals from Metop-A IASI and AVHRR using the EUMETSAT OSI-SAF matchup dataset. *Remote Sens. Environ.* **2012**, *126*, 184–194.

62. Massonnet, F.; Bellprat, O.; Guemas, V.; Doblas-Reyes, F.J. Using climate models to estimate the quality of global observational data sets. *Science* **2016**, *354*, 452–455.

63. Bulgin, C.E.; Eastwood, S.; Embury, O.; Merchant, C.J.; Donlon, C. The sea surface temperature climate change initiative: Alternative image classification algorithms for sea-ice affected oceans. *Remote Sens. Environ.* **2015**, *162*, 396–407.

remote sensing

MDPI

Article

Confirmation of ENSO-Southern Ocean Teleconnections Using Satellite-Derived SST

Brady S. Ferster [1,*], Bulusu Subrahmanyam [1] and Alison M. Macdonald [2]

[1] School of the Earth, Ocean and Environment, University of South Carolina, 701 Sumter Street, Columbia, SC 29208, USA; sbulusu@geol.sc.edu
[2] Physical Oceanography Department, Woods Hole Oceanographic Institution, MS 21, 266 Woods Hole Rd., Woods Hole, MA 02543, USA; amacdonald@whoi.edu
*   Correspondence: bferster@seoe.sc.edu; Tel.: +1-803-777-4529

Received: 26 January 2018; Accepted: 19 February 2018; Published: 23 February 2018

Abstract: The Southern Ocean is the focus of many physical, chemical, and biological analyses due to its global importance and highly variable climate. This analysis of sea surface temperatures (SST) and global teleconnections shows that SSTs are significantly spatially correlated with both the Antarctic Oscillation and the Southern Oscillation, with spatial correlations between the indices and standardized SST anomalies approaching 1.0. Here, we report that the recent positive patterns in the Antarctic and Southern Oscillations are driving negative (cooling) trends in SST in the high latitude Southern Ocean and positive (warming) trends within the Southern Hemisphere sub-tropics and mid-latitudes. The coefficient of regression over the 35-year period analyzed implies that standardized temperatures have warmed at a rate of 0.0142 per year between 1982 and 2016 with a monthly standard error in the regression of 0.0008. Further regression calculations between the indices and SST indicate strong seasonality in response to changes in atmospheric circulation, with the strongest feedback occurring throughout the austral summer and autumn.

Keywords: Southern Ocean; sea surface temperature; teleconnections; Antarctic Oscillation; El Niño-Southern Oscillation; AVHRR

1. Introduction

Southern Ocean is a highly dynamic component of the global ocean circulation that plays a key role in the transport of heat, the uptake of carbon, and the global climate system [1–8]. The Southern Ocean circulation is largely wind driven. Changes in the Southern Hemisphere wind field drive sea surface temperature (SST) gradients that can support a feedback mechanism and influence both the latitude of the Antarctic Circumpolar Current (ACC) and the distributions of heat and nutrients [9–13]. Southern Hemisphere atmospheric variability exhibits a large number of modes, mostly influenced by large-scale low-frequency patterns [14] and has been known to play a major role in Southern Hemisphere weather and climate [15,16]. Due to the wave patterns associated with large-scale teleconnections, the largest temperature anomalies occur in the Amundsen-Bellingshausen Sea [17]. For a more in-depth analysis into the relationship between large-scale atmospheric teleconnections and Southern Ocean SST, refer to [17]. To analyze trends of Southern Hemisphere air-sea interactions, two patterns of atmospheric variability are compared with SST: the Antarctic Oscillation (AAO) and the Southern Oscillation (SO).

The AAO (also referred as the Southern Annular Mode) is a large-scale low-frequency pattern and is the dominant mode of atmospheric variability in the Southern Hemisphere Westerlies [16]. Previous studies found that the westerly winds have shifted south due to the increasing frequency of the AAO and the growing Antarctic ozone hole [18,19] and significant changes have been observed in both the temperature and salinity of the Southern Ocean [4,20,21]. However, despite the increasing frequency of the AAO and the shifting Westerlies, statistical evidence indicates an insensitivity of the ACC and sloping isopycnals to decadal changes in wind stress [22]. Within the Southern Ocean and Antarctic waters, previous studies have compared SST with the AAO [23–25] on weekly to monthly scales, noting the significant anomalies induced by the atmospheric pattern.

The SO is a measure of the Walker circulation in the tropical South Pacific. This circulation relates to variations in ocean temperatures and atmospheric pressure across the broad expanse of the tropical Pacific. The SO is also a metric for the El Niño-Southern Oscillation (ENSO) index. Negative (positive) SO index signifies warm (cold) SST across the eastern tropical Pacific and can therefore describe patterns similar to El Niño (La Niña) events. Additionally, the SO and El Niño (La Niña) were found to be negatively (positively) related to the Pacific-South American (PSA) wave pattern [26], which can have strong influences on SST, sea-ice, and atmospheric temperatures along Antarctica [27–29]. Reference [29] describes the strengthened influence of the AAO and SO during in phase periods, indicating the strong relationship between the atmospheric circulation and the effect on ocean dynamics. In this analysis, the statistical significance of the influence of teleconnections on Southern Ocean SST is investigated.

Hypothesizing that the SO has a greater influence than the AAO in the Southern Ocean and that both teleconnections drive warming (cooling) trends during positive (negative) phases of the oscillations, this analysis compares atmospheric teleconnections to Southern Ocean SST in both space and time. Investigation of the relationships between SST patterns and atmospheric variability with which they are associated provides a basis for improved understanding of Antarctic sea-ice and air-sea dynamics within the Southern Ocean.

2. Materials and Methods

2.1. Observational Data

Historically, in situ observations in the Southern Ocean have been sparse and made difficult due to harsh austral winter conditions. Southern Ocean in situ observations were particularly limited prior to the Argo float program, with most of the high-quality collections derived from repeat hydrography programs [30,31], such as the World Ocean Circulation Experiment (WOCE). Compared to in situ observations, satellites have relatively high spatial resolution and for SST a long temporal record extending back to the 1980's. The Optimal Interpolated Sea Surface Temperatures version 2 (OISST v2) product [32], utilizes statistically blended Advanced Very High Resolution Radiometer (AVHRR) data and in situ measurements to accurately represent the sea surface. OISST v2 is obtained from the National Oceanic and Atmospheric Administration (NOAA) Earth Science Research Laboratory Physical Science Division. OISST v2 [32] has 0.25° resolution spatially and temporal coverage dating back to September 1981. For this analysis, full-year only, high-resolution AVHRR data were used (1982–2016) to prevent seasonal bias, and all data points with satellite-derived fractional sea-ice in the high latitudes were removed.

Variability in SST anomalies is compared against proxies for large-scale, low frequency climate patterns of AAO and SO. Both the AAO and SO indices come from the National Centers of Environmental Prediction, Climate Prediction Center (NCEP, CPC). AAO index is computed from the leading empirical orthogonal function of 700 hPa height anomalies between mid- and high latitudes from NCEP/NCAR (National Center for Atmospheric Research) reanalysis data [33]. SO index is computed from standardized observed sea level pressure anomalies between Tahiti, French Polynesia and Darwin, Australia, divided by monthly standard deviations [34].

2.2. Methods

To analyze the teleconnections and standardized SST anomalies, multiple statistical tests are performed. To calculate standardized anomalies, the monthly mean climatology from OISST v2 (1982–2016) is subtracted from the monthly SST record and divided by the monthly standard deviation of OISST v2, effectively removing seasonality. Standardized anomalies typically provide a better representation on the magnitude of anomalies since the influences of dispersion have been removed. The linear regressions of standardized anomalies are calculated using whole years for the duration of OISST v2 SST, 1982–2016. Within the regressions and correlations, *p*-values are used to determine significance (alpha = 0.05). Positive (negative) indices are defined as those above (below) the 70th (30th) percentile, which is approximately 0.5 (−0.5). We define neutral years as those with indices between −0.5 and 0.5. Additional seasonal comparisons define austral summer as January through March and austral winter as July through September.

3. Results

We begin by addressing the relationship between SST and the large-scale AAO and SO teleconnections during the period 1982–2016. This period is used as it includes all full years in the OISST v2 data product. As described above, to reduce seasonal bias, standardized anomalies are compared against the large-scale patterns of AAO and SO indices. Correlations between SST and the teleconnections (Figure 1a,b) relate positive anomalies in the mid-latitudes (30°S–50°S) and negative anomalies in the high latitude regions during positive phases. A positive AAO is associated with a poleward shift in the Westerlies, while a positive SO describes anomalously cold temperatures in the eastern Pacific (similar to La Niña events).

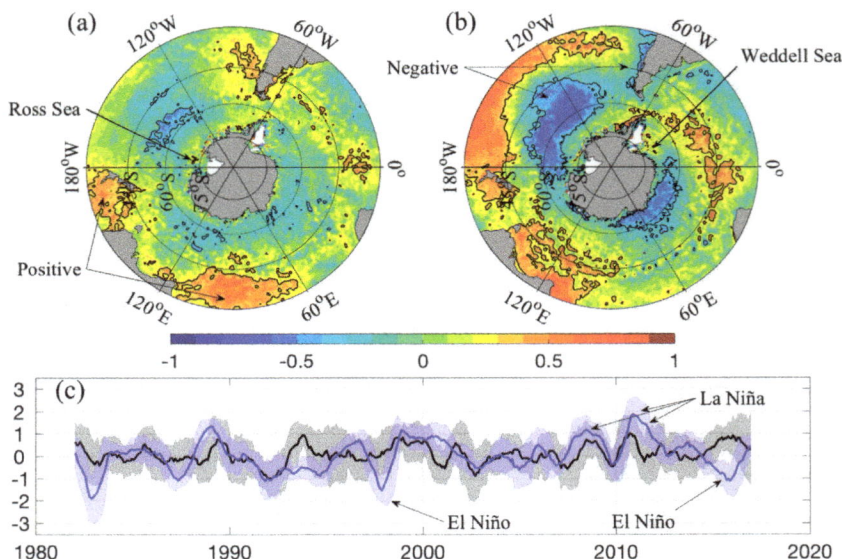

Figure 1. Pearson's correlation coefficient between standardized SST anomalies and the (**a**) Antarctic Oscillation (AAO) and (**b**) Southern Oscillation (SO). Negative (positive) coefficients are blue (red), and indicate decreased (increased) standardized SST anomalies. Coefficients interior to the black contour are significant (alpha = 0.05). (**c**) The 12-month running mean of AAO (black) and SO (blue) indices between 1982 and 2016, the shaded regions indicate the uncertainty.

During positive AAO years, the shift in the Westerlies correlates with anomalously warm surface waters in the mid-latitude Atlantic and Indian basins and a broad negative anomaly in the South Pacific (Figure 1a). A similar spatial pattern exists during positive SO years (Figure 1b). Positive SO years correlate with large-scale negative standardized SST anomalies in the high latitude Pacific and positive anomalies in the mid-latitude Pacific, along the west coast of Australia, and south of 45°S in the Atlantic. The spatial extent of significant SST anomaly correlations with the SO is larger than the correlations between SST and the AAO. Between the mid-2000's and 2014, both indices have been mostly in a positive phase (Figure 1c). When both indices are in the positive phase, spatial correlations relate increasing negative anomalies along the Antarctic coast with increasing positive anomalies in portions of the Ross and Weddell Seas and the sub-tropical Southern Ocean, similar to [24].

The influences of teleconnections are further explored through a more detailed analysis of the standardized SST anomalies (Figure 2). Spatial linear regressions from 1982 to 2016 in the Southern Ocean (Figure 2a) show significant large-scale surface warming (positive) in the mid-latitude Southern Ocean, the southern Indian Basin (60°E–120°E), and the Amundsen Sea sector. In addition, significant negative trends in the South Pacific Basin and Drake Passage region are seen. An estimate of the Southern Ocean trends suggests a mean value of 0.0142 per year, a median of 0.0154 per year, and a distribution negatively skewed towards negative trends.

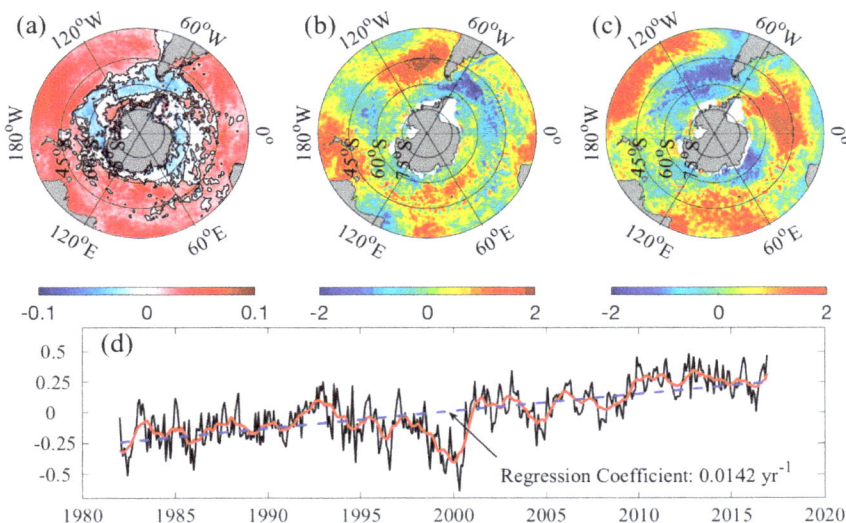

Figure 2. The 1982–2016 sea surface temperature (SST) coefficient of regression (year^{-1}) (**a**), mean standardized SST anomalies during 2016 (**b**), and 2010 (**c**). In (**a**), values interior to black contour lines represent significant trends (alpha = 0.05). (**d**) The monthly averaged standardized SST anomalies (black) in the Southern Ocean (30°S–70°S), 12-month running mean (red), and the linear regression (dashed blue). The coefficient of regression is 0.0142 year^{-1} and the coefficient of determination (r^2) is 0.436. The temporal monthly standard error in regression is 0.0008.

Comparing one-year averaged anomalies, we look to analyze two recent examples: a year with opposing index values and another with similar in-phase index values. In doing so, we explore the spatial regions and magnitudes of positive and opposing index years. Using Figure 1c, the years 2016 and 2010 were selected for each comparison respectively. The 2016 indices are described by a strong negative SO in the beginning of the year, changing to a strong positive SO by the end of the year with a consistently positive AAO (see Figure 1c). That is, the phases of the indices are mostly out of sync. The pattern of standardized SST anomalies (Figure 2b) from 2016 does show the strong similarities to the positive phases of both AAO and SO (i.e., negative sub-polar and positive sub-tropical anomalies), but also from the negative SO phase. The potency of the strong negative SO value (Figure 1b) is seen in the anomalously warm SSTs in the eastern and central sub-tropical Pacific Ocean (Figure 2b), both of which overshadow the values forced by the positive SO months.

The 2010 mean standardized SST anomalies (Figure 2c) were the start of successive La Niña years (denoted in Figure 1c) and a positive AAO. Most notable are the anomalously warm temperatures in the sub-tropical Pacific, South Atlantic, and western Indian Basins. Moreover, there are strong negative anomalies in the South Pacific and in the South Indian Ocean (near 60°E). The pattern of 2010 positive and negative anomalies (Figure 2c) shares strong similarities with the positive and negative SST and SO correlations (Figure 1b), showing the strong influence of the in-phase oscillations on SST.

Temporal analysis of the mean Southern Ocean SST (Figure 2d) suggests warming at a rate of 0.0142 per year between 1982 and 2016 (0.0092 °C per year if not standardized), with a standard error of 0.0008. The average was taken between 30°S and 70°S for each month. Throughout the 35-year period, there was a net warming of approximately 0.50 based on standardized anomalies (approximately 0.32 °C if not standardized). A spatial comparison indicates that positive index months correlate to broad-scale warming in the mid-latitudes that supports the warming trend. To show the more recent relationship of in phases indices, Figure 3 depicts the standardized SST anomalies for AAO positive and SO neutral, AAO neutral and SO positive, and both AAO and SO positive indices. In this comparison, the opposing relationships of neutral-phase indices mitigate anomalies. However, during in-phase months of AAO and SO indices, the anomalies are greater in magnitude and broader spatially compared to either out-of-phase relationship, similar to the results of [29].

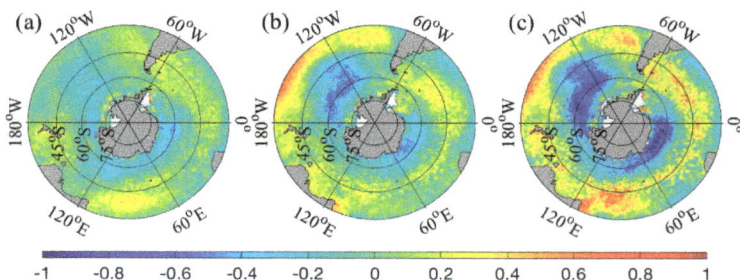

Figure 3. Monthly mean standardized sea surface temperature (SST) anomalies (°C) during (**a**) positive Antarctic Oscillation (AAO) and neutral Southern Oscillation (SO) months, (**b**) neutral AAO and positive SO months, and (**c**) both positive AAO and SO months. In each instance, a positive (negative) index is defined as greater (less) than 0.5 (−0.5) and neutral between −0.5 to 0.5.

A comparison of monthly and yearly averaged standardized SST anomalies (Figure 4) mark contrasting differences between the temporal scales of AAO and the SO. The monthly (Figure 4a) and yearly (Figure 4b) averaged standardized SST anomalies during positive AAO events both have negative anomalies in the Pacific Basin and positive anomalies in the Atlantic and Indian Basins. On longer time-scales, this yearly pattern of anomalous temperatures is stronger than monthly averaged. Differences in the yearly and monthly AAO anomalies (Figure 4c) show the stronger (red) yearly signal, particularly east of the Greenwich Meridian and south of Australia to the dateline. Yearly anomalies are weaker (blue) in the high latitude Indian, the mid to high-latitude Atlantic and mid-latitude Central Pacific.

Although global air-sea interactions are heavily influenced by large-scale teleconnections, seasonal regressions derive a response in SST to changes in index values (Figure 5). In both instances, the largest coefficients of regression occur in austral summer and autumn, while the weakest coefficients arise in austral winter. Similar results for SST and ENSO were previously described in [27] and [29]. Based on these findings, spatial changes in austral summer indices can be used to depict linear changes in temperature. However, minimally significant regressions between austral winter anomalies and the indices suggest a potential non-linear or lag relationship.

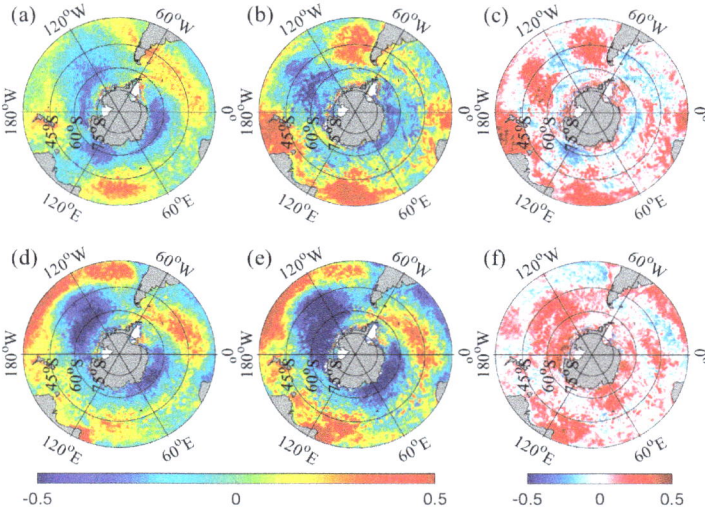

Figure 4. Monthly mean sea surface temperature (SST) standardized anomalies during positive Antarctic Oscillation (AAO) (**a**) and Southern Oscillation (SO) (**d**) months and (**b,e**) are standardized temperature anomalies during positive AAO and SO years respectively. (**c,f**) are the absolute value of yearly averaged anomalies minus the absolute value of monthly averaged anomalies. Red (blue) depicts yearly averages are greater (weaker) than monthly. In each instance, a positive index is defined as greater than 0.5.

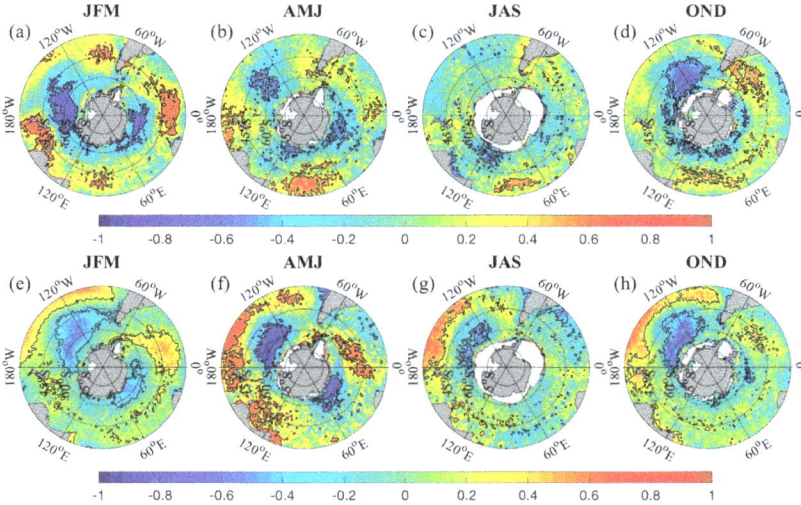

Figure 5. Coefficients of regression between the Southern Oscillation (AAO) (**a–d**) and standardized sea surface temperature (SST) anomalies (°C) from 1982 to 2016. (**a**) is monthly anomalies averaged over January to March (austral summer), (**b**) April through June, (**c**) July through September (austral winter), and (**d**) October to December. The coefficients of regression between the SO (**e–h**) SST anomalies are through the same temporal scale as (**a–d**) respectively. The largest coefficients occur with AAO and SO during the austral summer and autumn, while the smallest coefficients occur in austral winter.

4. Discussion

Analyzing correlations and mean standardized anomalies allow for the interpretation of how large-scale teleconnections influence the Southern Ocean. Strong positive indices of both the AAO and SO (La Niña) induce broad-scale cooling in the high latitudes of the Southern Ocean and warming in the mid-latitudes and Weddell Sea. Each index is seen to drive spatial anomalies on a magnitude of 0.5 and anomalies approaching 1 when both indices are in-phase. Within Figure 1a,b, the spatial correlations of the SO are more highly correlated spatially and based on magnitude, displaying the overlaying influence of the SO on the Southern Ocean SST. This result is important as the mid-2000's through 2012 were largely in-phase positive oscillations, associated with large-scale negative anomalies in the high latitudes and positive anomalies in the mid-latitudes and Weddell sea. More importantly, the spatial pattern in the temporal linear regression of SST is spatially similar to SST anomalies in 2010, a year marked by a strong La Niña period and a positive AAO index value, showing the importance of in-phase oscillations on SST. The continual positive indices within the 21st century would therefore be strongly contributing to the positive trend in Southern Ocean mid-latitude SST and negative trend within the Southern Hemisphere high latitudes.

The patterns of SST anomalies seen during positive SO monthly (Figure 4d) and yearly (Figure 4e) periods are similar, with intensification throughout the year in both the Indian and Pacific Basins (Figure 4f). This temporal disparity is most likely due to the time required for an Equatorial-Tropical Pacific phenomenon to influence large regions. A distinguishing feature of these patterns is that the magnitude of positive SO events in the Pacific Basin is greater than those associated with the AAO. The potential implication is that the SO plays a dominant role in the Southern Ocean. Furthermore, differences in yearly and monthly averaged anomalies are comparatively large. Thus, the longer duration of AAO and SO events could be an effective means of supporting or driving long-term trends.

SST anomalies can still be influenced by localized processes and feedback mechanisms. In years where the AAO and SO are increasingly positive, the Westerlies shift poleward and anomalously cooler temperatures are found in the high latitudes, with warmer temperatures in the mid-latitudes. There are breaks in the trend when strong El Niño events occur (negative SO), driving temperature changes that oppose the existing trend. Although this analysis supports the long-term increase in mean Southern Ocean surface temperature found by earlier analyses, it suggests that while the Atlantic and Indian Basins continue to warm, there has been significant cooling in the South Pacific as a result of the most recent patterns in the AAO and SO.

5. Conclusions

In summary, an analysis of satellite-derived SST observations provides statistical grounds for measuring and understanding the spatial correlations between SSTs and global atmospheric teleconnection patterns. We find SSTs to be significantly correlated to both the AAO and the SO, with larger magnitude of anomalies associated with the SO events. Large-scale spatial patterns of both the AAO and the SO are significantly correlated to sea surface temperatures in Southern Ocean, driving significant cooling in Antarctic sub-polar regions and warming in the Southern Hemisphere subtropics. The recent in phase positive AAO and SO patterns are simultaneously driving significant cooling in the high latitude Pacific basin, despite broad-scale warming throughout the Southern Ocean at a rate of 0.0142 per year. We further find that the strong in phase austral summer and autumn relationships are driving the most significant changes. Our analysis, which suggests the potential for a continuous warming trend should the AAO and SO spend extended periods in their positive phases, provides strong grounds for the promotion of continued monitoring of the high latitude SST.

Acknowledgments: B.S.F. is supported by the NASA/South Carolina Space-grant Graduate Assistantship. NOAA's OISST v2 AVHRR data is obtained through the Earth Science Research Laboratory Physical Science Division (NOAA/ESRL/PSD) (https://www.esrl.noaa.gov/psd/data/gridded/data.noaa.oisst.v2.highres.html). The indices for the Southern Oscillation and Antarctic Oscillation are distributed by the Climate Prediction

Remote Sens. **2018**, *10*, 331

Center (CPC) of the National Centers of Environmental Prediction (NCEP), National Oceanic and Atmospheric Administration (NOAA). A. Macdonald acknowledges support from NOAA Grant #NA160AR4310172.

Author Contributions: B.S.F., B.S., and A.M.M. conceived and designed the data analysis and interpretation of the results. B.S.F. prepared all of the figures and prepared the manuscript and B.S. and A.M.M. guided this work and corrected the article.

Conflicts of Interest: The authors declare no conflict of interest. The founding sponsors had no role in the design of the study; in the collection, analyses, or interpretation of data; in the writing of the manuscript, and in the decision to publish the results.

References

1. Marshall, J.; Speer, K. Closure of the meridional overturning circulation through Southern Ocean upwelling. *Nat. Geosci.* **2012**, *5*, 171–180. [CrossRef]
2. Cheng, L.; Trenberth, K.E.; Fasullo, J.; Boyer, T.; Abraham, J.; Zhu, J. Improved estimates of ocean heat content from 1960 to 2015. *Sci. Adv.* **2017**, *3*, e1601545. [CrossRef] [PubMed]
3. Landschützer, P.; Gruber, N.; Haumann, F.A.; Rödenbeck, C.; Bakker, D.C.; Van Heuven, S.; Hoppema, M.; Metzl, N.; Sweeney, C.; Takahashi, T.; et al. The reinvigoration of the Southern Ocean carbon sink. *Science* **2015**, *349*, 1221–1224. [CrossRef] [PubMed]
4. Sarmiento, J.L.; Hughes, T.M.C.; Stouffer, R.J.; Manabe, S. Simulated response of the ocean carbon cycle to anthropogenic climate warming. *Nature* **1998**, *393*, 245–249. [CrossRef]
5. Gille, S.T. Warming of the Southern Ocean Since the 1950's. *Science* **2002**, *295*, 1275–1277. [CrossRef] [PubMed]
6. McNeil, B.I.; Matear, R.J. Southern Ocean Acidification: A Tipping Point at 450-ppm Atmospheric CO_2. *Proc. Natl. Acad. Sci. USA* **2008**, *105*, 18860–18864. [CrossRef] [PubMed]
7. Montes-Hugo, M.; Doney, S.C.; Ducklow, H.W.; Fraser, W.; Martinson, D.; Stammerjohn, S.E.; Schofield, O. Recent Changes in Phytoplankton Communities Associated with Rapid Regional Climate Change Along the Western Antarctic Peninsula. *Science* **2009**, *323*, 1470–1473. [CrossRef] [PubMed]
8. Liu, J.; Curry, J.A. Accelerated Warming of the Southern Ocean and its Impacts on the Hydrological Cycle and Sea Ice. *Proc. Natl. Acad. Sci. USA* **2010**, *107*, 14987–14992. [CrossRef] [PubMed]
9. Dong, S.; Gille, S.T.; Sprintall, J. An Assessment of the Southern Ocean Mixed Layer Heat Budget. *J. Clim.* **2007**, *20*, 4425–4442. [CrossRef]
10. Downes, S.M.; Budnick, A.S.; Sarmiento, J.L.; Farneti, R. Impacts of wind stress on the Antarctic Circumpolar Current fronts and associated subduction. *Geophys. Res. Lett.* **2011**, *38*. [CrossRef]
11. Sun, C.; Watts, D.R. Heat flux carried by the Antarctic Circumpolar Current mean flow. *J. Geophys. Res.* **2002**, *107*. [CrossRef]
12. Dong, S.; Sprintall, J.; Gille, S.T. Location of the Antarctic polar front from AMSR-E satellite sea surface temperature measurements. *J. Phys. Oceanogr.* **2006**, *36*, 2075–2089. [CrossRef]
13. Gille, S.T. Meridional displacement of the Antarctic Circumpolar Current. *Philos. Trans. R. Soc. A.* **2014**, *372*, 20130273. [CrossRef] [PubMed]
14. Simmonds, I.; King, J.C. Global and hemispheric climate variations affecting the Southern Ocean. *Antarct. Sci.* **2004**, *16*, 401–413. [CrossRef]
15. Turner, J. The El Niño–Southern Oscillation and Antarctica. *Int. J. Climatol.* **2004**, *24*, 1–31. [CrossRef]
16. Hendon, H.H.; Lim, E.-P.; Nguyen, H. Seasonal variations of subtropical precipitation associated with the Southern Annular Mode. *J. Clim.* **2014**, *27*, 3446–3460. [CrossRef]
17. Hendon, H.H.; Thompson, D.W.J.; Wheeler, M. Australian rainfall and surface temperature variations associated with the Southern Hemisphere Annular Mode. *J. Clim.* **2006**, *20*, 2452–2467. [CrossRef]
18. Thompson, D.W.J.; Solomon, S. Interpretation of Recent Southern Hemisphere Climate Change. *Science* **2002**, *296*, 895–899. [CrossRef] [PubMed]
19. Thompson, D.W.J.; Solomon, S.; Kushner, P.J.; England, M.H.; Grise, K.M.; Karoly, D.J. Signatures of the Antarctic ozone hole in Southern Hemisphere surface climate change. *Nat. Geosci.* **2001**, *4*, 741–749. [CrossRef]
20. Gille, S.T. Yearly-Scale Temperature Trends in the Southern Hemisphere Ocean. *J. Clim.* **2008**, *21*, 4749–4765. [CrossRef]

21. Durack, P.J.; Wijffels, S.E. Fifty-year trends in global ocean salinities and their relationship to broad-scale warming. *J. Clim.* **2010**, *23*, 4342–4362. [CrossRef]

22. Böning, C.W.; Dispert, A.; Visbeck, M.; Rintoul, S.R.; Schwarzkopf, F.U. The response of the Antarctic Circumpolar Current to recent climate change. *Nat. Geosci.* **2008**, *1*, 864–869. [CrossRef]

23. Ciasto, L.M.; Thompson, D.W. Observations of large-scale ocean-atmosphere interaction in the Southern Hemisphere. *J. Clim.* **2008**, *21*, 1244–1259. [CrossRef]

24. Ciasto, L.M.; Alexander, M.A.; Deser, C.; England, M.H. On the persistence of cold-season SST anomalies associated with the Annular Mode. *J. Clim.* **2011**, *24*, 2500–2515. [CrossRef]

25. Xiao, B.; Zhang, Y.; Yang, X.Q.; Nie, Y. On the role of extratropical air-sea interaction in the persistence of the Southern Annular Mode. *Geophys. Res. Lett.* **2016**, *43*, 8806–8814. [CrossRef]

26. Mo, K.C. Relationships between low-frequency variability in the Southern Hemisphere and sea surface temperature anomalies. *J. Clim.* **2000**, *13*, 3599–3610. [CrossRef]

27. Welhouse, L.J.; Lazzara, M.A.; Keller, L.M.; Tripoli, G.J.; Hitchman, M.H. Composite analysis of the effects of ENSO events on Antarctica. *J. Clim.* **2016**, *29*, 1797–1808. [CrossRef]

28. Cerrone, D.; Fusco, G.; Simmonds, I.; Aulicino, G.; Budillon, G. Dominant Covarying Climate Signals in the Southern Ocean and Antarctic Sea Ice Influence during the Last Three Decades. *J. Clim.* **2017**, *30*, 3055–3072. [CrossRef]

29. Fogt, R.L.; Bromwich, D.H. Decadal variability of the ENSO teleconnection to the high-latitude South Pacific governed by coupling with the southern annular mode. *J. Clim.* **2006**, *19*, 979–997. [CrossRef]

30. Lyman, J.M.; Johnson, G.C. Estimating annual global upper-ocean heat content anomalies despite irregular in situ ocean sampling. *J. Clim.* **2008**, *21*, 5629–5641. [CrossRef]

31. Talley, L.D.; Feely, R.A.; Sloyan, B.M.; Wanninkhof, R.; Baringer, M.O.; Bullister, J.L.; Carlson, C.A.; Doney, S.C.; Fine, R.A.; Firing, E.; et al. Changes in Ocean Heat, Carbon Content, and Ventilation: A Review of the First Decade of GO-SHIP Global Repeat Hydrography. *Annu. Rev. Mar. Sci.* **2016**, *8*, 185–215. [CrossRef] [PubMed]

32. Reynolds, R.W.; Smith, T.M.; Liu, C.; Chelton, D.B.; Casey, K.S.; Schlax, M.G. Daily High-Resolution-Blended analyses for sea surface temperature. *J. Clim.* **2007**, *20*, 5473–5496. [CrossRef]

33. Kalnay, E.; Kanamitsu, M.; Kistler, R.; Collins, W.; Deaven, D.; Gandin, L.; Iredell, M.; Saha, S.; White, G.; Woollen, J.; et al. The NCEP/NCAR 40-Year Reanalysis Project. *Bull. Am. Meteorol. Soc.* **1996**, *77*, 437–471. [CrossRef]

34. Trenberth, K.E.; Caron, J.M. The Southern Oscillation revisited: Sea level pressures, surface temperatures, and precipitation. *J. Clim.* **2000**, *13*, 4358–4365. [CrossRef]

remote sensing

MDPI

Article

Spatio-Temporal Interpolation of Cloudy SST Fields Using Conditional Analog Data Assimilation

Ronan Fablet [1,*], Phi Huynh Viet [1], Redouane Lguensat [1], Pierre-Henri Horrein [1] and Bertrand Chapron [2]

[1]   IMT Atlantique, Lab-STICC, UBL, Brest 29238, France; vietphi3892@gmail.com (P.H.V.); redouane.lguensat@imt-atlantique.fr (R.L.); ph.horrein@imt-atlantique.fr (P.-H.H.)
[2]   Ifremer, LOPS, Brest 29200, France; bchapron@ifremer.fr
*   Correspondence: ronan.fablet@imt-atlantique.fr; Tel.: +33-229-001-287

Received: 27 November 2017; Accepted: 6 February 2018; Published: 17 February 2018

Abstract: The ever increasing geophysical data streams pouring from earth observation satellite missions and numerical simulations along with the development of dedicated big data infrastructure advocate for truly exploiting the potential of these datasets, through novel data-driven strategies, to deliver enhanced satellite-derived gapfilled geophysical products from partial satellite observations. We here demonstrate the relevance of the analog data assimilation (AnDA) for an application to the reconstruction of cloud-free level-4 gridded Sea Surface Temperature (SST). We propose novel AnDA models which exploit auxiliary variables such as sea surface currents and significantly reduce the computational complexity of AnDA. Numerical experiments benchmark the proposed models with respect to state-of-the-art interpolation techniques such as optimal interpolation and EOF-based schemes. We report relative improvement up to 40%/50% in terms of RMSE and also show a good parallelization performance, which supports the feasibility of an upscaling on a global scale.

Keywords: ocean remote sensing data; data assimilation; optimal interpolation; analog models; multi-scale decomposition; patch-based representation

---

1. Introduction

Long records of high-resolution Sea Surface Temperature (SST) are of high importance for a wide range of applications including among others weather and climate forecasting, ocean-atmosphere exchanges, the monitoring of tropical cyclones [1]. SST is an example of essential variables derived from remote sensing data [2–6], which play a critical role in climate models as well as numerical weather forecasts. SST field time series are for instance among the key satellite-derived data assimilated in ocean-atmosphere models [7,8] and hurricane dynamics [9]. Spaceborne sensors provide invaluable data to reconstruct satellite-derived high-resolution SST fields (typically, up to a few kilometers) on a global scale. Such SST fields may however comprise high rates of missing data. Optical sensors [10] may depict the highest missing data rates, as they cannot sense the ocean surface through clouds. Though less sensitive to atmospheric conditions [11], radiometers are also affected by thick clouds and heavy rain conditions.

The reconstruction of gap-free high-resolution SST fields from satellite-derived SST measurement has long been a critical issue [12–18]. Operational products typically rely on the Optimal Interpolation (OI). Amon others, cloud-free OSTIA [12], ODYSSEA [19], AMSR-E [17] products are examples of operational products which rely on OI. It produces the Best Linear Unbiased Estimator (BLUE) of the field given irregularly sampled observations. This model-driven approach requires selecting a covariance prior of the SST fields, most often exponential and Gaussian covariance models [12,18]. The parameterization of this covariance prior involves a trade-off between the size of the gaps to be filled and the fine-scale variability of the SST fields. Physically-driven data assimilation models [20] may outperform OI if relevant dynamical priors can be defined [21]. The trade-off to be considered between the complexity and genericity of this physical prior remains however complex, especially when considering the assimilation of a single sea surface tracer as SST.

Besides model-driven schemes, the ever increasing availability of satellite-derived data and of simulation data from high-resolution ocean models has paved the way for the development of data-driven methods. EOF-based models were among the early and perhaps most popular data-driven methods applied to the reconstruction of SST fields from cloudy SST data [14,15,22] as well of other sea surface tracers such as ocean colour [22]. EOF-based approaches are particularly appealing for ocean remote sensing as they relate to a model of the covariance structure of the considered fields and may adapt to any type of geometry of missing data and interpolation grid. Their use is also motivated by their ability to decompose the spatiotemporal variability of the sea surface fields according to different modes, which may be interpreted geophysically. A renewed interest can also be noticed for analog schemes and applications to forecasting and assimilation issues [23,24]. Analog schemes, proposed a long time ago in geoscience [25], rely on the idea that the dynamics of a given system may repeat to some extent. Given a set of previously observed or analysed data, one may retrieve examples similar to a current state in this set, such that the future of this current state may be forecasted from the known evolution of these similar situations. The lack of large-scale dataset along with the computational complexity of analog methods has long limited their applicability. In this context, we recently introduced the analog data assimilation (AnDA) and demonstrated its relevance for the reconstruction of complex dynamical systems for partial observations, including sea surface dynamics [26,27]. Here, as stated in the next section, we further explore and evaluate AnDA schemes for the reconstruction of cloud-free SST fields from satellite-derived measurements.

The remainder of the paper is organized as follows. Section 2 briefly reviews the related work on data assimilation and introduces the main contributions of this work. Section 3 presents the considered data and case-study region. Section 4 describes the proposed AnDA methods for the reconstruction of cloud-free SST fields. Section 5 presents experimental results. Section 6 further discusses our key contributions and future work.

2. Problem Statement and Related Work

Data assimilation is the classic framework for the reconstruction of sea surface geophysical fields from partial satellite observation series [20,28]. Two main categories of data assimilation methods may be distinguished: variational and statistical data assimilation. Variational methods rely on a continuous setting and states data assimilation as the minimization of a variational cost. Statistical methods involve state-space models [20,28]. They formulate data assimilation as the maximization or estimation of the posterior likelihood of the state series given an observation series. The state refers to the geophysical parameter of interest, here a cloud-free SST field at a given time. In this work, we focus on statistical data assimilation methods, which provides a greater flexibility to model state dynamics as well as the relationship between the state series and the observation series [20]. They also avoid determining the adjoint of the dynamic operator, which may be complex while reaching state-of-the-art reconstruction performance [29].

The state-space model typically comprises two key components:

- A dynamical model which states the time evolution of the state. Within a discrete statistical framework, it comes to define the likelihood of the state at a given time given the state at the previous time;
- An observation model which relates the state to the observation, here the cloud-free SST field to the SST observation with missing data.

Among the variety of algorithms proposed to solve for statistical data assimilation issues, Ensemble Kalman filters and smoothers (EnKF and EnKS) are particularly popular. They demonstrate both good assimilation performance and a high modeling flexibility [20]. It may also be noted that the optimal interpolation can be regarded as a statistical assimilation model, where the dynamical prior involves a Gaussian distribution, such that an analytical and numerical solution can be derived [12,20]. EnKS and EnKF may provide relevant solutions to implement optimal interpolation schemes for high-dimensional fields. The definition of the dynamical prior is a critical aspect of such model-driven assimilation scheme. Regarding ocean dynamics, the balance between modeling complexity and uncertainty is particularly complex. Especially, simplified models such as advection-diffusion or QG (Quasi-Geostrophic) priors [21] may only be valid approximations for specific space-time regions.

These issues have motivated the development of data-driven frameworks as an alternative to the definition of model-driven dynamical priors. We may for instance cite EOF-based (Empirical Orthogonal Function) interpolation techniques [14,15], which state interpolation issues as a matrix completion problem and iterates successive projections onto an EOF basis under the constraint of the observed data. Such techniques have been proven relevant for the reconstruction of large-scale SST fields. They may however lack some mathematically-sound interpretation in terms of data assimilation issue. Interestingly, the combination of analog forecasting operators and classic statistical assimilation schemes has led to the introduction of novel data-driven schemes for data assimilation, referred to as Analog Data Assimilation (AnDA) [24,26]. Especially, our previous work [26] presents an application of AnDA to the space-time interpolation of SST fields using patch-based and EOF decompositions. (We use in this paper the term patch to refer to a subset of $K \times K$ pixels centered on given pixel. K is the width of the patch. The term patch is widely used in image processing with the emergence of patch-based image representations [30,31]. Patch-based representations [30–33] provide means to encode the spatial structure of the images while providing a simple and computationally-efficient framework.

Here, we further extend AnDA for the spatiotemporal interpolation of SST fields to improve both reconstruction performance and computational efficiency. This work involves three main contributions:

- the introduction of conditional analog forecasting operators with a view to explicitly accounting for dependencies between the state to be reconstructed and auxiliary variables. In [24,26], the considered analog forecasting operators implicitly assumed the high-resolution component dX to be independent on the low-resolution component $\bar{X}$. Both theoretical and statistical studies [34,35] advocate for considering inter-scale dependencies, which relate to the multi-scale characteristics of ocean turbulence [35,36]. We show here that analog strategies are highly flexible to consider such conditioning;
- the introduction of an analog forecasting operator embedding physically-sound priors. We further benefit from the flexibility of analog operators to exploit the synergy between SSH (Sea Surface Height) [35,37] and SST. We investigate locally-linear analog forecasting operators where SSH is used as a complementary regressor;
- the reduction of the computational complexity of AnDA using a clustering-based analog forecasting operator. To improve the scalability of the proposed methodology, we show that we can significantly reduce the computational complexity of the analog data assimilation with no impact on reconstruction performance.

We demonstrate the relevance of these contributions through numerical experiments for real cloudy SST patterns and evaluate the computational complexity of the proposed models and their parallelization performance for future large-scale case-studies.

3. Data and Study Area

With a view to evaluating the proposed analog assimilation methodology detailed in the next section, we use reference gap-free L4 SST time series from which we create SST datasets with missing data using real missing data masks. We consider two gap-free L4 SST products:

- OSTIA SST: the OSTIA product delivered daily by the UK Met Office [12] with a $0.05° \times 0.05°$ spatial resolution (approx. 5 km). The OSTIA analysis combines satellite data provided by infrared sensors (AVHRR, AATSR, SEVIRI), microwave sensors (AMSR-E, TMI) and in situ data from drifting and moored buoys.

- MW SST: the microwave optimally-interpolated product distributed by REMSS (http://www.remss.com/measurements/sea-surface-temperature/oisst-description/). This product combines daily microwave satellite measurements (TMI, AMSR-E, AMSR2, WindSat sensors) for a 0.25° × 0.25° resolution.

From a spectral analysis of the SST fields, it may be noted that the MW SST dataset involves greater energy level for scales below 100km than OSTIA SST dataset. For both datasets, we consider SST time series from January 2007 to December 2015 (January 2008 to December 2015) in a region off South Africa (150 × 300 pixels) from 0°E to 7°E and 22.5°S to 60°S. This region comprises the Aghulas current and combines highly-dynamic areas and periods off South Africa and not as active areas in the northern part of the case-study region which is characterized by warmer waters. This region is also characterized by a significant variability of the cloud cover up to very high missing data rates (e.g., above 70%). These characteristics make this region a relevant and representative testbed for SST interpolation issues.

As real cloud masking time series, we consider the cloud masks associated with the METOP-AVHRR SST time series (Ocean and Sea Ice Satellite Application Facility (OSI SAF) (2016). GHRSST L3C global sub-skin Sea Surface Temperature from the Advanced Very High Resolution Radiometer (AVHRR) on Metop satellites (currently Metop-B) (GDS V2) produced by OSI SAF (GDS version 2). NOAA National Centers for Environmental Information. Dataset) METOP-AVHRR product is a high-resolution infrared sensor, which may involve very high missing data rates in the case-study area (see Figure 1 for an example of cloud mask pattern).

As detailed in the next section, the proposed analog data assimilation models may benefit from multi-source data. More particularly, in the considered case-study, we explore the extent to which SSH data may be useful to improve the interpolation of the SST. As gridded and interpolated SSH field, we consider daily SSH data with a 0.25° × 0.25° resolution distributed by the CMEMS (Copernicus Marine Environment Monitoring Service, marine.copernicus.eu).

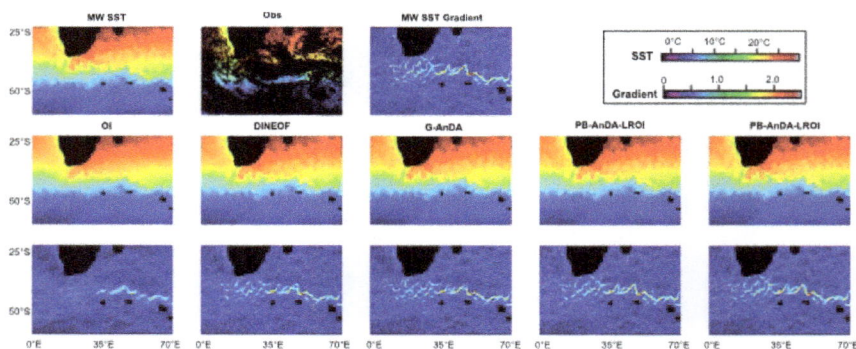

Figure 1. Reconstructed SST fields using OI, DINEOF, G-AnDA, PB-AnDA-LROI + dX + Z, PB-AnDA-LRM + dX + Z on day 150th for MW SST case-study: the first row depicts the reference SST field, the cloudy observation and the gradient magnitude; the second and third rows depict respectively the SST fields and their gradient magnitude for OI, DINEOF, G-AnDA, PB-AnDA-LROI + dX + Z, PB-AnDA-LRM + dX + Z. It may be noticed that PB-AnDA-LROI + dX + Z and PB-AnDA-LRM + dX + Z better reconstructs fine-scale structures for instance along the Aghulas return current as well as south of Madagascar island.

4. Method

4.1. Patch-Based Analog Data Assimilation

We consider the following scale-based decomposition of the SST fields:

$$X = \bar{X} + dX + \zeta \tag{1}$$

$\bar{X}$ refers to a background field. It may be a given as a mean field as well as optimally-interpolated fields. dX refers to high-resolution component to be estimated.

Following [26], we consider an analog data assimilation for the high-resolution component dX. It involves a patch-based state-dependent dynamical operator. Let us denote by $\mathcal{P}_s$ the patch centered on grid site s and $X(\mathcal{P}_s, t)$ the patch-level state for field X on $\mathcal{P}_s$ at time t. The considered dynamical operator for $\mathcal{P}_s$ at time t is stated as

$$dX\left(\mathcal{P}_s, t\right) = \mathcal{M}_{X(\mathcal{P}_s, t-1)}\left(X(\mathcal{P}_ps, t-1), \eta(\mathcal{P}_s, t-1)\right) \tag{2}$$

where $\mathcal{M}_{X(\mathcal{P}_s, t-1)}$ is the state-dependent operator at time t for grid site s. η is a random perturbation. We further constrain this patch-based model through an EOF-based decomposition of each patch-level state $dX\left(\mathcal{P}_s, t\right)$.

$$dX\left(\mathcal{P}_s, t\right) = \sum_{n=1}^{N} \alpha_n(s, t) B_n \tag{3}$$

With B_n the nth principal component of the EOF and $\alpha_n(s, t)$ the associated EOF expansion coefficient for patch $\mathcal{P}_s$ at time t. N_{EOF} refers to the number of vectors of the EOF basis. B will denote the matrix formed by all principal components.

The state-dependent dynamical operator $\mathcal{M}_{X(\mathcal{P}_s, t-1)}$ is stated as a locally-linear analog forecasting operator [24,26]. We assume that we are provided with a reference dataset, referred to as catalog $\mathcal{C}$ which comprises pairs of states $\{X(\mathcal{P}_{s_i}, t_i - 1), X(\mathcal{P}_{s_i}, t_i)\}_i$ at two consecutive time steps, referred to respectively as analogs and successors. For a given kernel denoted by $\mathcal{K}$, let us denote by $a_k(s, t)$ the kth analog (i.e., nearest-neighbor) of state $X(\mathcal{P}_s, t)$ in catalog $\mathcal{C}$ and $s_k(s, t)$ its successor. The locally-linear analog operator is stated as a multivariate linear regression in the EOF space between the analogs $\{a_k(s, t))\}_k$ and their successors $\{s_k(s, t))\}_k$ with a zero-mean Gaussian perturbation. The linear regression is fitted using a weighted least-square estimate with weights $\{\mathcal{K}(a_k(s, t), X(\mathcal{P}_s, t-1))\}_k$. The covariance of the Gaussian perturbation is estimated from the residual of the linear regression for the K pairs of analogs and successors. In [24,26], the regression variables are directly the states projected onto the EOF space. Here, as detailed in the next section, we consider different parameterization of the regression variables as well as of the kernel $\mathcal{K}$ to explore the potential conditioning of the dynamics of state dX by other variables (e.g., the low-resolution component $\bar{X}$ or a velocity field).

Given the observation model associated with the considered cloudy SST observations

$$Y(t, s) = X(t, s) + \epsilon(t, s), \ \forall s \in \Omega_t \tag{4}$$

With Ω_t the cloud-free region at time t, the reconstruction of high-resolution component dX given observation time series Y relies on the ensemble Kalman smoother (EnKS) associated with the considered analog dynamical operators. The EnKS is a forward-backward sequential algorithm. It represents the state at each time step from the mean and covariance of a set of members, which are evolved in time based on the analog forecasting operator and updated at each time step from the available observations using a Kalman-based recursion. We refer the reader to [24] for the details of the analog EnKS. Here, the analog EnKS is applied independently to overlapping patch locations. We then reconstruct field dX as a mean over overlapping patches. An additional postprocessing step is applied to remove possible patch-related blocky artifacts using a patch-based and global EOF filtering. The resulting workflow is sketched in Figure 2.

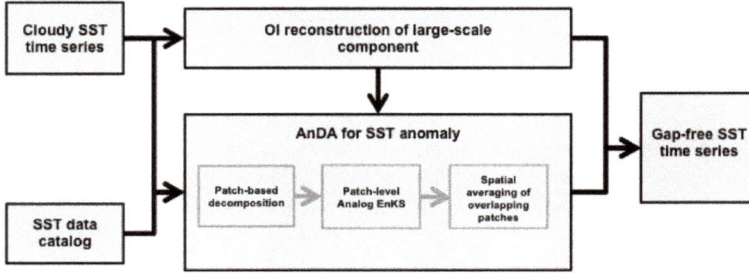

Figure 2. Workflow of the proposed framework for the reconstruction of gap-free SST time series from cloudy SST data: given a cloudy SST field time series, it first applies an optimal interpolation to reconstruct the large-scale component $\hat{X}$ and second the analog data assimilation (AnDA) of the anomaly dX (cf. (Equation (1))). This second step exploits a reference SST catalog and is constrained by the reconstructed large-scale component. The resulting gap-free SST time series is the sum of the large-scale component $\hat{X}$ and of the anomaly dX. We also sketch the main steps involved in the AnDA scheme.

4.2. Conditional and Physically-Derived Analog Forecasting Operators

Let us denote by U a co-variable with the same space-time resolution as field X. The Conditioning of analog forecasting operator $\mathcal{M}_{X(\mathcal{P}_s,t-1)}$ at time t and patch $\mathcal{P}_s$ may be issued:

- from the selection of analogs based on both variables dX and U, and not solely based on dX as in [24,26]. This comes to take into account variable Z in kernel $\mathcal{K}$. We typically consider a parameterization of kernel $\mathcal{K}$ as $\mathcal{K}_{dX} \cdot \mathcal{K}_U$ using kernels applied respectively to fields dX and Z. Here, we will consider a Gaussian kernel for $\mathcal{K}_{dX}$ and a correlation-based kernel for kernel $\mathcal{K}_U$. It may be noted that the considered kernels only exploit the spatial dimensions;
- from the fit of a multivariate linear regression using both dX and U, or transformed version of U, as regression variables and not solely based on dX as in [24,26]. For instance, following previous studies [34,38], one may consider the low-resolution field $\bar{X}$ as a potentially-relevant information to improve the forecasting of the high-resolution field dX.

It may be emphasized that a given co-variable may be used only for one of these two types of conditioning.

We also explore the potential relationship between locally-linear analog forecasting operator and physical operator. As a sea surface geophysical tracer, advection-diffusion priors may be regarded as relevant first-order approximations [21]. The advection-diffusion prior is given by

$$\partial_t X + \langle \omega, \nabla X \rangle = \kappa \Delta X \tag{5}$$

With ω the sea surface velocity field and κ the diffusion coefficient. Given that satellite-derived altimeter fields, denoted here by Z, provide a low-resolution estimate of the sea surface velocity fields, field ω may be written as $\nabla^{\perp} Z + \delta \omega$ with $\delta \omega$ the unresolved velocity component. Using decomposition Equation (1), advection-diffusion prior Equation (5) suggests considering a locally-linear patch-based analog forecasting operator Equation (2) where both variables dX, $\bar{X}$ and Z are considered as regression variables. Similarly to EOF decomposition Equation (3) for field dX, we also consider patch-based EOF decompositions for fields $\bar{X}$ and Z to constrain the estimation of the analog locally-linear operator.

It may be noted that this inference of such multi-modal analog forecasting operators relate to the approximation of the underlying unresolved velocities from local analogs.

4.3. Computationally-Efficient Analog Assimilation Strategies

An important goal of this study is to evaluate the computational complexity of the proposed analog assimilation strategies and their parallelization properties. By construction, the considered patch-level decomposition relies on the independent processing of each patch location for the considered grid. This ensures the computational complexity to evolve linearly with the number of grid points. The considered EnKS procedure involves two main steps: the forecasting and the analysis step. Given the relatively low-dimensional EOF-based representation of each patch, the computational complexity of the analog EnKS mainly relates to the ananlog forecasting step. In the standard version used in [24], the computational complexity may be decomposed as $N_M \cdot (C_{search} + C_{fit} + C_{forecasting})$ with N_M the number of members used to represent the state at each time step, C_{search} the computational cost of the search for analogs, C_{fit} the computational cost of the fit of the analog forecasting operator and $C_{forecasting}$ the computational cost of the application of the fitted forecasting operator. The first two ones are obviously the most important ones.

To speed up the search for local analogs, we can benefit from large research effort dedicated to nearest-neighbor search, especially approximate nearest-neighbor search [39,40]. Here, we consider FLANN (Fast Library for Approximate Nearest Neighbors) frameworkavailable at http://www.cs.ubc. ca/research/flannforadditionaldetails), which is among the state-of-the-art schemes for approximate nearest-neighbor search. It relies on an offline computation of a tree-based indexing structure. We let the reader [40,41] and FLANN (Fast Library for Approximate Nearest Neighbors) library.

Importantly, it may be noted that at a given time step many members can be expected to share similar dynamics. Therefore, fitting a local analog forecasting operator for each member as in [24,26] is expected to be computationally-redundant. To reduce this computational redundancy, we introduce a clustering-based strategy. For a given time step, we constrain the computational complexity to a given number of analog forecasting operator fit, denoted by N_{Fit}. We first clusterized the members into N_{Fit} using a K-means procedure [42]. We then fit an analog forecasting model for each cluster using the analogs and successors to the center of the cluster. For the forecasting step, we apply to each member the analog forecasting operator of the cluster it is assigned to. Overall, this clustering-based strategy leads to cost $N_{Fit} \cdot C_{fit}$ to compare to the original $N_M \cdot C_{fit}$. Here, we typically set N_{Fit} to 3 whereas $N_M = 100$.

4.4. Experimental Setting

Computational setting: The considered experiments, especially regarding the evaluation of the computational complexity of the proposed methods, have been implemented onto Teralab platform (https://www.teralab-datascience.fr/fr/) using a virtual machine with the following setting: 30 CPUs with a 64 G RAM (24 CPUs are used for processing tasks and others as backup or for background tasks). All experiments were run using Python and PB-AnDA Python library available at https://github.com/rfablet/PB_ANDA. We used Multiprocessing Python module to implement AnDA onto the considered multi-core platform.

Benchmarked models and algorithms: We consider different patch-based analog assimilation models, referred to as PB-AnDA, corresponding to different parameterizations of the analog forecasting operators:

- for the low-resolution component $\bar{X}$, we consider two options: (i) optimally-interpolated fields projected onto a region-level EOF decomposition with 20 components which resolve spatial scales up to approximately 100 km, (ii) the mean field. The first one is referred to LROI and second one to LRM;

- for the search for analogs, we explored both a simple kernel with no conditioning by the low-resolution component, such that $\mathcal{K} = \mathcal{K}_{dX}$ and a kernel $\mathcal{K} = \mathcal{K}_{dX} * \mathcal{K}_Z$ with $Z = \|\nabla \bar{X}\|$ to introduce a conditioning of the analog forecasting operators by the low-resolution gradient magnitude as suggested in [34]. As both settings resulted in very similar interpolation performance (e.g., RMSE of 0.24 for MW SST dataset for both settings), we only report results for the simplest kernel choice (i.e., $\mathcal{K} = \mathcal{K}_{dX}$) in the subsequent analysis.

- three types of regression variables were evaluated: locally-linear operators using only dX as regression variables (dX), using dX and $\bar{X}$ as regression variables ($dX + \bar{X}$) and using dX and Z as regression variables ($dX + Z$). A fully-developed locally-linear approximation of an advection-diffusion prior would consist in considering both dX, $\bar{X}$ and Z as regression variables. It resulted in the same performance as considering only dX and Z (see Table 1) and was not included in the reported results. We might recall that all locally-linear models are fitted within EOF subspaces.

Table 1. Interpolation performance of PB-AnDA models for the MW SST case-study for three zones of interest in the case-study region: we refer the reader to the main text for the description of the different PB-AnDA parameterizations.

PB-AnDA	LROI			LRM		
	dX	$dX + Z$	$dX + \bar{X}$	dX	$dX + Z$	$dX + \bar{X}$
Zone 1	0.35 ± 0.06	$\mathbf{0.34 \pm 0.05}$	0.35 ± 0.06	0.34 ± 0.06	$\mathbf{0.33 \pm 0.05}$	0.34 ± 0.06
Zone 2	0.33 ± 0.09	$\mathbf{0.32 \pm 0.08}$	0.33 ± 0.09	0.32 ± 0.08	$\mathbf{0.30 \pm 0.07}$	0.32 ± 0.08
Zone 3	$\mathbf{0.18 \pm 0.04}$	$\mathbf{0.18 \pm 0.04}$	$\mathbf{0.18 \pm 0.04}$	0.19 ± 0.04	0.19 ± 0.04	0.19 ± 0.04

Overall, we refer to a specific model as follows. For instance, model $PB\text{-}AnDA + LROI + dX + \bar{X} + Z$ implements the patch-based analog assimilation with: (i) optimally-interpolated fields as low-resolution component and (ii) locally-linear analog forecasting operators with auxiliary variables $\bar{X}$ and Z. All patch-based analog assimilation models involve similar parameter setting regarding the number of members, $N_M = 100$, and the patch-based EOF decomposition with $N_{EOF} = 50$, which account for 96% of the total variance of the SST datasets.

For benchmarking purposes, we considered two state-of-the-art interpolation techniques and a global AnDA model:

- a classic optimal interpolation with a Gaussian space-time covariance structure: the spatial and time correlation lengths were tuned from cross-validation experiments for the considered SST datasets to respectively 3 days and 100 km. This interpolation is referred to as OI and implemented using [43];
- a DINEOF interpolation [14]: the EOF-based interpolation comes to iteratively project the reconstructed field onto the EOF basis while modifying only SST values for missing data areas. We use 40 EOF components to account for about 95% of the total variance. This interpolation referred to as DINOEF is applied globally onto the entire case-study region.
- a direct application of AnDA over the entire region: this interpolation referred to as G-AnDA exploits the same EOF decomposition as DINEOF and $N_M = 100$ members in the implemented AnDA.

5. Results

5.1. Interpolation Performance

We first compare interpolation performance of the considered methods, namely OI, DINEOF, G-AnDA and PB-AnDA + LROI + dX for the both OSTIA and MW case-studies (Tables 2 and 3). Similar conclusions can be drawn from these experiments with a significant gain of the proposed patch-based AnDA model compared to the three other approaches. For instance, We report a relative gain up to 50% in RMSE w.r.t. OI and of 40% w.r.t. DINEOF. Though the direct application of AnDA to the entire region lead to a slight improvement (e.g., mean RMSE of 0.38 for G-AnDA w.r.t. 0.40

for DINEOF and 0.43 for OI on MW SST dataset), the additional relative gain greater than 35% in RMSE of the patch-based version PB-AnDA + LROI + dX emphasizes the relevance of the proposed multi-scale and patch-based decomposition to account for fine-scale structures. The analysis of the mean correlation coefficients between the interpolated fields and the reference fields for scales below 100 km leads to the same conclusion.

Table 2. Interpolation performance for the MW SST case-study: mean root mean square error (RMSE) and correlation coefficients with the MW SST for OI, DINEOF, G-AnDA and PB-AnDA methods. We refer the reader to the main text for the details on the considered parameterizations.

Criterion	RMSE	Correlation
OI	0.48 ± 0.05	0.69 ± 0.07
DINEOF	0.40 ± 0.04	0.79 ± 0.04
G-AnDA	0.38 ± 0.04	0.81 ± 0.03
PB-AnDA + LROI + dX	0.24 ± 0.03	0.93 ± 0.02

Table 3. Interpolation performance for the OSTIA SST case-study: mean root mean square error (RMSE) and correlation coefficients with the OSTIA SST for OI, DINEOF, G-AnDA and PS-MS-AnDA methods. We refer the reader to the main text for the details on the considered parameterizations.

Criterion	RMSE	Correlation
OI	0.42 ± 0.11	0.83 ± 0.07
DINEOF	0.40 ± 0.10	0.86 ± 0.06
G-AnDA	0.38 ± 0.08	0.87 ± 0.04
PB-AnDA + LROI + dX	0.22 ± 0.04	0.90 ± 0.03

Based on the above results, we further compared the performance of the different parameterization of the proposed Pb-AnDA models. The higher energy level of MW SST fields for scales below 100 km made the MW SST case-study more appropriate for this analysis. We evaluate the interpolation performance for PB-AnDA modes using respectively dX, $dX + Z$ and $dX + \bar{X}$ variables using both the optimally-interpolated field (LROI) and the yearly mean (LRM) as low-resolution background. We report in Table 1 the RMSE for three specific zones: a first zone from (10°E, 36.25°S) to (56.25°E, 45°S), a second zone from (55°E, 38.75°S) to (75°E, 47.5°S) and a third zone from (35°E, 26.25°S) to (55°E, 35°S). The first two zones depict highly-dynamical patterns, whereas the dynamics on the third one are not as intense. Whereas auxiliary variables do not bring any improvement for Zone 3 for both LROI and LRM settings, a slight mean improvement is reported when considering Z for the two other zones (i.e., the EOF-based decomposition of the SSH field) as auxiliary variable (e.g., RMSE values of 0.32 vs. 0.30 for PB-AnDA-LRM-dX and PB-AnDA-LRM-$dX + Z$ in Zone 2). Surprisingly, the exploitation of the optimally-interpolated background (LROI setting) may be outperformed by LRM setting (e.g., RMSE of 0.32 for PB-AnDA-LROI-dX and 0.3 for PB-AnDA-LRM-dX in Zone 2). This may suggest that the space-time smoothing of the optimal interpolation for highly-dynamical situations may result in local biases. To check for this hypothesis, we run a complementary experiment using 5-daily SST field in order to simulate even higher-dynamical situations. As reported in Table 4, these experiments further pinpoint the relevance of PB-AnDA-LRM-$dX + Z$ setting to deal with highly-dynamical situations. Example reported in Figures 1 and 3 illustrates these conclusions. Visually, the use of the SSH as auxiliary variable (Z) leads to a better reconstruction of the fine-scale structures. This is further illustrated in Figure 4 for Zone 2.

As a synthesis, we report in Figure 5 the time series of the mean RMSE and correlation for the MW SST case-study for OI (blue), DINEOF, G-AnDA, PB-AnDA-LROI-$dX + Z$, PB-AnDA-LRM-$dX + Z$. These results clearly emphasize the relevance of PB-AnDA models. Besides lower RMSE values and higher correlation coefficients, PB-AnDA models also depict a much lower variability. Similarly to the zone-specific results discussed above, LRM and LROI settings lead to a very similar performance (mean RMSE of 0.239 for PB-AnDA-LRM-$dX + Z$ and of 0.241 for PB-AnDA-LROI-$dX + Z$). Though this may not appear significant when considering a spatiotemporal mean, this is the case when considering specific dates and zones corresponding to highly-dynamical situations. I may be noted that the overall computational complexity of PB-AnDA-LRM-$dX + Z$ is significantly lower as it did not require the computation of the OI field as low-resolution background.

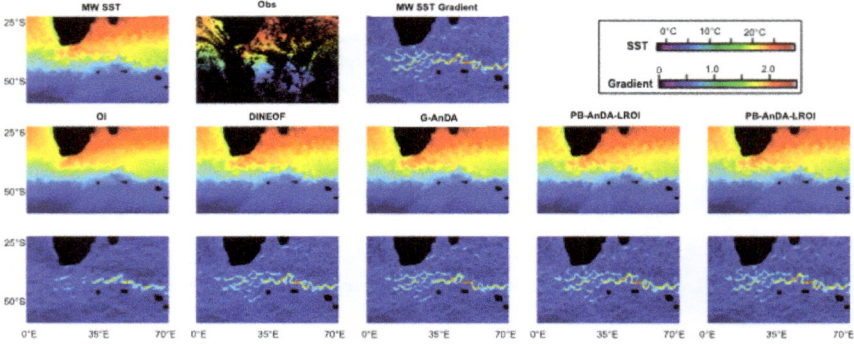

Figure 3. Reconstructed SST fields using OI, DINEOF, G-AnDA, PB-AnDA-LROI + dX + Z, PB-AnDA-LRM + dX + Z on day 150 for MW SST case-study: refer to Figure 1. It may be noticed that only PB-AnDA-LROI + dX + Z and PB-AnDA-LRM + dX + Z retrieves the fine-scale eddy-like structure off South Africa on this particular date.

Table 4. Interpolation performance of PB-AnDA models for a 5-daily MW SST case-study for three zones of interest in the case-study region: we refer the reader to the main text for the description of the different PB-AnDA parameterizations.

PB-AnDA	LROI			LRM		
	dX	$dX + Z$	$dX + \tilde{X}$	dX	$dX + Z$	$dX + \tilde{X}$
Zone 1	0.49 ± 0.10	$\mathbf{0.47 \pm 0.10}$	0.49 ± 0.10	0.51 ± 0.12	$\mathbf{0.45 \pm 0.09}$	0.51 ± 0.12
Zone 2	0.48 ± 0.14	$\mathbf{0.43 \pm 0.12}$	0.48 ± 0.14	0.49 ± 0.17	$\mathbf{0.38 \pm 0.10}$	0.48 ± 0.17
Zone 3	0.23 ± 0.06	$\mathbf{0.22 \pm 0.06}$	0.23 ± 0.06	0.23 ± 0.06	$\mathbf{0.22 \pm 0.06}$	0.23 ± 0.06

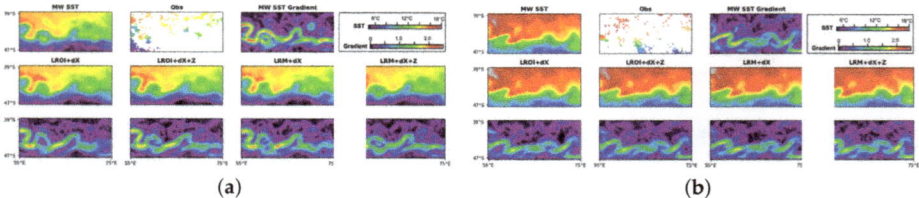

(a) (b)

Figure 4. Comparison of assimilation results for Zone 2 and case-study MW SST for different parameterization of the PB-AnDA models: we report the example for two dates, respectively the 104th and 309th of the MW SST time series, in panels (**a,b**). For each panel, the first row depicts the MW SST field (MW SST), the cloudy SST observation (Obs) and the gradient field (MW SST Gradient). The second and third displays respectively the reconstructed SST fields and their gradient magnitude for Pb-AnDA using LROI + dX, dX + Z, LROI + LRM + dX, LRM + dX + Z parameterizations.

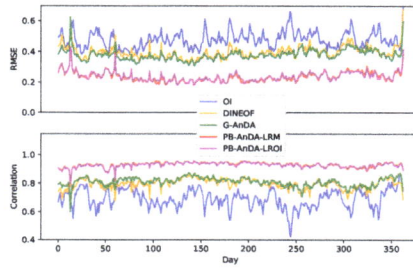

Figure 5. Mean RMSE and correlation time series for MW SST case-study: OI (blue), DINEOF (orange), G-AnDA (green), PB-MS-AnDA-LROI (red), PB-MS-AnDA-LRM (purple) methods.

5.2. Computational Complexity and Scalability

Regarding computational complexity issues, we evaluate the computational time of the PB-AnDA models with respect to the number processing cores. Figure 6 emphasizes the scalability of PB-AnDA models with good parallelization performance, as the computational time almost reaches the optimal linear decrease w.r.t. the number of cores in logarithmic scale. This parallelization performance directly relates to the proposed patch-based setting which leads to the independent sequential assimilation of the considered patches.

These experiments also stress the significant reduction of the computational complexity resulting from the clustering-based analog forecasting operators. When considering a 12-core architecture, the computational time is for instance reduced by a factor of about 4 between the proposed clustering-based scheme compared with the original one [24].

Overall, for the considered case-study region with 194 patches and the considered multi-core architecture, the overall computation time required by PB-AnDA is significantly less than that of G-AnDA (23 min vs. 82 min), though higher than that of DINEOF (23 min vs. 4 min). These results support an operational application of the proposed AnDA models on a global scale for high-resolution SST fields using state-of-the-art multi-core architecture.

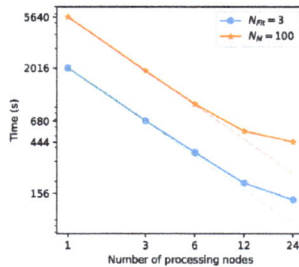

Figure 6. Parallelization performance of the proposed PB-AnDA setting: we report in logarithmic scale the computational time of the assimilation of 24 patches using the standard PB-AnDA with $N_M = 100$ members (orange,-) and using the clustering-based version with $N_{Fit} = 3$ clusters (blue,-). The dashed lines indicate the computational time of a theoretically-optimal multi-core parallelization.

6. Conclusions

We presented an application of the analog data assimilation [24] to the interpolation of SST fields. Using patch-based and EOF-based decompositions as in [26], the main contributions of this study are three-fold: (i) the introduction of conditional and physically-driven analog forecasting operators, (ii) the reduction of the computational complexity of PB-AnDA models with clustering-based analog forecasting operators, (iii) the demonstration of the scalability of PB-AnDA models to scale up to large scale-datasets. Overall, this study supports the investigation of the operational application of PB-AnDA models for an improved spatio-temporal interpolation of SST fields compared with optimal interpolation [12] and EOF-based schemes [14]. Among the issues to be delt with, the size and nature of the SST catalogs to be archived is certainly a critical question. Future work should further explore the extent to which purely-observation-based catalog may be self-sufficient or appropriately complemented by numerical simulation datasets. Preliminary results suggest that purely-observation-based catalogs might be a relevant option. Future should also investigate how AnDA may also provide a flexible framework to combine multi-source and multi-scale SST data through adapted observation models [10].

Beyond the reconstruction of gapfilled SST fields, we believe that the reported experiments illustrate the potential of PB-AnDA models for the reconstruction of geophysical products from remote sensing data, especially other sea surface tracers such as SSH (Sea Surface Height), SSS (Sea Surface Salinity) and ocean colour, as well as atmospheric variables. It might be noted that our recent application on the interpolation of altimeter-derived SSH fields further supports this potential [27]. For such applications, the relevance of PB-AnDA models is expected to strongly depend on one hand on the availability of large-scale simulation or observation-driven datasets to build representative catalogs of exemplars of the range of space-time scales of interests, and, on the other hand, on the validity of the assumption that state dynamics are locally-linear with respect to the considered regressors.

Acknowledgments: This work was supported by ANR (Agence Nationale de la Recherche, grant ANR-13-MONU-0014), Labex Cominlabs (grant SEACS), Teralab (grant TIAMSEA) and CNES (grant OSTST-MANATEE).

Author Contributions: R.F., R.L. and B.C. conceived and designed the experiments; P.V. performed the experiments; R.F. P.V. and R.L. analyzed the data; P.V., R.L. and PH.H. contributed reagents/materials/analysis tools; R.F. wrote the paper." Authorship must be limited to those who have contributed substantially to the work reported.

Conflicts of Interest: The authors declare no conflicts of interest.

References

1. Donlon, C.J.; Minnett, P.J.; Gentemann, C.; Nightingale, T.J.; Barton, I.J.; Ward, B.; Murray, M.J. Toward Improved Validation of Satellite Sea Surface Skin Temperature Measurements for Climate Research. *J. Clim.* **2002**, *15*, 353–369.
2. Merchant, C.J.; Embury, O.; Roberts-Jones, J.; Fiedler, E.; Bulgin, C.; Corlett, G.K.; Good, S.; McLaren, A.; Rayner, N.; Morak-Bozzo, S.; et al. Sea surface temperature datasets for climate applications from Phase 1 of the European Space Agency Climate Change Initiative (SST CCI). *Geosci. Data J.* **2014**, *1*, 179–191.
3. Hollmann, R.; Merchant, C.J.; Saunders, R.; Downy, C.; Buchwitz, M.; Cazenave, A.; Chuvieco, E.; Defourny, P.; de Leeuw, G.; Forsberg, R.; et al. The ESA Climate Change Initiative: Satellite Data Records for Essential Climate Variables. *Bull. Am. Meteorol. Soc.* **2013**, *94*, 1541–1552.
4. Filipponi, F.; Valentini, E.; Taramelli, A. Sea Surface Temperature changes analysis, an Essential Climate Variable for Ecosystem Services provisioning. In Proceedings of the 9th International Workshop on the Analysis of Multitemporal Remote Sensing Images (MultiTemp), Brugge, Belgium, 27–29 June 2017; pp. 1–8.
5. Li, J.; Heap, A.D. Spatial interpolation methods applied in the environmental sciences: A review. *Environ. Model. Softw.* **2014**, *53*, 173–189.
6. Li, J.; Wang, P.; Han, H.; LI, J.; Zhang, J. On the assimilation of satellite sounder data in cloudy skies in numerical weather prediction models. *J. Meteorol. Res.* **2016**, *30*, 169–182.
7. Penny, S.G.; Hamill, T.M. Coupled Data Assimilation for Integrated Earth System Analysis and Prediction. *Bull. Am. Meteorol. Soc.* **2017**, *98*, ES169–ES172.
8. Waters, J.; Lea, D.J.; Martin, M.J.; Mirouze, I.; Weaver, A.; While, J. Implementing a variational data assimilation system in an operational 1/4 degree global ocean model. *Q. J. R. Meteorol. Soc.* **2015**, *141*, 333–349.

9. Wada, A.; Kunii, M. The role of ocean-atmosphere interaction in Typhoon Sinlaku (2008) using a regional coupled data assimilation system. *J. Geophys. Res. Oceans* **2017**, *122*, 3675–3695.

10. Guan, L.; Kawamura, H. Merging Satellite Infrared and Microwave SSTs: Methodology and Evaluation of the New SST. *J. Oceanogr.* **2004**, *60*, 905–912.

11. Wentz, F.; Gentemann, C.; Smith, D.; Chelton, D. Satellite Measurements of Sea Surface Temperature through Clouds. *Science* **2000**, *288*, 847–850.

12. Donlon, C.J.; Martin, M.; Stark, J.; Roberts-Jones, J.; Fiedler, E.; Wimmer, W. The Operational Sea Surface Temperature and Sea Ice Analysis (OSTIA) system. *Remote Sens. Environ.* **2012**, *116*, 140–158.

13. Donlon, C. The Next Generation of Multi-Sensor Merged Sea Surface Temperature Data Sets for Europe. In *Remote Sensing of the European Seas*; Barale, V., Gade, M., Eds.; Springer: Berlin, Germany, 2008; pp. 177–188, doi:10.1007/978-1-4020-6772-3_14.

14. Ping, B.; Su, F.; Meng, Y. An Improved DINEOF Algorithm for Filling Missing Values in Spatio-Temporal Sea Surface Temperature Data. *PLoS ONE* **2016**, *11*, e0155928.

15. Beckers, J.M.; Rixen, M. EOF Calculations and Data Filling from Incomplete Oceanographic Datasets. *J. Atmos. Ocean. Technol.* **2003**, *20*, 1839–1856.

16. Reynolds, R.W.; Smith, T.M. Improved Global Sea Surface Temperature Analyses Using Optimum Interpolation. *J. Clim.* **1994**, *7*, 929–948.

17. Reynolds, R.W.; Smith, T.M.; Liu, C.; Chelton, D.B.; Casey, K.S.; Schlax, M.G. Daily High-Resolution-Blended Analyses for Sea Surface Temperature. *J. Clim.* **2007**, *20*, 5473–5496.

18. Tandeo, P.; Ailliot, P.; Autret, E. Linear Gaussian state-space model with irregular sampling: Application to sea surface temperature. *Stoch. Environ. Res. Risk Assess.* **2010**, *25*, 793–804.

19. Dash, P.; Ignatov, A.; Martin, M.; Donlon, C.; Brasnett, B.; Reynolds, R.W.; Banzon, V.; Beggs, H.; Cayula, J.F.; Chao, Y. Group for High Resolution Sea Surface Temperature (GHRSST) analysis fields inter-comparisons—Part 2: Near real time web-based level 4 SST Quality Monitor (L4-SQUAM). *Deep Sea Res. Part II Top. Stud. Oceanogr.* **2012**, *77*, 31–43.

20. Evensen, G. *Data Assimilation*; Springer: Berlin/Heidelberg, Germany, 2009.

21. Ubelmann, C.; Klein, P.; Fu, L.L. Dynamic Interpolation of Sea Surface Height and Potential Applications for Future High-Resolution Altimetry Mapping. *J. Atmos. Ocean. Technol.* **2014**, *32*, 177–184.

22. Sirjacobs, D.; Alvera-Azcárate, A.; Barth, A.; Lacroix, G.; Park, Y.; Nechad, B.; Ruddick, K.; Beckers, J.M. Cloud filling of ocean colour and sea surface temperature remote sensing products over the Southern North Sea by the Data Interpolating Empirical Orthogonal Functions methodology. *J. Sea Res.* **2011**, *65*, 114–130.

23. Zhao, Z.; Giannakis, D. Analog Forecasting with Dynamics-Adapted Kernels. *arXiv* **2014**, arXiv:physics1412.3831.

24. Lguensat, R.; Tandeo, P.; Aillot, P.; Fablet, R. The Analog Data Assimilation. *Mon. Weather Rev.* **2017**, doi:10.1175/MWR-D-16-0441.1.

25. Lorenz, E.N. Deterministic Nonperiodic Flow. *J. Atmos. Sci.* **1963**, *20*, 130–141.

26. Fablet, R.; Viet, P.H.; Lguensat, R. Data-Driven Models for the Spatio-Temporal Interpolation of Satellite-Derived SST Fields. *IEEE Trans. Comput. Imag.* **2017**, *3*, 647–657.

27. Lguensat, R.; Huynh Viet, P.; Sun, M.; Chen, G.; Fenglin, T.; Chapron, B.; Fablet, R. Data-Driven Interpolation of Sea Level Anomalies Using Analog Data Assimilation. Available online: https://hal.archives-ouvertes.fr/hal-01609851 (accessed on 4 October 2017).

28. Asch, M.; Bocquet, M.; Nodet, M. *Data Assimilation*; Fundamentals of Algorithms, Society for Industrial and Applied Mathematics: Heidelberg, Germany, 2016; doi:10.1137/1.9781611974546.

29. Fairbairn, D.; Pring, S.R.; Lorenc, A.C.; Roulstone, I. A comparison of 4DVar with ensemble data assimilation methods. *Q. J. R. Meteorol. Soc.* **2014**, *140*, 281–294.

30. Buades, A.; Coll, B.; Morel, J.M. A non-local algorithm for image denoising. In Proceedings of the CVPR'05 IEEE Conference on Computer Vision and Pattern Recognition, San Diego, CA, USA, 20–25 June 2005; Volume 2, pp. 60–65.

31. Mairal, J.; Bach, F.; Ponce, J. Sparse Modeling for Image and Vision Processing. *Found. Trends Comput. Graph. Vis.* **2014**, *8*, 85–283.

32. Freeman, W.T.; Liu, C. Markov Random Fields for Super-Resolution. In *Advances in Markov Random Fields for Vision and Image Processing*; Blake, A., Kohli, P., Rother, C., Eds.; MIT Press: Cambridge, MA, USA, 2011.

33. Criminisi, A.; Perez, P.; Toyama, K. Region filling and object removal by exemplar-based image inpainting. *IEEE Trans. Image Process.* **2004**, *13*, 1200–1212.

34. Fablet, R.; Rousseau, F. Joint Interpolation of Multisensor Sea Surface Temperature Fields Using Nonlocal and Statistical Priors. *IEEE J. Sel. Top. Appl. Earth Obs. Remote Sens.* **2016**, *9*, 2665–2675.

35. Klein, P.; Isern-Fontanet, J.; Lapeyre, G.; Roullet, G.; Danioux, E.; Chapron, B.; Le Gentil, S.; Sasaki, H. Diagnosis of vertical velocities in the upper ocean from high resolution sea surface height. *Geophys. Res. Lett.* **2009**, *36*, L12603.

36. Bernard, D.; Boffetta, G.; Celani, A.; Falkovich, G. Inverse Turbulent Cascades and Conformally Invariant Curves. *Phys. Rev. Lett.* **2007**, *98*, 024501.

37. Tandeo, P.; Ailliot, P.; Ruiz, J.; Hannart, A.; Chapron, B.; Cuzol, A.; Monbet, V.; Easton, R.; Fablet, R. Combining Analog Method and Ensemble Data Assimilation: Application to the Lorenz-63 Chaotic System. In *Machine Learning and Data Mining Approaches to Climate Science*; Lakshmanan, V., Gilleland, E., McGovern, A., Tingley, M., Eds.; Springer: Berline, Germany, 2015; pp. 3–12.

38. Fablet, R.; Rousseau, F. Missing data super-resolution using non-local and statistical priors. In Proceedings of the 2015 IEEE International Conference on Image Processing (ICIP), Quebec City, QC, Canada, 27–30 September 2015; pp. 676–680.

39. Iwamura, M.; Sato, T.; Kise, K. What Is the Most Efficient Way to Select Nearest Neighbor Candidates for Fast Approximate Nearest Neighbor Search. In Proceedings of the 2013 IEEE International Conference on Computer Vision (ICCV), Sydney, Australia, 1-8 December 2013; pp. 3535–3542.

40. Muja, M.; Lowe, D.G. Scalable Nearest Neighbor Algorithms for High Dimensional Data. *IEEE Trans. Pattern Anal. Mach. Intell.* **2014**, *36*, 2227–2240.

41. Muja, M.; Lowe, D.G. Fast approximate nearest neighbors with automatic algorithm configuration. In Proceedings of the VISAPP'09 International Conference on Computer Vision Theory and Applications, Lisboa, Portugal, 5–8 February 2009; Volume 2, p. 2.

42. Duda, R.O.; Hart, P.; Stork, D. *Pattern Classification*; John Wiley & Sons: New York, NY, USA, 2012.

43. Escudier, R.; Bouffard, J.; Pascual, A.; Poulain, P.M.; Pujol, M.I. Improvement of coastal and mesoscale observation from space: Application to the northwestern Mediterranean Sea. *Geophys. Res. Lett.* **2013**, *40*, 2148–2153.

remote sensing

MDPI

Article

Optimal Estimation of Sea Surface Temperature from AMSR-E

Pia Nielsen-Englyst [1,*], Jacob L. Høyer [1], Leif Toudal Pedersen [2], Chelle L. Gentemann [3], Emy Alerskans [1], Tom Block [4] and Craig Donlon [5]

[1]   Danish Meteorological Institute, Lyngbyvej 100, DK-2100 Copenhagen Ø, Denmark; jlh@dmi.dk (J.L.H.); ea@dmi.dk (E.A.)
[2]   DTU-Space, Technical University of Denmark, DK-2800 Lyngby, Denmark; ltp@eolab.dk
[3]   Earth and Space Research, Seattle, WA 98121, USA; cgentemann@gmail.com
[4]   Brockmann Consult GmbH, Max-Planck-Str. 2, 21502 Geesthacht, Germany; tom.block@brockmann-consult.de
[5]   European Space Agency/European Space Research and Technology Centre (ESA/ESTEC), 2201 AZ Noordwijk, The Netherlands; craig.donlon@esa.int
*   Correspondence: pne@dmi.dk; Tel.: +45-3915-7424

Received: 26 November 2017; Accepted: 30 January 2018; Published: 2 February 2018

Abstract: The Optimal Estimation (OE) technique is developed within the European Space Agency Climate Change Initiative (ESA-CCI) to retrieve subskin Sea Surface Temperature (SST) from AQUA's Advanced Microwave Scanning Radiometer—Earth Observing System (AMSR-E). A comprehensive matchup database with drifting buoy observations is used to develop and test the OE setup. It is shown that it is essential to update the first guess atmospheric and oceanic state variables and to perform several iterations to reach an optimal retrieval. The optimal number of iterations is typically three to four in the current setup. In addition, updating the forward model, using a multivariate regression model is shown to improve the capability of the forward model to reproduce the observations. The average sensitivity of the OE retrieval is 0.5 and shows a latitudinal dependency with smaller sensitivity for cold waters and larger sensitivity for warmer waters. The OE SSTs are evaluated against drifting buoy measurements during 2010. The results show an average difference of 0.02 K with a standard deviation of 0.47 K when considering the 64% matchups, where the simulated and observed brightness temperatures are most consistent. The corresponding mean uncertainty is estimated to 0.48 K including the in situ and sampling uncertainties. An independent validation against Argo observations from 2009 to 2011 shows an average difference of 0.01 K, a standard deviation of 0.50 K and a mean uncertainty of 0.47 K, when considering the best 62% of retrievals. The satellite versus in situ discrepancies are highest in the dynamic oceanic regions due to the large satellite footprint size and the associated sampling effects. Uncertainty estimates are available for all retrievals and have been validated to be accurate. They can thus be used to obtain very good retrieval results. In general, the results from the OE retrieval are very encouraging and demonstrate that passive microwave observations provide a valuable alternative to infrared satellite observations for retrieving SST.

Keywords: remote sensing; sea surface temperature (SST); microwave; optimal estimation

---

1. Introduction

Sea surface temperature (SST) is an essential climate variable that is fundamental for climate monitoring, understanding of air–sea interactions, and numerical weather prediction. It has been observed from thermal infrared (IR) satellite instruments since the early 1980s, but these observations are limited by their inability to retrieve SST under clouds and biasing from aerosols [1–3]. SST observations from passive microwave (PMW) sensors are widely recognized as an important alternative to the IR observations [4,5]. PMW SST retrievals are not prevented by non-precipitating clouds and the impact of aerosols is small [6,7]. The first PMW SST retrieval algorithm was developed for the Scanning Multichannel Microwave Radiometer (SMMR) on NIMBUS-7 [8,9]. The algorithm used a linear combination of brightness temperatures and the earth incidence angle. It was a two-stage algorithm, first calculating wind speed then selecting the SST regression coefficients if winds were greater than or less than 7 m·s^{-1}. SMMR suffered from significant calibration problems (solar contamination of the hot-load calibration target) that resulted in large errors in the retrieved SST, limiting its usefulness [10].

Since December 1997, a series of satellites have carried well-calibrated PMW radiometers capable of accurately retrieving SST. The first accurate PMW SST data was from the high inclination orbit Tropical Rainfall Measuring Mission (TRMM) Microwave Imager (TMI) that observed equatorward of 40 degrees latitude (limited by the high inclination orbit of the spacecraft and the weak sensitivity of x-band channels (10.65 GHz) at SST less than ~285 K). This was followed in 2002 by the first global PMW SST data from AQUA's Advanced Microwave Scanning Radiometer—Earth Observing System (AMSR-E) and more recently the follow-on instrument AMSR-2 flown on the Global Change Observing Mission (GCOM-1, e.g., [11]) launched in May 2012. The first spaceborne polarimetric microwave radiometer, WindSat, launched in January 2003, also provides measurements for retrieving SST [12]. These data demonstrated how through-cloud SST measurements offer unique opportunities for research into air–sea interactions, improved coverage in persistently cloudy regions, and could improve the existing operational SST products [6,7,13].

AMSR-E is a partnership, with a JAXA instrument carried on a NASA satellite. Both JAXA and NASA have routinely produced SSTs from AMSR-E; each using different retrieval algorithms developed and refined over years of experience with PMW data. The JAXA algorithm is just focused on producing SST using the 6V channel [14,15]. The NASA algorithm determines SST and wind speed at the same time using a radiative transfer based two-step algorithm [16]. The JAXA AMSR-E SSTs were compared to collocated IR satellite SSTs, for 7 months of data in 2003, and found a 0.0 K mean difference and standard deviation of 0.71/0.60 K for day/night [17]. The NASA AMSR-E SSTs from June 2002 through October 2011 were compared to global drifter SST data, and results showed a mean difference of −0.05 K and standard deviation of 0.48 K [18]. Both products are widely used in the operational ocean and atmospheric modeling communities (e.g., [19]) as well as for scientific research and applications [20–23].

SST retrievals using the optimal estimation (OE) principle have been developed several years ago for IR retrievals (see e.g., [24–26]). The OE retrieval methodology differs from standard regression models, in that it utilizes a forward model that includes a priori information about the ocean and atmospheric state to calculate simulated brightness temperatures. This methodology leads to improvements in the accuracy of IR SST retrievals using OE rather than a Non Linear SST (NLSST) algorithm and was used creating the first IR SST climate data record from the European Space Agency Climate Change Initiative (ESA-CCI) project [24,27]. Although an OE algorithm incurs additional computational costs because of the required forward modeling, it has significant advantages as the optimal estimator can be designed to estimate both retrieval uncertainty and sensitivity [28].

Climate quality global PMW SST retrievals are challenging due to the need to exclude retrievals that include observations from channels affected by wind, atmospheric attenuation and emission, sun-glint, land contamination, and Radio Frequency Interference (RFI). Different groups (e.g., [16,17]) use different exclusion criteria to flag affected data. The use of simulations and the OE methodology for the retrieval is therefore tempting as the algorithm automatically shifts away from the compromised channels and provides additional information which can be used to filter erroneous retrievals.

OE has previously been applied to multi-frequency PMW data [29] and results from AMSR-E retrievals were reported by [30,31]. In these studies, SST was among the retrieved parameters but the focus was to retrieve sea ice parameters. The OE technique has also been applied for WindSat retrievals [32], where SST also was among the retrieved parameters but the focus was on wind vector retrievals. The objective of this study is to develop a retrieval algorithm that provides an optimal and physically consistent retrieval with a specific focus on retrieving SST to be used for generation of a climate data record.

The paper is structured such that Section 2 describes the satellite and in situ data, the matchup database used for developing the OE retrieval and the OE processor. The results are presented in Section 3, and discussion and conclusions are in Sections 4 and 5, respectively.

2. Materials and Methods

2.1. Data

2.1.1. AMSR-E Brightness Temperatures

JAXA's AMSR-E instrument was launched in May 2002 on NASA's Aqua satellite. The AMSR-E instrument is a conical scanning microwave imaging radiometer that measures both vertical and horizontal linear polarizations at 6.9, 10.7, 18.7, 23.8, 36.5, and 89.0 GHz channels using an antenna diameter of 1.6 m. This suite of channels was chosen to support accurate retrievals of ocean, ice and atmospheric parameters. AMSR-E data are available from June 2002 through October 2011 when the antenna rotating mechanism on the instrument failed. This study uses the spatially resampled L2A swath data product AMSR-E V12 [33], produced by Remote Sensing Systems (RSS) and distributed by NASA's National Snow and Ice Data Center (NSIDC; https://nsidc.org/data/ae_l2a). The spatial resampling is generated by applying the Backus–Gilbert method to the L1A data. The RSS L2A product includes brightness temperatures for all AMSR-E channels that have been calibrated to the RSS version 7 standard, which includes inter-calibration with other satellite radiometers, and a correction to the AMSR-E hot load used during the calibration [34]. Brightness temperatures are re-sampled to the resolution of other channels and the location where the reflection vector intersects the geostationary sphere, used for development of RFI flagging, is included in the dataset. Sun glint angles are also calculated as a part of the RSS L2A AMSR-E V12 files. For this analysis, we use the re-sampling to 6.9 GHz resolution (75 × 43 km) for the five lowest frequencies.

2.1.2. In Situ Observations

The in situ dataset used for algorithm testing and validation is composed of quality-controlled measurements taken from the International Comprehensive Ocean-Atmosphere Dataset (ICOADS) version 2.5.1 [35], and the Met Office Hadley Centre (MOHC) Ensembles dataset version 4.2.0 (EN4) [36]. Observations from drifting buoys constitute the main source of observations. The temperature sensor on a drifting buoy is placed at around 20 cm depth in calm water, although the depth in perturbed conditions is poorly known. Temperature measurements are typically made hourly with an uncertainty from sensor calibration inferred to be about 0.2 °C [37]. MOHC quality control (QC) flags and track flags are provided with the data. See Atkinson et al. [38] for more information on the quality control.

In addition, temperature observations are used from the Argo profiling floats (see e.g., [39]). The data and quality control of these observations are described in Good et al. [36]. In this study, the uppermost temperature observations from the Argo observations have been used, which have a typical depth of 5 m [40] and a very high accuracy, with uncertainties of 0.002 °C [41,42]. Both Argo and drifting buoy observations have previously been used for algorithm development and validation studies [43–46].

2.1.3. Ancillary Data Fields

The OE method utilizes a priori information about the state of the ocean and atmosphere as first guess. For this study, we used Numerical Weather Prediction (NWP) information from the ERA-Interim NWP data as first guess on the atmospheric and oceanic state [47]. The ERA-Interim SST fields are from the Operational Sea Surface Temperature and Ice Analysis (OSTIA) level 4 SST analysis, which is generated from IR and PMW satellite observations blended with in situ data from drifting buoys [19,21]. The larger number of IR satellites and the higher spatial resolution compared to the PMW satellite observations means that the OSTIA analysis is dominated by the IR satellite observations. An independent validation showed a global mean difference (OSTIA—Argo floats) of −0.05 K and standard deviation of 0.55 K (see Section 3.3), which was found when the OSTIA SST analysis fields were compared against observations from Argo floats for 2009–2011 (using the data filtering methods described in Section 2.2.2). For Sea Surface Salinity (SSS), we used the monthly, 0.25 degrees spatial resolution, objectively analyzed mean fields, SSS from the World Ocean Atlas (WOA) 2013 version 2 [48,49]. This climatology was determined from historical salinity data from a wide variety of sources that were carefully quality controlled and objectively analyzed into monthly globally complete maps of SSS. These data were linearly interpolated in time and space to the matchup location.

2.2. Matchup Database

2.2.1. ESA-CCI Multi-Sensor Matchup Dataset

The basis for the retrieval algorithmic development and tuning is a Multi-sensor Matchup Dataset (MMD) pioneered by the ESA-CCI SST project [50]. The MMD has been constructed as a general dataset for algorithm development and not specifically for OE development. It includes AMSR-E orbital data matched to in situ measurements (drifting buoys and Argo floats) constrained by a maximally allowed geodesic distance and a maximal time difference.

The MMD was created using a Multi-sensor Matchup System (MMS) software that reads in all the in situ observations and finds the corresponding matching satellite observations throughout the full dataset. Matches were only included within a maximal geodesic distance of 20 km and a time difference of maximally 4 h. The spatial distance ensures that the in situ measurement is located within an AMSR-E footprint. The temporal distance balances the need for accurate collocated data with the need for a large number of useable matches.

The collocated AMSR-E data include a 21 by 21 pixel window with the matchup location in the center as well as all variables of the corresponding in situ measurement. The ERA-Interim NWP data were referenced to each AMSR-E pixel and each in situ measurement and spatially interpolated to the data raster using Climate Data Operators (CDO) [51]. This ancillary information includes a subset of the available ERA-Interim variables, covering a time range of −60 h to +36 h around the matchup time. Processing of the matchup dataset has been performed on the Climate and Environmental Monitoring from Space Facility (CEMS) computing facility at the Centre for Environmental Data Analysis (CEDA).

2.2.2. Data Filtering Methods

Developing an accurate retrieval algorithm relies on the quality of the satellite observations and auxiliary fields used for the retrieval and validation. It is therefore essential to flag erroneous matchups as accurately as possible and produce a clean dataset for the development.

Data have been flagged if any of the following quality flags were set to fail: AMSR-E pixel data quality, AMSR-E scan data quality, MOHC QC flag and MOHC track flag. If the brightness temperature was outside the normal range (0–320 K) for any of the channels, the data were flagged. To discard cases where the atmospheric contribution (largest for the 18–36 GHz channels) exceeds the information from the surface, data were flagged if the difference between the H and V brightness temperatures for the 18–36 GHz channels <0 K (for valid oceanic retrievals V should always be larger than H). Data were also flagged based on the spatial standard deviation of the 23V, 23H, 36V and 36H GHz brightness temperatures in the 21 × 21 pixel extracts surrounding each matchup. If the standard deviation of these channels were higher than 55, 35, 25 and 25 K, respectively, the data were flagged to remove obviously bad observations. Additionally, matchups were excluded if the ancillary data seemed to be erroneous. If in situ or NWP SSTs were less than −2 °C or greater than 40 °C or if NWP wind speeds were greater than 20 m·s^{-1} then the matchup was flagged. All these flags have been combined into a gross error flag, which in total removes 13.1% of the drifter matchups.

Additional filtering was added to account for various situations where the SST retrieval could be compromised, namely during: land and ice contamination (due to antenna side-lobe contamination), sun glitter contamination, geostationary satellite and ground-based RFI, diurnal warming, and rain contamination. To avoid contamination due to land and ice, the AMSR-E land/ocean flag and NWP sea ice fraction were used to flag data. Applying this filter alone resulted in a flagging of 13.1% of the drifter matchups, which had already passed the gross error check. To avoid sun glitter contamination, data with sun glint angle <25° were flagged (9.6%). Potential contamination due to RFI was detected using the observation location (for ground based RFI) and reflection vector (for geo-stationary RFI) using Table 2 presented in Gentemann et al. [52] (6.5%). To avoid diurnal warming effects, daytime data with NWP wind speeds <4 m·s^{-1} were flagged (8.0%). Rain contamination was accounted for by flagging data if the brightness temperature of the 18V channel >240 K (0.4%). Applying all these checks at once leads to an elimination of 41.1% of the total drifter matchups.

Finally, to obtain a more equal latitudinal distribution of the drifter matchups, a limit of 40,000 matchups per degree of latitude was imposed, removing an additional 12.2% of the drifter matchups. The summary statistics for different steps in the flagging process are shown in Table 1. The outcome is a high-quality, globally representative, final drifter matchup database. The focus for this study is the year 2010, which consists of 3,764,798 filtered drifter matchups that are used in the following.

Table 1. Summary of 2010 data discarded due to the different filters applied.

Flagging	N	% Removed
All matchups	7,278,035	
Gross error flag	6,323,288	13.1
- Land/ice mask [1]		13.1
- Sun glitter [1]		9.6
- RFI [1]		6.5
- Diurnal warming [1]		8.0
- Rain [1]		0.4
All above checks	4,286,354	41.1
Even out by latitude	3,764,798	12.2
Total	3,764,798	48.3

[1] Percentage of gross error checked matchups removed by applying each filter individually.

The final number of matchups per month during 2010 is shown in Figure 1. Figure 2a shows the geographical distribution of the matchups per square kilometer after all checks have been applied. Figure 2b shows the latitudinal distribution of matchups before and after the even-out-by-latitude filter has been applied, where the red line denotes the maximum allowed matchups per latitude.

Several subsets of the filtered drifter matchups have been applied throughout this study. The subsets were randomly selected from the filtered drifter matchups. This approach was chosen to reduce computational efforts and very small effects were seen on the final results, when the size of the subset was increased or reduced.

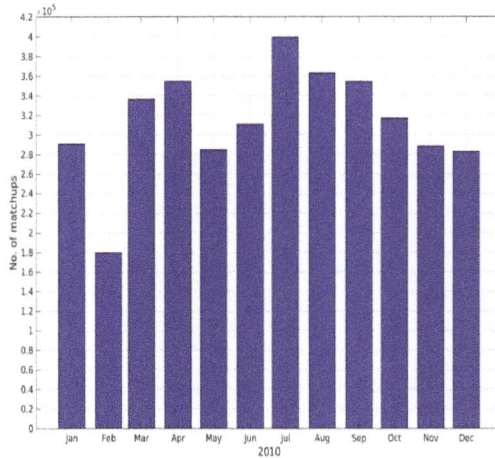

Figure 1. Number of matchups per month.

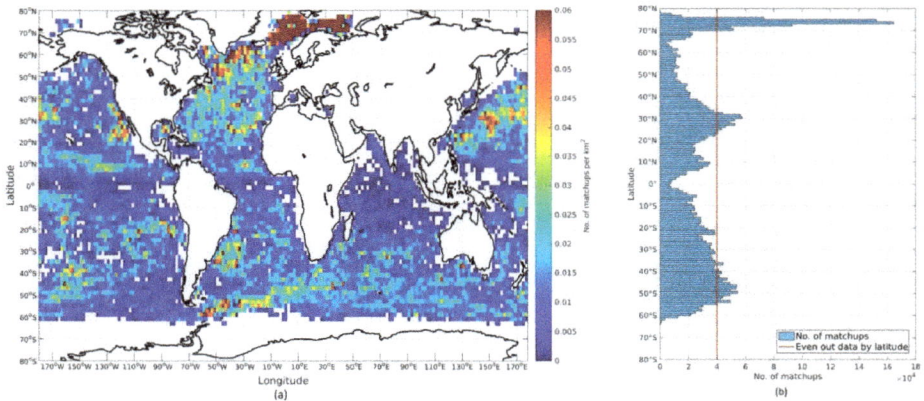

Figure 2. (**a**) Geographical distribution of drifter matchups per square kilometer during 2010; (**b**) the latitudinal distribution of drifter matchups before and after the number of matchups have been evened out by latitude. The red line denotes the maximum allowed matchups per latitude.

The matchups with Argo floats are limited to 95,240, prior to quality filtering. The gross error check removes 18.9% of the Argo matchups. Removing matchups contaminated by land/ice, RFI, rain, and diurnal warming effects sums up to a total removal of 39.3%, leaving 57,810 Argo matchups to be used in the following. Due to a more equal latitudinal matchup distribution and the limited number of matchups, the even-out-by-latitude filter has not been applied here. Figure 3a shows the geographical distribution of the Argo matchups per square kilometer and Figure 3b shows the latitudinal distribution of the Argo matchups after all checks have been applied.

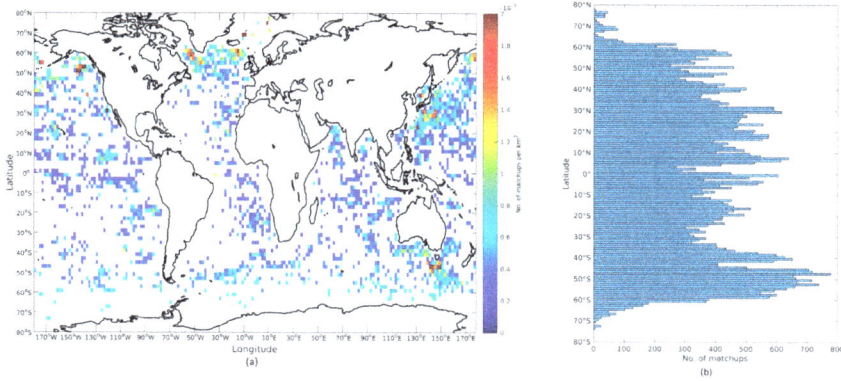

Figure 3. (**a**) Geographical distribution of Argo matchups per square kilometer during 2009–2011; (**b**) the latitudinal distribution of Argo matchups.

2.3. Optimal Estimation Development

The OE method can be used to retrieve geophysical parameters (e.g., SST) from PMW observations [29]. The relationship between the geophysical parameters and the measured brightness temperatures can be generalized to the following expression [53]:

$$\mathbf{y} = \mathbf{F}(\mathbf{x}) + e, \tag{1}$$

where $\mathbf{y}$ is the measurement vector (observed microwave brightness temperatures); $\mathbf{F}(\mathbf{x})$ is the non-linear forward model approximating the physics of the measurement, including the surface emissivity and the radiative transfer through the atmosphere [54]; $\mathbf{x}$ is the state vector containing the relevant geophysical properties of the ocean and atmosphere; and e is a residual uncertainty term containing uncertainties due to the measurement noise and uncertainties in the forward model.

The forward model predicts the top-of-atmosphere microwave brightness temperatures that should be measured by the individual channels of a radiometer given knowledge of the relevant geophysical parameters ($\mathbf{x}$) of the ocean and atmosphere. The forward model used in this study is based on the physical surface emissivity and Radiative Transfer Model (RTM) described in Wentz et al. [54]. The RTM consists of an atmospheric absorption model for oxygen, water vapor and cloud liquid water and a sea surface emissivity model that determines the emissivity as a function of SST, SSS, sea surface wind speed and direction. Some components have been adjusted with respect to Wentz et al. [54]. These include the wind directional signal of sea surface emissivity, which has been suppressed as it did not improve the retrievals; and the fact that we only use the V- and H-polarizations for the 5 lower frequencies: 6.9, 10.7, 18.7, 23.8, 36.5 GHz.

The aim of this study is to invert Equation (1) to retrieve the most likely state vector $\mathbf{x}$ that can reproduce the observed microwave brightness temperatures, $\mathbf{y}$. In this study, the inversion approach follows the OE technique by Rodgers [53] and we broadly follow his conventions.

In OE, a priori information about the expected mean and covariance of the geophysical parameters can be used to put restrictions on the variances of the estimated geophysical parameters and thereby improve the retrieval. In this case, the prior information is NWP fields as described in Section 2.1.3.

Assuming the forward model is a general function of the state, the measurement + model error has a Gaussian distribution, and there is a prior estimate with a Gaussian uncertainty distribution, the maximum probability state $\mathbf{x}$ can be found by minimizing the cost function, $\mathbf{J}$:

$$\mathbf{J} = [\mathbf{y} - \mathbf{F}(\mathbf{x})]^{\mathsf{T}} \mathbf{S}_{\epsilon}^{-1} [\mathbf{y} - \mathbf{F}(\mathbf{x})] + (\mathbf{x} - \mathbf{x_a})^{\mathsf{T}} \mathbf{S}_a^{-1} (\mathbf{x} - \mathbf{x_a}), \tag{2}$$

where S_ϵ is a covariance matrix for the measurement and forward model uncertainties, S_a is the covariances of the a priori state x_a (the a priori guess of the ocean and atmospheric state x). The cost function is a measure of the goodness of the fit to both the measurements (first term on the right) and the a priori state (second term on the right) balanced by the inverse of their relative uncertainties (S_ϵ and S_a).

In this nonlinear case, Newtonian iteration is a straightforward numerical method for finding the zero gradient of the cost function, J. Using Newtonian iteration, the state x that minimizes the cost function can be found by:

$$x_{i+1} = x_i + S_x \left[K_i^T S_\epsilon^{-1}(y - F(x_i)) - S_a^{-1}(x_i - x_a) \right], \tag{3}$$

where S_x is the error covariance matrix of the retrieved parameters:

$$S_x = \left(S_a^{-1} + K_i^T S_\epsilon^{-1} K_i \right)^{-1}. \tag{4}$$

The matrix **K** expresses the sensitivity of the forward model to a perturbation in the retrieved parameters, i.e., it is a matrix consisting of the partial derivatives of the brightness temperatures in a particular channel with respect to each parameter of the state vector. Due to non-linearity, these partial derivatives need to be computed at each iteration (state).

2.3.1. Initial OE Setup

The measurement vector, **y**, used in our forward model consists of dual polarization observations (v-pol and h-pol) at the 5 lower frequencies: 6.9, 10.7, 18.7, 23.8, 36.5 GHz. Four geophysical parameters are considered to be the leading terms controlling the observed microwave brightness temperatures in the measurement situation (considering open-ocean only):

$$x = [WS, TCWV, TCLW, SST], \tag{5}$$

where WS is the wind speed, TCWV is the integrated columnar atmospheric water vapor content, TCLW is the integrated (columnar) cloud liquid water content, and SST is the sea surface temperature.

The variations of the retrieved geophysical parameters are restricted by the use of a priori information from NWP about the mean (a priori state) and covariances of the parameters. OE can be considered to be an adjustment of the a priori state vector based on the difference between simulated and observed brightness temperatures. The method takes appropriate account of errors by combining the a priori state vector and the information content in the observed brightness temperatures. The covariance matrix of the geophysical parameters related to x is fixed to:

$$S_a = \begin{bmatrix} e_{WS}^2 & 0 & 0 & 0 \\ 0 & e_{TCWV}^2 & 0 & 0 \\ 0 & 0 & e_{TCLW}^2 & 0 \\ 0 & 0 & 0 & e_{SST}^2 \end{bmatrix}, \tag{6}$$

where $e_{WS} = 2$ m·s^{-1}, $e_{TCWV} = 0.9$ mm, $e_{TCLW} = 1$ mm and $e_{SST} = 0.50$ K. The uncertainties on the WS, TCVW and TCLW are best estimates based upon published validation results (see e.g., [47,55–58]). The SST uncertainty is derived from a comparison against Argo drifting buoys, using the MMD (see Section 3.3). The measurement covariance matrix, S_ϵ is initially set to a diagonal matrix with all diagonal elements equal to 0.1 K [54]. The retrieved state vector is obtained by performing the Newtonian iteration, as described in Equation (3).

2.3.2. Testing for Convergence

For each iteration, the quality can be assessed by comparing the simulated and observed brightness temperatures and requiring these to be consistent within a certain uncertainty. This idea is quantified by the root-mean-square error (RMSE_{TB}):

$$\text{RMSE}_{\text{TB}} = \sqrt{\frac{1}{n}\sum_{i=1}^{n}(\text{TB}_{\text{obs}} - \text{TB}_{\text{calc}})^2}, \tag{7}$$

which is a confidence indicator of how well the simulated brightness temperatures, TB_{calc} fit the observed ones, TB_{obs}. The RMSE_{TB} criteria is chosen here over e.g., χ^2 as it provides an almost linear relationship with the performance of the OE, which will be shown in Section 3. Figure 4a illustrates the mean RMSE_{TB} difference for each iteration using a subset of the drifter MMD. The uncertainty bars mark one standard deviation. A strong reduction in RMSE_{TB} and standard deviation is found by performing the first iteration. The second iteration similarly leads to a decrease in RMSE_{TB} and standard deviation, while the following iterations show no significant improvement on the mean RMSE_{TB}. The usefulness of RMSE_{TB} as a confidence indicator will be further illustrated in Section 3.

To reduce computational cost, a convergence analysis is performed to decide whether a retrieval process has converged to sufficient precision or if further iterations are required. According to Rodgers [53] the most straightforward convergence test is to ensure that the cost function (Equation (2)) is actually being minimized. The change in the cost function between two subsequent iterations will always be small near a cost minimum. Noting that the expected value of the cost function at the minimum is equal to m degrees of freedom ($m = 10$) an appropriate test would be to require the change between iterations of $\Delta J = J_i - J_{i+1} \ll m$ or $\Delta J = J_i - J_{i+1} < 0.1$. In addition, ΔJ is required to be positive at the final solution.

A maximum of 10 iterations are allowed and a failure to meet the above convergence criterion within 10 iterations leads to an exclusion of the data (<0.1%). Figure 4b shows the number of iterations performed for all drifter matchups during 2010 by applying the convergence criterion. Usually, convergence is reached by iteration 3–4. This setup will be referred to as the initial optimal estimator.

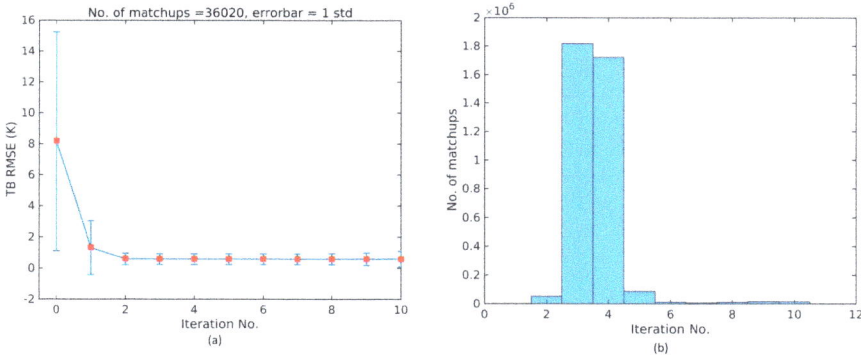

Figure 4. (**a**) The mean RMSE_{TB} for all channels is plotted for each iteration number. Uncertainty bars show one standard deviation; (**b**) number of iterations performed for all drifter matchups during 2010.

2.3.3. Improving the Forward Model

The PMW observations in the 6–10 GHz observe the subskin SSTs at ~1 mm depth, which is different from the IR skin SST observations where a cool skin effect is included [59]. For conditions free of diurnal variability signals, which is the case for the filtered MMD used here, we can assume that the PMW observed subskin SSTs are similar to the in situ drifting buoy observations at the nominal depth of ~20 cm [23]. We can thus use the in situ observations from the drifting buoys to validate the algorithm and for improving the forward model. Similarly, we can assume that the OSTIA SST fields, which are foundation temperatures, are similar to the subskin SSTs and the SSTs at 20 cm.

The OE method assumes an unbiased prior and forward model, which is not necessarily the case. The comparison against Argo SSTs has already addressed the bias in the prior NWP SST field with a mean difference of -0.05 K, which has been adjusted for. In order to get a measure of the forward model deficiency, we compare TB_{calc} with TB_{obs}. If we assume optimal forward model input variables and unbiased TB_{obs}, we can regard the TB_{obs}–TB_{calc} differences as inefficiencies in the forward model that we would like to correct for. In other words, we want to use the best available input variables. Therefore, retrieved WS, TCWV and TCLW and in situ SST values are used in the forward model calculations, to bring us as close to true oceanic and atmospheric conditions as possible. The retrieved variables are obtained by running the initial optimal estimator for a subset of 37,242 matchups. These input variables are used to run the forward model once and the difference TB_{obs}–TB_{calc} is calculated for each channel. Part of the observed channel biases may be a result of the difference between the RTM used here and the one used in calibration of the RSS L2A product [60]. In addition, the RTM used here, does not include wind directional effects. Following Merchant et al. [25] cells with TB_{obs}–TB_{calc} differences falling outside the range given by the median ± 3 robust standard deviations (RSD) in any of our 10 channels are discarded. Furthermore, only matchups that have passed the convergence test (Section 2.3.2) are included. The derived average TB_{obs}–TB_{calc} differences of the 10 channels range from -0.75 K on 10 GHz H to 0.62 K on 18.7 GHz V and are subsequently used as a constant bias correction of the forward model. In addition to the constant bias correction, an updated error covariance matrix $\mathbf{S_e}$ is calculated from the TB_{obs}–TB_{calc} subset. The updated $\mathbf{S_e}$ used in the following has an average of square root diagonals of 0.20 K, smallest for 10.7 GHz H and 36.5 GHz H (0.09 K) and largest for 6.9 GHz H (0.31 K).

Further steps are taken to improve the forward model used in the retrieval. The updated optimal estimator has been run for the 3,724,216 drifter matchups in 2010. The retrieved WS, TCWV and TCLW values have subsequently been used together with in situ SST to run the forward model once. Similar to the approach used for the constant bias correction, the simulated brightness temperatures are then compared with satellite observations for each channel. To improve the forward model, we use a correction scheme based upon a multivariate regression model. The regression model applies an empirical fit of TB_{calc}–TB_{obs} to analytic functions of in situ SST, retrieved WS and NWP wind direction relative to the azimuthal look, φ_r. The fitting is done on averaged TB_{calc}–TB_{obs} values for binned data with respect to SST, WS and φ_r and with binning intervals of: 1 °C, 2 m·s^{-1} and 15°, respectively. Only average values from bins with more than 50 members are used when the regression coefficients are determined. Four sinusoidal terms were found to be the most optimal in representing the wind direction biases. The optimal regression model used for the forward model residuals is:

$$a_1 + b_1 SST + b_2 SST^2 + c_1 WS + c_2 WS^2 + d_1 \cos(\varphi_r) + d_2 \sin(\varphi_r) + d_3 \cos\left(\tfrac{\varphi_r}{2}\right) +$$
$$d_4 \sin\left(\tfrac{\varphi_r}{2}\right) + d_5 \cos\left(\tfrac{\varphi_r}{3}\right) + d_6 \sin\left(\tfrac{\varphi_r}{3}\right) + d_7 \cos\left(\tfrac{\varphi_r}{4}\right) + d_8 \sin\left(\tfrac{\varphi_r}{4}\right), \tag{8}$$

with individual coefficients calculated for each channel. This correction is added to the simulated brightness temperatures individually every time the forward model is called using retrieved SST and WS from the latest iteration.

Figure 5a–d show the average (solid lines) and standard deviation (dashed lines) of the difference TB_{calc}–TB_{obs} (final iteration) for all channels, and all matchups during 2010, before (black) and after (blue) the empirical bias correction scheme has been applied. Figure 5a indicates a positive bias at high latitudes, no bias at mid-latitudes and a slightly positive bias at the equatorial regions before the empirical bias correction has been applied. The black line of Figure 5b shows an almost linear trend in bias ranging from a positive bias of about 0.5 K in cold waters, no bias at temperatures ~20–25 °C and a slightly positive bias for warmer waters, which is in good agreement with Figure 5a. Figure 5c also reveals a systematic bias in the TB_{calc}–TB_{obs} difference with the NWP WS before the empirical bias correction is applied. At low wind speeds little bias is present but with increasing wind speeds the bias rapidly becomes larger. Also, the binned φ_r statistics reveal a dependency with a positive bias around $\varphi_r = 250°$ that might be related to wind direction effects not included in the forward model. The bottom plots show the number of matchups in each bin (blue curve) and the cumulative percentage of matchups (red curve).

The blue lines of Figure 5a–d are the updated residuals after the empirical bias correction scheme has been added to TB_{calc} for all channels, each time the forward model is called. The application of the empirical bias correction improves the behavior of the residuals against each of the four factors by flattening their bias curves and bringing them closer to zero. The standard deviation of the TB_{calc}–TB_{obs} difference also decreases with the application of the empirical bias correction scheme.

The retrieved parameters have been compared with and without including the empirical bias correction in the retrieval. The distributions are very similar and mean differences between the two retrievals are: -0.02 m·s^{-1}, -0.04 mm and 7.19×10^{-4} mm for WS, TCWV and TCLW, respectively.

The empirical bias correction of the forward model completes the steps taken towards the final OE setup. The final OE configuration is briefly summarized in Table 2 and Figure 6 illustrates the different processes performed in the final Danish Meteorological Institute (DMI) OE algorithm. First, the algorithm reads in the predefined $\mathbf{S}_\epsilon$, $\mathbf{S}_a$ and e values, where e is the perturbation used to calculate the Jacobians. The observation loop is started for each satellite observation pixel or matchup by reading the observed brightness temperatures and the first guess values. Thereafter, the iteration process is initiated. For each iteration, the forward model is used to calculate the simulated brightness temperature from the state vector (in the first step: state vector = first guess). Moreover, the Jacobians ($\mathbf{K}$), cost function ($\mathbf{J}$), uncertainty ($\mathbf{S_x}$) and sensitivity ($\mathbf{A}$, Section 3.1) are calculated. The change in the cost function between two iterations is used to test for convergence and a maximum of 10 iterations are allowed. Until convergence is met, the state vector is updated for each iteration step and the iteration continues. When the iteration process is stopped the state vector is saved together with the uncertainties, corresponding averaging kernels and simulated brightness temperatures.

Table 2. Final optimal estimator configuration.

Aspect of Optimal Estimator	Configuration
Initial forward model, iF(x)	Modified from Wentz et al. [54] (Section 2.3.1)
Channels used in retrieval	6.9, 10.7, 18.7, 23.8, 36.5 GHz (V/H)
First guess fields, x_a	NWP (Section 2.1.3)
Prior error covariance for SST, e_{SST}	0.5 K
Error covariance of observations and model, S_ϵ	Full matrix (Section 2.3.2)
Convergence criterion	$\Delta J = J_i - J_{i+1} < 0.1$. (Section 2.3.2)
IterationsImproved forward model, F(x)	Max iterations = 10iF(x) + corrections (Section 2.3.3)

The drifter matchups covering 2010 have been processed and the final OE SST is presented and validated in the following section. Section 3.3 presents the results using the independent observations from Argo floats covering 2009–2011.

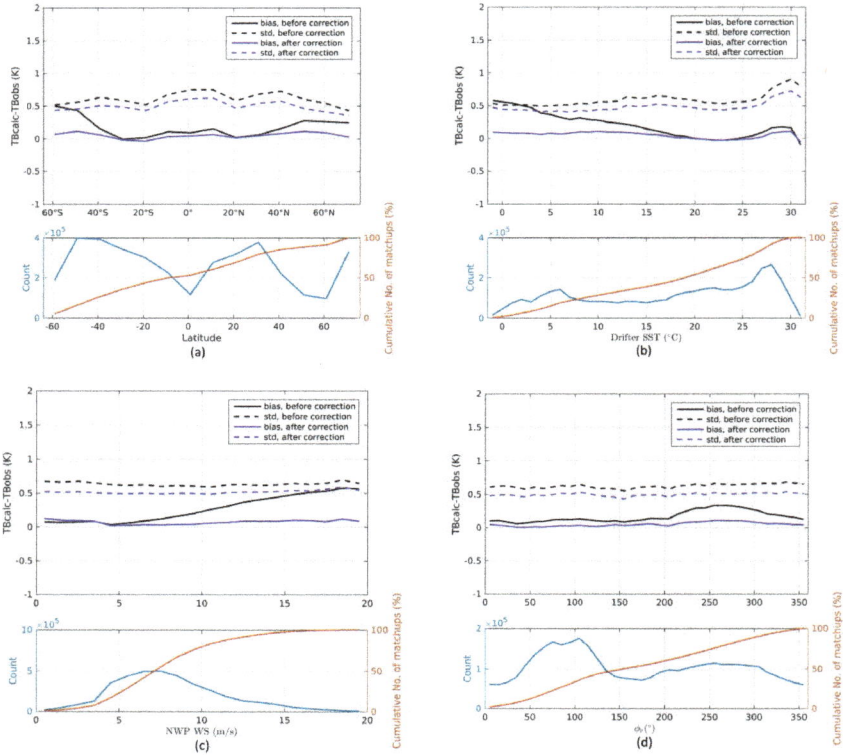

Figure 5. Mean simulated minus observed brightness temperatures for all channels against (**a**) Latitude; (**b**) Drifter SST; (**c**) NWP WS; (**d**) NWP wind direction relative to azimuthal satellite look. Dashed lines are standard deviations and solid lines are biases. The black and blue colors denote differences before/after the empirical bias correction has been applied. The bottom plots show the number of matchups (blue) and the cumulative percentage of matchups (red) for each data bin.

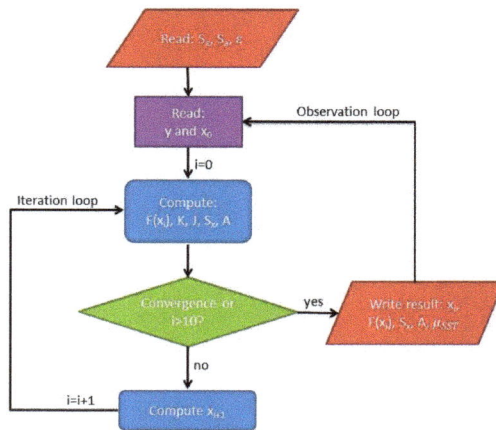

Figure 6. Flowchart of the DMI OE algorithm.

3. Results

The DMI OE retrieval scheme has been run with the final configuration described in Section 2.3.3 for the total number of filtered drifter matchups: 3,764,798 for 2010 (Section 2.2.2). The summary statistics of the OE SSTs against drifters are given in Table 3. The retrievals that did not manage to fulfill the convergence criterion described in Section 2.3.2 within the 10 maximum allowed iterations have been eliminated (<0.1%). The OE retrievals that have passed the convergence criterion give SSTs with an initial bias of 0.02 K and a standard deviation of 0.57 K. We have not checked that the retrievals that did pass the convergence test have actually converged to the required solution. This can be done in several ways. One way is to apply a gross error check to throw away cases with unrealistic conditions. These include: (1) temperature conditions outside the accepted range between −2 °C and 35 °C; (2) retrieved wind speeds outside the range 0–30 m·s^{-1}; and (3) retrieved cloud liquid water outside the range: 0–1.5 kg/kg (mass of condensate/mass of moist air). Applying this gross error check removes 9% of the retrievals and reduces the standard deviation to 0.54 K. Another approach is to check that the retrieval is consistent with the satellite observations by evaluating the RMSE$_{TB}$ value as described in Section 2.3.2. The practical usefulness of the quality indicator, RMSE$_{TB}$, is shown in Figure 7. All retrievals that have passed the convergence test have been binned with respect to RMSE$_{TB}$ with a bin size of 0.1 K. The number of members in each bin is shown in the bottom plot (blue curve) together with the cumulative percentage (red curve). The middle plot displays the binned distribution of OE SST minus drifter SST (with bin size of 1 K) as a function of binned RMSE$_{TB}$, where the color bar is the number of matchups in each bin. The top plot shows the mean (solid) and standard deviation (dashed) of OE SST minus drifter SST as a function of the binned RMSE$_{TB}$ statistic. We notice a large increase in scatter as RMSE$_{TB}$ increases. This makes the RMSE$_{TB}$-value an efficient indicator of the quality of the OE SST retrieval. Limiting RMSE$_{TB}$ to 1 K removes only 8% of the converged retrievals and leaves the remaining 92% with a bias of 0.02 K and standard deviation of 0.51 K. These results reflect that the RMSE$_{TB}$ quality indicator provides a better discrimination of quality compared to the gross error check.

Around 64% of converged retrievals have a RMSE$_{TB}$ value below 0.5 K and a corresponding bias of 0.02 K and standard deviation of 0.47 K, while 42% have a RMSE$_{TB}$-value less than 0.35 K and a corresponding bias of 0.02 K and standard deviation of 0.45 K. The validation results of the NWP SSTs are included here for reference, but note that drifting buoy observations and PMW observations have already been included in the generation of the NWP fields, as explained earlier. In the following we will only consider the 64% "good" retrievals, which have a corresponding RMSE$_{TB}$ < 0.5 K.

Table 3. Comparison of retrieved SSTs and NWP SSTs against drifter SSTs for various subsets.

Filter	Bias/K OE-Drifter	std/K OE-Drifter	Bias/K NWP-Drifter	std/K NWP-Drifter	N (10^6)	
Convergence test passed	0.02	0.57	−0.04	0.50	3.7429	=100%
Gross error check	0.04	0.54	−0.04	0.50	3.4071	=91%
RMSE$_{TB}$ < 1 K	0.02	0.51	−0.04	0.50	3.4329	=92%
RMSE$_{TB}$ < 0.50 K	0.02	0.47	−0.04	0.48	2.3953	=64%
RMSE$_{TB}$ < 0.35 K	0.02	0.45	−0.04	0.47	1.5681	=42%

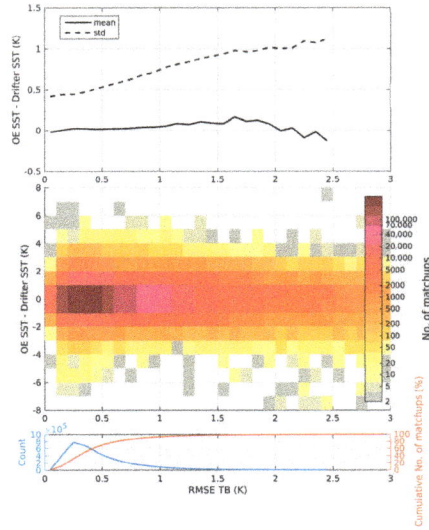

Figure 7. OE SST minus drifter SST as a function of binned RMSE$_{TB}$. The dashed line is standard deviation while the solid line is bias in the upper figure. The surface plot in the middle figure shows the number of matchups in each bin, while the bottom plot shows the number of matchups (blue) and the cumulative percentage of matchups (red) in each RMSE$_{TB}$ bin.

Figure 8a,b show global maps of the gridded (with grid size of 5 degrees) mean and standard deviation of OE SST minus drifter SST, respectively, for the 64% best retrievals with a corresponding RMSE$_{TB}$ < 0.5 K. The geographical distribution of the mean OE SST minus drifter SST reveals a dependency on latitude, with positive bias at mid-latitudes and negative bias in high latitudes and the equatorial region, likely linked to surface emissivity issues (dependent on wind speed and direction) and atmospheric effects. We notice areas with high standard deviations in e.g., the Gulf Stream Extension, the Kuroshio Current and the Aghulas Retroflection areas. These western boundary current regions are known to be very dynamical with high mesoscale activity and large SST gradients over smaller scales [61,62]. The mesoscale SST gradients will result in enhanced differences when the large (64 × 32 km native instantaneous field of view at 6.9 GHz) satellite footprints are compared with in situ observations. The elevated variability in these regions is therefore not related to the quality of the OE SST retrieval.

Figure 9a,b show the OE SST performance, considering the retrievals with RMSE$_{TB}$ < 0.5 K, as a function of binned drifter SST and NWP WS, respectively. The OE SST displays a warm bias for drifter SSTs in the range of 15–25 °C and similar for the small fraction of very high (>28 °C) SSTs. Figure 9b shows that the OE SST has a bias dependency on the NWP WS with a warm bias for low (<6 m·s^{-1}) wind speeds and a cold bias for higher wind speeds.

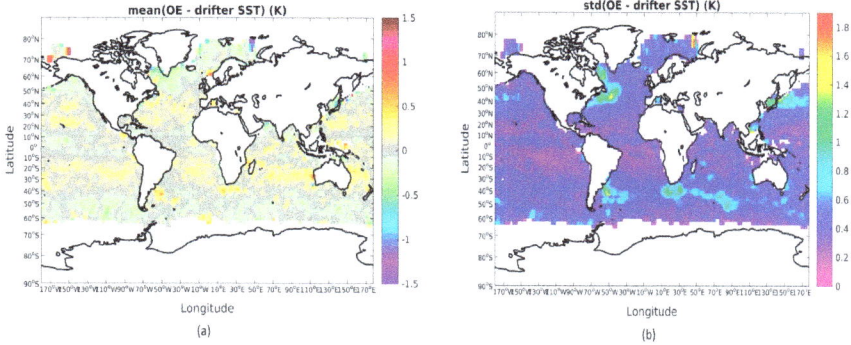

Figure 8. (**a**) Mean OE SST minus drifter SST; (**b**) Mean standard deviation of the OE retrieved minus drifter SST difference. Only retrievals with a corresponding RMSE_TB < 0.5 K are plotted in the figures.

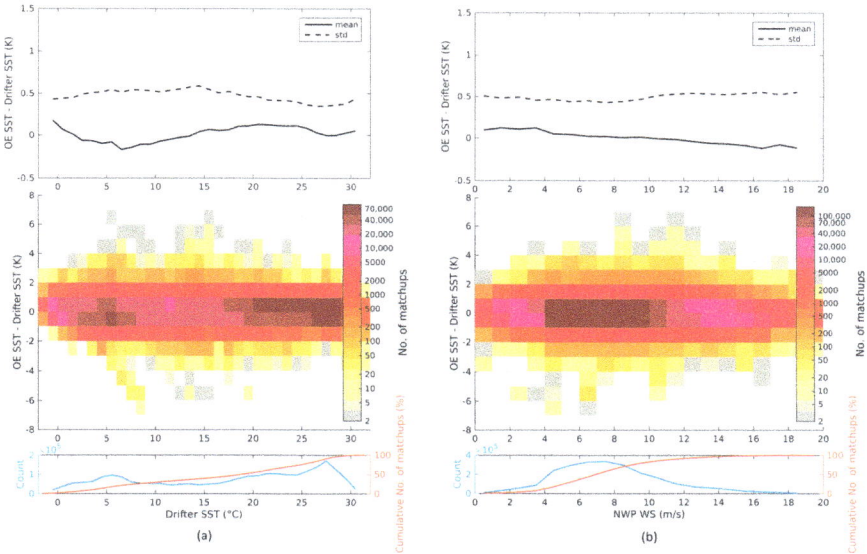

Figure 9. OE SST minus drifter SST as a function of binned (**a**) Drifter SST; (**b**) NWP WS. Solid lines are bias and dashed lines are standard deviation in the upper figures. The surface plots in middle figures show the number of matchups in each bin, while the bottom plots show the total number of matchups (blue) and the cumulative percentage of matchups (red) in each drifter SST and NWP WS bin, respectively. Only retrievals with a corresponding RMSE_TB < 0.5 K are plotted in the figures.

3.1. SST Sensitivity

One of the benefits of using the OE framework is that it naturally provides several diagnostics for assessing the quality and sensitivity of the retrieval. One of these diagnostics is the averaging kernel matrix, **A**, which contains the sensitivities of the retrieved parameters to the true state on its diagonal (and cross-sensitivities between parameters on the off-diagonals):

$$\mathbf{A}_{ij} = \frac{dx_i}{dx_j^t},$$ (9)

where x^t is the true state. If the averaging kernel was equal to the identity matrix the a priori state would have no influence on the retrieved state, which instead would be obtained purely from the information content of the measured brightness temperatures. The mean SST sensitivity for all drifter matchups during 2010 is found to be 0.50 with above OE setup and Figure 10a shows the geographical distribution of SST sensitivity. The SST sensitivity is lowest in high latitudes and increases towards the equatorial region, which is consistent with the fact that ∂ TB/ ∂ SST is smaller for cold waters (especially for X-band 10.65 GHz channels) [63]. The equatorial region reveals sensitivities of ~0.6 while high latitudes have sensitivities around 0.4. Sensitivities from 0.39 to 0.65 were reported in Gentemann et al. [64] for 0 °C and 30 °C SST, respectively, for an AMSR-E regression type retrieval. In addition, Prigent et al. [63] used simulations to derive channel sensitivities ∂ TB/ ∂ SST of ~0.3 to 0.6 for the 6 GHz V. These results are in good agreement with the sensitivities obtained here.

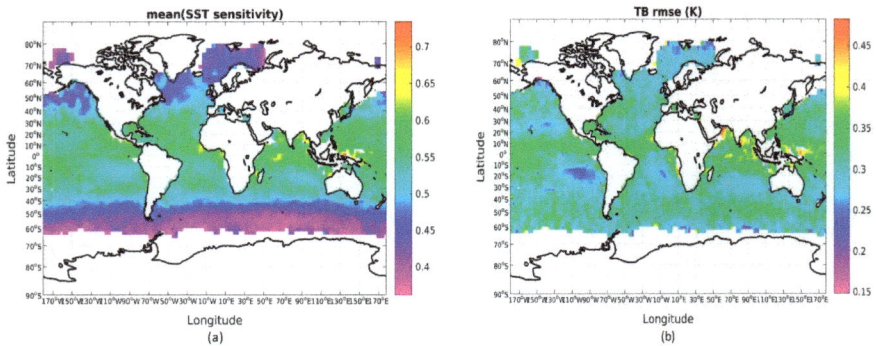

Figure 10. Gridded statistics of (**a**) mean sensitivity and (**b**) mean RMSE$_{TB}$. Only retrievals with a corresponding RMSE$_{TB}$ < 0.5 K are plotted.

3.2. Retrieval Uncertainty

The OE technique offers several options to estimate an uncertainty for each individual retrieval. The OE methodology directly provides an estimate of the retrieval uncertainty, S_x, due to uncertainties in the measurements, forward model, and in the a priori state vector (see Equation (4)). Considering all converged drifter matchups during 2010, the global mean uncertainty is 0.35 K. From Figure 7, it is evident that the quality of the SST retrieval is closely connected to the RMSE$_{TB}$ value from the retrieval. For that reason, we have set up an additional uncertainty indicator based on a scaled RMSE$_{TB}$ value, using a scaling factor of 0.55. Figure 11 shows the validation results for the uncertainties of the converged matchups, where the actual SST retrieval differences against drifter observations are displayed versus the theoretical uncertainties obtained from the RMSE$_{TB}$ values. The dashed line represents the ideal uncertainty under the assumptions that drifting buoys have a total uncertainty of 0.2 K and that the sampling uncertainty is 0.3 K. The point to satellite footprint sampling difference is estimated based on the results in Høyer et al. [44]. It is evident from the figure that there is a good agreement between the observed uncertainty and the modeled uncertainty estimates that are based on the RMSE$_{TB}$ and an integrated part of every OE retrieval. The mean modeled uncertainty is estimated to 0.48 K including the in situ and sampling uncertainty. Figure 10b shows the geographical distribution of RMSE$_{TB}$ considering the best 64% retrievals with a corresponding RMSE$_{TB}$ < 0.5 K.

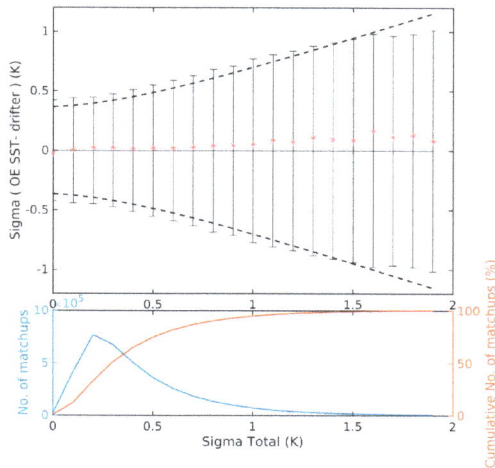

Figure 11. OE SST uncertainty validation with respect to drifter SST. Dashed lines show the ideal uncertainty model accounting for uncertainties in drifter SST and the sampling error. Solid black lines show one standard deviation of the retrieved minus drifter differences for each 0.1 K bin and the red symbols mark the mean bias. The bottom plot shows number of matchups (blue) and the cumulative percentage of matchups for each bin (red).

3.3. Validation against Independent Argo Floats

The validation against drifting buoys favors the NWP SST validation results as the drifting buoys are included in the OSTIA fields. For that reason, the OE retrieval algorithm has also been run for the 57,810 Argo matchups covering the period 2009–2011 (Section 2.2.2). The matchups with Argo floats are fewer than for drifting buoys; however, they are independent and can thus be used to compare the performance of the OE SST and NWP SST. Table 4 shows the validation of the OE SSTs and NWP SSTs against Argo floats for various subsets based on the listed filters. The Argo validation results resemble what was found for the drifting buoy observations with a very clear relation between the $RMSE_{TB}$ and the quality of the OE retrieval, and the highest quality OE retrievals performing better than the NWP SSTs. Note that the standard deviation of differences also includes the point to footprint sampling effects that are larger for the OE retrievals than for the NWP, which has an original spatial resolution of 0.05 degrees in latitude and longitude.

Table 4. Comparison of OE SSTs and NWP SSTs against Argo SSTs for various subsets.

Filter	Bias/K OE-Argo	std/K OE-Argo	Bias/K NWP-Argo	std/K NWP-Argo	N	
Convergence test passed	0.01	0.61	−0.05	0.55	57789	=100%
Gross error check	0.03	0.58	−0.05	0.54	51846	=90%
$RMSE_{TB} < 1$ K	0.01	0.55	−0.06	0.54	53150	=92%
$RMSE_{TB} < 0.50$ K	0.01	0.50	−0.06	0.51	36639	=63%
$RMSE_{TB} < 0.35$ K	0.01	0.49	−0.05	0.50	23410	=41%

Similar to drifters, two uncertainty estimates are given. The OE uncertainty $\mathbf{S_x}$ has an average value of 0.35 K, while the modeled estimate has an uncertainty of 0.47 K. The modeled uncertainty is calculated using a scaled $RMSE_{TB}$ with a scaling factor of 0.65. The modeled uncertainty has been evaluated for the Argo matchups and the result is shown in Figure 12. The dashed lines represent the ideal uncertainty under the assumptions that Argo floats have an accuracy of 0.002 K [42] and that the sampling uncertainty is 0.3 K [44].

Remote Sens. **2018**, *10*, 229

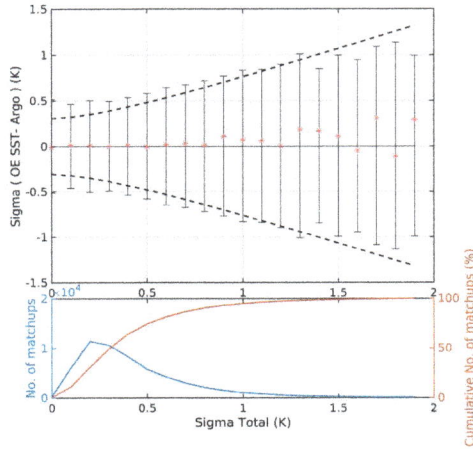

Figure 12. OE SST uncertainty validation with respect to Argo SSTs. Dashed lines show the ideal uncertainty model accounting for uncertainties in Argo SST and the sampling error. Solid black lines show one standard deviation of the retrieved minus Argo differences for each 0.1 K bin and the red symbols mark the mean bias. The bottom plot shows number of matchups (blue) and the cumulative percentage of matchups (red) for each bin.

3.4. OE vs. NWP Latitudinal Performance

Overall the OE SST performs similar to the NWP SST when compared to SST from drifters and Argo floats. However, there are regional differences in the performance of the two SST estimates. Figure 13 shows the latitudinal difference in standard deviations of OE and NWP SST compared against the same set of drifters and Argo floats, respectively. The figure shows that the OE SST performs better than NWP SST in both northern and southern mid-latitudes, while NWP SST performs better in the tropics. The latitudinal pattern in the relative performance is remarkably similar for both the drifting buoys and the Argo floats.

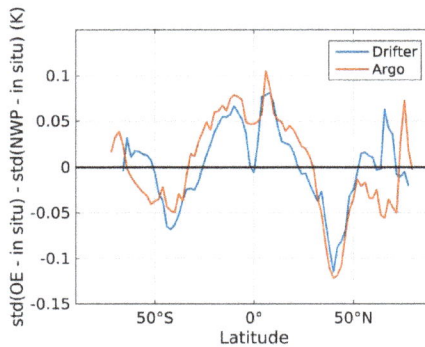

Figure 13. The latitudinal difference in standard deviations of OE and NWP SST compared against in situ SST. The blue curve is the comparison against drifters, while the red curve shows the comparison against Argo floats.

4. Discussion

The OE algorithm developed here is an attempt to develop a dedicated SST OE algorithm based on multi-channel microwave radiometer satellite observations. The OE methodology with the integrated forward modeling based on physical properties that provide simulated brightness temperature estimates for every retrieval contains valuable information on pixel level about the quality of the satellite observations. In addition, the OE technique offers a possibility for identifying and discarding RFI information in a very efficient way that is not possible with statistical retrieval methods, such as regression models.

The basis for the OE setup, the development and improvements in the forward model, as well as the validation is the comprehensive MMD that contains all the required fields for a retrieval matched with in situ observations. The MMD is a very powerful tool for algorithm development and makes testing and dependency assessment straightforward. The results presented here demonstrate that updating the state variable through several iterations is crucial for the quality of the retrievals, with a clear reduction in $RMSE_{TB}$ for the first two iterations. This approach resembles what is used for OE sea ice concentration retrieval [30,31] but differs from the OE IR SST retrievals, where one inversion is typically performed [24,25]. The need for several iterations is probably a result of the non-linear behavior of the forward model.

The forward model is an essential part of the OE retrievals. The OE performance therefore depends on the performance of the forward model and any deficiencies in the model to simulate the observed brightness temperature will propagate into the retrieval. During development, it was evident that using the MMD to improve the forward model is an essential step. The main improvements in the forward models ability to simulate the observed brightness temperature were found for the wind speed and SST dependency. Using results from a regression model to correct the forward model led to significant improvements in both the bias and the standard deviation between the simulated and observed brightness temperatures.

In the microwave part of the spectrum, several geophysical factors contribute to changes observed by the satellite, such as the ocean state, water vapor and cloud liquid waters etc. [63]. This means that the sensitivity in e.g., the 6 GHz channel to the actual SST variations is not as high as reported for the thermal IR part of the spectrum, where TCLW and TCWV cases are discarded before assessing the sensitivity [65]. In Prigent et al. [63] a regression based retrieval model is used to derive maximum SST sensitivities in the order of 0.65 for the 6 GHz V channel and decreasing for lower SSTs. These results are in good agreement with what we find here, with an overall sensitivity of 0.50 to the true SST. In addition, the global pattern shows a higher sensitivity at lower latitudes with warmer waters, which is also in agreement with the modeled results.

The global validation results with a mean and standard deviation of OE SST minus drifter SST of 0.02 ± 0.47 K for the best 64% of the matchups are very encouraging. These results are comparable or better than the previous validation results for PMW SST retrievals [66,67], which showed a degradation in the SST performance in cold and moist conditions. Similarly, Gentemann [18] reported on a latitudinal and SST dependency in the standard deviations, when compared to drifting buoys. The OE SST performance shows an increase in standard deviations in regions with large mesoscale activity and strong SST gradients. The reason for this is probably the temporal and spatial sampling errors when satellite observations are compared to pointwise in situ observations. Despite an enhanced spatial sampling effect from the larger PMW satellite footprint, the OE SST retrievals perform better than NWP SST in regions with mesoscale activity at both northern and southern mid-latitudes when validated against drifting buoys and Argo floats. In the tropics NWP SST performs better. It is worth to notice the consistency between drifters and Argos.

The algorithm has been developed to retrieve subskin SSTs in conditions not affected by diurnal warming, as these matchups were filtered out. This allowed the use of drifter observations for improving the forward model and for a detailed validation against drifting buoys with a nominal depth of 20 cm and Argo observations at 5 m. SSTs can also be retrieved during daytime conditions with diurnal warming, but accurate validation of the performance in these conditions against in situ would require a model or method for the subskin to 20 cm or 5 m correction.

The OE retrieval provides optimal estimates of four state variables: WS, TCWV, TCLW and SST. The focus in this paper has been on the SST retrieval and on validating this state variable against accurate in situ observations. To validate all state variables against accurate in situ observations is not straightforward for the other variables. A detailed evaluation and validation of the three other parameters are outside the scope of this paper.

The main strength of OE is that it is able to select channels with the most information and provides an independent estimate for the individual retrievals on the uncertainty of the retrievals. The estimated uncertainties were validated against independent in situ observations. The accurate and reliable uncertainty estimates increase the applicability of the SST dataset. A need for very accurate SST retrievals will of course reduce the number of SST retrievals, but a realistic uncertainty estimate guides the users to select the combination of accuracy versus data coverage that is optimal for their application.

5. Conclusions

In this study, the optimal estimation (OE) method has been used to retrieve subskin sea surface temperature (SST) from passive microwave (PMW) satellite observations. The results indicate that the OE SST has an overall bias (OE SST—drifter SST) of 0.02 K and standard deviation of 0.47 K when considering the 64% matchups, where the simulated and observed brightness temperatures are most consistent. The corresponding mean modeled uncertainty is 0.48 K including the in situ and sampling uncertainty. An independent validation against Argo observations shows a mean difference of 0.01 K, standard deviation of 0.48 K and modeled uncertainty of 0.47 K considering the 62% best matchups. The modeled uncertainty estimates, available for each retrieval, have proven to be accurate and reliable, when compared to in situ observations. The main advantage of the OE technique is its capability to provide valuable information on pixel level about the quality of the satellite observations, which can be used directly to identify and discard erroneous retrievals (e.g., contamination from extreme wind, atmospheric attenuation and emission, sun-glint, land/ice, rain and RFI).

Future work on the OE methodology can include more focus on improving the forward model with regard to the other state variables and on the estimation of the S_a and the S_e statistical parameters, as these are key parameters in the retrieval process. Alternative forward models could be tested in the retrieval process, but few accurate forward models exist at present that are suitable for use in an OE context. More work should therefore be put into improving the forward models with the aim of PMW OE SST retrievals. In addition, more work could be done on assessing the role of the first guess values and the impact of these observations on the retrieved SSTs. In the present work, we have disregarded observations in the vicinity of sea ice, but considering the results obtained within the ESA-CCI Sea Ice project, a future development could include the development of an integrated ocean and sea ice OE processor, that is able to estimate the sea ice concentration and SST at the same time and thus allowing for PMW SSTs closer to the marginal ice zone.

In the context of developing an SST climate data record, it is important to note that microwave radiometer provides an independent technique to conventional infrared (IR) radiometer satellite retrievals, potentially adding robustness to the resulting multi-mission satellite SST record. Furthermore, as some areas of the global ocean are quasi-permanently obscured by clouds, few IR measurements are available: microwave radiometry mitigates this negative situation meaning that multi-mission SST datasets provide a better representation of the SST with close to daily coverage with a single wide swath satellite instrument. We note that the GCOM-W AMSR-2 is now continuing the legacy of AMSR-E with an improved capability until the early 2020's. The rotating joint for the antenna scan mechanism of AMSR-E degraded within ~9.5 years of launch. The design lifetime of AMSR-2 was 5 years so a replacement is urgently needed. However, there is currently no follow-on mission either planned or in development to provide continuity of the 6–7 GHz frequency imaging capability that is fundamental for the retrieval of SST from microwave satellite radiometry. In future, we strongly advocate that the following issues are urgently addressed:

1. That the future 6–7 GHz frequency wide swath imaging capability currently provided by GCOM-W AMSR-2 is sustained by flying a new mission. This implies immediate initiation of satellite development for a potential launch in 2025.
2. That the spatial resolution of the 6–7 GHz frequency channels is significantly improved to provide a ~10 km native instantaneous spatial resolution (which may imply using a large 6–9 m rotating deployable mesh antenna). This is required to minimize significant loss of data in the coastal zone and marginal sea ice zone due to side lobe contamination, currently there are no valid measurements within 100 km of these areas.
3. That the radiometric quality of future satellite microwave radiometers is significantly improved over current capability. This is required because in many areas imaging microwave radiometer measurements are the only measurements available for the SST climate data record in areas characterized by quasi-permanent cloud cover that confounds thermal IR satellite SST retrievals.
4. That appropriate combination with one or more higher-frequency channels (10, 18, 37 GHz) is provided in order to resolve the ambiguities in the microwave measurements if too few channels are available.

In terms of current capability, we conclude that overall, the OE SST retrieval results for AMSR-E are very promising and demonstrate that the OE algorithm is complementary to the standard SST retrieval algorithms that are available today.

Acknowledgments: This study was funded as part of the ESA-CCI SST. The MMS was developed under the ESA-CCI and the FIDUCEO project Horizon 2020 research and innovation program under grant agreement No 638822. We thank Christopher J. Merchant for providing valuable inputs and comments to this article.

Author Contributions: Pia Nielsen-Englyst, Jacob L. Høyer, and Leif Toudal Pedersen conceived and designed the experiments. Pia Nielsen-Englyst and Jacob L. Høyer performed the experiments. Pia Nielsen-Englyst, Jacob L. Høyer, Leif Toudal Pedersen, Chelle L. Gentemann, and Craig Donlon analyzed the data. Tom Block developed the matchup database. Emy Alerskans and Chelle L. Gentemann developed the data filtering methods.

Conflicts of Interest: The authors declare no conflict of interest. The founding sponsors had no role in the design of the study; in the collection, analyses, or interpretation of data; in the writing of the manuscript, or in the decision to publish the results.

References

1. Vázquez-Cuervo, J.; Armstrong, E.M.; Harris, A. The effect of aerosols and clouds on the retrieval of infrared sea surface temperatures. *J. Clim.* **2004**, *17*, 3921–3933. [CrossRef]
2. Reynolds, R.W. Impact of mount pinatubo aerosols on satellite-derived sea surface temperatures. *J. Clim.* **1993**, *6*, 768–774. [CrossRef]
3. Reynolds, R.W.; Rayner, N.A.; Smith, T.M.; Stokes, D.C.; Wang, W. An improved in situ and satellite sst analysis for climate. *J. Clim.* **2002**, *15*, 1609–1625. [CrossRef]
4. Donlon, C.; Rayner, N.; Robinson, I.; Poulter, D.J.S.; Casey, K.S.; Vazquez-Cuervo, J.; Armstrong, E.; Bingham, A.; Arino, O.; Gentemann, C.; et al. The global ocean data assimilation experiment high-resolution sea surface temperature pilot project. *Bull. Am. Meteorol. Soc.* **2007**, *88*, 1197–1213. [CrossRef]
5. Donlon, C.J.; Casey, K.S.; Gentemann, C.L.; Harris, A. Successes and challenges for the modern sea surface temperature observing system. In *Proceeding of OceanObs'09: Sustained Ocean Observations and Information for Society;* Hall, J., Harrison, D.E., Stammer, D., Eds.; ESA Publication WPP-306: Venice, Italy, 2010; Volume 2.
6. Wentz, F.J.; Gentemann, C.; Smith, D.; Chelton, D.B. Satellite measurements of sea surface temperature through clouds. *Science* **2000**, *288*, 847–850. [CrossRef] [PubMed]
7. Chelton, D.B.; Wentz, F.J. Global microwave satellite observations of sea surface temperature for numerical weather prediction and climate research. *Bull. Am. Meteorol. Soc.* **2005**, *86*, 1097–1115. [CrossRef]
8. Wilheit, T.T.; Chang, A.T.C. An algorithm for retrieval of ocean surface and atmospheric parameters from the observations of the scanning multichannel microwave radiometer. *Radio Sci.* **1980**, *15*, 525–544. [CrossRef]

9.   Lipes, R.G. Description of SEASAT radiometer status and results. *J. Geophys. Res.* **1982**, *87*. [CrossRef]
10.   Milman, A.S.; Wilheit, T.T. Sea surface temperatures from the scanning multichannel microwave radiometer on Nimbus 7. *J. Geophys. Res.* **1985**, *90*. [CrossRef]
11.   Maeda, T.; Taniguchi, Y.; Imaoka, K. GCOM-W1 AMSR2 Level 1R Product: Dataset of brightness temperature modified using the antenna pattern matching technique. *IEEE Trans. Geosci. Remote Sens.* **2016**, *54*, 770–782. [CrossRef]
12.   Gaiser, P.W.; St Germain, K.M.; Twarog, E.M.; Poe, G.A.; Purdy, W.; Richardson, D.; Grossman, W.; Jones, W.L.; Spencer, D.; Golba, G.; et al. The WindSat spaceborne polarimetric microwave radiometer: Sensor description and early orbit performance. *IEEE Trans. Geosci. Remote Sens.* **2004**, *42*, 2347–2361. [CrossRef]
13.   Xie, S.-P. Satellite observations of cool ocean—Atmosphere interaction*. *Bull. Am. Meteorol. Soc.* **2004**, *85*, 195–208. [CrossRef]
14.   Shibata, A. Calibration of AMSR-E SST toward a monitoring of global warming. In Proceedings of the 2005 IEEE International Geoscience and Remote Sensing Symposium, Seoul, Korea, 29 July 2005.
15.   Shibata, A. Features of ocean microwave emission changed by wind at 6 GHz. *J. Oceanogr.* **2006**, *62*, 321–330. [CrossRef]
16.   Wentz, F.J.; Meissner, T. *AMSR_E Ocean Algorithms*; Remote Sensing Systems: Santa Rosa, CA, USA, 2007; p. 6.
17.   Hosoda, K.; Murakami, H.; Shibata, A.; Sakaida, F.; Kawamura, H. Difference characteristics of sea surface temperature observed by GLI and AMSR aboard ADEOS-II. *J. Oceanogr.* **2006**, *62*, 339–350. [CrossRef]
18.   Gentemann, C.L. Three way validation of MODIS and AMSR-E sea surface temperatures. *J. Geophys. Res. Oceans* **2014**, *119*, 2583–2598. [CrossRef]
19.   Donlon, C.J.; Martin, M.; Stark, J.; Roberts-Jones, J.; Fiedler, E.; Wimmer, W. The Operational Sea Surface Temperature and Sea Ice Analysis (OSTIA) system. *Remote Sens. Environ.* **2012**, *116*, 140–158. [CrossRef]
20.   Hosoda, K.; Kawamura, H.; Sakaida, F. Improvement of New Generation Sea Surface Temperature for Open ocean (NGSST-O): A new sub-sampling method of blending microwave observations. *J. Oceanogr.* **2015**, *71*, 205–220. [CrossRef]
21.   Stark, J.D.; Donlon, C.J.; Martin, M.J.; McCulloch, M.E. OSTIA: An operational, high resolution, real time, global sea surface temperature analysis system. In Proceedings of the OCEANS, Aberdeen, UK, 18–21 June 2007; pp. 1–4.
22.   Reynolds, R.W.; Smith, T.M.; Liu, C.; Chelton, D.B.; Casey, K.S.; Schlax, M.G. Daily high-resolution-blended analyses for sea surface temperature. *J. Clim.* **2007**, *20*, 5473–5496. [CrossRef]
23.   Donlon, C.; Casey, K.; Robinson, I.; Gentemann, C.; Reynolds, R.; Barton, I.; Arino, O.; Stark, J.; Rayner, N.; LeBorgne, P.; et al. The GODAE high-resolution sea surface temperature pilot project. *Oceanography* **2009**, *22*, 34–45. [CrossRef]
24.   Merchant, C.J.; Le Borgne, P.; Marsouin, A.; Roquet, H. Optimal estimation of sea surface temperature from split-window observations. *Remote Sens. Environ.* **2008**, *112*, 2469–2484. [CrossRef]
25.   Merchant, C.J.; Le Borgne, P.; Roquet, H.; Marsouin, A. Sea surface temperature from a geostationary satellite by optimal estimation. *Remote Sens. Environ.* **2009**, *113*, 445–457. [CrossRef]
26.   Merchant, C.J.; Embury, O. Simulation and Inversion of Satellite Thermal Measurements. In *Experimental Methods in the Physical Sciences*; Elsevier: Amsterdam, The Netherlands, 2014; Volume 47, pp. 489–526. ISBN 978-0-12-417011-7.
27.   Merchant, C.J.; Embury, O.; Roberts-Jones, J.; Fiedler, E.; Bulgin, C.E.; Corlett, G.K.; Good, S.; McLaren, A.; Rayner, N.; Morak-Bozzo, S.; et al. Sea surface temperature datasets for climate applications from Phase 1 of the European Space Agency Climate Change Initiative (SST CCI). *Geosci. Data J.* **2014**, *1*, 179–191. [CrossRef]
28.   Merchant, C.J.; Le Borgne, P.; Roquet, H.; Legendre, G. Extended optimal estimation techniques for sea surface temperature from the Spinning Enhanced Visible and Infra-Red Imager (SEVIRI). *Remote Sens. Environ.* **2013**, *131*, 287–297. [CrossRef]
29.   Pedersen, L.T. Merging microwave radiometer data and meteorological data for improved sea ice concentrations. *EARSeL Adv. Remote Sens.* **1994**, *3*, 81–89.
30.   Melsheimer, C.; Heygster, G.; Mathew, N.; Toudal Pedersen, L. Retrieval of sea ice emissivity and integrated retrieval of surface and atmospheric parameters over the Arctic from AMSR-E data. *J. Remote Sens. Soc. Jpn.* **2009**, *29*, 236–241.

31. Scarlat, R.C.; Heygster, G.; Pedersen, L.T. Experiences with an Optimal estimation algorithm for surface and atmospheric parameter retrieval from passive microwave data in the arctic. *IEEE J. Sel. Top. Appl. Earth Obs. Remote Sens.* **2017**, *10*, 3934–3947. [CrossRef]

32. Bettenhausen, M.H.; Smith, C.K.; Bevilacqua, R.M.; Wang, N.Y.; Gaiser, P.W.; Cox, S. A nonlinear optimization algorithm for WindSat wind vector retrievals. *IEEE Trans. Geosci. Remote Sens.* **2006**, *44*, 597–610. [CrossRef]

33. Ashcroft, P.; Wentz, F.J. AMSR-E/Aqua L2A Global Swath Spatially-Resampled Brightness Temperatures (Tb), Version 3. 2013. Available online: https://cmr.earthdata.nasa.gov/search/concepts/C190757121-NSIDC_ECS.html (accessed on 30 July 2016).

34. Wentz, F.J. *SSM/I Version-7 Calibration Report*; Remote Sensing Systems: Santa Rosa, CA, USA; 2013; p. 46.

35. Woodruff, S.D.; Worley, S.J.; Lubker, S.J.; Ji, Z.; Eric Freeman, J.; Berry, D.I.; Brohan, P.; Kent, E.C.; Reynolds, R.W.; Smith, S.R.; et al. ICOADS Release 2.5: Extensions and enhancements to the surface marine meteorological archive. *Int. J. Climatol.* **2011**, *31*, 951–967. [CrossRef]

36. Good, S.A.; Martin, M.J.; Rayner, N.A. EN4: Quality controlled ocean temperature and salinity profiles and monthly objective analyses with uncertainty estimates: THE EN4 DATA SET. *J. Geophys. Res. Oceans* **2013**, *118*, 6704–6716. [CrossRef]

37. O'Carroll, A.G.; Eyre, J.R.; Saunders, R.W. Three-Way Error Analysis between AATSR, AMSR-E, and In Situ sea surface temperature observations. *J. Atmos. Ocean. Technol.* **2008**, *25*, 1197–1207. [CrossRef]

38. Atkinson, C.P.; Rayner, N.A.; Kennedy, J.J.; Good, S.A. An integrated database of ocean temperature and salinity observations. *J. Geophys. Res. Oceans* **2014**, *119*, 7139–7163. [CrossRef]

39. Roemmich, D.; Johnson, G.; Riser, S.; Davis, R.; Gilson, J.; Owens, W.B.; Garzoli, S.; Schmid, C.; Ignaszewski, M. The Argo Program: Observing the Global Oceans with Profiling Floats. *Oceanography* **2009**, *22*, 34–43. [CrossRef]

40. Gille, S.T. Decadal-Scale Temperature Trends in the Southern Hemisphere Ocean. *J. Clim.* **2008**, *21*, 4749–4765. [CrossRef]

41. Kennedy, J.J. A review of uncertainty in in situ measurements and data sets of sea surface temperature: In situ sst uncertainty. *Rev. Geophys.* **2014**, *52*, 1–32. [CrossRef]

42. Abraham, J.P.; Baringer, M.; Bindoff, N.L.; Boyer, T.; Cheng, L.J.; Church, J.A.; Conroy, J.L.; Domingues, C.M.; Fasullo, J.T.; Gilson, J.; et al. A review of global ocean temperature observations: Implications for ocean heat content estimates and climate change: Review of ocean observations. *Rev. Geophys.* **2013**, *51*, 450–483. [CrossRef]

43. Embury, O.; Merchant, C.J.; Corlett, G.K. A reprocessing for climate of sea surface temperature from the along-track scanning radiometers: Initial validation, accounting for skin and diurnal variability effects. *Remote Sens. Environ.* **2012**, *116*, 62–78. [CrossRef]

44. Høyer, J.L.; Karagali, I.; Dybkjær, G.; Tonboe, R. Multi sensor validation and error characteristics of Arctic satellite sea surface temperature observations. *Remote Sens. Environ.* **2012**, *121*, 335–346. [CrossRef]

45. Merchant, C.J.; Embury, O.; Rayner, N.A.; Berry, D.I.; Corlett, G.K.; Lean, K.; Veal, K.L.; Kent, E.C.; Llewellyn-Jones, D.T.; Remedios, J.J.; et al. A 20 year independent record of sea surface temperature for climate from Along-Track Scanning Radiometers: SST for climate from atsrs. *J. Geophys. Res. Oceans* **2012**, *117*. [CrossRef]

46. Udaya Bhaskar, T.V.S.; Rahman, S.H.; Pavan, I.D.; Ravichandran, M.; Nayak, S. Comparison of AMSR-E and TMI sea surface temperature with Argo near-surface temperature over the Indian Ocean. *Int. J. Remote Sens.* **2009**, *30*, 2669–2684. [CrossRef]

47. Dee, D.P.; Uppala, S.M.; Simmons, A.J.; Berrisford, P.; Poli, P.; Kobayashi, S.; Andrae, U.; Balmaseda, M.A.; Balsamo, G.; Bauer, P.; et al. The ERA-Interim reanalysis: configuration and performance of the data assimilation system. *Q. J. R. Meteorol. Soc.* **2011**, *137*, 553–597. [CrossRef]

48. Zweng, M.M.; Reagan, J.R.; Antonov, J.I.; Seidov, D.; Biddle, M.M. World Ocean Atlas 2013, Volume 2, Salinity. 2013. Available online: https://repository.library.noaa.gov/view/noaa/14848 (accessed on 26 June 2017).

49. Boyer, T.P.; Antonov, J.I.; Baranova, O.K.; Garcia, H.E.; Johnson, D.R.; Mishonov, A.V.; O'Brien, T.D.; Seidov, D.; Smolyar, I.; Zweng, M.M.; et al. *World Ocean Database 2013*; NOAA Printing Office: Silver Spring, MD, USA, 2013; 208p.

50. Block, T.; Embacher, S.; Merchant, C.J.; Donlon, C. High performance software framework for the calculation of satellite-to-satellite data matchups (MMS version 1.2). *Geosci. Model Dev. Discuss.* **2017**, 1–15. [CrossRef]

51. Schulzweida, U.; Kornblueh, L.; Quast, R. *CDO User's Guide—Climate Data Operators*; Max Planck Institute for Meteorology: Hamburg, Germany, 2010; pp. 1–173.

52. Gentemann, C.L.; Hilburn, K.A. In situ validation of sea surface temperatures from the GCOM-W1 AMSR2 RSS calibrated brightness temperatures: Validation of RSS GCOM-W1 SST. *J. Geophys. Res. Oceans* **2015**, *120*, 3567–3585. [CrossRef]

53. Rodgers, C.D. *Inverse Methods for Atmospheric Sounding—Theory and Practice*; Series on Atmospheric Ocanic and Planetary Physics; World Scientific Publishing Co. Pte. Ltd.: Singapore, 2000; Volume 2, ISBN 978-981-281-371-8.

54. Wentz, F.J.; Meissner, T. *AMSR Ocean Algorithm. Algorithm Theoretical Basis Document*; Remote Sensing Systems: Santa Rosa, CA, USA, 2000.

55. Chelton, D.B.; Freilich, M.H. Scatterometer-based assessment of 10-m wind analyses from the operational ECMWF and NCEP Numerical weather prediction models. *Mon. Weather Rev.* **2005**, *133*, 409–429. [CrossRef]

56. Jakobson, E.; Vihma, T.; Palo, T.; Jakobson, L.; Keernik, H.; Jaagus, J. Validation of atmospheric reanalyses over the central Arctic Ocean: Tara reanalyses validation. *Geophys. Res. Lett.* **2012**, *39*. [CrossRef]

57. Li, J.-L.F.; Waliser, D.; Woods, C.; Teixeira, J.; Bacmeister, J.; Chern, J.; Shen, B.-W.; Tompkins, A.; Tao, W.-K.; Köhler, M. Comparisons of satellites liquid water estimates to ECMWF and GMAO analyses, 20th century IPCC AR4 climate simulations, and GCM simulations. *Geophys. Res. Lett.* **2008**, *35*. [CrossRef]

58. Jiang, J.H.; Su, H.; Zhai, C.; Perun, V.S.; Del Genio, A.; Nazarenko, L.S.; Donner, L.J.; Horowitz, L.; Seman, C.; Cole, J.; et al. Evaluation of cloud and water vapor simulations in CMIP5 climate models using NASA "A-Train" satellite observations: Evaluation of IPCC Ar5 model simulations. *J. Geophys. Res. Amospheres* **2012**, *117*. [CrossRef]

59. Donlon, C.J.; Minnett, P.J.; Gentemann, C.; Nightingale, T.J.; Barton, I.J.; Ward, B.; Murray, M.J. Toward improved validation of satellite sea surface skin temperature measurements for climate research. *J. Clim.* **2002**, *15*, 353–369. [CrossRef]

60. Meissner, T.; Wentz, F.J. The emissivity of the ocean surface between 6 and 90 GHz over a large range of wind speeds and earth incidence angles. *IEEE Trans. Geosci. Remote Sens.* **2012**, *50*, 3004–3026. [CrossRef]

61. Pascual, A.; Faugère, Y.; Larnicol, G.; Le Traon, P.-Y. Improved description of the ocean mesoscale variability by combining four satellite altimeters. *Geophys. Res. Lett.* **2006**, *33*. [CrossRef]

62. Legeckis, R. A survey of worldwide sea surface temperature fronts detected by environmental satellites. *J. Geophys. Res.* **1978**, *83*. [CrossRef]

63. Prigent, C.; Aires, F.; Bernardo, F.; Orlhac, J.-C.; Goutoule, J.-M.; Roquet, H.; Donlon, C. Analysis of the potential and limitations of microwave radiometry for the retrieval of sea surface temperature: Definition of MICROWAT, a new mission concept: Sea Temperature from Microwaves. *J. Geophys. Res. Oceans* **2013**, *118*, 3074–3086. [CrossRef]

64. Gentemann, C.L.; Meissner, T.; Wentz, F.J. Accuracy of satellite sea surface temperatures at 7 and 11 GHz. *IEEE Trans. Geosci. Remote Sens.* **2010**, *48*, 1009–1018. [CrossRef]

65. Merchant, C.J.; Harris, A.R.; Roquet, H.; Le Borgne, P. Retrieval characteristics of non-linear sea surface temperature from the advanced very high resolution radiometer. *Geophys. Res. Lett.* **2009**, *36*. [CrossRef]

66. Castro, S.L.; Wick, G.A.; Jackson, D.L.; Emery, W.J. Error characterization of infrared and microwave satellite sea surface temperature products for merging and analysis. *J. Geophys. Res.* **2008**, *113*. [CrossRef]

67. Dong, S.; Gille, S.T.; Sprintall, J.; Gentemann, C. Validation of the advanced microwave scanning radiometer for the earth observing system (AMSR-E) sea surface temperature in the southern ocean. *J. Geophys. Res.* **2006**, *111*. [CrossRef]

remote sensing

MDPI

Article

Exploring Machine Learning to Correct Satellite-Derived Sea Surface Temperatures

Stéphane Saux Picart [1,*], Pierre Tandeo [2], Emmanuelle Autret [3] and Blandine Gausset [1]

[1] Météo-France/Centre de Météorologie Spatiale, Avenue de Lorraine, B.P. 50747, 22307 Lannion CEDEX, France; blandine.gausset@gmail.com
[2] IMT Atlantique, Lab-STICC, UBL, 29238 Brest, France; pierre.tandeo@imt-atlantique.fr
[3] Ifremer, Laboratoire d'Océanographie Physique et Spatiale, ZI Pointe du Diable CS 10070, 29280 Plouzané, France; emmanuelle.autret@ifremer.fr
*   Correspondence: stephane.sauxpicart@meteo.fr; Tel.: +33-296-05-67-07

Received: 20 November 2017; Accepted: 27 January 2018; Published: 1 February 2018

Abstract: Machine learning techniques are attractive tools to establish statistical models with a high degree of non linearity. They require a large amount of data to be trained and are therefore particularly suited to analysing remote sensing data. This work is an attempt at using advanced statistical methods of machine learning to predict the bias between Sea Surface Temperature (SST) derived from infrared remote sensing and ground "truth" from drifting buoy measurements. A large dataset of collocation between satellite SST and in situ SST is explored. Four regression models are used: Simple multi-linear regression, Least Square Shrinkage and Selection Operator (LASSO), Generalised Additive Model (GAM) and random forest. In the case of geostationary satellites for which a large number of collocations is available, results show that the random forest model is the best model to predict the systematic errors and it is computationally fast, making it a good candidate for operational processing. It is able to explain nearly 31% of the total variance of the bias (in comparison to about 24% for the multi-linear regression model).

Keywords: machine learning; systematic error; sea surface temperature; random forest

---

1. Introduction

Characterising the error associated with data, from observations or model outputs, is essential for correct use and analysis. It is, however, often a very complex problem requiring many assumptions. When large amounts of data are available, data-driven methods provide a convenient way to work around that complexity, especially for remote sensing data. Recently in this domain, authors proposed to use relevant machine learning methods (see, e.g., [1] or [2]). These methods are based on statistical models and are able to automatically solve lots of regression and/or classification problems. Here, we focus on a regression problem and we test various classical methods using linear and nonlinear assumptions.

Sea Surface Temperature (SST) is a variable which has been estimated over a long time period from infrared radiometers' acquisitions on board satellites. It is used in many domains such as meteorological, climatic or ecosystem studies. For many of these applications it is important to know the accuracy of the data being used. This has been recognised internationally by the community of satellite SST data producers and users and a formal recommendation has been put forward by the Group for High Resolution SST (GHRSST) in the GHRSST Data Specification version 2.0 [3] to include Sensor-Specific Error Statistics (SSES) in distributed products.

The error is the difference between the measured SST and the true SST [4]. Systematic errors in satellite estimation of SST may have various origins [5]. One of the sources is the retrieval algorithm itself. Other sources include the calibration of the sensor which may not be accurate and contamination by sea ice, undetected clouds or atmospheric Saharan dust [6]. These effects result in inaccuracies in retrieved SST.

There is no consensus on how to derive SSES, and each data producer uses their own methodology. Methodologies are often based on regression of the error against a set of explanatory variables using a large dataset of collocations between drifting buoys and satellite retrievals. One approach, based on look-up tables (LUT) established from comparisons with in situ observations, provides discrete values of SSES. This is the case for the SSES hypercube of bias and standard deviation [7] used for observations from the Moderate Resolution Imaging Spectroradiometer (MODIS). Castro et al. [8] also proposed, for many infrared and microwave SST products, bias corrections from LUT representing bias and standard deviation dependencies on retrieval conditions such as wind speed, water vapor, view angle and SST. In Petrenko and Ignatov [9] and in Petrenko et al. [10], the SSES method is based on the segmentation of the SST domain and local regression coefficients are applied for each segment. In order to avoid possible discontinuities and noise introduced by these methods, other definitions of continuous SSES are proposed by Tandeo et al. [11] and Xu et al. [12]. A different approach is to model and propagate the uncertainties independently of in situ data [13], as used by the SST Climate Change Initiative [14].

The objective of this study is to make a first evaluation of the potential of advanced statistical methods of machine learning to model and predict the bias between satellite derived SST products and drifting buoy measurements, considered to be ground truth.

This study is based on a large data set of collocations between satellite and in situ measurements, described in Section 2. The methods are explained in Section 3 and interpretation of the results are presented in Section 4. Discussion and conclusion are in Section 5.

2. Data

The Spinning Enhanced Visible and Infrared Imager (SEVIRI) on board the Geostationary satellite Meteosat Second Generation (MSG) operates in the thermal infrared channels, enabling SST retrieval. Hourly SST products are computed operationally at the Centre de Météorologie Spatiale (CMS) in the framework of the EUMETSAT Ocean and Sea Ice (OSI SAF) project. The basis of SST retrieval is a split-window algorithm using the 10.8 and 12 μm channels. A complete description of the retrieval methodology can be found in Le Borgne et al. [15].

As part of the operational processing of MSG data, SST products are created as well as a Match-up DataSet (MDS) by collocating satellite information and in situ measurements (drifting buoys) collected from the Global Telecommunication System with a 5 days delay to ensure sufficient coverage. A satellite match-up is searched for within ±1 hour and information is extracted for a 5 × 5 pixel box around the position of the measurements. The MDS includes satellite SST but also includes some other variables used in the SST processing such as Numerical Weather Prediction (NWP) model output (e.g., total atmospheric content of water vapour) and level 1 SEVIRI brightness temperature in the channels of interest for SST retrieval.

Thereafter, only night-time data are used to minimize the difference between skin SST retrieved by infrared sensors like SEVIRI and bulk SST measured by drifting buoys. The quality level (QL) is a confidence indicator designed to help users filter out data that are not sufficiently good for their application. As per recommendations formulated in the Product User Manual [16] only data with a higher QL are considered (3, 4 or 5) because for QL < 3 the SST retrieved is likely to be contaminated by undetected clouds.

The MDS used in this work covers two years (August 2013–July 2015) and contains 485,600 match-ups. The period from August 2013 to July 2014 is used as the learning sample (with 249,318 match-ups) to train the statistical models and the other period from August 2014 to July 2015 is used as test sample (with 236,282 match-ups) to validate statistical models.

Infrared sensors are sensitive to the skin temperature of the sea, whereas drifting buoy measurements are taken at a depth of between 20 to 30 cm which can lead to significant differences [17]. However, since this study focuses on night-time data only, it is expected that these differences are small and therefore we make the assumption that the drifting buoy measurements constitute the sea truth. The accuracy of SEVIRI SST, ΔSST, is therefore defined as:

$$\Delta SST = SST_{sat} - SST_{buoys} \tag{1}$$

where SST_{sat} is the SST estimation given in SEVIRI products and SST_{buoys} is the temperature measured by drifting buoys.

Figure 1a shows the spatial distribution of ΔSST averaged over the training dataset for $5 \times 5°$ boxes and Figure 1b represents the number of data points in each box. A strong negative bias is noticeable in the intertropical zone, primarily due to the high atmospheric water vapour content in this region [18], and secondarily due to the presence of Saharan dust in the atmosphere [6]. On the other hand positive biases are observed in the southern hemisphere and around the Mediterranean Sea due to the drier atmosphere. Note that due to both cloud coverage and geographical distribution of drifting buoys, the spatial distribution of the match-ups is not homogeneous at all, as shown by Figure 1b.

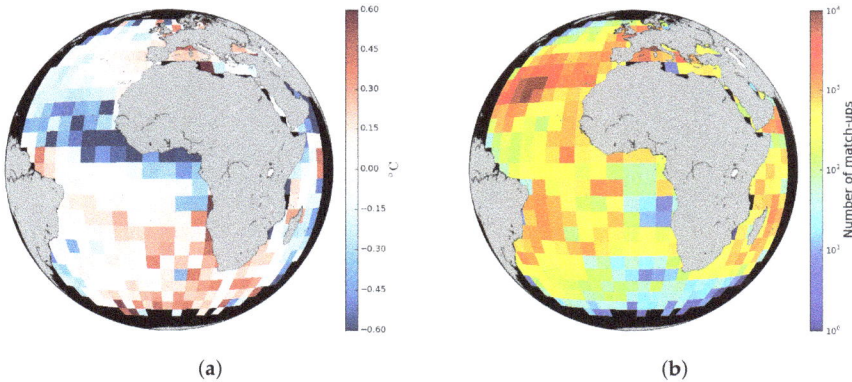

(a)   (b)

Figure 1. Training dataset (August 2013–July 2014, 249 318 match-ups): (a) mean difference between satellite and drifting buoys (ΔSST); (b) number of match-ups.

In this work, we use ten variables to model the bias in satellite-derived SST. These are listed in Table 1. Atmospheric water vapour is the primary cause of error in SST retrieval algorithms and therefore the integrated water vapour from the ECMWF model is one of the most important variables to model the bias between satellite and in situ SST. The differences between the 3.9 and the 8.7 µm channels and between the 10.8 and the 12.0 µm channels provides information on the presence of atmospheric Saharan dust [6] which can strongly affect the quality of the retrieval. These differences are averaged over 5×5 pixel boxes (of the MDS) in order to smooth out the effect of radiometric noise. Despite the fact that this study focuses only on night-time data, it is important to include the wind speed and solar zenith angle to take into account possible residual diurnal warming effects at the beginning of the night. The number of valid (clear sky) pixels in the 5 × 5 pixel boxes is also used as it provides information on the level of cloudiness around the central pixel. Additionally, the standard deviation of SST in the boxes is also informative of the spatial variability (presence of thermal fronts) which may be a source of error in satellite to in situ comparison. Finally the SST is also used as a model input variable in order to account for SST dependent error in the algorithm. Since the algorithm is calibrated on a global scale, it may indeed show weaknesses for retrieving extreme temperatures (low or high).

Note that all this information is available without delay. This would allow an online statistical model to produce SST error in real time so long as the model is already trained.

Table 1. Description of the Match-up DataSet (MDS) ancillary variables.

Name	Description
Latitude	Latitude of in situ measurements
Wind speed	Near surface wind speed (ECMWF)
Solar zenith angle	Angle between zenith and sun position
Satellite zenith angle	Angle between zenith and satellite position
Integrated water vapour	Integrated water vapour in the atmosphere (ECMWF)
IR_039 – IR_087 averaged in box	Difference between channel 3.9 µm and 8.7 µm averaged in 5 × 5 pixels box
IR_108 – IR_120 averaged in box	Difference between channel 10.8 µm and 12.0 µm averaged in 5 × 5 pixels box
Number of valid pixels	Number of valid retrievals (quality level 3, 4 or 5) in 5 × 5 pixels box
SST STD	Standard deviation of SST in 5 × 5 pixels box
SST	SST retrieved from SEVIRI

3. Methods

The goal of this study is to estimate ΔSST defined in Equation (1) in order to operationally adjust SST_{sat} measurements. This will be achieved by modelling the impact of simultaneous covariates presented in Table 1 and denoted as $\{X_i,\ i = 1, \ldots, p\}$. The relationship between ΔSST and covariates can be either nonlinear as for the latitude between 20°N and 60°N (Figure 2a) or linear with a negative slope (Figure 2b) for the integrated water vapour. Note that nonlinear interactions of covariates can also be detected (not shown). In this paper, we compare 4 regression models classically used in machine learning. They are described below.

- The first model used is a simple linear regression expressed as

$$\Delta SST = \alpha_0 + \sum_{i=1}^{p} \alpha_i X_i + \sum_{i=1}^{p}\sum_{j=1}^{p} \alpha_{i,j} X_i X_j \tag{2}$$

  where the α parameters correspond to the intercept, the linear and quadratic effects of covariates, and interactions between covariates.
- The second model, LASSO (Least Absolute Shrinkage and Selection Operator, see Tibshirani et al. [19]), is similar to the first one, except with a sparsity constraint on the α parameters. Thus, it is a subversion of Equation (2), where some of the α values are null. The LASSO model is based on a numerical optimization to find the alpha parameters that minimize the following expression:

$$\min_{\alpha} \frac{1}{2N} \sum^{N} ||\Delta SST - \alpha_0 + \sum_{i=1}^{p} \alpha_i X_i + \sum_{i=1}^{p}\sum_{j=1}^{p} \alpha_{i,j} X_i X_j||^2 + \lambda \sum^{p} |\alpha| \tag{3}$$

where N corresponds to the number of training samples used to learn the model, $P = 1 + p + p^2$ is the total number of alpha parameters and λ is estimated by cross validation.

- The third model, GAM (Generalized Additive Model, see Hastie et al. [20]), uses nonlinear functions to model the impact of the covariates such as

$$\Delta SST = \alpha_0 + \sum_{i=1}^{p} f_i(X_i) + \sum_{i=1}^{p} \sum_{j=1}^{p} f_{i,j}(X_i X_j) \tag{4}$$

where functions f_i and $f_{i,j}$ are adjusted using local linear regressions, as in Figure 2a,b.

- The last model, random forest (see Breiman et al. [21]), applies N random samplings with replacement such as

$$\Delta SST = \frac{1}{N} \sum_{i=1}^{N} t_i(X_1, \ldots, X_p) \tag{5}$$

where t_i are the different regression trees (see Breiman et al. [22]). A tree is based on simple decision criteria on the X covariates such as: if X < threshold then $\Delta SST = $ value1 else $\Delta SST = $ value2. The threshold value is learned from the data, maximizing the difference between value1 and value2. Then, we recursively split the dataset in 2 leaves at each node of the tree. In this paper, we use trees with a maximum of 1000 nodes and the forest is based on 100 trees.

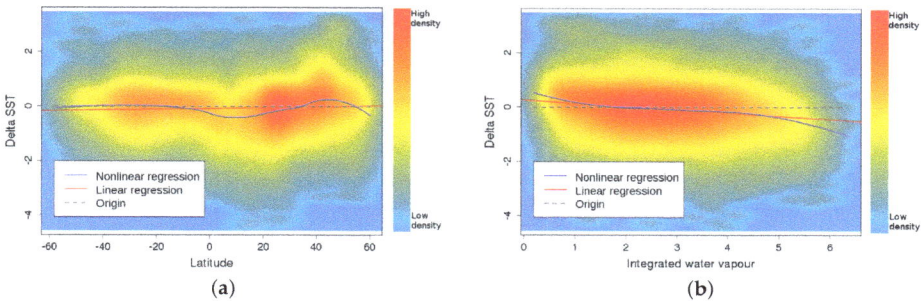

Figure 2. Density scatterplot of ΔSST as function of (**a**) latitude and (**b**) integrated water vapour fitted using linear and nonlinear (lowess) regressions.

Hereinafter, the statistical estimates of ΔSST proposed by the presented 4 models will be denoted by $\widehat{\Delta SST}$. In order to select the best model we use the adjusted R^2 (denoted as R_{adj}) which is negatively impacted by the number of parameters in the model and the Root Mean Square Error (RMSE). The R_{adj} is given by the following equation:

$$R_{adj}^2 = 1 - \frac{\sum^N ||\Delta SST - \widehat{\Delta SST}||^2}{\sum^N ||\Delta SST - \overline{\Delta SST}||^2} \frac{N-1}{N-P-1} \tag{6}$$

where $\widehat{\Delta SST}$ corresponds to the estimations given by one of the 4 models presented above and $\overline{\Delta SST}$ the mean value computed with the N training samples. The use of the adjusted R_{adj}^2 enables a model to be found with a good fit and a low number of parameters to avoid over-fitting.

4. Results

Using the training dataset (249,318 match-ups) the four regression models presented above are determined. These models are then applied to the test sample (236,282 match-ups) to predict the bias in the satellite estimate of SST (Equation (1)). The performance of each model is assessed using R_{adj}^2 and RMSE. Results are presented in Table 2.

Table 2. Results of R_{adj}^2 and RMSE for the different models.

	Linear Regression	LASSO	GAM	Random Forest
R_{adj}^2	24.65%	24.43%	28.44%	30.96%
RMSE	0.576 K	0.576 K	0.562 K	0.554 K

These results show that the link between ΔSST and X covariates is clearly nonlinear. Indeed, R_{adj}^2 of GAM and random forest are 4 and 6% better than linear regression (30.96% and 28.44% against 24.65%). LASSO results are very similar to linear regression. The RMSE of nonlinear models are also improved in comparison to linear models.

We denote corrected SST$_{corr}$ as the satellite SST to which the predicted bias has been removed: SST$_{corr}$ = SST$_{sat}$ − $\widehat{\Delta SST}$, and we compare the corrected SST to in situ SST. Figure 3 illustrate the zonal performances of the four models on the test sample (August 2014 to July 2015) in comparison with uncorrected SST difference (in black line). All four models are able to reduce the strong negative bias between 5°S and 35°N. Linear regression and LASSO models both over estimate the bias around 5°S and amplify it North of 50°N. On the contrary GAM and random forest give better results consistently. At high latitude (where fewer match-ups are available) the random forest estimates the bias more accurately than the GAM model.

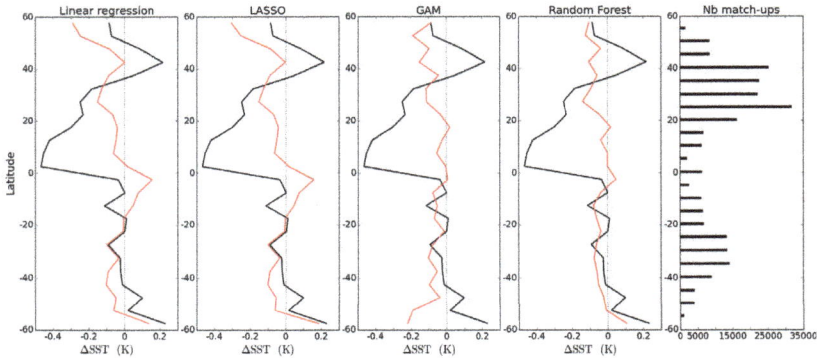

Figure 3. ΔSST= SST$_{sat}$ − SST$_{buoys}$ (black line) and $\widehat{\Delta SST}$ = SST$_{corr}$ − SST$_{buoys}$ (red line) averaged per latitude bins of 5° for linear regression, Least Square Shrinkage and Selection Operator (LASSO), Generalised Additive Model (GAM) and random forest models (from left to right) for the test sample. Far right plot shows the number of match-ups for each latitude bin.

In the following development we focus on the random forest model because it gives better results and is faster to apply (once trained) than the GAM procedure, making it a better choice for operational applications.

The global mean and standard deviation of ΔSST for the test dataset (August 2014 to July 2015) before correction are equal to −0.082 and 0.664 K respectively, and after correction they are reduced to −0.071 and 0.547 K respectively. This improvement of the standard deviation can be visualized using

Figure 4 which represents the probability density functions of ΔSST and $\widehat{\Delta\text{SST}}$: when the predicted bias is applied the distribution is narrower and more Gaussian.

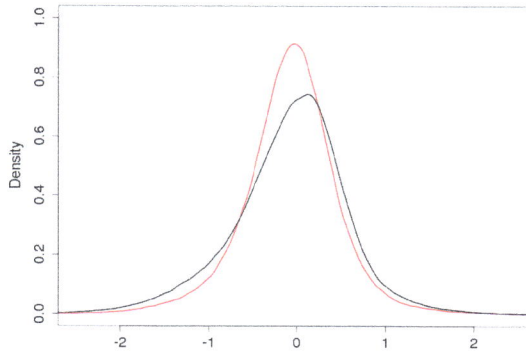

Figure 4. Histograms of $\Delta\text{SST} = \text{SST}_{\text{sat}} - \text{SST}_{\text{buoys}}$ (black line) and $\widehat{\Delta\text{SST}} = \text{SST}_{\text{corr}} - \text{SST}_{\text{buoys}}$ with correction by random forest model on test sample (red line).

Figure 5 shows the geographical distribution of the mean and standard deviation of the difference between SST_{sat} and $\text{SST}_{\text{buoys}}$ (Figure 5a,b respectively) and between SST_{corr} and $\text{SST}_{\text{buoys}}$ (Figure 5c,d respectively). Comparison of Figure 5a,c illustrate the overall reduction in the bias after subtracting the predicted bias to the satellite SST. The standard deviation is also reduced (Figure 5b,d) but high values are still observed in a number of regions. For instance, around the coast of West Africa, in the Mediterranean Sea and in the Red Sea: these regions are subject to atmospheric mineral dust events occurring only during a few months every year. High standard deviation is also visible in the Gulf Stream region and south of South Africa where strong SST gradients are observed.

Here we focus on a case study: the random forest model is applied to a satellite scene (30 April 2015 at 12 a.m., see Figure 6). This scene is composed of 792 408 clear-sky pixels (QL>2). This scene was chosen because it corresponds to a large event of Saharan dust visible on Figure 6c which represents the Saharan Dust Index (SDI, a dimensionless quantity correlated to the concentration of mineral dust particles, [6]). SDI above 0.1 indicates an amount of mineral dust particles in the atmosphere that significantly affect SST retrieval. Figure 6a,b show the integrated water vapour content and the wind speed at 10m from ECMWF NWP respectively. The predicted error from the random forest model is shown in Figure 6d.

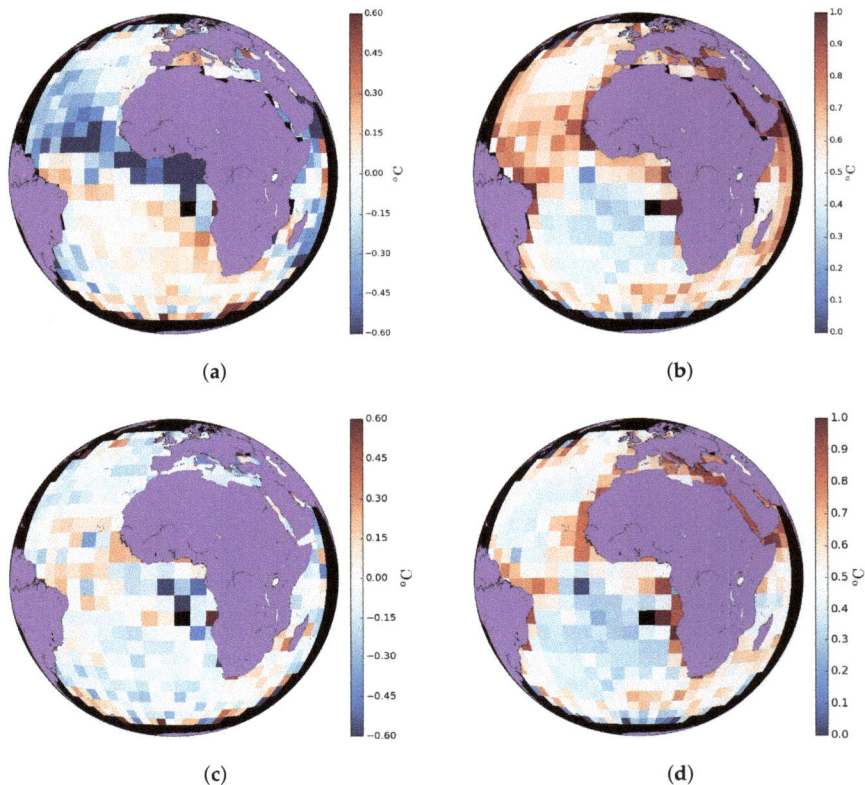

Figure 5. (**a**) Mean and (**b**) standard deviation of $SST_{sat} - SST_{buoys}$ on test sample; (**c**) Mean and (**d**) standard deviation of $SST_{corr} - SST_{buoys}$ with correction by random forest model on test sample.

Large scale features in the predicted error can be visually correlated to atmospheric features. The predicted error is largely negative where the atmosphere is humid (integrated water vapour above $4.5 \, g \, cm^{-2}$) in combination with high a SDI (of the order of 0.3 or above): this is the case around the Southern coast of West Africa. On the other hand, positive errors occur when the atmosphere is drier than average (integrated water vapour below $3 \, g \, cm^{-2}$) in combination with low SDI values (below 0.2): this is the case off the coast of Brazil where a south eastward thong of drier atmosphere leads to positive predicted error. It is interesting to note that where there is a combination of dry atmosphere and higher than normal SDI, the predicted error is often positive: this is the case in the Mediterranean Sea and off the coast of Namibia. There is no visible correlation between predicted error and wind speed, which is not altogether surprising since at 12 am residual diurnal warming would be minimal (and probably only observed in the western part of the domain).

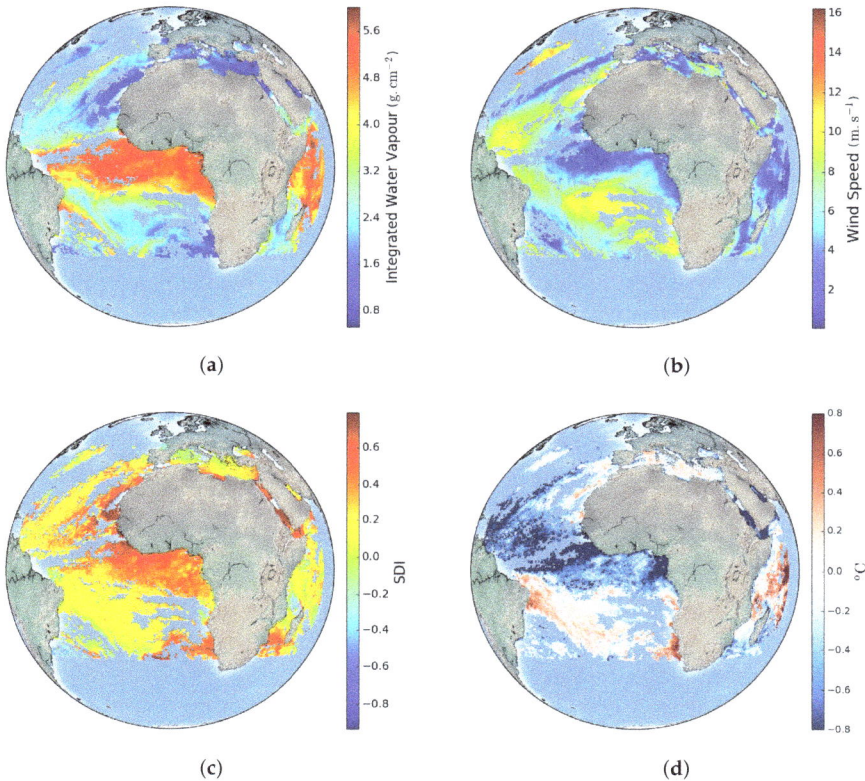

Figure 6. (**a**) Integrated water vapour; (**b**) Wind speed; (**c**) Saharan Dust Index; (**d**) ΔSST predicted by random forest model (30 April 2015 at 12 a.m.)

5. Discussion and Conclusions

The Match-up DataSet (MDS) of Météo-France Centre de Météorologie Spatiale (CMS) which collocates satellite and in situ Sea Surface Temperature (SST) measurements has been used to define statistical models of SST bias (ΔSST predicted by a model) for the Spinning Enhanced Visible and Infrared Imager (SEVIRI) sensor on-board Meteosat Second Generation (MSG) geostationary satellite. Linear regression, LASSO, GAM and the random forest model were used to fit ΔSST using the information of ten covariates.

It was shown that the nonlinear models (GAM and random forest) perform better in predicting the bias of satellite SST retrieval than linear models. They clearly manage to reduce the zonal biases associated with high water vapour content.

The random forest model was preferred over the GAM because of its slightly better results but mostly because it is quicker to run once the model has been trained making it a better choice for operational application. The random forest has then been studied further and applied on a study case (one 15-min of acquisition of the SEVIRI instrument).

Ocean and Sea Ice Satellite Application Facility (OSI SAF) operational processing of SST performed at CMS uses a very basic principle to derive SST bias: a MDS is used to compute the bias per quality level. The bias is then attributed to each pixel according to its respective quality level. This method provides discrete fields of bias which are to be avoided according the Group for High Resolution SST (GHRSST) Data Specification version 2.0 [3]. Above all this method provides a R^2_{adj} equals to 16.74 %, so a statistical model like random forest, with a R^2_{adj} equals to 30.96 %, is twice as efficient in capturing the variance of the error and would therefore be a valuable update to the processing chain.

The main limitation of the use of statistical models to predict the error in SST retrieval is scarceness of the in situ data. Despite the large number of match-ups some areas or phenomena are not sampled sufficiently to have a significant impact on the model. This can be seen when comparing Figure 5a,c. The model is well able to estimate the bias in the satellite retrieval at large spatial scales, which is largely due to the inability of the SST algorithm to cope with a varying atmosphere. However the high standard deviations in localised areas on Figure 5d may suggest that the model does not fully capture the spatial variability of the error.

Currently, the methodology has been proven successful in predicting bias in SST derived from a geostationary satellite which provides a large number of collocations with in situ data due to high temporal resolution of acquisitions (15 min). Although no test has been done on the minimum size of MDS required to train the random forest model, it is anticipated that for polar orbiting satellites a longer period would be needed. This is certainly a limitation associated with data-driven methodologies.

The large amount of data required to train the random forest model properly, means that an accurate model cannot be built to estimate the SST error in the first few month of the life of a satellite because too few match-ups are available. This is certainly true for polar orbiting satellite which do not provide as many match-ups as geostationary satellite. A temporary model could be built even prior to the launch of a satellite by using radiative transfer simulations of brightness temperature to train the model and updated later when sufficient in situ match-ups become available.

More work could be done to assess the random forest performance during daytime or to determine whether inclusion of simulations of brightness temperature as a covariate would be beneficial. Nevertheless, advanced statistical models such as random forest are promising for evaluation of the systematic error in SST retrieval from space with respect to in situ measurements. It is worth noting that these techniques may be applied in many other remote sensing contexts as long as large match-up datasets are available.

Acknowledgments: The data from the EUMETSAT Satellite Application Facility on Ocean and Sea Ice used in this study are accessible through the SAF's homepage http://osi-saf.eumetsat.int. The authors would like to thank Météo-France for funding Blandine Gausset's master thesis. The authors also wish to thank the reviewers for their very constructive comments.

Author Contributions: The idea of this study was initially designed by Pierre Tandeo, Emmanuelle Autret and Stéphane Saux Picart. Blandine Gausset did most of the work during her master thesis at Météo-France and Pierre Tandeo performed some extra analysis. The paper was mostly written by Stéphane Saux Picart and Blandine Gausset with help of the the author co-authors.

Conflicts of Interest: The authors declare no conflict of interest.

References

1. Lary, D.J.; Alavi, A.H.; Gandomi, A.H.; Walker, A.L. Machine learning in geosciences and remote sensing. *Geosci. Front.* **2016**, *7*, 3–10.
2. Zhang, L.; Zhang, L.; Du, B. Deep learning for remote sensing data: A technical tutorial on the state of the art. *IEEE Geosci. Remote Sens. Mag.* **2016**, *4*, 22–40.

3. GHRSST Science Team. The Recommended GHRSST Data Specification (GDS) 2.0. 2011. Available online: https://www.ghrsst.org/about-ghrsst/governance-documents/ (accessed on 29 January 2018).
4. Merchant, C.; Paul, F.; Popp, T.; Ablain, M.; Bontemps, S.; Defourny, P.; Hollmann, R.; Lavergne, T.; Laeng, A.; de Leeuw, G.; et al. Uncertainty information in climate data records from Earth observation. *Earth Syst. Sci. Data* **2017**, *9*, 511–527.
5. Merchant, C.J.; Horrocks, L.A.; Eyre, J.R.; O'Carroll, A.G. Retrievals of sea surface temperature from infrared imagery: origin and form of systematic errors. *Q. J. R. Meteorol. Soc.* **2006**, *132*, 1205–1223.
6. Merchant, C.J.; Embury, O.; Le Borgne, P.; Bellec, B. Saharan dust in nighttime thermal imagery: Detection and reduction of related biases in retrieved sea surface temperature. *Remote Sens. Environ.* **2006**, *104*, 15–30.
7. Evans, R.; Kilpatrick, K. The MODIS hypercube. In Proceedings of the 8th International GHRSST Science Team meeting, Melbourne, Australia, 14–17 May 2007. Available online: https://www.ghrsst.org/meetings/8th-international-ghrsst-science-team-meeting-ghrsst-viii/(accessed on 29 January 2018).
8. Castro, S.L.; Wick, G.A.; Jackson, D.L.; Emery, W.J. Error characterization of infrared and microwave satellite sea surface temperature products for merging and analysis. *J. Geophys. Res. Ocean.* **2008**, *113*, doi:10.1029/2006JC003829.
9. Petreko, B.; Ignatov, A. SSES in ACSPO. In Proceedings of the GHRSST XV Science Team Meeting, Cape Town, South Africa, 2–6 June 2014. Available online: https://www.ghrsst.org/meetings/15th-international-ghrsst-science-team-meeting-ghrsst-xv/(accessed on 29 January 2018).
10. Petrenko, B.; Ignatov, A.; Kihai, Y.; Dash, P. Sensor-Specific Error Statistics for SST in the Advanced Clear-Sky Processor for Oceans. *J. Atmos. Ocean. Technol.* **2016**, *33*, 345–359.
11. Tandeo, P.; Autret, E.; Piolle, J.; Tournadre, J.; Ailliot, P. A Multivariate Regression Approach to Adjust AATSR Sea Surface Temperature to In Situ Measurements. *IEEE Geosci. Remote Sens. Lett.* **2009**, *6*, 8–12.
12. Xu, F.; Ignatov, A.; Liang, X. Towards continuous error characterization of sea surface temperature in the advanced clear-sky processor for oceans. In Proceedings of the 89th AMS Annual Meeting, 16th Conference on Satellite Meteorology and Oceanography, Phoenix, AZ, USA, 10–16 January 2009; pp. 11–15.
13. Bulgin, C.; Embury, O.; Corlett, G.K.; Merchant, C. Independent uncertainty estimates for coefficient based sea surface temperature retrieval from the Along-Track Scanning Radiometer instruments. *Remote Sens. Environ.* **2016**, *178*, 213–222.
14. SST-CCI. SST CCI Uncertainty Characterisation Report v2; SST-CCI-UCR-UOE-002. 2013. Available online: www.esa-sst-cci.org (accessed on 29 January 2018).
15. Le Borgne, P.; Roquet, H.; Merchant, C. Estimation of Sea Surface Temperature from the Spinning Enhanced Visible and Infrared Imager, improved using numerical weather prediction. *Remote Sens. Environ.* **2011**, *115*, 55–65.
16. OSI SAF. *Geostationary Sea Surface Temperature Product User Manual*; Technical Report; EUMETSAT: Berlin, Germany, 2011.
17. Donlon, C.J.; Minnett, P.J.; Gentemann, C.; Nightingale, T.J.; Barton, I.J.; Ward, B.; Murray, M.J. Toward Improved and Validation of Satellite and Sea Surface and Skin Temperature and Measurements and for Climate and Research. *J. Clim.* **2002**, *15*, 353–359.
18. Marsouin, A.; Le Borgne, P.; Legendre, G.; Péré, S.; Roquet, H. Six years of OSI-SAF METOP-A AVHRR sea surface temperature. *Remote Sens. Environ.* **2015**, *159*, 288–306.
19. Tibshirani, R.J. Regression Shrinkage and Selection via the lasso. *J. R. Stat. Soc.* **1996**, *58*, 267–288.
20. Hastie, T.J.; Tibshirani, R.J. *Generalized Additive Models*; Chapman & Hall/CRC: London, UK, 1990.
21. Breiman, L. Random Forests. *Mach. Learn.* **2001**, *45*, 5–32.
22. Breiman, L.; Friedman, J.H.; Olshen, R.A.; Stone, C.J. *Classification and Regression Trees*; Wadsworth: Belmont, CA, USA, 1984.

remote sensing

MDPI

Article

The Accuracies of Himawari-8 and MTSAT-2 Sea-Surface Temperatures in the Tropical Western Pacific Ocean

Angela L. Ditri [1,*,†], **Peter J. Minnett** [2], **Yang Liu** [2], **Katherine Kilpatrick** [2] and **Ajoy Kumar** [1]

[1]   Meteorology and Ocean Sciences & Coastal Studies, College of Science and Technology,
      Millersville University, Millersville, PA 17551, USA; Ajoy.Kumar@millersville.edu
[2]   Ocean Sciences, Rosenstiel School of Marine and Atmospheric Science, University of Miami,
      Miami, FL 33149, USA; pminnett@miami.edu (P.J.M.); yliu@rsmas.miami.edu (Y.L.);
      kkilpatrick@rsmas.miami.edu (K.K.)
*   Correspondence: alditri@udel.edu; Tel.: +1-215-313-8708
†   Current affiliation: Center for Remote Sensing, College of Earth, Ocean, and Environment,
    University of Delaware, Newark, DE 19716, USA.

Received: 20 November 2017; Accepted: 24 January 2018; Published: 1 February 2018

Abstract: Over several decades, improving the accuracy of Sea-Surface Temperatures (SSTs) derived from satellites has been a subject of intense research, and continues to be so. Knowledge of the accuracy of the SSTs is critical for weather and climate predictions, and many research and operational applications. In 2015, the operational Japanese MTSAT-2 geostationary satellite was replaced by the Himawari-8, which has a visible and infrared imager with higher spatial and temporal resolutions than its predecessor. In this study, data from both satellites during a three-month overlap period were compared with subsurface in situ temperature measurements from the Tropical Atmosphere Ocean (TAO) array and self-recording thermometers at the depths of corals of the Great Barrier Reef. Results show that in general the Himawari-8 provides more accurate SST measurements compared to those from MTSAT-2. At various locations, where in situ measurements were taken, the mean Himawari-8 SST error shows an improvement of ~0.15 K. Sources of the differences between the satellite-derived SST and the in situ temperatures were related to wind speed and diurnal heating.

Keywords: sea surface temperatures; geostationary satellite; infrared; tropical western Pacific Ocean; the Great Barrier Reef; accuracy

---

1. Introduction

Sea-surface temperature (SST) is a key variable for the study of the climate, weather, and ocean. The tropical western Pacific Ocean and eastern Indian Ocean, often referred to as the Tropical Warm Pool (TWP), have some of the highest SSTs (e.g., [1]). High SSTs throughout the tropical belt lead to meridional convergence in the lower troposphere and convection producing the clouds of the intertropical convergence zone (ITCZ) [2] which, being the ascending arm of the Hadley Cells to the north and south, is a driver of the large scale atmospheric circulation. As such, it is also a major part of the earth's hydrological cycle [3]. The bright cloud tops in the ITCZ also influence the regional planetary albedo and thus influence the radiative heat budget of the earth. The vertical atmospheric motion is driven by the high SSTs in the equatorial regions [4]. The SST is also an indicator of the upper ocean heat content that is closely connected to the generation and intensification of cyclones in the TWP [5,6]. The cyclones frequently make landfall to the west, where damage and loss of life can be extreme (e.g., [7]). Improved accuracy of SSTs is critical for better forecasts of such events.

A further aspect of high temperatures in the tropics is the risk to coral reefs, which are damaged by elevated temperatures both when occurring episodically, such as on diurnal time scales [8,9] and over several days [10,11]. Dire consequences follow when elevated temperatures are sustained over weeks and months [12]. If the temperatures revert to the range to which they are acclimated, the corals can recover, however extended periods of high temperatures can lead to extensive coral mortality [9,13]. The defense mechanism of corals when subjected to elevated temperatures is to expel the symbiotic algae (zooxanthellae) living in their tissues causing the coral to lose their color, an effect widely referred to as coral bleaching. Coral bleaching can also result from anomalously low temperatures [14], especially where the corals are exposed to cold air temperatures at low tide [15]. The corals are also stressed by increasing ocean acidification [16,17].

The western Pacific and eastern Indian Oceans are home to extensive coral reefs, including those in the so-called Coral Triangle that encompasses the waters between the island of Borneo in the west and the Solomon Islands in the east, and the northern extent of the Philippines in the north, and the Timor, Arafura, and Coral Seas to the south [18]. The Coral Triangle contains the highest diversity of corals and of the species that are associated with them, including reef fishes [19]. The Coral Triangle does not exist in isolation, but is embedded in a much larger area of corals and high marine biodiversity that includes the Great Barrier Reef (GBR) off Queensland, Australia, to the south. At present, the GBR is experiencing extensive and severe bleaching, especially in the northern part where mortality is very high (>50%); the GBR coral bleaching is the worst on record [20,21]. The episode began in 2014 with record high SSTs through much of the Coral Triangle and GBR, and is part of a global event that is especially severe in the Pacific and Indian Oceans [22]. The intensity and spatial extent of this, and past severe bleaching occurrences, are clearly linked to the spatial patterns of elevated SSTs [12], and are expected to become worse as the oceans warm [12,23]. Thus, the areas of the Coral Triangle and GBR present a very pressing need for accurate measurements of ocean temperature over long periods and over large areas.

A valuable source of global near-surface ocean temperatures are those measured from surface drifting buoys [24] deployed to provide measurements for weather forecasting and studying surface currents. The temperature measurements, taken at a depth of about 20 cm in calm seas, are in widespread use [25,26], but equatorial upwelling and surface current divergence tends to remove the drifters from the tropics [27,28]. Thus, there is a paucity of measurements in the Coral Triangle and GBR regions.

Near-surface temperatures are measured in the tropics by thermometers of the Global Tropical Moored Buoy Array, which, in the Tropical Western Pacific Ocean, comprises the Triangle Trans-Ocean Buoy Network (TRITON; [29]). This is a deep-water mooring array, and so does not extend into the shallow waters where the corals are found. In contrast, temperature measurements have been made by the Australian Institute of Marine Science (AIMS) for many years by self-recording thermometers deployed at the coral depths by divers. These measurements are very good indicators of the thermal stress experienced by the corals and, being recorded during our analysis period with a 10-min resolution, provide data that resolve rapid changes, such as those associated with diurnal heating [30]. However, they are relatively sparse in space, and the data loggers have to be recovered before the temperatures can be analyzed.

Thus, the surface temperature fields derived from satellites are a very attractive source of information to study the potential threats to the wellbeing of the corals, especially as a recent study has shown they are an accurate proxy for temperatures at the depths of the corals [31], even though the satellite-derived temperatures are skin temperatures (SST_{skin}; [32,33]). Satellite derived SSTs cover large areas and those from geostationary satellites positioned over the Equator of the Pacific Ocean provide frequent measurements over the Coral Triangle and GBR. However, the appropriate application of satellite-derived SSTs to assessing the thermal conditions experienced by corals depends on knowledge of the errors and uncertainties in the SSTs retrieved from the satellite measurements. Our objective is to determine the accuracies of SSTs derived from geostationary satellites in the Tropical Western Pacific Ocean, including the Coral Triangle and the GBR. Our focus will be on SSTs derived from infrared radiometers on geostationary satellites, as these provide more rapid sampling than the polar orbiters which offers the possibility of capturing short period heating events [31]. Given that clouds obscure the surface in the infrared and thus prevent the derivation of SSTs, the rapid sampling by geostationary sensors increases the likelihood of determining SSTs at given locations as clouds pass. The accuracies of the satellite-derived SSTs will be established by comparisons with in situ measurements from the Triangle Trans-Ocean buoy Network (TRITON) moored buoys in the western part of the TAO array and from the GBR temperature loggers.

The paper is organized as follows: the next section introduces the data, beginning with the satellite retrieved SSTs and analysis methods, followed by a presentation of the results. A discussion of the results comes before the conclusions, which includes suggestions for further work.

2. Materials and Methods

Himawari-8, the first of a new generation of geostationary meteorology satellites of the Japan Meteorological Agency (JMA), began their operation on 7 July 2015. Himawari-8 replaced the MTSAT-2 (Multifunctional Transport Satellite-2, also referred to as Himawari-7). Though Himawari-8 became operational in July, MTSAT-2 continued operation until 4 December 2015 [34]. Himawari-8 is located at 140.7°E above the equator while the MTSAT-2 is located at 145°E above the equator. Himawari-8 carries the Advanced Himawari Imager (AHI), which has significant improvements in comparison to the imager onboard MTSAT-2. The AHI is capable of generating full disk images with a 10-min sampling frequency. The AHI has 16 spectral bands of which four are infrared (IR), λ = 8.60, 10.45, 11.20, and 12.35 μm [35,36], that are used for SST retrievals. These IR bands have a spatial resolution of 2 km at nadir. The temporal and spatial resolutions of Himawari-8 AHI are improved from those of the MTSAT-2 imager, which has only five spectral bands. Three of the five spectral bands used for IR SST retrievals include λ = 3.75, 10.8, and 12.0 μm [37] which have a 4 km spatial resolution and sampling intervals of 60 min [38,39]. Table 1 summarizes the satellite characteristics. SST fields from both satellites are provided by the National Oceanic and Atmospheric Administration (NOAA) in GHRSST Level-2 Pre-processed format (L2P; [32]). A recent study by Kramar et al. [40] found that Himawari-8 AHI SSTs derived using NOAA's Advanced Clear-sky Processor for Oceans (ACSPO; [41]) show better accuracy than those produced by the Japan Aerospace Exploration Agency (JAXA). Based on this result, the MTSAT-2 and Himawari-8 SSTs used here are those produced by the NOAA Office of Satellite Products and Operations (OSPO).

Table 1. Details of satellite data used in this study.

Satellite Name	Spatial Resolution	Temporal Resolution	Position	No. of Spectral Bands	Operation Period
MTSAT-2	4 km	Hourly	0°N, 145.0°E	5	2010 to 2015
Himawari-8	2 km	10 min	0°N, 140.7°E	16	2015 to 2022

The MTSAT-2 SSTs were derived using the long-established Non-Linear SST atmospheric correction algorithm [37] that uses measurements taken in the thermal infrared centered at λ = 10.8 and 12.0 μm:

$$SST = a_0 + a_1 \times T_{11} + a_2 \times (T_{11} - T_{12}) \times T_{sfc} + a_3 \times (T_{11} - T_{12}) \times (\sec(\theta) - 1) \qquad (1)$$

where T_n are brightness temperatures, in K, measured at n = rounded integer values of λ, θ is the satellite zenith angle and T_{sfc} is a prior estimate of the surface temperature. Equation (1) can be used during both day and night. The coefficients are derived from a correlation analysis between the satellite brightness temperature measurements and coincident subsurface temperatures from buoys. The night-time algorithm, also due to Walton, Pichel, Sapper, and May [37], includes measurements from a third channel centered at λ = 3.75 μm:

$$SST = a_0 + a_1 \times T_{11} + a_2 \times (T_{3.7} - T_{12}) \times T_{sfc} + a_3 \times (\sec(\theta) - 1) \qquad (2)$$

The measurements in the mid-infrared atmospheric transmission window, $T_{3.7}$, suffer from contamination from scattered and reflected solar radiation that occurs during the day, so these measurements can only be used at night. The metadata in the MTSAT-2 files indicate that the retrieved SSTs are a skin temperature.

The ACSPO SST atmospheric correction algorithm applied to Himawari-8 AHI uses measurements from four infrared bands, labeled 11, 13, 14, and 15 at central wavelengths, λ = 8.60, 10.45, 11.20, and 12.35 μm [35,36]; it takes the form [40]:

$$SST = a_0 + a_1 \times T_{10} + a_2 \times (T_{10} - T_{12}) + a_3 \times (T_{10} - T_9) \times \sec(\theta) + a_4 \times (T_{10} - T_{11}) \times \sec(\theta)$$
$$+ a_5 \times (T_{10} - T_9) \times T_{sfc} + a_6 \times (T_{10} - T_{11}) \times T_{sfc} + a_7 \times (T_{10} - T_{12}) \times T_{sfc}$$
(3)

where here T_{sfc} is an estimate of the surface temperature taken from the daily Canadian Meteorological Center (CMC) L4 SST analysis [42]. The coefficients a_i are derived from regression analysis of collocated, contemporaneous brightness temperature measurements of the satellite radiometer with those of quality-controlled subsurface temperatures measured from drifting and moored buoys in the iQuam data set (in situ SST Quality Monitor; [43]). Thus the Himawari-8 AHI SSTs are considered a "subskin" temperature [40]. Since Equation (3) does not use brightness temperature measurements in the mid-infrared atmospheric transmission window it can be used for both daytime and night-time SST retrievals. For successful retrieval of SST, the brightness temperatures have to be screened to remove all measurements that include a component of emission from clouds. The Himawari-8 AHI data were taken from ftp://ftp.star.nesdis.noaa.gov/pub/sod/sst/acspo_data/l2/ahi/.

In this study, satellite SSTs were compared to multiple subsurface in situ temperature measurements. These in situ stations include seven TRITON moored buoys from the TAO array and seven self-recording thermometers attached to corals in the GBR (Table 2). The accuracy of the near-surface thermometers on the TRITON buoys is 0.05 K [44], and that of the GBR thermometers is better than ±0.1 K [11].

Table 2. Locations and depth of in situ measurements.

Station	Lat	Lon	Thermometer Depth
TAO 1	0°N	147°E	1.5 m
TAO 2	0°N	156°E	1.5 m
TAO 3	2°N	137°E	1.5 m
TAO 4	2°N	147°E	1.5 m
TAO 5	8°N	137°E	1.5 m
TAO 6	2°N	156°E	1.5 m
TAO 7	2°S	156°E	1.5 m
GBR 1	21.87°S	152.52°E	10.4 m
GBR 2	16.64°S	146.11°E	7.0 m
GBR 3	18.83°S	147.63°E	3.3 m
GBR 4	18.49°S	146.87°E	1.9 m
GBR 5	21.03°S	150.85°E	7.1 m
GBR 6	21.41°S	151.64°E	7.1 m
GBR 7	21.11°S	152.55°E	8.3 m

The TAO data used here were provided by NOAA/PMEL (Pacific Marine Environmental Laboratory (https://www.pmel.noaa.gov/tao/drupal/disdel/) and in situ data from the GBR were provided by the Australian Institute of Marine Science (AIMS). SSTs derived from Himawari-8 are subskin SSTs, whereas the in situ measurements are at 1.5 m depth on the TRITON buoys of the TAO array (http://www.jamstec.go.jp/jamstec/TRITON/real_time/overview/po-t1), the depths of the thermometers on the GBR vary. The depths of the GBR thermometers are given below the lowest astronomical tide and thus the depths below the surface will depend on the state of the tide; typically the tidal amplitudes are 5 m for spring tides and 2 m for neap tides [45]. Three months

(1 August 2015–31 October 2015) of data were compiled during a period when both satellites were operational. The selection of in situ stations (Table 2, Figure 1) required data to be available in this period. A 5 × 5 pixel box of data was extracted from both satellites' images around each in situ location. This was to ensure there would an adequate amount of satellite data of the best quality (quality flag 5). The best quality data were spatially averaged to give one satellite measurement for each in situ location for each comparison.

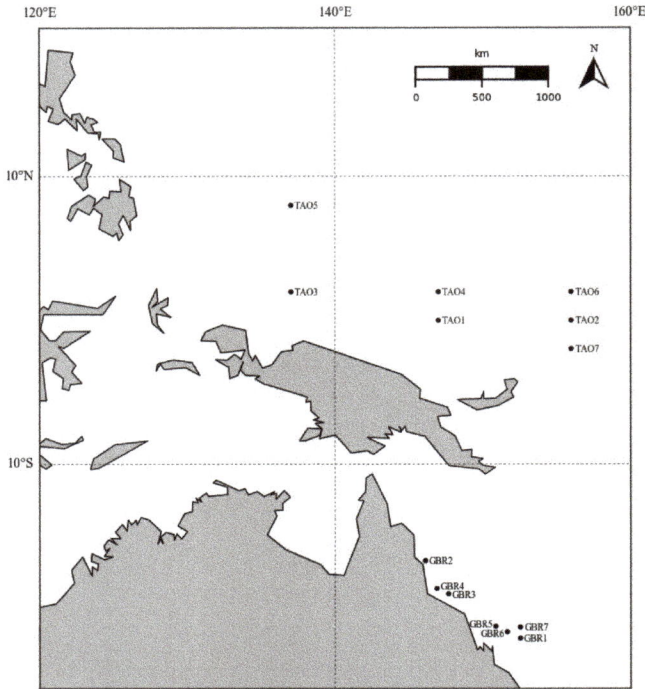

Figure 1. Locations of in situ stations providing subsurface temperatures for this study.

Himawari-8 AHI and the thermometers in the GBR had a temporal resolution of 10 min, but the measurements were not synchronized. Data from the GBR were linearly interpolated to match the times of the Himawari-8 data. MTSAT-2 and the TAO array had an hourly temporal resolution; data from the GBR and Himawari-8 were averaged to match this hourly temporal resolution; the averaging process is summarized in Table 3. The temperature differences were calculated by subtracting the in situ subsurface temperature from the satellite SST. To facilitate an analysis of diurnal heating patterns, both satellite and in situ data were converted from coordinated universal time (UTC) to local time. When separating day and night data, the time interval was limited to 10 h of daylight and nighttime centered on local noon and midnight to avoid issues around dusk and dawn because of difficulties in cloud screening of MTSAT-2 SST data close to sunrise and sunset [46].

Wind speed is a critical parameter in determining the amplitude of diurnal heating and cooling [47] and in this study, wind speeds provided in the satellite files were used. The wind speeds are derived from the National Centers for Environmental Prediction (NCEP) Global Forecast System (GFS) fields [48] and are representative of a value at 10 m height at 1° spatial resolution and are produced every six hours. They are interpolated to the times and positions of the satellite-derived SST fields.

Table 3. Sampling and averaging summary for generating matchups between satellite and in situ measurements.

		GBR	TAO
MTSAT-2	1 h sampling interval	In situ temperatures within the hour following the satellite SST measurement were averaged.	Hourly in situ temperatures were paired with corresponding hour of the satellite SST.
Himawari-8 AHI	10 min sampling interval	Although same sampling intervals, they were not synchronized. In situ temperatures were interpolated to satellite sample times.	Satellite SST samples that were within the hour of the in situ sample were averaged.

3. Results

The statistics of the differences between the satellite-derived SST and the in situ temperatures for each of the TAO and GBR stations are shown in Table 4. The statistics of the differences between the satellite-derived SSTs for the Himawari-8 and MTSAT-2 for the entire data set used here are shown in Table 5, and for day and night conditions at the two sets of in situ measurements in Table 6. In general, the mean and median values of the differences are smaller for the Himawari-8 AHI SSTs compared to those of MTSAT-2, but there are exceptions. The standard deviations of the differences do not show the expected improvements in Himawari-8 AHI SSTs compared to those of MTSAT-2, especially at the GBR stations; these are shown graphically in Figure 2 as box-whisker plots. The central bar in the box indicates the median value, and the lower and upper borders of the box represent the first and third quartiles of the distribution of values; the extreme values of the whiskers are the minimum and maximum values, excluding outliers. Outliers were considered those to be beyond 1.5 times the upper and lower quartiles, and are not shown in this and other figures.

Table 4. Statistics of satellite Sea-Surface Temperature (SST)—in situ temperatures (K). Upper row for each station is for Himawari-8 SSTs, and the lower row for MTSAT-2 SSTs.

Station	θ	N	N of Outliers	Min	Max	Day Mean	Night Mean	Day Median	Night Median	Day STD	Night STD	Day RSD	Night RSD
TAO 1	7°	1464	19	−1.294	1.759	0.034	0.009	0.033	0.022	0.368	0.353	0.349	0.330
	2°	848	15	−1.385	1.670	0.141	0.268	0.194	0.300	0.447	0.341	0.409	0.250
TAO 2	18°	909	6	−1.045	0.977	−0.048	−0.058	−0.057	−0.060	0.336	0.324	0.349	0.349
	13°	480	13	−1.480	1.970	0.113	0.224	0.157	0.283	0.453	0.353	0.427	0.263
TAO 3	5°	1768	64	−1.810	2.605	0.119	−0.036	0.105	−0.007	0.554	0.439	0.473	0.395
	10°	1105	45	−2.000	2.230	0.160	0.165	0.161	0.189	0.500	0.358	0.413	0.315
TAO 4	8°	1255	46	−2.039	1.419	0.005	−0.037	0.046	0.010	0.441	0.420	0.355	0.362
	3°	639	22	−1.565	2.020	0.053	0.197	0.094	0.243	0.467	0.321	0.416	0.331
TAO 5	10°	1004	26	−1.791	0.758	−0.158	−0.233	−0.124	−0.184	0.359	0.327	0.319	0.299
	13°	550	23	−1.535	1.833	0.181	0.226	0.203	0.252	0.457	0.291	0.400	0.295
TAO 6	18°	1008	6	−1.286	2.520	−0.011	−0.047	−0.027	−0.055	0.385	0.346	0.403	0.333
	13°	507	20	−1.980	1.825	0.096	0.220	0.121	0.225	0.489	0.292	0.399	0.295
TAO 7	18°	880	14	−2.184	0.967	−0.279	−0.310	−0.243	−0.286	0.498	0.431	0.441	0.374
	13°	458	14	−2.240	1.530	−0.179	−0.025	−0.128	−0.008	0.513	0.423	0.478	0.474
GBR 1	29°	1886	58	−2.523	2.003	0.089	−0.239	0.121	−0.177	0.545	0.497	0.491	0.383
	27°	1136	46	−2.271	1.763	0.147	−0.137	0.159	−0.042	0.437	0.429	0.367	0.343
GBR2	20°	1844	21	−1.300	2.873	0.117	−0.163	0.095	−0.174	0.490	0.353	0.504	0.360
	19°	977	38	−1.513	2.174	0.231	0.079	0.205	0.108	0.433	0.380	0.375	0.341
GBR3	23°	1980	22	−1.665	2.922	0.474	0.116	0.507	0.132	0.545	0.447	0.575	0.437
	22°	1260	46	−1.625	3.007	0.512	0.298	0.530	0.382	0.402	0.357	0.327	0.293
GBR 4	23°	1838	20	−1.445	2.287	0.082	−0.193	0.086	−0.208	0.483	0.365	0.498	0.381
	22°	1138	44	−1.037	2.485	0.249	0.155	0.169	0.236	0.380	0.265	0.306	0.222
GBR 5	27°	1996	19	−1.428	1.252	0.133	−0.050	0.150	−0.030	0.381	0.349	0.395	0.356
	25°	1334	41	−1.365	1.578	0.122	0.069	0.141	0.101	0.311	0.251	0.273	0.235
GBR6	28°	1944	17	−1.304	1.700	0.334	0.127	0.346	0.143	0.433	0.388	0.460	0.400
	26°	1163	38	−1.609	1.360	0.280	0.187	0.288	0.218	0.331	0.288	0.282	0.232
GBR 7	28°	1787	9	−0.821	2.451	0.885	0.675	0.930	0.736	0.548	0.530	0.575	0.548
	26°	954	7	−1.325	2.379	0.849	0.829	0.894	0.889	0.449	0.385	0.466	0.414

θ is the satellite zenith angle; N is the number of matchups (Column 3) and of outliers (Column 4); RSD is Robust Standard Deviation.

Table 5. Statistics of satellite SST—in situ temperatures (K).

	N	Mean	Median	STD	RSD
Himawari-8	21563	0.180	0.155	0.534	0.492
MTSAT-2	12549	0.261	0.269	0.480	0.402

Table 6. Statistics of satellite SST—in situ temperatures (K) for day and night conditions at the positions of the Tropical Atmosphere Ocean (TAO) moorings and the Great Barrier Reef (GBR) stations.

	N	Day Mean	Night Mean	Day Median	Night Median	Day STD	Night STD	Day RSD	Night RSD
TAO/Himawari-8	8288	−0.022	−0.086	−0.015	−0.075	0.454	0.399	0.393	0.366
TAO/MTSAT-2	4587	0.099	0.189	0.137	0.230	0.487	0.351	0.420	0.321
GBR/Himawari-8	13,275	0.299	0.037	0.283	0.006	0.561	0.510	0.543	0.461
GBR/MTSAT-2	7962	0.329	0.196	0.302	0.189	0.452	0.424	0.392	0.325

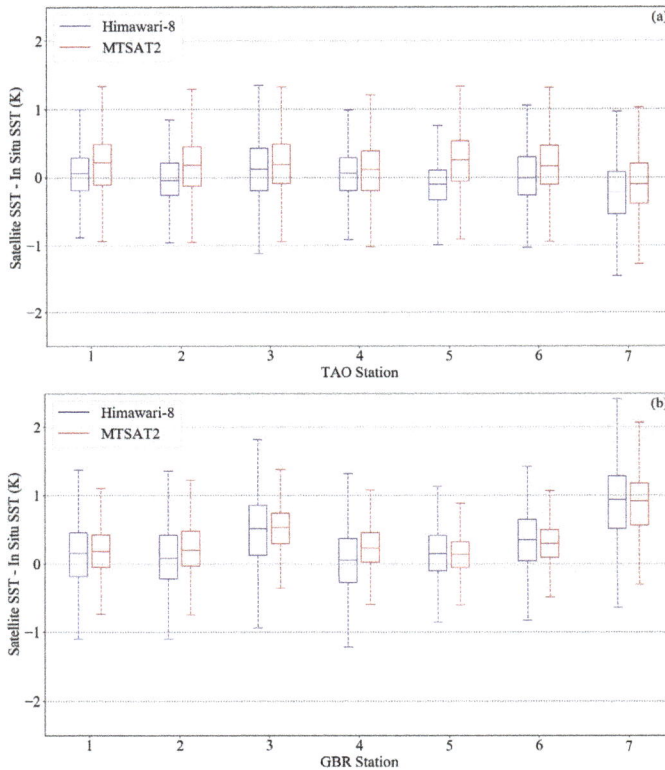

Figure 2. Box plots for the temperature difference between the satellite-derived SST and subsurface temperature at each TAO mooring (**a**) and GBR station (**b**). Blue boxes and whiskers represent the temperature differences for the Himawari-8 SSTs. Red represents the temperature differences for the MTSAT-2 SSTs. Outliers are not plotted.

Time series of temperatures measured by the satellites and in situ thermometers for a sample TAO station is shown in Figure 3 and for a sample GBR station in Figure 4. Gaps in the satellite-derived SSTs are where clouds have obscured the surface. At both stations, there is a marked diurnal heating signal in both the in situ sub-surface temperature measurements, and in the satellite-derived SST; this is seen in the times series of measurements at all stations. The days with the largest signals are those with high insolation and low wind speed. In the absence of significant wind-driven vertical mixing, the subsurface temperature signal characteristic of diurnal heating decays with depth [49] and this is revealed in the larger amplitudes of the diurnal temperature signals in the satellite-derived SSTs than in the subsurface temperatures. What is also apparent in these time series is the better agreement between MTSAT-2 SSTs and the subsurface temperatures at night than those derived from Himawari-8, which shows colder SSTs at night. During the day, the Himawari-8 AHI SSTs are generally colder than those of MTSAT-2. The systematic day-night characteristics of the differences between the satellite-derived SSTs and the subsurface temperatures are shown in Figures 5 and 6. The larger median differences, as shown by the bar in the boxes, and the length of the whiskers, between the satellite-derived SSTs and the subsurface temperatures during the day than during the night can be explained by the effects of diurnal heating introducing thermal gradients between the SSTs and the temperatures below.

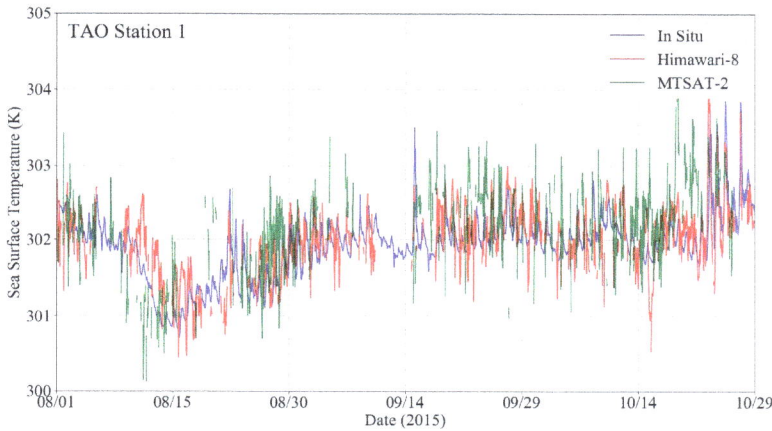

Figure 3. Time series of near-surface temperature measurements at the TAO-1 mooring at 0°N, 147°E (blue) and the satellite-derived SST from Himawari-8 (red) and MTSAT-2 (green).

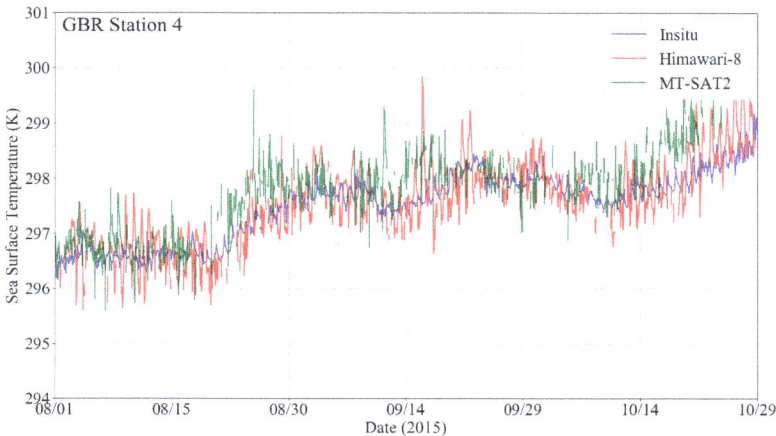

Figure 4. Time series of temperature measurements at the depth (1.9 m) of the corals GBR-4 station at 18.49°S, 146.87°E (blue), and the satellite-derived SST from Himawari-8 (red) and MTSAT-2 (green).

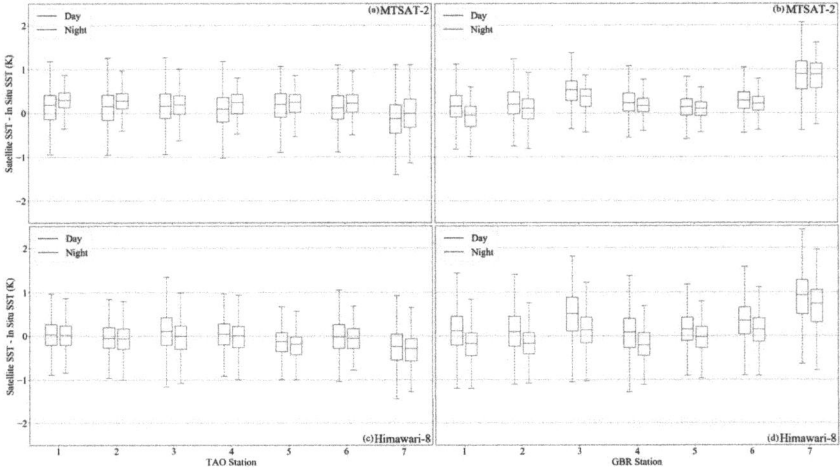

Figure 5. Box plots of the differences between satellite and in situ temperatures for day (blue) and night (red). The top row (**a,b**) is for MTSAT-2 SSTs, and the bottom row (**c,d**) for Himawari-8. The left column (**a,c**) is for the TAO moorings and the right column (**b,d**) for the GBR stations. Outliers are not shown.

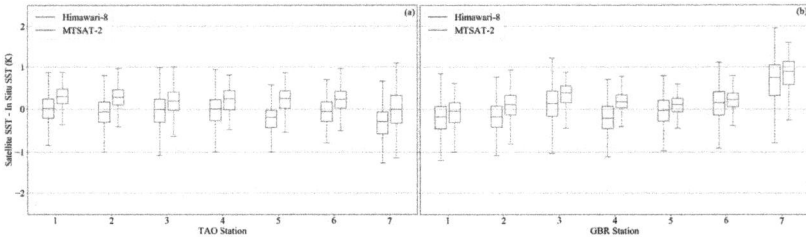

Figure 6. Box plots of the night-time differences for the satellite-derived SST and the in situ temperatures for Himarari-8 (blue) and MTSAT-2 (red). TAO moorings are at left (**a**); GBR stations at right (**b**). Outliers are not plotted.

4. Discussion

The results presented in the previous section show discrepancies that can be explained by many different points of view. This section first introduces the main summary of the results, followed by an explanation of the errors and uncertainties. Since the results show generally smaller SST differences during night-time measurements for both satellites, the diurnal heating effect is an important aspect to investigate. In this discussion, we present data that indeed shows a diurnal signal, along with its effects on the SST differences, and the statistics. Although the diurnal heating effect is one perspective used to help explain discrepancies, other sources such as footprint size and calibration errors are mentioned.

The comparison between the satellite-derived SSTs and the subsurface temperatures show better agreement in the mean for the Himawari-8 AHI SSTs in the areas of both the western TAO array and GBR. For the SST derived from both MTSAT-2 and Himawari-8, results show that the variation within the SST differences decrease during the night. For night-time measurements, the Himawari-8 AHI SSTs show better agreement in the area of the western TAO array. Though the standard deviation of the temperature differences for the MTSAT-2 is smaller than the Himawari-8, the Himawari-8 AHI SSTs are in general more accurate at night. Overall, Himawari-8 shows a mean SST difference of 0.18 K for all stations, while MTSAT-2 shows a mean difference of 0.26 K. The most reduced discrepancy was 0.17 K at GBR Station 4 for the Himawari-8 AHI SSTs. Differences between the stations can be related to the proximity to land, and the depth of the in situ measurement.

The comparisons between SSTs derived from satellite data and in situ measurements are often interpreted as an assessment of the accuracy of the satellite SSTs, but this interpretation assumes the in situ measurements are perfect and accurate, and there are no contributions to the differences from the method of comparison itself [50].

The terms "error" and "uncertainty" have distinct meanings. Error is the difference between a measured value and the true value (generally not known) and uncertainty is the dispersion, or spread, of a group of measurements of the same quantity. Thus, uncertainty is a quantification of the doubt about the measurement result [51,52]. The nature of errors and uncertainties are often described as either systematic or random. Systematic errors and uncertainties can be reduced significantly, and possibly eliminated, through an understanding of their sources and by averaging multiple measurements. In contrast, those that are random cannot be eliminated, but can be reduced by repeating the same measurement, or by taking multiple measurements under the same conditions.

Typically, the accuracy of a satellite-derived SST is expressed as a mean error, or bias, and a scatter, or standard deviation, but these are based on the assumption of a Gaussian distribution. In reality, the symmetry of a Gaussian distribution in studies such as this is rarely seen due to the effects of undetected clouds, which are nearly always colder than the underlying sea surface. This introduces a negative skewness to the distribution. The use of the median and robust standard deviation, which reduces the sensitivity to outliers in the distribution, has become a more accepted method of estimating the central value and dispersion of the differences between satellite-derived and in situ temperatures [53–55].

The differences between satellite-derived SSTs and in situ temperature measurements, within acceptable spatial and temporal interval for coincidence [54,56] are not simply an estimate of the accuracy of the satellite SST retrievals as, it is clear that there are multiple contributors to these temperature differences. Some of these contributors include inaccuracies in the in situ measurements and imperfections in the cloud screening algorithms. In addition, because of the finite intervals in space and time between the satellite and the in situ measurement, there is a contribution from the variability in the ocean temperature fields, e.g., [56]. Many of the contributors are independent of each other and can be summed in quadrature to determine the total uncertainty in the differences, which, when combined with the errors and uncertainties in the satellite radiometric measurements can provide the desired estimate of the accuracy for the satellite SST retrievals.

As stated above, the accuracy of the GBR thermometers is better than ±0.1 K and the measurements are recorded with a precision of 0.02 K [11]. The near-surface thermometers on the TRITON moored buoys is given as 0.05 K [44]. Thus, the accuracies of the in situ thermometer, while non-zero, are not likely to be the major cause of the discrepancies.

Given the Robust Standard Deviation (RSD) of the differences between the satellite-derived SSTs and the subsurface temperatures at each of the stations are less sensitive to outliers, these were expected to be smaller than the Standard Deviation, but there are several cases where this is not so. Examination of the histograms of differences revealed that stations where the RSD is larger resulted from distributions that are bimodal, or at least without a clear single peak. Those bimodal histograms may indicate a factor that if identified could be used to determine better estimates of the differences with in situ temperatures, and eventually to better estimates of the accuracies of the satellite-derived SSTs.

A possible cause of a bimodal distribution in the differences between satellite-derived SSTs and the subsurface temperatures is diurnal heating, and strong diurnal signals in the discrepancies of SST from both satellites are apparent as many stations. Figure 7 shows box-whisker plots of the discrepancies at TAO Station 3, which is quite typical of data from the TAO stations. The characteristics of the discrepancies with Himawari-8 SSTs are better behaved than the comparisons using MTSAT-2 SSTs, in that the pattern is less variable in the median, but the negative median error in the evening and early part of the night is unexpected. The negative median discrepancies in the Himawari-8 comparisons, are more pronounced at the GBR Station 4 (Figure 8).

Large temperature differences occurred when there were low wind speeds during the time of the highest insolation around local noon (Figures 9 and 10). This is related to the thermal stratification within the upper ocean. For wind speeds >6 ms^{-1}, the variation within the temperature differences decreased, approaching 0 K, presumably because higher wind speeds mix the upper part of the water column, decreasing the change in the temperature between the SST and the temperature at the depth of the in situ measurements. When the water is well mixed, the difference between measurements will be smaller; making the satellite derived SSTs closer to the in situ measurements and decreasing the discrepancies. SSTs were separated by day and night time to assess the effects of diurnal heating. For Himawari-8, the median temperature difference for each station is generally close to 0 K. The standard deviations in the discrepancies decreased for both satellites at all stations during the night compared to during the day. When comparing night-time differences for both satellites, there is a smaller variation seen within the MTSAT-2 for all in situ stations. Both satellites show better results when the diurnal heating effect is eliminated during the night-time samples and for wind speeds >6 ms^{-1}.

Additional uncertainties within the different mean discrepancies with in situ temperatures for each satellite could be related to the footprint sizes. Though a 5 × 5 pixel array from both satellites were used to compare with the in situ measurements, the differences in resolution causes the array to cover a different total area. The lower spatial resolution of the MTSAT-2 has higher probability of incorporating in situ errors when compared to the spatial resolution of Himawari-8.

Apart from the physical differences between satellite-measured SSTs and buoy measured temperatures at depth, in situ measurements have uncertainties that contribute to the differences. By considering the differences in temperatures measured by pairs of buoys at times of close approach, Emery et al. [57] concluded that the buoy temperatures have an uncertainty of 0.15 K. A subsequent analysis using three-way comparisons between two satellite-derived SSTs and temperatures from drifting buoys, a technique that allows an estimate of the uncertainty to be made for each data set, produced an estimate of the buoy temperature uncertainties of 0.23 K [58]. Other estimates of the uncertainties in temperatures measured from drifting buoys span the range of 0.12 K to 0.67 K [59] (Table 2).

Generally, the in situ measurements are prescreened to remove or note low quality observations. It was previously found that moored buoys had lower measurement uncertainties, whereas drifting buoys and ships introduce more noise [59]. When comparing satellite measurements of SST to in situ measurements, not all of the discrepancies can be assigned to errors in the satellite retrievals. However, the contributions from sources other than the satellite retrieval error should be very similar for both MTSAT-2 and Himawari-8 comparisons with in situ measurements, so the differences in the discrepancies are an indication of the changes in error and uncertainties in the SSTs derived from each satellite.

The data showed that Himawari-8 had an average SST difference of 0.18 ± 0.53 K, with an average median of 0.16 K. The MTSAT-2 had an average SST difference of 0.26 ± 0.48 K, with an average median of 0.27 K. Overall, the Himawari-8 AHI SSTs had smaller discrepancies with the in situ temperatures by an average of 0.08 K. When analyzing only night-time measurements, in which the effects of diurnal heating in the upper ocean should be small, the SST differences had a smaller variation for both satellites at all in situ locations. At times of higher wind speeds, there were also smaller variations within the SST discrepancies.

The large variations in the SST discrepancies were likely related to diurnal thermal stratification that occurs when the water column is not being mixed. High insolation during the day and wind speeds <6 ms^{-1} are conducive to the formation of thermal stratification [33]. Areas with wind speeds <6 ms^{-1} cover about 30% of the global ocean surface [33], but in areas of the TRITON moorings the fraction is much larger [60]; the amplitude of diurnal variability of wind speed is generally <0.4 ms^{-1} [61]. Thus, in the area of the TRITON buoys of the TAO moorings, it is likely that the conditions for the generation of diurnal heating will be met. Thus, the results of this study are consistent with the comparisons being influenced by diurnal heating, leading to an increase in the magnitudes of the SST discrepancies during the day. The expected wind-speed dependence in the diurnal heating signal is apparent in the satellite—in situ temperatures, with smaller discrepancies being seen at higher wind speeds during the day; this is most apparent in Himawari-8 AHI SST comparisons with GBR temperatures (Figure 10). However, distortion of temperature field in the upper ocean by TRITON buoys was found to cause temperature differences at a depth of 0.2 m below the water line on opposite sides of the TRITON buoy of up to 1 K in conditions of large diurnal heating, i.e., a diurnal heating amplitude of 2–3 K [62]. The effects of flow distortion around the buoys are strongly time dependent and thus contribute to the differences found here, but in a manner that is very difficult to quantify.

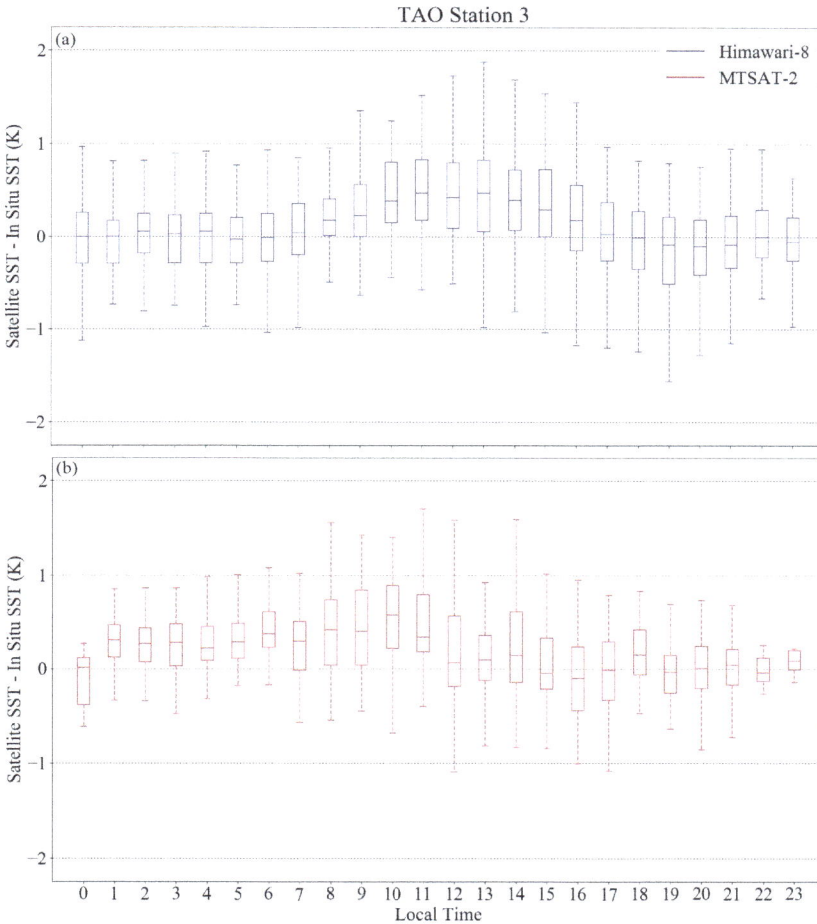

Figure 7. Box plots of the hourly differences between the satellite-derived SSTs and the subsurface temperatures measured at TAO Station 3 at 2°N, 137°E. Differences of SSTs from Hiawari-8 are shown in blue in (**a**), and from MTSAT-2 in red in (**b**). Outliers are not plotted. The data are of best quality level from 1 August 2015–31 October 2015.

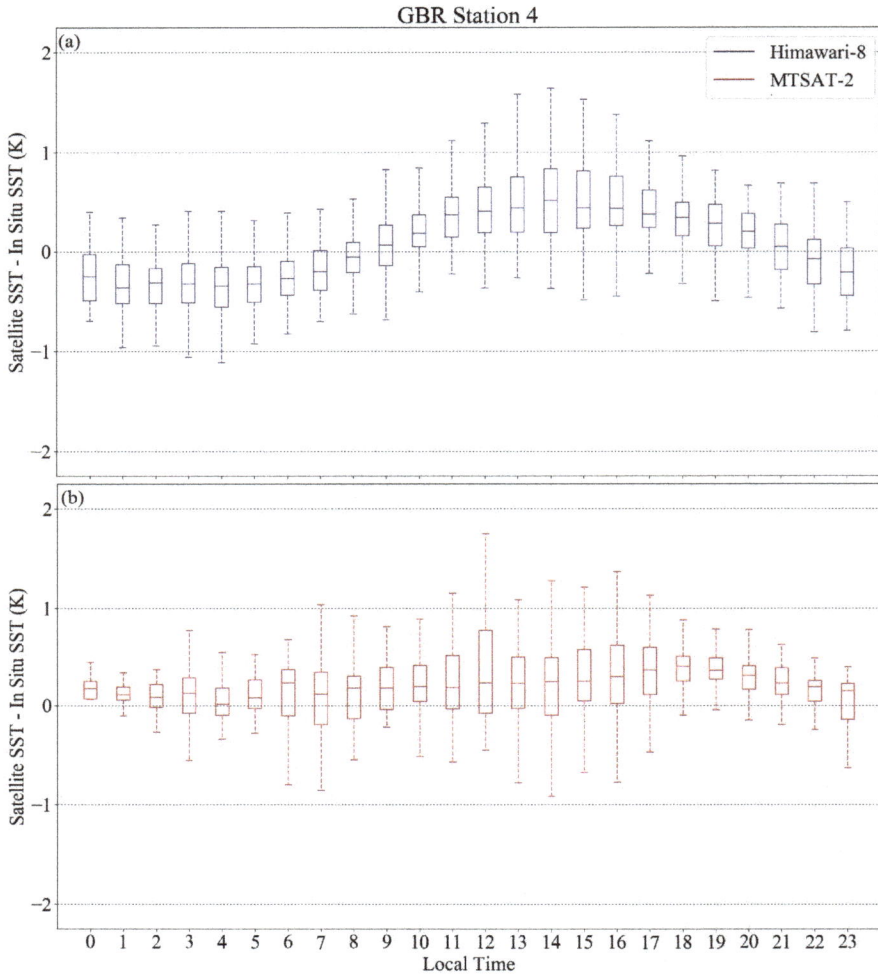

Figure 8. As Figure 7, but for GBR Station 4 at 18.49°S, 146.87°E with in situ temperatures measured at a depth of 1.9 m. Differences of SSTs from Hiawari-8 are shown in blue in (**a**), and from MTSAT-2 in red in (**b**).

Inaccuracies in the calibration of the infrared measurements on both MTSAT-2 and Himawari-8 radiometers could lead to bias errors in the satellite-data as errors in the brightness temperatures propagate through the atmospheric correction algorithms. Similarly, brightness temperature errors could compromise the effectiveness of the cloud screening algorithms leading to classification errors allowing pixels with cloud contamination to be misidentified as cloud-free. The objective of the Global Space-based Inter-Calibration System (GSICS) program is to assess the calibration accuracy of thermal infrared (IR) channels of imaging radiometers on geostationary satellites [63]. The reference sensors are the hyperspectral Infrared Atmospheric Sounding Interferometers (IASI) on the European polar-orbiting MetOp satellites [64]. The high resolution spectral measurements of IASI are convolved with the relative spectral response functions of the channels on the satellites on the geostationary satellites to allow comparison between the measurements [65].

GSICS comparisons of MTSAT-2 brightness temperatures and IASI spectral measurements indicate average differences of +0.08 K in the 10.8 μm channel and +0.10 K for the 12 μm channel, with small seasonal variations in the differences [65]. The seasonal fluctuations in the 3.8 μm channel differences are much more pronounced, reaching ~+0.4K around the vernal equinox and somewhat smaller at the autumnal equinox; around the solstices the differences are close to zero [65]. MTSAT-2 is a three-axis stabilized satellite and there is evidence of larger errors about local midnight when the entrance aperture of the imager faces to the sun, and stray solar radiation appears to degrade the calibration of the MTSAT-2 infrared channels [66]. The behavior of these midnight errors are similar to those found for the infrared channels of the imager on the Geostationary Operational Environmental Satellite (GOES -11 and -12; [67]). However, there is no significant evidence of this effect in our analysis.

Comparisons have been made between Himawari-8 AHI brightness temperatures and measurements of IASI on MetOp-A and MetOp-B, of the Atmospheric InfraRed Sounder (AIRS) on Aqua [68] and of the Cross-Track Infrared Sounder (CrIS) on the Suomi-NPP satellite [69]. Preliminary results indicate errors of <±0.1 K for the four bands used in the ACSPO atmospheric correction, Equation (3) [66]. However, the errors are positive for the measurements at 10.4 μm and 11.2 μm, but negative at 8.6 μm and at 12.4 μm. Thus, some of the brightness temperature difference terms in Equation (3) will have small effects on the retrieved temperatures, while those that combine the measurements with errors of opposite signs will make larger contributions. These calibration errors are based on analysis of only one month of data, from early in the Himawari-8 mission, and analysis of longer time series may lead to more confident estimates of the calibration errors.

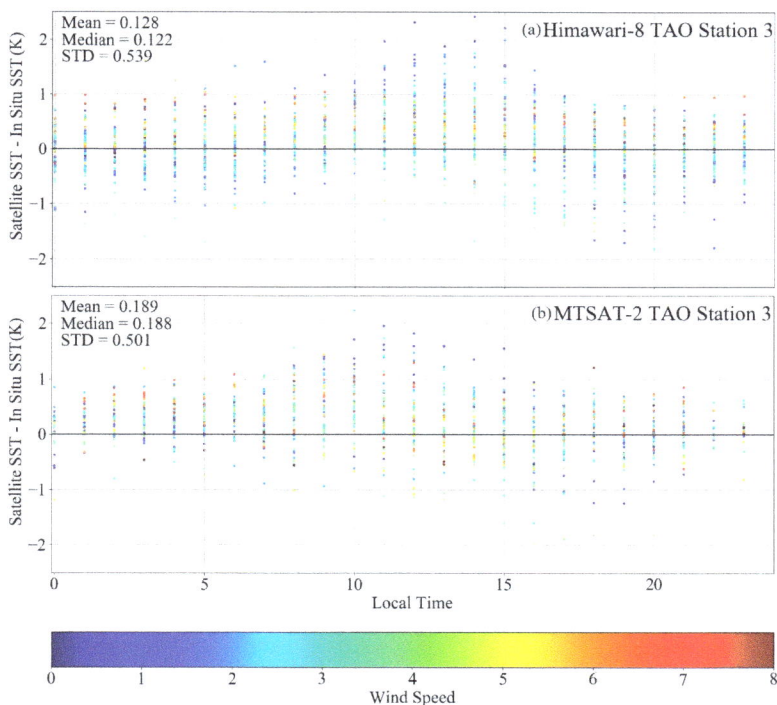

Figure 9. Differences between satellite-derived SST and in situ temperatures for each hour of the day at the TAO mooring 3 at 2°N, 137°E. The colors represent wind speed in ms^{-1}. The dots correspond to highest quality data from 1 August 2015–31 October 2015. The top panel is for SSTs from Himawari-8 and the lower panel for SSTs from MTSAT-2.

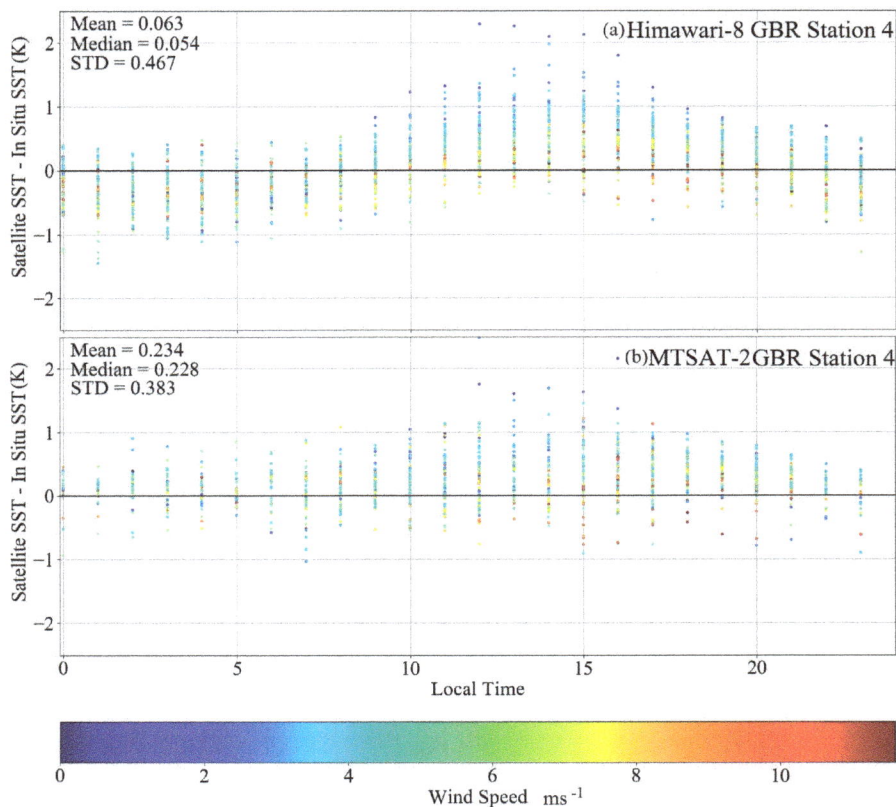

Figure 10. As Figure 9 but for GBR Station 4 at 18.49°S, 146.87°E with in situ temperatures measured at a depth of 1.9 m.

In a separate study comparing AHI observations and model simulations of brightness temperatures using radiative transfer simulations, Zou, Zhuge and Weng [35] found larger errors in some of the channels. These authors considered a wide geographic area over land as well as ocean, for satellite zenith angles up to 60°, and found a scene dependence of the bias errors, which may be indicative of an imperfect detector non-linearity correction in the calibration process [70]. But, Zou, Zhuge and Weng [35] acknowledge the imperfections in the radiative transfer models could have contributed to the larger estimates of the calibration errors.

5. Conclusions

This study was motivated by an interest to quantify the accuracies of SSTs in the tropical western Pacific Ocean derived from the AHI on Himawari-8, in particular how the SSTs derived from the new sensor have improved compared to those of the heritage sensor on MTSAT-2. The study was facilitated by a period of concurrent operation of both satellites. A particular application of the Himawari-8 AHI SSTs is the contribution they can make to monitoring and studying the health of the corals of the Great Barrier Reef, especially in conditions applying thermal stress to the corals, possibly leading to bleaching and mortality. Earlier work [31] had shown how well SSTs derived from a number of satellite infrared radiometers can represent the temperatures at the depths of the corals, and demonstrated the benefit of SSTs derived from geostationary satellites not only to better resolve the diurnal heating and cooling, but also to reduce obscuration by clouds. Based on the better spectral, temporal, and spatial resolution of the AHI, it was expected to show improved SST accuracy when compared with in situ measurements.

Based on the results of Kramar, Ignatov, Petrenko, Kihai, and Dash [40] we selected AHI SSTs derived using the NOAA ASCPO cloud screening and atmospheric algorithms, and the comparable MTSAT-2 SST data produced by NOAA. Comparison of SSTs from both satellites with subsurface temperatures from thermometers on the TRITON buoys of the TAO moorings and on the Great Barrier Reef, showed the Himawari-8 AHI produces SSTs that agree better with the in situ measurements, but the improvement is relatively modest. The Himawari-8 AHI SSTs appear better, in a qualitative sense, but the quantitative improvements were less pronounced when gross statistics are considered. However, the representation of diurnal variability in the Himawari-8 AHI SSTs is more physical than those in the SSTs from MTSAT-2. In addition to the expected regional bias errors inherent in atmospheric correction algorithms derived for application over larger geographic areas [71], there is evidence in the literature of on-board calibration issues that lead to inaccuracies in the Himawari-8 AHI brightness temperatures that no doubt contribute to the discrepancies reported here.

As more experience is gained with the Himawari-8 AHI data, corrections for the calibration problems will no doubt be found and we can expect improved accuracies in the AHI SSTs; this will benefit many scientific studies in this area. The Himawari-8 AHI SSTs will lead to better forecasts of typhoon activity off Japan's coast, helping protect the land and its people. In the Great Barrier Reef area, more accurate SSTs will lead to better monitoring the changing ocean temperatures that have harmful effects on the surrounding coral and ecosystems [11].

Acknowledgments: Angela Ditri undertook this research as a summer undergraduate intern at the Ocean Sciences Department of the Rosenstiel School. The GBR temperature time series were provided by the Australian Institute of Marine Science (AIMS), Townsville, Queensland, Australia, and the TRITON mooring data from the Pacific Marine Environmental Laboratory (PMEL) of the National Oceanic and Atmospheric Administration (NOAA). The MTSAT-2 and Himawari-8 AHI SSTs were extracted from the Center for Satellite Applications and Research (STAR) of the National Environmental Satellite, Data, and Information Service (NESDIS) (ftp://ftp.star.nesdis.noaa.gov). This study was funded in part by the NASA Physical Oceanography Program (NNX13AE30G). Angela Ditri is currently supported by the NASA Space Grant College and Fellowship Program (NASA Grant NNX15AI19H).

Author Contributions: Angela Ditri undertook the analysis and contributed to the writing of the manuscript. Peter Minnett conceived and directed the study and wrote most of the manuscript. Yang Liu and Katherine Kilpatrick provided guidance during the analyses. Ajoy Kumar was Angela Ditri's undergraduate supervisor and provided guidance during the analyses and the writing of the paper.

Conflicts of Interest: The authors declare no conflict of interest.

References

1. Robinson, I.S. *Measuring the Oceans from Space: The Principles and Methods of Satellite Oceanography*; Springer Science & Business Media: Berlin, Germany; New York, NY, USA, 2004; p. 670.
2. Schneider, T.; Bischoff, T.; Haug, G.H. Migrations and dynamics of the intertropical convergence zone. *Nature* **2014**, *513*, 45–53. [CrossRef] [PubMed]
3. Oki, T.; Kanae, S. Global Hydrological Cycles and World Water Resources. *Science* **2006**, *313*, 1068–1072. [CrossRef] [PubMed]
4. Clement, A.C. The Role of the Ocean in the Seasonal Cycle of the Hadley Circulation. *J. Atmos. Sci.* **2006**, *63*, 3351–3365. [CrossRef]
5. Wada, A.; Chan, J.C.L. Relationship between typhoon activity and upper ocean heat content. *Geophys. Res. Lett.* **2008**, *35*, 36–44. [CrossRef]
6. Benestad, R.E. On tropical cyclone frequency and the warm pool area. *Nat. Hazards Earth Syst. Sci.* **2009**, *9*, 635–645. [CrossRef]
7. Zhang, Q.; Liu, Q.; Wu, L. Tropical Cyclone Damages in China 1983–2006. *Bull. Am. Meteorol. Soc.* **2009**, *90*, 489–495. [CrossRef]

8. Dunn, S.R.; Thomason, J.C.; Le Tissier, M.D.A.; Bythell, J.C. Heat stress induces different forms of cell death in sea anemones and their endosymbiotic algae depending on temperature and duration. *Cell Death Differ.* **2004**, *11*, 1213–1222. [CrossRef] [PubMed]

9. Fitt, W.; Brown, B.; Warner, M.; Dunne, R. Coral bleaching: Interpretation of thermal tolerance limits and thermal thresholds in tropical corals. *Coral Reefs* **2001**, *20*, 51–65. [CrossRef]

10. Rogers, J.S.; Monismith, S.G.; Koweek, D.A.; Torres, W.I.; Dunbar, R.B. Thermodynamics and hydrodynamics in an atoll reef system and their influence on coral cover. *Limnol. Oceanogr.* **2016**, *61*, 2191–2206. [CrossRef]

11. Berkelmans, R.; De'ath, G.; Kininmonth, S.; Skirving, W.J. A comparison of the 1998 and 2002 coral bleaching events on the Great Barrier Reef: Spatial correlation, patterns, and predictions. *Coral Reefs* **2004**, *23*, 74–83. [CrossRef]

12. Hughes, T.P.; Kerry, J.T.; Álvarez-Noriega, M.; Álvarez-Romero, J.G.; Anderson, K.D.; Baird, A.H.; Babcock, R.C.; Beger, M.; Bellwood, D.R.; Berkelmans, R.; et al. Global warming and recurrent mass bleaching of corals. *Nature* **2017**, *543*, 373–377. [CrossRef] [PubMed]

13. Donner, S.D.; Skirving, W.J.; Little, C.M.; Oppenheimer, M.; Hoegh-Guldberg, O.V.E. Global assessment of coral bleaching and required rates of adaptation under climate change. *Glob. Chang. Biol.* **2005**, *11*, 2251–2265. [CrossRef]

14. Kemp, D.W.; Colella, M.A.; Bartlett, L.A.; Ruzicka, R.R.; Porter, J.W.; Fitt, W.K. Life after cold death: Reef coral and coral reef responses to the 2010 cold water anomaly in the Florida Keys. *Ecosphere* **2016**, *7*, e01373. [CrossRef]

15. Hoegh-Guldberg, O.; Fine, M. Low temperatures cause coral bleaching. *Coral Reefs* **2004**, *23*, 444. [CrossRef]

16. Kavousi, J.; Parkinson, J.E.; Nakamura, T. Combined ocean acidification and low temperature stressors cause coral mortality. *Coral Reefs* **2016**, *35*, 903–907. [CrossRef]

17. Langdon, C.; Atkinson, M.J. Effect of elevated pCO_2 on photosynthesis and calcification of corals and interactions with seasonal change in temperature/irradiance and nutrient enrichment. *J. Geophys. Res.* **2005**, *110*, C09S07. [CrossRef]

18. Veron, J.E.N.; Devantier, L.M.; Turak, E.; Green, A.L.; Kininmonth, S.; Stafford-Smith, M.; Peterson, N. Delineating the Coral Triangle. *Galaxea J. Coral Reef Stud.* **2009**, *11*, 91–100. [CrossRef]

19. Barber, P.H. The challenge of understanding the Coral Triangle biodiversity hotspot. *J. Biogeogr.* **2009**, *36*, 1845–1846. [CrossRef]

20. Great Barrier Reef Marine Park Authority. *Interim Report: 2016 Coral Bleaching Event on the Great Barrier Reef. Preliminary Findings of a Rapid Ecological Impact Assessment and Summary of Environmental Monitoring and Incident Response*; Great Barrier Reef Marine Park Authority: Townsville, Australia, 2016; p. 27.

21. Cave, D.; Gillis, J. Large Sections of Australia's Great Reef Are Now Dead, Scientists Find. *New York Times*, 15 March 2017.

22. Eakin, C.M.; Liu, G.; Gomez, A.M.; De La Cour, J.L.; Heron, S.F.; Skirving, W.J.; Geiger, E.F.; Tirak, K.V.; Strong, A.E. Global Coral Bleaching 2014–2017: Status and an Appeal for Observations. *Reef Encount.* **2016**, *43*, 20–26.

23. Ainsworth, T.D.; Heron, S.F.; Ortiz, J.C.; Mumby, P.J.; Grech, A.; Ogawa, D.; Eakin, C.M.; Leggat, W. Climate change disables coral bleaching protection on the Great Barrier Reef. *Science* **2016**, *352*, 338–342. [CrossRef] [PubMed]

24. Reverdin, G.; Boutin, J.; Martin, N.; Lourenco, A.; Bouruet-Aubertot, P.; Lavin, A.; Mader, J.; Blouch, P.; Rolland, J.; Gaillard, F.; et al. Temperature Measurements from Surface Drifters. *J. Atmos. Ocean. Technol.* **2010**, *27*, 1403–1409. [CrossRef]

25. Zhang, H.-M.; Reynolds, R.W.; Lumpkin, R.; Molinari, R.; Arzayus, K.; Johnson, M.; Smith, T.M. An Integrated Global Observing System for Sea Surface Temperature Using Satellites and in Situ Data: Research to Operations. *Bull. Am. Meteorol. Soc.* **2009**, *90*, 31–38. [CrossRef]

26. Lumpkin, R.; Özgökmen, T.; Centurioni, L. Advances in the application of surface drifters. *Annu. Rev. Mar. Sci.* **2017**, *9*, 59–81. [CrossRef] [PubMed]

27. Lumpkin, R.; Maximenko, N.; Pazos, M. Evaluating Where and Why Drifters Die. *J. Atm. Ocean. Technol.* **2012**, *29*, 300–308. [CrossRef]

28. Elipot, S.; Lumpkin, R.; Perez, R.C.; Lilly, J.M.; Early, J.J.; Sykulski, A.M. A global surface drifter data set at hourly resolution. *J. Geophys. Res. Ocean.* **2016**, *121*, 2937–2966. [CrossRef]

29. Ando, K.; Kuroda, Y.; Fujii, Y.; Fukuda, T.; Hasegawa, T.; Horii, T.; Ishihara, Y.; Kashino, Y.; Masumoto, Y.; Mizuno, K.; et al. Fifteen years progress of the TRITON array in the Western Pacific and Eastern Indian Oceans. *J. Oceanogr.* **2017**, *73*, 403–426. [CrossRef]

30. Zhu, X.; Minnett, P.J.; Berkelmans, R.; Hendee, J.; Manfrino, C. Diurnal warming in shallow coastal seas: Observations from the Caribbean and Great Barrier Reef regions. *Cont. Shelf Res.* **2014**, *82*, 85–98. [CrossRef]

31. Zhu, X.; Minnett, P.J.; Beggs, H.; Berkelmans, R. Thermal features and diurnal warming at the Great Barrier Reef derived from satellite data. *Remote Sens. Environ.* **2018**. in review.

32. Donlon, C.J.; Robinson, I.; Casey, K.S.; Vazquez-Cuervo, J.; Armstrong, E.; Arino, O.; Gentemann, C.; May, D.; LeBorgne, P.; Piollé, J.; et al. The Global Ocean Data Assimilation Experiment High-resolution Sea Surface Temperature Pilot Project. *Bull. Am. Meteorol. Soc.* **2007**, *88*, 1197–1213. [CrossRef]

33. Donlon, C.J.; Minnett, P.J.; Gentemann, C.; Nightingale, T.J.; Barton, I.J.; Ward, B.; Murray, J. Toward improved validation of satellite sea surface skin temperature measurements for climate research. *J. Clim.* **2002**, *15*, 353–369. [CrossRef]

34. Kurihara, Y.; Murakami, H.; Kachi, M. Sea surface temperature from the new Japanese geostationary meteorological Himawari-8 satellite. *Geophys. Res. Lett.* **2016**, *43*, 1234–1240. [CrossRef]

35. Zou, X.; Zhuge, X.; Weng, F. Characterization of Bias of Advanced Himawari Imager Infrared Observations from NWP Background Simulations Using CRTM and RTTOV. *J. Atmos. Ocean. Technol.* **2016**, *33*, 2553–2567. [CrossRef]

36. Petrenko, B.; Ignatov, A.; Kihai, Y.; Dash, P. Sensor-Specific Error Statistics for SST in the Advanced Clear-Sky Processor for Oceans. *J. Atmos. Ocean. Technol.* **2016**, *33*, 345–359. [CrossRef]

37. Walton, C.C.; Pichel, W.G.; Sapper, J.F.; May, D.A. The development and operational application of nonlinear algorithms for the measurement of sea surface temperatures with the NOAA polar-orbiting environmental satellites. *J. Geophys. Res.* **1998**, *103*, 27999–28012. [CrossRef]

38. Liang, X.; Ignatov, A.; Kramar, M.; Yu, F. Preliminary Inter-Comparison between AHI, VIIRS and MODIS Clear-Sky Ocean Radiances for Accurate SST Retrievals. *Remote Sens.* **2016**, *8*, 203. [CrossRef]

39. Bessho, K.; Date, K.; Hayashi, M.; Ikeda, A.; Imai, T.; Inoue, H.; Kumagai, Y.; Miyakawa, T.; Murata, H.; Ohno, T.; et al. An Introduction to Himawari-8/9—Japan's New-Generation Geostationary Meteorological Satellites. *J. Meteorol. Soc. Jpn. Ser. II* **2016**, *94*, 151–183. [CrossRef]

40. Kramar, M.; Ignatov, A.; Petrenko, B.; Kihai, Y.; Dash, P. Near Real Time SST Retrievals from Himawari-8 at NOAA Using ACSPO System. In Proceedings of the Ocean Sensing and Monitoring VIII, Baltimore, MD, USA, 17–21 April 2016; Arnone, R.A., Hou, W.W., Eds.; SPIE: Baltimore, MD, USA, 2016; p. 98270L.

41. Liang, X.-M.; Ignatov, A.; Kihai, Y. Implementation of the Community Radiative Transfer Model in Advanced Clear-Sky Processor for Oceans and validation against nighttime AVHRR radiances. *J. Geophys. Res. Atmos.* **2009**, *114*. [CrossRef]

42. Brasnett, B.; Surcel-Colan, D. Assimilating Retrievals of Sea Surface Temperature from VIIRS and AMSR2. *J. Atmos. Ocean. Technol.* **2016**, *33*, 361–375. [CrossRef]

43. Xu, F.; Ignatov, A. In situ SST Quality Monitor (iQuam). *J. Atmos. Ocean. Technol.* **2014**, *31*, 164–180. [CrossRef]

44. Kawai, Y.; Kawamura, H.; Tanba, S.; Ando, K.; Yoneyama, K.; Nagahama, N. Validity of sea surface temperature observed with the TRITON buoy under diurnal heating conditions. *J. Oceanogr.* **2006**, *62*, 825–838. [CrossRef]

45. Wolanski, E. *Physical Oceanographic Processes of the Great Barrier Reef*; CRC Press: Boca Raton, FL, USA, 1994; p. 208.

46. Zhang, H.; Beggs, H.; Majewski, L.; Wang, X.H.; Kiss, A. Investigating sea surface temperature diurnal variation over the Tropical Warm Pool using MTSAT-1R data. *Remote Sens. Environ.* **2016**, *183*, 1–12. [CrossRef]

47. Gentemann, C.L.; Minnett, P.J. Radiometric measurements of ocean surface thermal variability. *J. Geophys. Res.* **2008**, *113*, C08017. [CrossRef]

48. Saha, S.; Moorthi, S.; Wu, X.; Wang, J.; Nadiga, S.; Tripp, P.; Behringer, D.; Hou, Y.-T.; Chuang, H.-Y.; Iredell, M.; et al. The NCEP Climate Forecast System Version 2. *J. Clim.* **2014**, *27*, 2185–2208. [CrossRef]

49. Ward, B. Near-Surface Ocean Temperature. *J. Geophys. Res.* **2006**, *111*, C02005. [CrossRef]

50. Corlett, G.K.; Merchant, C.J.; Minnett, P.J.; Donlon, C.J. Assessment of Long-Term Satellite Derived Sea Surface Temperature Records. In *Experimental Methods in the Physical Sciences, Optical Radiometry for Ocean Climate Measurements*; Zibordi, G., Donlon, C.J., Parr, A.C., Eds.; Academic Press: Cambridge, MA, USA, 2014; Volume 47, pp. 639–677.

51. Working Group 1 of the Joint Committee for Guides in Metrology. *Evaluation of Measurement Data—Guide to the Expression of Uncertainty in Measurement*; BIPM: Sèvres, France, 2008; p. 134.

52. Bell, S. *A Beginner's Guide to Uncertainty of Measurement*; National Physical Laboratory: Teddington, UK, 2001; p. 41.

53. Kilpatrick, K.A.; Podestá, G.; Walsh, S.; Williams, E.; Halliwell, V.; Szczodrak, M.; Brown, O.B.; Minnett, P.J.; Evans, R. A decade of sea surface temperature from MODIS. *Remote Sens. Environ.* **2015**, *165*, 27–41. [CrossRef]

54. Embury, O.; Merchant, C.J.; Corlett, G.K. A reprocessing for climate of sea surface temperature from the along-track scanning radiometers: Initial validation, accounting for skin and diurnal variability effects. *Remote Sens. Environ.* **2012**, *116*, 62–78. [CrossRef]

55. Merchant, C.J.; Harris, A.R. Toward the elimination of bias in satellite retrievals of skin sea surface temperature. 2: Comparison with in situ measurements. *J. Geophys. Res.* **1999**, *104*, 23579–23590. [CrossRef]

56. Minnett, P.J. Consequences of sea surface temperature variability on the validation and applications of satellite measurements. *J. Geophys. Res.* **1991**, *96*, 18475–18489. [CrossRef]

57. Emery, W.J.; Baldwin, D.J.; Schlüssel, P.; Reynolds, R.W. Accuracy of in situ sea surface temperatures used to calibrate infrared satellite measurements. *J. Geophys. Res.* **2001**, *106*, 2387–2405. [CrossRef]

58. O'Carroll, A.G.; Eyre, J.R.; Saunders, R.W. Three-Way Error Analysis between AATSR, AMSR-E, and In Situ Sea Surface Temperature Observations. *J. Atmos. Ocean. Technol.* **2008**, *25*, 1197–1207. [CrossRef]

59. Kennedy, J.J. A review of uncertainty in in situ measurements and data sets of sea surface temperature. *Rev. Geophys.* **2014**, *51*, 1–32. [CrossRef]

60. Woods, S.; Minnett, P.J.; Gentemann, C.L.; Bogucki, D. Influence of the oceanic cool skin layer on global air–sea CO_2 flux estimates. *Remote Sens. Environ.* **2014**, *145*, 15–24. [CrossRef]

61. Dai, A.; Deser, C. Diurnal and semidiurnal variations in global surface wind and divergence fields. *J. Geophys. Res. Atmos.* **1999**, *104*, 31109–31125.

62. Kawai, Y.; Ando, K.; Kawamura, H. Distortion of Near-Surface Seawater Temperature Structure by a Moored-Buoy Hull and Its Effect on Skin Temperature and Heat Flux Estimates. *Sensors* **2009**, *9*, 6119–6130. [CrossRef] [PubMed]

63. Goldberg, M.; Ohring, G.; Butler, J.; Cao, C.; Datla, R.; Doelling, D.; Gärtner, V.; Hewison, T.; Iacovazzi, B.; Kim, D.; et al. The Global Space-Based Inter-Calibration System. *Bull. Am. Meteorol. Soc.* **2011**, *92*, 467–475. [CrossRef]

64. Blumstein, D.; Chalon, G.; Carlier, T.; Buil, C.; Hebert, P.; Maciaszek, T.; Ponce, G.; Phulpin, T.; Tournier, B.; Simeoni, D.; et al. IASI Instrument: Technical Overview and Measured Performances. In Proceedings of the Optical Science and Technology, the SPIE 49th Annual Meeting, Denver, CO, USA, 2–6 August 2004; Strojnik, M., Ed.; Volume 5543, pp. 196–207.

65. Hewison, T.J.; Wu, X.; Yu, F.; Tahara, Y.; Hu, X.; Kim, D.; Koenig, M. GSICS inter-calibration of infrared channels of geostationary imagers using Metop/IASI. *IEEE Trans. Geosci. Remote Sens.* **2013**, *51*, 1160–1170. [CrossRef]

66. Okuyama, A.; Andou, A.; Date, K.; Hoasaka, K.; Mori, N.; Murata, H.; Tabata, T.; Takahashi, M.; Yoshino, R.; Bessho, K. Preliminary Validation of Himawari-8/AHI Navigation and Calibration. In Proceedings of the Earth Observing Systems XX, San Diego, CA, USA, 10–13 August 2015; Butler, J.J., Jack, X., Gu, X., Eds.; p. 96072E.

67. Yu, F.; Wu, X.; Raja, M.R.V.; Li, Y.; Wang, L.; Goldberg, M. Diurnal and scan angle variations in the calibration of GOES imager infrared channels. *IEEE Trans. Geosci. Remote Sens.* **2013**, *51*, 671–683. [CrossRef]

68. Aumann, H.H.; Chahine, M.T.; Gautier, C.; Goldberg, M.D.; Kalnay, E.; McMillin, L.M.; Revercomb, H.; Rosenkranz, P.W.; Smith, W.L.; Staelin, D.H.; et al. AIRS/AMSU/HSB on the Aqua Mission: Design, science objectives, data products, and processing systems. *IEEE Trans. Geosci. Remote Sens.* **2003**, *41*, 253–264. [CrossRef]

69. Han, Y.; Revercomb, H.; Cromp, M.; Gu, D.; Johnson, D.; Mooney, D.; Scott, D.; Strow, L.; Bingham, G.; Borg, L.; et al. Suomi NPP CrIS measurements, sensor data record algorithm, calibration and validation activities, and record data quality. *J. Geophys. Res. Atmos.* **2013**, *118*, 12734–12748. [CrossRef]
70. Saunders, R.W.; Blackmore, T.A.; Candy, B.; Francis, P.N.; Hewison, T.J. Monitoring Satellite Radiance Biases Using NWP Models. *IEEE Trans. Geosci. Remote Sens.* **2013**, *51*, 1124–1138. [CrossRef]
71. Minnett, P.J. The regional optimization of infrared measurements of sea-surface temperature from space. *J. Geophys. Res.* **1990**, *95*, 13497–13510. [CrossRef]

remote sensing

MDPI

Article

Role of El Niño Southern Oscillation (ENSO) Events on Temperature and Salinity Variability in the Agulhas Leakage Region

Morgan L. Paris * and Bulusu Subrahmanyam

School of the Earth, Ocean, and Environment, University of South Carolina, 701 Sumter Street,
Columbia, SC 29208, USA; sbulusu@geol.sc.edu
* Correspondence: mparis@email.sc.edu; Tel.: +01-937-974-1515

Received: 17 November 2017; Accepted: 16 January 2018; Published: 18 January 2018

Abstract: This study explores the relationship between the Agulhas Current system and El Niño Southern Oscillation (ENSO) events. Specifically, it addresses monthly to yearly variations in Agulhas leakage where the Agulhas Current sheds waters into the Atlantic Ocean, in turn affecting meridional overturning circulation (MOC). Sea surface temperature (SST) data from the National Oceanic and Atmospheric Administration's (NOAA) Advanced Very High Resolution Radiometer (AVHRR) combined with sea surface salinity (SSS) from Soil Moisture Ocean Salinity (SMOS) and Simple Ocean Data Assimilation (SODA) reanalysis are used to explore changes in Agulhas leakage dynamics. Agulhas leakage is anomalously warm in response to El Niño and anomalously cool in response to La Niña. The corresponding SSS signal shows both a primary and secondary signal response. At first, the SSS signal of Agulhas leakage is anomalously fresh in response to El Niño, but this primary signal is replaced by a secondary anomalously saline signal. In response to La Niña, the primary SSS signal of Agulhas leakage is anomalously saline, while the secondary SSS signal is anomalously fresh. The lag between the peak of ENSO and the response in SST and the corresponding primary SSS signal of Agulhas leakage is about 20 months, followed by the secondary SSS signal at a lag of about 26 months. In general, increasing ENSO strength increases the extremes of the resulting anomalous SST and SSS signal and impacts the Agulhas leakage region earlier during El Niño and slightly later during La Niña.

Keywords: Agulhas Current; Indian Ocean; sea surface temperature; sea surface salinity; El Niño Southern Oscillation; Simple Ocean Data Assimilation (SODA); Soil Moisture Ocean Salinity (SMOS); Advanced Very High Resolution Radiometer (AVHRR)

---

1. Introduction

The Agulhas Current, a western boundary current, is a limb of the wind-driven anti-cyclonic circulation of the South Indian Ocean. The current originates south of Madagascar forming a narrow flow stabilized by east Africa's steep continental slope. Past the southern tip of Africa, the flow retroflects eastward as the Agulhas Return Current forming the southern arm of the Indian Ocean subtropical gyre which is a part of the Southern Hemisphere super gyre [1]. A phenomenon known as Agulhas leakage occurs at the area of retroflection and transports warm saline water into the Atlantic through the shedding of Agulhas rings, cyclones, and filaments. This system feeds the upper arm of the Atlantic meridional overturning circulation (AMOC). Variability in leakage may impact the strength of overturning sequentially altering climate patterns [2]. Fluctuations in the strength of Agulhas leakage are controlled by long-term and short-term fluctuations in the Agulhas Current and source current dynamics [3]. This paper uses sea surface temperature (SST) and sea surface salinity (SSS) to explore the influence of the El Niño-Southern Oscillation (ENSO) on the Agulhas Current system and, ultimately, Agulhas leakage. Previous work by Biastoch et al. [4] established a link between Agulhas leakage and changes in heat and salt transports into the Atlantic. Specifically, Agulhas rings are distinguishable from surrounding waters by their high salt content derived from strong evaporation occurring in the retroflection region which boosts the salinity within the rings [5]. In other words, a close relationship exists between SST, SSS, and circulation supporting our use of SST and SSS as a proxy for Agulhas leakage.

The aim of the study is to define the relationship between ENSO and Agulhas leakage in terms of SST and SSS response. This relationship cannot be fully understood without first connecting the influence of the ENSO signal across the three ocean basins involved: the Pacific Ocean, the Indian Ocean, and the Atlantic Ocean. In terms of global circulation, the Indian Ocean acts as the link between the Pacific and the Atlantic Ocean, contributing nearly 12.6 Sv to Agulhas leakage, of which, about 7.9 Sv originates from the Pacific moving into the Atlantic [6]. Nearly half of the Indian Ocean contribution to Agulhas leakage comes from the Indonesian Throughflow (ITF) with a smaller portion originating south of Australia by Tasman leakage [6]. The ITF has been found to increase during La Niña and decrease during El Niño [7]. An analysis by Le Bars et al. [8] suggests changes to ITF strength influences Agulhas leakage because the two currents are codependent. Within the Indian Ocean basin, the westward flowing South Equatorial Current (SEC) circulates water from the Indian Ocean subtropical gyre and ITF to the Madagascar coast. This westward transport of water between 60°E and 100°E is modeled at mean speeds of ~0.1 m·s^{-1}, taking ~1.3 years for waters from the ITF to reach 77°E [6]. Upon reaching the Madagascar coast, the SEC splits at 17°S into a northern and southern branch. The southern branch feeds into the East Madagascar Current (EMC) while the northern branch bifurcates against the African coast into the Mozambique Channel (MC) [9]. A shift in the intensity and position of the tropical and subtropical gyre in response to positive (negative) SSH anomalies associated with ENSO wind anomalies (see next paragraph) changes the intensity of the SEC, thus altering flow through the MC and EMC [10]. The EMC sheds eddies near the tip of Madagascar [11], contributing ~25 Sv to the Agulhas Current [9]. The MC consists of a train of westward flowing eddies [11] which contribute ~5 Sv to the Agulhas Current [3].

Furthermore, the formation of eddies in the EMC and MC can be related to incoming Rossby waves crossing the Indian Ocean (see next paragraph) [12], and during El Niño years more eddies are released [11]. While the exact mechanics driving Agulhas leakage are still highly debated, a robust link has been identified between these eddies and the westward shift of the retroflection loop as well as the generation of a "Natal Pulse", a large solitary meander in the current that progresses downstream to influence retroflection dynamics. Recent research suggests that although Agulhas Current meanders may not be the dominant mode of variance, they destabilize the flow, causing increased Agulhas leakage events [12,13]. It is important to note that De Ruijter et al. [11] traced the propagation of eddies from south of Madagascar at 5–10 cm·s^{-1}. Therefore, it takes approximately 6 months after formation for eddies from the MC and EMC to influence Agulhas leakage.

A clear connection established by circulation patterns links the Pacific Ocean to the Indian Ocean into the Atlantic Ocean. We are interested in the processes that alter this system to explain why we are seeing the anomalous SST and SSS patterns highlighted in this paper. However, the mechanisms involved in ENSO signal propagation have yet to be deciphered. For this study, the work of Putrasahan et al. [10] is used to define the proposed process by which an ENSO signal originating in the Pacific Ocean propagates into the Indian Ocean basin and ultimately alters the properties of Agulhas leakage. During the mature season of El Niño (La Niña), fluctuations of Walker circulation cause anomalous easterly winds (strong westerly) winds to form over Indonesia, generating upwelling (downwelling) Kelvin waves. Anomalous easterly (westerly) winds actively suppress (enhance) convection, causing a basin-wide warming (cooling) trend [14]. The wind anomalies over Indonesia combined with Ekman pumping generate off-equatorial Rossby waves that travel westward. Note, this process explains the previously mentioned ENSO-associated SSH anomalies that Palastanga et al. [15] found to be influencing the SEC, further impacting the MC and EMC. The ENSO signal also enters the Indian Ocean along the western Australian coast by a pathway known as the subtropical North Pacific ray-path. North Pacific Rossby waves generated during ENSO events impinge on the western boundary and move equatorward along the "ray-path" of Kelvin–Munk waves to reflect as equatorial Kelvin waves. When the reflected Kelvin waves impinge upon the Australian continent they become coastally trapped and move poleward along the coast, where they radiate Rossby waves into the south Indian Ocean [16]. Ultimately, an ENSO event triggers two sets of westward-propagating Rossby waves at 12°S and 25°S from wind forcing and Kelvin waves, respectively. As previously mentioned, the eddy activity of the EMC and MC are influenced by Rossby-wave propagation at 25°S and 12°S, respectively [12].

Rossby waves alone cannot explain signal propagation. A second parameter, wind stress, also plays an important role. The previously described anomalous wind and SST conditions in the Indian Ocean basin that form in response to ENSO are correlated with a wind-stress anomaly along the equator [17]. This is further supported by the strong correlation present between weakened trade winds in the Pacific, a characteristic of an El Niño, and strengthened trade winds in the tropical Indian Ocean. Strengthened trade winds along the tropical Indian Ocean create a zonal band of positive wind stress curl over the tropics, forcing the continued propagation of the Rossby waves at 25°S and 12°S [10]. Using SSH, Putrasahan et al. [10] was able to correlate SST anomalies of Agulhas leakage to wind stress and found a lag of approximately 2 years. In other words, it takes approximately 2 years for tropical warm anomalies formed from El Niño-associated wind anomalies to reach the Agulhas leakage region. This is relatively consistent with the earlier mentioned time scales of ocean circulation, where it takes a little more than ~1.3 years for waters to cross the Indian Ocean basin [6] and then ~6 months for eddies from EMC and MC to interact with Agulhas leakage [11].

Our study aims to define the relationship between ENSO events and SST and SSS variability in the Agulhas leakage region. In other words, the results presented in this paper are intended to describe the observed effects of ENSO on Agulhas leakage, focusing on defining the relationship itself rather than determining the various driving mechanisms of signal propagation. The previous paragraphs highlight the potential mechanisms of signal propagation serving as evidence and support the notion that such a relationship between Agulhas leakage and ENSO exists. Our study is predominantly important with respect to SSS because the response of SSS in the Agulhas leakage region to ENSO is a novel topic yet to be understood. Newly launched satellite-derived salinity missions used in this study, such as the National Aeronautics and Space Administration's (NASA) Soil Moisture Active Passive (SMAP) and the European Space Agency's (ESA) Soil Moisture and Ocean Salinity (SMOS), are an innovative approach to studying SSS. In respect to SST, Putrasahan et al. [10] established a link between the interannual variability of SST of Agulhas leakage and ENSO, then determined the lag in response of Agulhas leakage to be about 2 years. However, the Putrasahan et al. [10] study does not specifically investigate the difference between El Niño and La Niña but rather relies on a correlation analysis to distinguish between the phases and the influence of ENSO strength. The results presented in this paper further the work done by Putrasahan et al. [10] by evaluating the SST signal propagation from a different perspective and distinguishing El Niño events from La Niña events. This paper uses spatial plots to illustrate the entire propagation of an ENSO signal, from where the SST and SSS signal originates to movement of the signal across the Indian Ocean basin to surround the source currents and ultimately signal interaction with the Agulhas current system, changing Agulhas leakage SST and SSS properties.

2. Materials and Methods

ENSO events were determined using the Oceanic Niño Index (ONI) obtained from the National Weather Service and Climate Prediction website. The SST anomalies used to calculate the ONI are from the Extended Reconstructed Sea Surface Temperature (ERSST) version 4 dataset derived from the International Comprehensive Ocean-Atmosphere Dataset (ICOADS). Threshold values were calculated from the anomalies in the Niño 3.4 region (5°N–5°S, 120°–170°W) with an applied 3-month running mean and are based on a centered 30-year base period updated every 5 years.

Satellite-derived measurements from the Advanced Very High Resolution Radiometer (AVHRR) by the National Oceanic and Atmospheric Administration (NOAA) were the sole source of SST data used to interpret temperature trends in the Indian Ocean and Agulhas leakage region. The data included in this study is entitled NOAA NCEI OISST (version 2) daily SST data and was obtained from http://iridl.ldeo.columbia.edu/SOURCES/.NOAA/.NCDC/.OISST/.version2/.AVHRR/.sst/. The data includes a combination of both AVHRR and in-situ data for optimal interpolation. The data set spans from October 1981 to December 2015 with daily intervals at 0.25° × 0.25° spatial resolution that we converted to monthly averages. AVHRR is appropriate for the purposes of this study because of this long time span of data coverage that allows the majority of previous ENSO events to be evaluated; 21 ENSO events occur between 1981 and 2015. Furthermore, the data set is bias-corrected to achieve a uniform performance throughout a wide range of atmospheric and oceanic conditions. Comparisons with in-situ buoys indicate that the global accuracy of current Pathfinder algorithm is 0.02° ± 0.5 °C [18]. Satellite-derived salinity measurements from Soil Moisture Ocean Salinity (SMOS) and model-based products from Simple Ocean Data Assimilation (SODA) reanalysis were used to evaluate SSS. SMOS version 2.0 level 3 monthly SSS data at a 0.25° × 0.25° spatial resolution was obtained from Barcelona Expert Centre (http://bec.icm.csic.es/). This data set spans from January 2010 to June 2016 and is still operational. SODA version 3.3.1 reanalysis is obtained from the Asia Pacific Data Research Center (APDRC) at a monthly temporal resolution and a 0.25° × 0.25° spatial resolution spanning from 1980–2015.

This study classifies all ENSO events between 1981 and 2015 as an El Niño if the ONI value was at or exceeding a +0.5° anomaly threshold for 3 consecutive months, and a La Niña if the ONI value was at or below a −0.5° anomaly for 3 consecutive months. Any remaining years are considered to be neutral. This classification process is consistent with that used by NOAA's Climate Prediction Center (http://www.cpc.ncep.noaa.gov/). The threshold is further divided into weak events with a 0.5°–0.9° anomaly range, moderate events with 1.0°–1.4° anomaly range, and strong events with anomalous values greater or equal to 1.5° established by Jan Null at Golden Gate Weather Services (http://ggweather.com/enso/oni.htm). Table 1 represents the strong, weak, and moderate classifications of La Niña and El Niño. The months of January–March were selected to represent the relationship between ENSO and leakage dynamics because it is during the mature season of ENSO (December–March) [19].

Table 1. El Niño Southern Oscillation (ENSO) phase years between 1981–2015.

El Niño	La Niña	Neutral	
1982–1983 [1]	1983–1984 [2]	1980	2003
1986–1987	1984–1985 [2]	1981	2005
1987–1988	1988–1989 [1]	1985	2008
1991–1992	1995–1996 [2]	1989	2011
1994–1995 [2]	1998–1999	1990	2012
1997–1998 [1]	1999–2000	1992	2013
2002–2003	2000–2001 [2]	1993	2014
2004–2005 [2]	2007–2008	1995	
2006–2007 [2]	2010–2011 [2]	1996	
2009–2010	2011–2012	1998	
2015–2016 [1]		2001	

[1] Strong ENSO event; [2] weak ENSO event.

The SST interannual anomalies used throughout the study were obtained by computing the average monthly SST from the full AVHRR data set (1981–2015) and subtracting them from the monthly average of a given year. The same process was used for SSS anomalies except using SODA reanalysis from 1980–2015 and SMOS data from 2010–2016. In other words, the mean seasonal cycle was removed.

We defined the Agulhas leakage region based on the location in which prevalent transport of warm saline waters is observed. This region spans from the tip of the African continental shelf to the oceanic subtropical front (37°–45°S) [5] and has a western limit established by the Good Hope transect [20] and an eastern limit at the point of retroflection (10°–20°E) [14]. The box-averaged SST and SSS of this region were obtained to create a time series of SST and SSS changes in the Agulhas leakage region to further represent possible changes in Agulhas leakage dynamics. Note that Dencausse et al. [21] defines the Agulhas retroflection loop as having an average position at 18°E meaning that the retroflection loop is present within our defined box for the Agulhas leakage region. However, the results of box-averaged time series are not observed as being largely skewed by the retroflection signal because the position of the retroflection loop is highly variable, and using the box-average mitigates the influence of extreme values that may come from retroflection interaction within the defined Agulhas leakage region.

A Pearson Product-Moment Correlation Coefficient analysis was performed to obtain Figure 11 representing the lag between the peak of ENSO signals and box-averaged SST and SSS at the point of Agulhas retroflection. The peak in ENSO signal was defined as the average ONI value during the peak (December–March) of defined El Niño or La Niña years. This was correlated with the 3-month running mean of box-averaged SST and SSS at monthly lag intervals from corresponding El Niño or La Niña years.

3. Results

3.1. Sea Surface Temperature (SST) and Sea Surface Salinity (SSS) Signal Response to El Niño and La Niña

Figures 1 and 2 illustrate the average AVHRR SST and SODA SSS anomalies during the peak, the following year, and two years after the peak (January to March) of all La Niña years and El Niño years, respectively (listed in Table 1). Figure 3 is an extension of Figures 1 and 2 showing the continuation of the SSS signal in the Agulhas leakage region in the months beyond two years following the peak of ENSO to up to three years following the peak of ENSO.

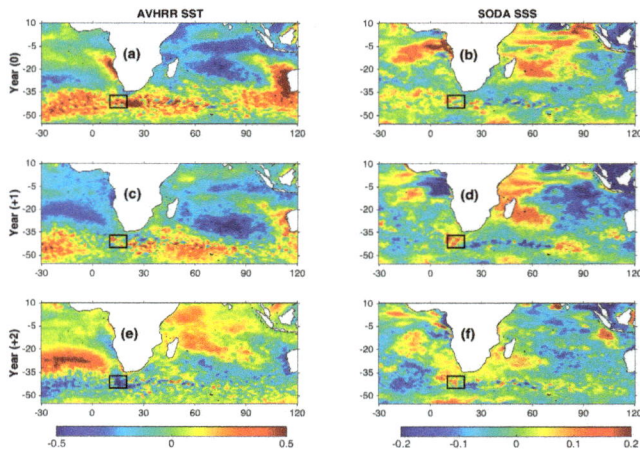

Figure 1. Composite mean for all La Niña events listed in Table 1 from January–March for both Advanced Very High Resolution Radiometer (AVHRR) sea surface temperature (SST) anomalies (left column) and Simple Ocean Data Assimilation (SODA) sea surface salinity (SSS) anomalies (right column) during the peak (**a,b**), the following year (**c,d**), and two years after (**e,f**). The boxes represent the established leakage region (37°–45°S, 10°–20°E).

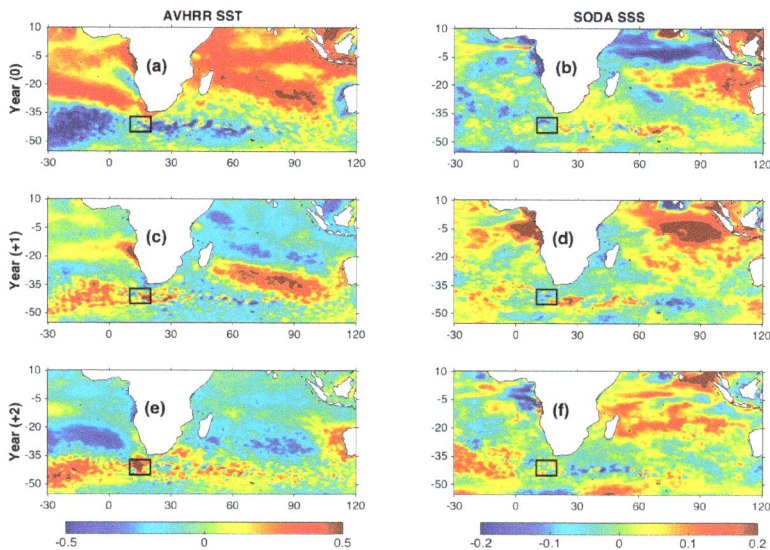

Figure 2. Same as Figure 1 except for all El Niño events listed in Table 1 during the peak (**a,b**), the following year (**c,d**), and two years after (**e,f**).

During the peak of La Niña, SST anomalies in the Indian Ocean basin are dominantly cool, excluding warming off the west coast of Australia (Figure 1a). This is opposite of El Niño where SST anomalies are dominantly warm excluding cooling off the coast of Australia (Figure 2a). In respect to SST, there is a single dominant warming (cooling) signal during El Niño (La Niña). The corresponding SSS is composed of two simultaneously occurring but contrasting signals each reflecting an area of either dominantly fresh or saline SSS. Furthermore, the original region of these anomalously fresh and saline signals switch during El Niño and La Niña. In other words, La Niña shows high salinity in the north-west in the Equatorial Indian Ocean with low salinity in the south-east near the coast of Sumatra (Figure 1b), and El Niño shows low salinity in the north-west with high salinity in the south-east (Figure 2b). Throughout the paper, this dual SSS signal is referred to as a primary signal and a secondary signal. The primary SSS signal is the region of anomalously low (high) SSS originating in the north-west Equatorial Indian Ocean during El Niño (La Niña). The secondary SSS signal refers to the anomalously high (low) SSS occurring simultaneously in the south-east Indian Ocean near the coast of Sumatra. If a zero-lag correlation were to exist between Agulhas leakage and ENSO, Figures 1a,b and 2a,b would show that La Niña is related to warm saline leakage and El Niño is related to cool fresh leakage, respectively. However, the SST and SSS signal directly surrounding the African coast during the peak of ENSO does not correspond to the SST and SSS signal observed in the leakage region at this same time. Near the tip of the African coast, waters are warm and saline (cool and fresh) during the peak of El Niño (La Niña).

When evaluating the SST anomalies for one year following an ENSO event, the established warming (cooling) trend identified during the peak of El Niño (La Niña) is now isolated near the region east of Madagascar at ~15°–35°S (Figures 1c and 2c). The spatial variation of SSS identified during a peak ENSO event remains consistent into the next year, but the secondary signal is pushed further west and the primary signal surrounds Madagascar (Figures 1d and 2d). One year after the peak of El Niño (La Niña), the dominant positive (negative) SST and the primary fresh (saline) SSS signal surround Madagascar.

The initial impact on Agulhas leakage from the propagating dominant SST signal and primary SSS signal is not strongly observed until two years after the peak of ENSO (Figures 1e,f and 2e,f). At this time, the SST and SSS signal of Agulhas leakage exhibits opposite temperature and salinity trends for El Niño compared to La Niña. Two years after the peak of El Niño, Agulhas leakage is anomalously warm and fresh (Figure 2e,f). In contrast, two years after the peak of La Niña Agulhas leakage is anomalously cool and saline (Figure 1e,f). Note, this is the same signal that surrounded Madagascar the year prior. The initial primary fresh (saline) SSS signal in the Agulhas leakage region is replaced by the secondary saline (fresh) SSS signal, but the SST signal of Agulhas leakage remains consistent. The secondary SSS signal did not reach the Madagascan coast until April–June of two years after La Niña (Figure 3a) and January–March of two years after El Niño (Figure 2f). Therefore, interaction with Agulhas leakage by the secondary saline (fresh) SSS signal occurs as late as the end of two years (start of three years) following El Niño (La Niña) (Figure 3d,f,g).

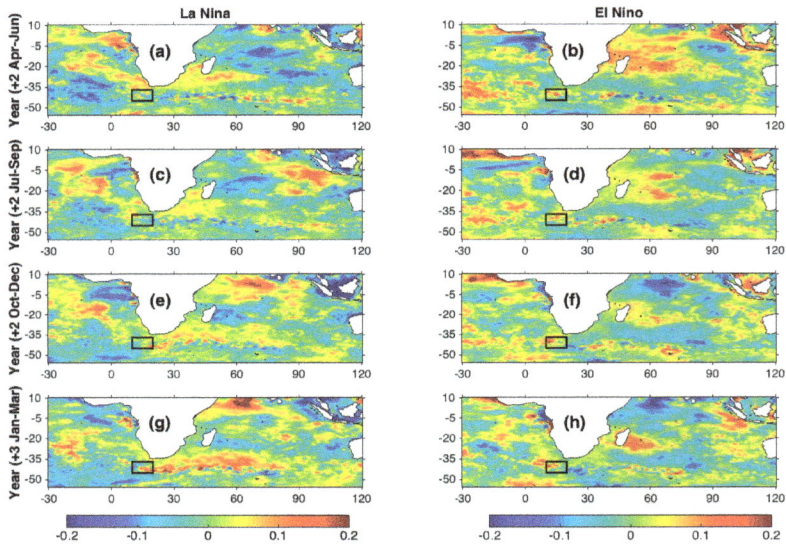

Figure 3. Composite mean of SODA SSS anomalies for all La Niña events (left column) and for all El Niño events (right column) listed in Table 1 two years after the peak for April–June (a,b), July–September (c,d), October–December (e,f), and three years after the peak for January–March (g,h). The boxes represent the established leakage region (37°–45°S, 10°–20°E).

3.2. Variability of Agulhas Leakage in Response to Strength of El Niño Southern Oscillation (ENSO) Events

During the 36-year period between 1980–2016, a total of 11 El Niño and 10 La Niña events occurred (Table 1). The events varied in strength according to ONI values, with the strongest El Niño events occurring in 1982–1983, 1997–1998, 2015–2016 and the strongest La Niña event occurring in 1988–1989. Three El Niño years and five La Niña years are classified as weak events, while the majority of ENSO events are classified as moderate strength. Furthermore, the majority of El Niño events (6 events) directly transitioned to a La Niña episode the following year. However, between 2003–2007, three separate El Niño events occurred without a complete La Niña event in between. There were four occurrences in which a La Niña episode was directly followed by a second episode (Table 1). Both a strong and weak event for El Niño and La Niña was selected to explore the influence of ENSO strength on Agulhas leakage, as represented in Figures 4–7. First and foremost, we analyze whether the individual ENSO events follow the same patterns of SST and SSS signal expression and propagation that was found in Section 3.1 for the average of all El Niño and La Niña events.

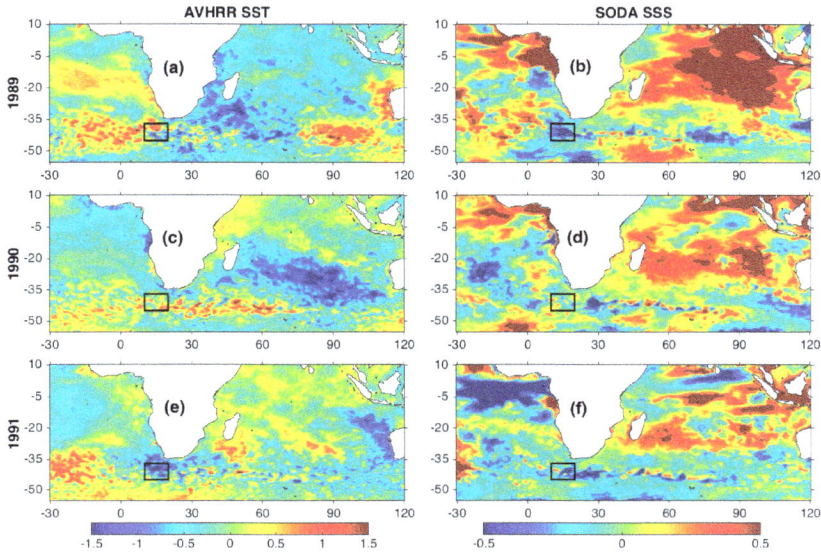

Figure 4. Composite mean for strong 1988–1989 La Niña event from January-March for both AVHRR SST anomalies (left column) and SODA SSS anomalies (right column) during the peak (**a**,**b**), the following year (**c**,**d**), and two years after (**e**,**f**). The boxes represent the established leakage region (37–45°S, 10–20°E).

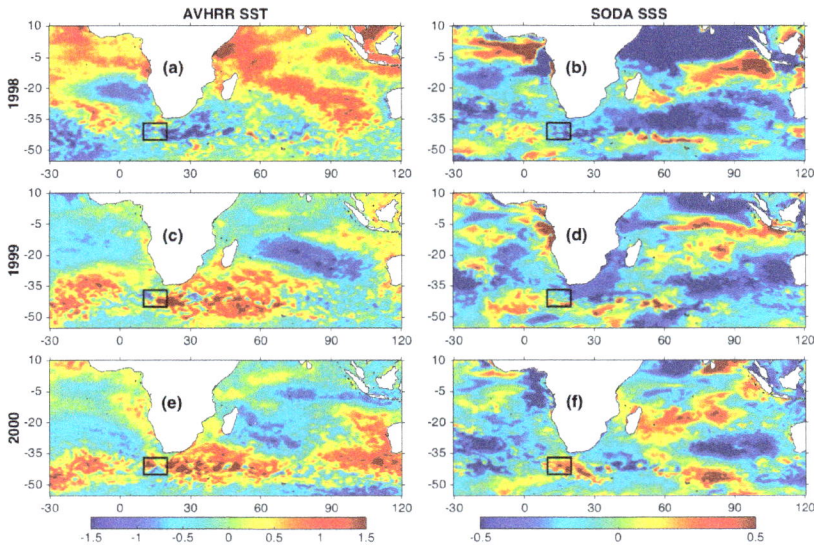

Figure 5. Same as Figure 4 except in reference to the strong 1997–1998 El Niño event during the peak (**a**,**b**), the following year (**c**,**d**), and two years after (**e**,**f**).

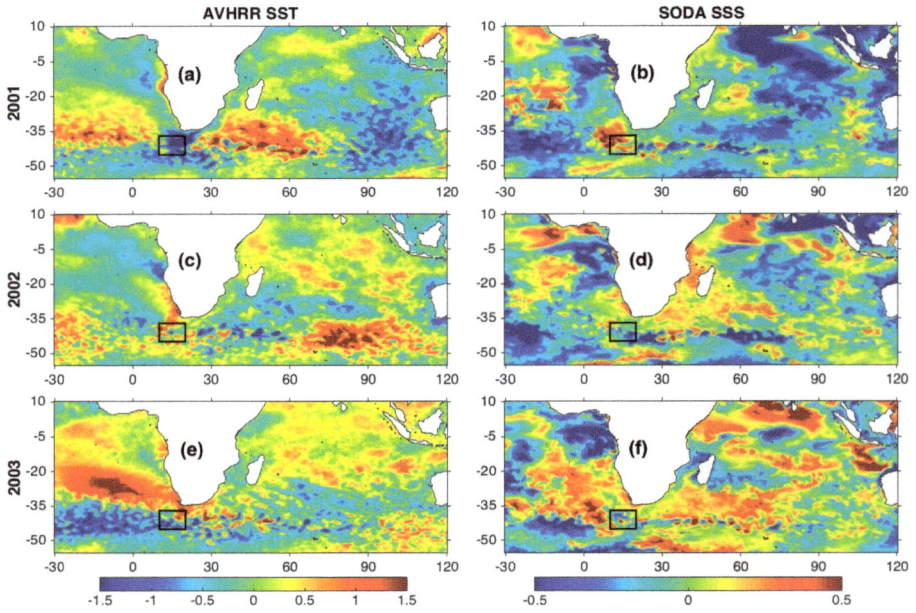

Figure 6. Same as Figure 4 except in reference to the weak 2000–2001 La Niña event during the peak (**a**,**b**), the following year (**c**,**d**), and two years after (**e**,**f**).

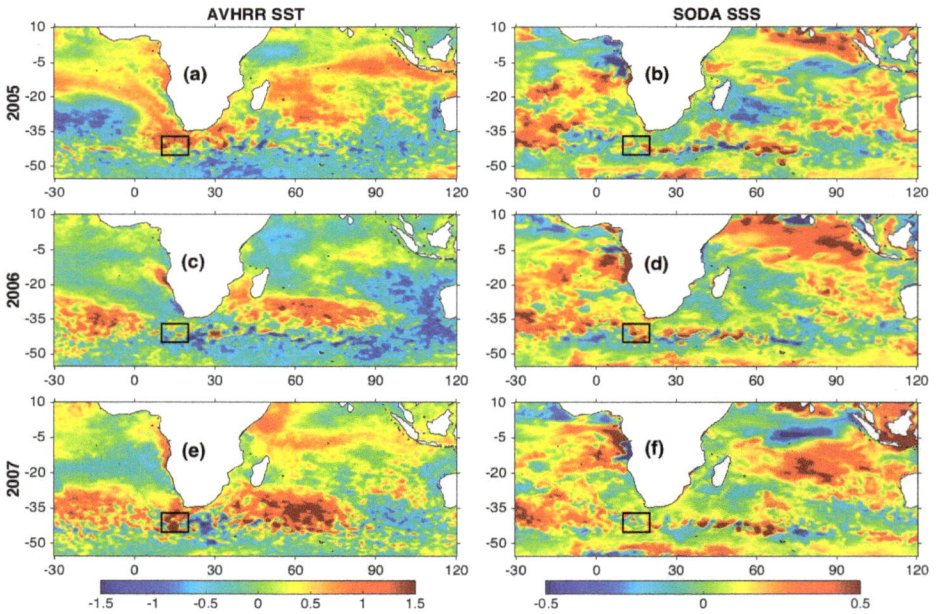

Figure 7. Same as Figure 4 except in reference to the weak 2004–2005 El Niño event during the peak (**a**,**b**), the following year (**c**,**d**), and two years after (**e**,**f**).

Beginning with the SST signal, we expect to see basin-wide warming (cooling) during the peak of El Niño (La Niña) that propagates westward the following year to surround Madagascar and east of Madagascar between 15°–35°S. Ultimately, two years after the peak of El Niño (La Niña) Agulhas leakage responds by releasing anomalously warm (cool) water. The SST signals of all four individual ENSO events during the peak year show the basin-wide temperature trend that can also be identified the following year near the coast of Madagascar (Figures 4a,c, 5a,c, 6a,c and 7a,c). A response in Agulhas leakage two years after the peak of ENSO is evident in both strong events as well as the weak El Niño (Figures 4e, 5e, and 7e). Note, warm waters from the strong El Niño event are seen in the Agulhas leakage region as early as one year after the peak (Figure 5c). The weak La Niña, however, shows a mix of both warm and cool waters in 2003, two years after the peak of La Niña. The expected cool waters dominate the leakage box itself, but an evident warm pool has formed off the southern tip of Africa moving into the Atlantic (Figure 6e).

Moving on to the SSS signal, we expect that during the peak of an El Niño (La Niña) the Indian Ocean basin should be anomalously fresh (saline) in the north-west and saline (fresh) in the south-east i.e., express the primary and secondary SSS signal respectively. The following year, the dual pattern of primary and secondary SSS signals should persist, but shifted to the west so that the primary fresh (saline) waters surround Madagascar with the secondary saline (fresh) waters to the east moving westward. Finally, two years following the peak of El Niño (La Niña) the Agulhas leakage region should initially express the primary fresh (saline) waters and later transition to the secondary saline (fresh) waters. The dual SSS signals of the Indian Ocean basin is evident in the strong El Niño (Figure 5b) and to a lesser extent in the weak El Niño (Figure 7b) but the strong and weak La Niña express only a single saline or fresh signal respectively (Figures 4b and 6b). Surrounding or approaching the Madagascar coast a year after the peak of ENSO, we see the expected primary saline signal for La Niña and primary fresh signal for El Niño for all events (Figures 4d, 5d, 6d, and 7d). Finally, two years following the peak of ENSO, the strong La Niña shows anomalously fresh SSS because the expected primary saline signal is still near the Madagascan coast. It is important to note that this observed fresh signal is not indicative of the expected secondary signal due to the original dominant saline response in the Indian Ocean basin (Figure 4f). The weak La Niña, however, does show isolated areas of the expected primary saline leakage in 2003 with the secondary fresh signal observed surrounding Madagascar at the same time (Figure 6f). In comparison, the strong El Niño event expresses the expected primary fresh SSS signal as early as 1999 (Figure 5d). By the year 2000, the secondary saline signal is released in Agulhas leakage and continues to surround Madagascar (Figure 5f). The Agulhas leakage two years after the weak El Niño is a mix of the expected primary fresh SSS with some high-salinity waters. The expected high-saline secondary SSS signal is propagating toward Madagascar at this time (Figure 7f).

In addition, Figures 8 and 9, showing the weak ENSO anomalies subtracted from the strong ENSO anomalies, support the observed differences between the strong and weak events and highlight other important variations. Positive (red) regions in these figures indicate that the stronger event expressed warmer SST or more saline SSS relative to the corresponding weak event. Negative (blue) values in these figures are indicative of cooler SST or fresher SSS anomalies during the strong event with respect to the corresponding weak event. The most noticeable difference between the strong and weak SST signal is the positive SST values in Figure 9c and negative SSS values in Figure 9d of the Agulhas leakage region. Furthermore, two years after the peak El Niño, the Agulhas leakage of a strong El Niño is warmer than that of the weak El Niño excluding a small region where an anomalously warm core eddy was present in leakage during the weak event (Figure 9d). The intensity of the SST anomalies of Agulhas leakage is also greater during a strong La Niña than the weak La Niña, as shown by the negative anomalies in Figure 8e. The same can be said for the primary saline SSS signal during La Niña (Figure 8f). In general, the SSS during the strong La Niña event is more saline across the entire Indian Ocean than the weak event (Figure 8b), and the strong El Niño more strongly expresses the dual SSS signal (saline in the south-east and fresh in the north-west) compared to the weak El Niño event (Figure 9b).

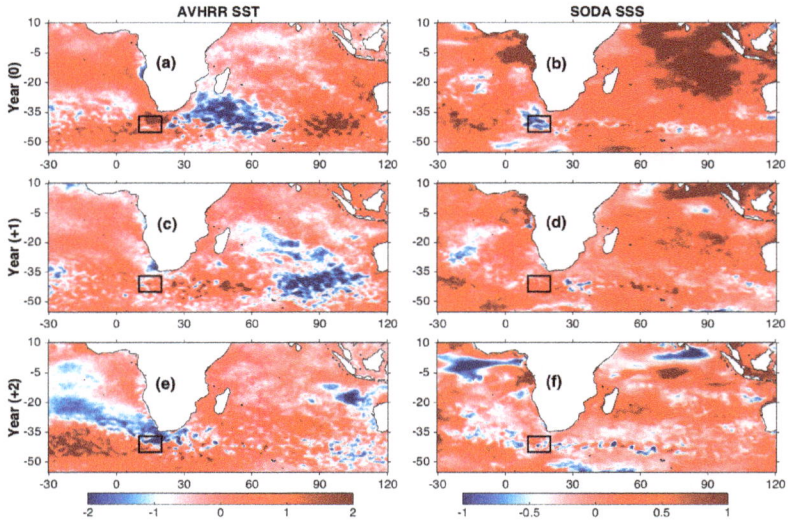

Figure 8. Composite mean for strong 1988–1989 La Niña minus the weak 2000–2001 La Niña from January–March for both AVHRR SST anomalies (right column) and SODA SSS anomalies (left column) during the peak (**a,b**), the following year (**c,d**), and two years after (**e,f**). The boxes represent the established leakage region (37°–45°S, 10°–20°E).

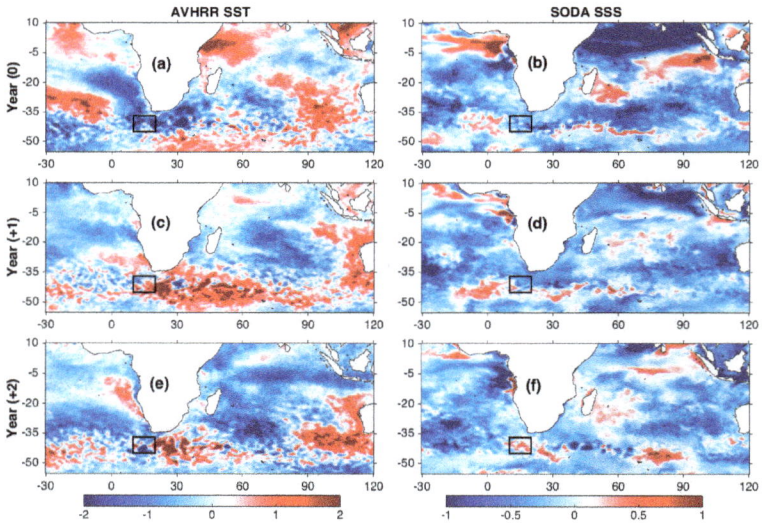

Figure 9. Same as Figure 8 except in reference to the strong 1997–1998 El Niño minus the weak 2004–2005 El Niño during the peak (**a,b**), the following year (**c,d**), and two years after (**e,f**).

3.3. *Temporal Variability of ENSO Signal*

A time series of box-averaged SST and SSS anomalies in the Agulhas leakage region is shown in Figure 10. We used this time series to evaluate the changes in the SST and SSS of Agulhas leakage in response to all ENSO events that have occurred between 1981–2016. From the previous sections, we have defined the type of SST and SSS response observed for La Niña events compared to El Niño. In this section, we are primarily interested in further identifying the lag between the peak of an ENSO event and the SST and SSS response of Agulhas leakage. A lead-lag correlation analysis for both SST and SSS to the ONI index for ENSO was performed, and the results are presented in Figure 11. This figure serves as a quantitative analysis of the following trends from the time series.

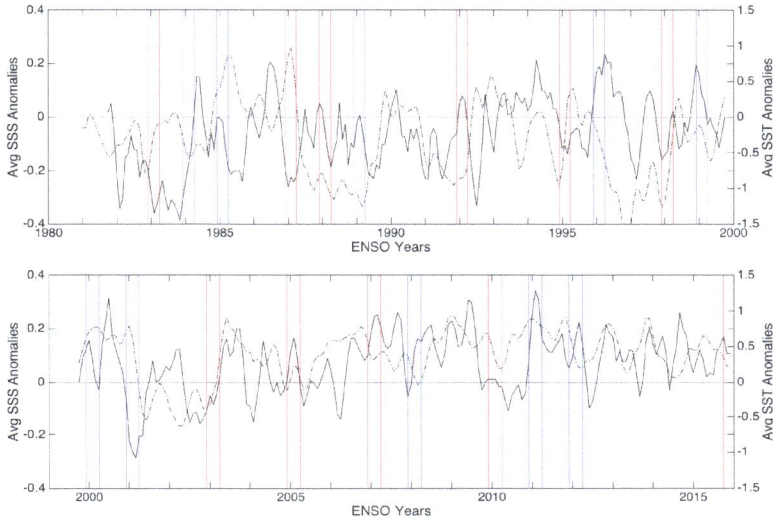

Figure 10. Time series from 1981–2016 for monthly box-averaged (37°–45°S, 10°–20°E) AVHRR SST (solid line) and SODA SSS (dashed line) anomalies. ENSO events are represented during their peak (January–March) by vertical shaded bars (red for El Niño and blue for La Niña).

There are a few seasonal and decadal trends of Agulhas leakage that should be mentioned first. First and foremost, there is a natural yearly high to low fluctuation in SST and SSS. Additionally, after 2005 the SST and SSS show an increasing trend in which the anomalies rarely drop below zero. In general, the time series for SST and SSS closely mimic each other in shape, meaning that for observed peaks and dips in the SST time series there are corresponding peaks and dips in the SSS time series. A correlation analysis quantifies this association as having a correlation of 0.243. While this may seem low, it is important to notice that the SSS anomalies have a smaller range than SST anomalies, from −0.4 to 0.4 and −1.5 to 1.5, respectively. Also, the SSS trend line is shifted slightly right when compared to SST, thus explaining the low calculated correlation compared to what is observed in Figure 10. In the next paragraph, when we isolate trends with respect to ENSO events, it will become evident that SSS response shows the primary and secondary signal response, explaining why corresponding SSS response occurs later than that of SST.

When evaluating the time series with respect to ENSO events, we identify the anomalous SST and SSS values two years after the marked ENSO years. For all El Niño events, the SST trend line is strongly positive at this time and for all La Niña events, the SST trend line is strongly negative. Additionally, extreme values of SST are found to correspond to the strongest ENSO events. In particular in 2000, 2 years after a strong El Niño, the SST increases to nearly 1.2 °C. With respect to SSS, the same patterns emerge but, as previously noted, the corresponding positive (negative) peaks (dip) occur slightly after the two-year mark of El Niño (La Niña) because these are related to the secondary SSS signal. At the two-year mark itself, the opposite primary negative (positive) SSS signal is evident. Notably, more than two years after a strong El Niño event (1982–1983) an extreme SSS dip of nearly −0.2 is observed.

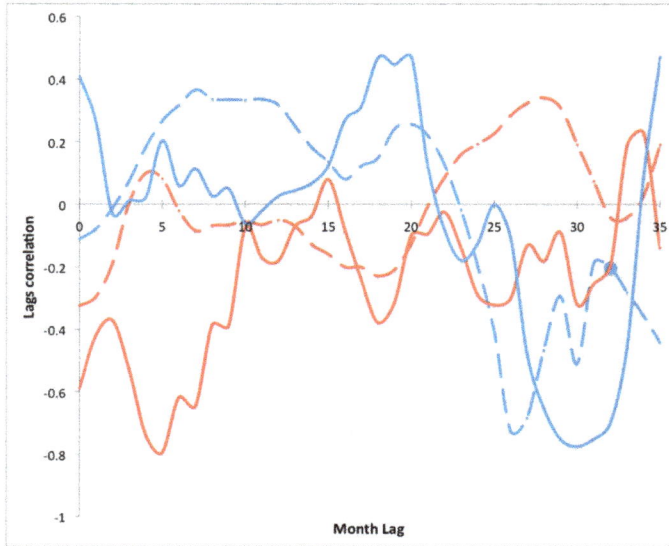

Figure 11. Correlation analysis between absolute values of Oceanic Niño Index (ONI) index and box-averaged AVHRR SST anomalies (solid line) and SODA SSS anomalies (dashed line) at monthly lag intervals for El Niño years in red and La Niña years in blue.

An additional lead-lag correlation analysis of the SST and SSS to the ONI ENSO index is included in order to analyze the time series with more precise quantitative measurement of the delay in response between the peak of ENSO and Agulhas leakage response. A positive correlation indicates that as the ENSO index strengthens, the anomalous SST and SSS is also increasing, meaning the Agulhas leakage is anomalously warm and saline. The opposite can be said for a negative correlation, as ENSO strength increases the SST and SSS are decreasing, meaning the Agulhas leakage is anomalously cool and fresh. During a La Niña event, it is expected that SST should have a negative correlation and the SSS should first show a positive correlation followed by a negative correlation, indicative of its primary and secondary SSS signal. Figure 11 indicates that the La Niña SST of Agulhas leakage has a positive correlation with the ONI index until 24 months after the peak of La Niña. A decrease in correlation begins as early as 20 months and continues on until the greatest negative correlation at 30 months after the peak of La Niña events. The corresponding SSS signal during La Niña is also dominantly positive, with the strongest positive correlation at a 20-month lag. After this peak, the correlation drastically decreases until the greatest negative correlation at a 26-month lag continuing to a 30-month lag. The same monthly lags apply to the El Niño SSS correlation but with the opposite sign: a negative correlation transitioning to a positive correlation. The SST correlation with El Niño is also opposite to that observed during La Niña. Most importantly, an increasing correlation is present at a 24-month lag for El Niño SST, which peaks with the strongest positive correlation at about a 34-month lag.

4. Discussion

4.1. SST and SSS Signal Response to El Niño and La Niña

According to Putrasahan et al. [10] and others, changes to circulation patterns coupled with Rossby wave propagation and wind-stress curl carry an ENSO signal from the Pacific, across the Indian Ocean, to alter Agulhas leakage. Putrasahan et al. [10] proved this connection by correlating warmer waters in the Agulhas leakage region to El Niño. We use this established connection to discuss the differences in the SST and SSS signal with respect to El Niño events compared to La Niña events, as presented in Section 3.1.

The observed SST signal of basin-wide warming (cooling) during El Niño (La Niña) from Figure 2a (Figure 1a) is consistent with the findings of Kug et al. [22] and Kilpatrick et al. [18] corresponding to fluctuations in Walker circulation and easterly (westerly) wind anomalies over Indonesia during the boreal winter/spring. The observed regions of positive and negative SSS in Figures 1b and 2b can be attributed to the influence of anomalous ENSO conditions as well as coincident positive (negative) Indian Ocean Dipole (IOD) with El Niño (La Niña). Primarily, easterly (westerly) winds drive upwelling (downwelling) along the Sumatra coast, explaining the observed secondary signal of positive (negative) SSS anomalies further enhanced by reduced (increased) rainfall created from the atmospheric circulation patterns of El Niño (La Niña). The negative primary SSS signal observed in the equatorial Indian Ocean during El Niño is explained by horizontal advection of low salinity waters from the Bay of Bengal, while the primary positive SSS in the equatorial Indian Ocean during La Niña is produced by eastward-flowing Wyrtki jets [23].

A zero-lag SST and SSS response of waters near the tip of Africa is most likely related to changes in southern African rainfall associated with ENSO-driven SST patterns in the Pacific rather than ENSO SST and SSS signal transmission. Most of the severe droughts have happened during the mature phase of El Niño and the wettest summers during La Niña [24]. Decreased (increased) rainfall could explain the observed salinization (freshening) along the coast during El Niño (La Niña). Additionally, weaker (stronger) upwelling favorable winds are present during El Niño (La Niña), and this ultimately creates conditions in which SST is anomalously warm (cool) during El Niño (La Niña). This positive correlation between ENSO and SST and SSS properties applies only to the south coast of Africa, while a negative correlation exists with the Agulhas Current system south of 36°S [25]. In other words, it is highly unlikely that ENSO influences the SST or SSS signal of Agulhas leakage during the peak of the ENSO event. Instead, it is possible that atmospheric-related weather patterns associated with ENSO alter the SST and SSS conditions near the southern tip of Africa, explaining the observed changes to that region in Figures 1a,b and 2a,b.

One year after the peak of ENSO, both the dominant SST signal and the dual SSS signal have shifted westward. The previous basin-wide trend in SST is isolated near the region east of Madagascar at ~15°–35°S. Note, this is the latitude where the SEC is found [15], supporting the idea that the SEC is involved in signal transmission. Grunseich et al. [23] attributes the westward movement of the SSS signal to Rossby waves associated with the westward flowing SEC. Most importantly, the SST signal and primary SSS signal surround Madagascar to interact with the source currents, which has the potential to influence eddy formation of the EMC and MC, thus continuing on to impact Agulhas leakage. This impact is initially seen two years after the peak of ENSO, as expected by the Putrasahan et al. [10] study. However, Putrasahan et al. [10] did not identify the continuation of this impact occurring for nearly 3 years after the peak of an El Niño (La Niña). During this time, the SST signal is consistently positive (negative) but the original primary fresh (saline) SSS signal is replaced by the secondary saline (fresh) SSS signal. Further evidence supporting the lag in response of Agulhas leakage is discussed in Section 4.3. From Figures 1–3, representing the average response of SST and SSS to ENSO events, we have established that both the SST and SSS signal of Agulhas leakage have a contrasting response to an El Niño versus a La Niña event. A single opposite SST signal occurs two years after the peak of El Niño as compared to La Niña, warming and cooling respectively. The SSS response is composed of the initial primary signal followed by the opposite secondary signal. First, the SSS of Agulhas leakage is anomalously fresh (saline) two years after the peak of El Niño (La Niña). In the following months, this anomalously fresh (saline) signal is replaced by the secondary saline (fresh) SSS signal in Agulhas leakage. Furthermore, these figures provide a general understanding of the transmission of an ENSO signal. The signal originates in the Pacific Ocean and during the peak of an ENSO event (December–March) it moves into the Indian Ocean basin, changing SST and SSS properties. The following year, the SST and SSS signals propagate westward across the Indian Ocean basin to surround Madagascar and begin to influence the Agulhas Current. Strong changes to Agulhas leakage are observed two years following an ENSO event and continue for nearly three years after ENSO. This process is consistent with the circulation, Rossby wave, eddy process, and wind curl explained in the Introduction. The next section explores how the strength of an ENSO event alters this signal formation and transmission.

4.2. Variability of Agulhas Leakage in Response to Strength of ENSO Events

Figures 4–7, showing individual ENSO events of either strong or weak strength, were analyzed in Section 3.2 for any deviations from the established SST and SSS trends identified in Section 3.1. Deviations from this established trend may be indicative of the interaction of ENSO strength on SST and SSS expression. With respect to the SST signal, the only deviation in signal propagation occurred near the Agulhas leakage region. During the strong El Niño episode, warm waters appeared a year earlier than expected and during the weak La Niña episode an evident warm pool formed off the southern tip of Africa contrasting the cool waters in the Agulhas leakage region. This warm pool occurred two years after the peak of the weak La Niña event in 2003. That year was also the peak of the 2002–2003 El Niño. In Section 4.1, we explained that atmospheric circulation associated with ENSO is known to cause warming near the tip of Africa, thus explaining this mixed signal during the weak La Niña.

There were several deviations from the established trend in the SSS signal. In particular, during the peak of La Niña for both the strong and weak event, there was only a saline or fresh signal, respectively, instead of the expected dual primary and secondary signals evident in both the El Niño episodes. The peak of the strong La Niña event, 1989, is also a negative IOD year that peaks in July [11], which may explain the strongly negative SSS signal. Furthermore, the clear dual SSS signal of the strong El Niño can partially be attributed to the positive IOD that peaked in November of 1998 [11]. The lack of any IOD influence may have inhibited the strong expression of an anomalous negative SSS in the north-western region during the weak La Niña. Regardless, the westward propagation of the SSS signal occurred as expected in all events. However, the influence of the primary SSS signal on Agulhas leakage, expected to occur two years following the peak ENSO event, was not consistent in all events. The primary signal was delayed in the strong La Niña event and appeared nearly a year early in the strong El Niño event. Interestingly, the primary signal was present in the Agulhas leakage two years after both the weak El Niño and La Niña. The mix of low salinity waters with the expected primary saline signal in Agulhas leakage two years after the weak El Niño may be the result of weak signal strength. While the primary fresh signal was present in the north-western Indian Ocean basin during the peak of this El Niño, it was not strong relative to the secondary saline signal, which may explain the corresponding weak expression in Agulhas leakage.

Trends identified in Figures 8 and 9, highlight important differences between the strong and weak events and provide further support for the signal transmission discussed. The warm fresh Agulhas leakage signal was identified previously as occurring nearly a year earlier during a strong El Niño. This is further supported by the positive and negative SSS values identified in the Agulhas leakage region (Figure 9c,d). The stronger anomalies identified in the Agulhas leakage region with respect to strong ENSO events indicate that ENSO strength is proportional to signal strength. Lastly, the trends referring to the expression of SSS in the Indian Ocean are most likely attributed to the IOD influence previously mentioned rather than differences in ENSO strength.

In conclusion, the established SST and SSS signal response to El Niño and La Niña is consistent for all events of varying ENSO strength, but the strength appears to influence the intensity of the signal and the time of its transmission, thereby explaining any variations from the established trends. In general, increasing ENSO strength increases the extremes of the resulting anomalous signal and impacts the Agulhas leakage region earlier during El Niño and slightly later during La Niña. As previously discussed, weakened trade winds in the Pacific and resultant strengthened trade winds in the tropical Indian ocean create a zonal band of positive wind-stress curl over the tropics which

Putrasahan et al. [10] correlated to the Agulhas leakage response. Knowing this, we can support this observed variation in time of signal propagation with ENSO phase and strength to a corresponding change in the strength of wind-stress curl [17]. As trade winds strengthen in the tropical Indian Ocean during stronger El Niño events, a corresponding increase in wind-stress curl would be expected across the tropical Indian Ocean, providing increased propagation of the SSS and SST signals. The opposite can be said for stronger La Nina events which correspond to further weakening of trade winds in the tropical Indian Ocean and a corresponding decrease in wind-stress curl slowing propagation of the SSS and SST signals. This temporal perspective is further explored in the next section.

4.3. Temporal Variability of ENSO Signal

In Section 3.3, we used the time series (Figure 10) to evaluate changes in the SST and SSS of Agulhas leakage for all ENSO events between 1981–2016. We also conducted a quantitative analysis of the trends from the time series using the lead–lag correlation analysis in Figure 11. Explaining the seasonal trends in the time series helps to better understand how the Agulhas leakage SST and SSS time series is changing with respect to ENSO. The naturally annual fluctuations in SST and SSS are likely attributed to a seasonal signal related to changes in current speed [26]. The increasing trend in the time series after 2005 could be indicative of a decadal warming trend observed in the Agulhas Current region [27] and an increase in leakage in response to shifting Southern Hemisphere westerlies [28].

When evaluating the time series with respect to ENSO events, we evaluated the anomalous SST and SSS values two years after the marked ENSO years, because the time series represents a box average of Agulhas leakage and in the previous sections the initial response of Agulhas leakage to the ENSO signal was found at this time. The trends in the time series at this time support the temporal perspective of both the primary and secondary SSS signal and the single dominant but lasting SST signal observed in the Agulhas leakage region.

The time series in Figure 10 only serves to provide a general qualitative analysis of the trends found in the previous sections and the time during which they are taking place. The lead–lag correlation analysis of the SST and SSS to the ONI ENSO index in Figure 11 supports this analysis with a more precise quantitative measure of the delay in response between the peak of ENSO and the Agulhas leakage response. The trends identified in Figure 11 support the existence of a single SST response of Agulhas leakage to ENSO, and this response has an average lag of ~20 months that persists and strengthens until a lag of nearly 30 months. Additionally, the dual primary and secondary signal response of the SSS of Agulhas leakage to ENSO is supported in Figure 11. The primary SSS signal has an average lag of ~20 months followed by the opposite secondary SSS signal at a lag of ~26 months.

5. Conclusions

Thus far, we have explored yearly fluctuations in Agulhas leakage using anomalous SST from AVHRR and SSS from SODA reanalysis. Figures 1 and 2 show that the response of Agulhas leakage to an El Niño event is opposite to that of a La Niña event for both SST and SSS. The single dominant SST response of Agulhas leakage occurs around 20 months and continues for nearly 30 months after the peak of ENSO and is anomalously high for El Niño and anomalously low for La Niña. At the same time, the corresponding SSS signal is anomalously saline for La Niña and fresh for El Niño. However, the SSS signal is composed of this initial primary signal, which reaches the region first, followed by the secondary signal that reaches Agulhas leakage about 26 months after the peak of ENSO. The secondary signal is fresh for La Niña and saline for El Niño. Additionally, Figures 3–10 suggest ENSO strength influences the intensity of both the signal and the time of its transmission. In general, increasing ENSO strength increases the extremes of the resulting anomalous signal and impacts the Agulhas leakage region earlier during El Niño and slightly later during La Niña. Correlation observations from Figure 11 support the Putrasahan et al. [10] proposed 2-year period for an ENSO signal to influence Agulhas leakage and also establishes the lags for the primary and secondary SSS signal response of Agulhas leakage.

The aim of this study was only to highlight that a relationship between ENSO and Agulhas leakage exists and define this relationship using SST and SSS, as well as to identify the time in which this response lags an ENSO event. We also explored the effect of ENSO strength on this relationship. Most notably, this study evaluates this process from start to finish to not only show that ENSO-generated SST and SSS signals influence Agulhas leakage, but also where these signals originate and how they propagate, providing a visual and qualitative representation of the connection between the Pacific, Atlantic, and Indian Ocean. Further investigation is required to understand the dynamics of the various systems involved in this relationship and signal propagation. Suggestions for future work include addressing how the eastern and central modes of El Niño may change this relationship. Furthermore, the depth signal of this described trend is another important parameter worth considering. This paper relied primarily on AVHRR SST and SODA SSS. Both data sets have been used in countless previous studies to explore the Indian Ocean basin or Agulhas current system and have been proven reliable for doing so [23,25]. However, it is important to note some issues with these data sets that may have influenced the results. Specifically, when conditions deviate from the mean atmosphere and ocean conditions, errors arise in AVHRR SST retrieval [18]. Therefore, AVHRR is unable to perform properly under cloudy conditions and direct sunlight, and we mentioned earlier that rainfall events are an expected response to La Niña events. The full intensity of SST readings may, therefore, be compromised due to interpolation to account for missed readings due to cloud or direct sunlight interactions. Future studies may benefit from using another data source for SST and compare the findings. Additionally, SSS from SODA is reanalysis-based. Observational comparisons would be beneficial to further support the findings in this paper.

We include an additional figure, Figure 12, to show that recent salinity missions will be valuable resources for future work in this field. Figure 12 highlights the SSS signal from SMOS of the most recent El Niño event (2015–2016). SMOS is able to provide a more accurate representation than previously available models. Also, recently launched (launch date: 31 January 2015 and data availability since April 2015) Soil Moisture Active Passive (SMAP) derived salinity will be useful in ENSO studies.

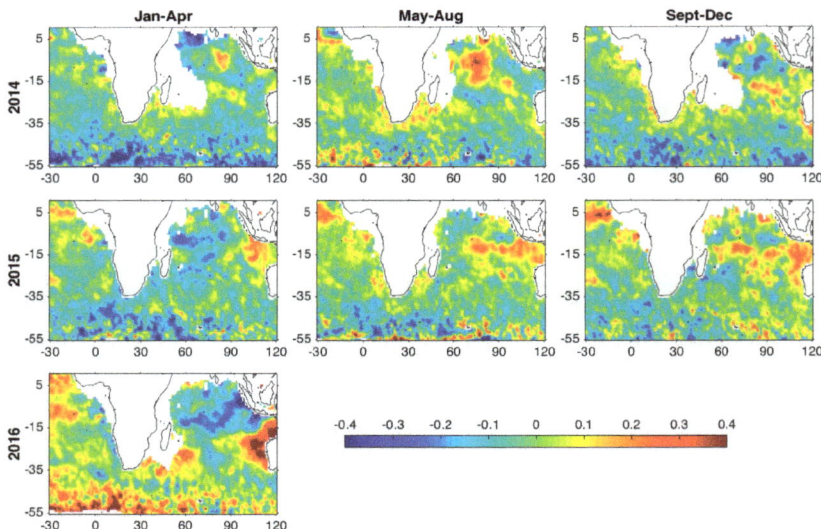

Figure 12. Composite mean of Soil Moisture Ocean Salinity (SMOS) SSS anomalies from January–April (**first column**), May–August (**middle column**) and September–December (**last column**) during the most recent El Niño event starting and stopping in 2014 (**top row**), restarting in 2015 (**middle row**), and reaching its peak in 2016 (**bottom row**).

Acknowledgments: This work is supported by the NASA/South Carolina Space Grant Consortium Mini-REAP grant awarded to BS. The Authors duly acknowledge the various data sources for the freely available data. AVHRR SST is downloaded from the National Center for Environmental Information (http://iridl.ldeo.columbia.edu/SOURCES/.NOAA/.NCDC/.OISST/.version2/.AVHRR/.sst/). The SMOS binned data used for this study is the level 3 Operational V2.0 provided by the ESA obtained from the SMOS Barcelona Expert Center Data distribution and visualization services (http://cp34-bec.cmima.csic.es/data/data-access/). The SMAP data are produced by Remote Sensing Systems, Santa Rosa, CA, and version 2.0 level 3 is obtained from NASA JPL PO.DAAC. SODA reanalysis product is a multi-institutional collaborative project between Texas A&M University and University of Maryland, which is obtained from the Asia-Pacific Data Research Center (APDRC) of the International Pacific Research Centre (IPRC). The authors would like to thank three anonymous reviewers and the editor, whose comments significantly contributed to the improvement of this paper.

Author Contributions: Morgan L. Paris and Bulusu Subrahmanyam conceived and designed the data analysis and interpretation of the results. Morgan L. Paris prepared all the figures and writing of this article, and Bulusu Subrahmanyam designed the project, guided this work and corrected the paper.

Conflicts of Interest: The authors declare no conflict of interest. The founding sponsors had no role in the design of the study; in the collection, analyses, or interpretation of data; in the writing of the manuscript, and in the decision to publish the results.

References

1. Ridgway, K.R.; Dunn, J.R. Observational evidence for a Southern Hemisphere oceanic supergyre. *Geophys. Res. Lett.* **2007**, 34. [CrossRef]
2. Beal, L.M.; De Ruijter, W.P.; Biastoch, A.; Zahn, R. On the role of the Agulhas system in ocean circulation and climate. *Nature* **2011**, *472*, 429–436. [CrossRef] [PubMed]
3. Simon, M.H.; Arthur, K.L.; Hall, I.R.; Peeters, F.J.; Loveday, B.R.; Barker, S.; Ziegler, M.; Zahn, R. Millennial-scale Agulhas Current variability and its implications for salt-leakage through the Indian–Atlantic Ocean Gateway. *Earth Planet. Sci. Lett.* **2013**, *383*, 101–112. [CrossRef]
4. Biastoch, A.; Durgadoo, J.V.; Morrison, A.K.; Van Sebille, E.; Weijer, W.; Griffies, S.M. Atlantic multi-decadal oscillation covaries with Agulhas leakage. *Nat. Commun.* **2015**, *6*. [CrossRef] [PubMed]
5. Ruijter, W.D.; Biastoch, A.; Drijfhout, S.S.; Lutjeharms, J.R.E.; Matano, R.P.; Pichevin, T.; Weijer, W. Indian-Atlantic interocean exchange: Dynamics, estimation and impact. *J. Geophys. Res. Oceans* **1999**, *104*, 20885–20910. [CrossRef]
6. Durgadoo, J.V.; Rühs, S.; Biastoch, A.; Böning, C.W. Indian Ocean sources of Agulhas leakage. *J. Geophys. Res. Oceans* **2017**, *122*, 3481–3499. [CrossRef]
7. Meyers, G. Variation of Indonesian throughflow and the El Niño-Southern Oscillation. *J. Geophys. Res. Oceans* **1996**, *101*, 12255–12263. [CrossRef]
8. Le Bars, D.L.B.; Dijkstra, H.A.; De Ruijter, W.P.M. Impact of the Indonesian Throughflow on Agulhas leakage. *Ocean Sci. Discuss.* **2013**, *10*, 353–391. [CrossRef]
9. Stramma, L.; Lutjeharms, J.R. The flow field of the subtropical gyre of the South Indian Ocean. *J. Geophys. Res. Oceans* **1997**, *102*, 5513–5530. [CrossRef]
10. Putrasahan, D.; Kirtman, B.P.; Beal, L.M. Modulation of SST Interannual Variability in the Agulhas Leakage Region Associated with ENSO. *J. Clim.* **2016**, *29*, 7089–7102. [CrossRef]
11. De Ruijter, W.P.; van Aken, H.M.; Beier, E.J.; Lutjeharms, J.R.; Matano, R.P.; Schouten, M.W. Eddies and dipoles around South Madagascar: formation, pathways and large-scale impact. *Deep Sea Res. Part I Oceanogr. Res. Pap.* **2004**, *51*, 383–400. [CrossRef]
12. Schouten, M.W.; De Ruijter, W.P.; Van Leeuwen, P.J. Upstream control of Agulhas Ring shedding. *J. Geophys. Res. Oceans* **2002**, *107*. [CrossRef]
13. Elipot, S.; Beal, L.M. Characteristics, energetics, and origins of Agulhas Current meanders and their limited influence on ring shedding. *J. Phys. Oceanogr.* **2015**, *45*, 2294–2314. [CrossRef]
14. Loveday, B.R.; Durgadoo, J.V.; Reason, C.J.; Biastoch, A.; Penven, P. Decoupling of the Agulhas leakage from the Agulhas Current. *J. Phys. Oceanogr.* **2014**, *44*, 1776–1797. [CrossRef]
15. Palastanga, V.; Van Leeuwen, P.J.; De Ruijter, W.P.M. A link between low-frequency mesoscale eddy variability around Madagascar and the large-scale Indian Ocean variability. *J. Geophys. Res. Oceans* **2006**, *111*. [CrossRef]
16. Cai, W.; Meyers, G.; Shi, G. Transmission of ENSO signal to the Indian Ocean. *Geophys. Res. Lett.* **2005**, *32*. [CrossRef]

17. Feng, M.; Meyers, G. Interannual variability in the tropical Indian Ocean: a two-year time-scale of Indian Ocean Dipole. *Deep Sea Res. Part II Top. Stud. Oceanogr.* **2003**, *50*, 2263–2284. [CrossRef]
18. Kilpatrick, K.A.; Podesta, G.P.; Evans, R. Overview of the NOAA/NASA advanced very high resolution radiometer Pathfinder algorithm for sea surface temperature and associated matchup database. *J. Geophys. Res. Oceans* **2001**, *106*, 9179–9197. [CrossRef]
19. Tokinaga, H.; Tanimoto, Y. Seasonal transition of SST anomalies in the tropical Indian Ocean during El Niño and Indian Ocean dipole years. *J. Meteorol. Soc. Jpn. Ser. II* **2004**, *82*, 1007–1018. [CrossRef]
20. Swart, S.; Speich, S.; Ansorge, I.J.; Goni, G.J.; Gladyshev, S.; Lutjeharms, J.R. Transport and variability of the Antarctic Circumpolar Current south of Africa. *J. Geophys. Res. Oceans* **2008**, *113*. [CrossRef]
21. Dencausse, G.; Arhan, M.; Speich, S. Spatio-temporal characteristics of the Agulhas Current retroflection. *Deep Sea Res. Part I Oceanogr. Res. Pap.* **2010**, *57*, 1392–1405. [CrossRef]
22. Kug, J.S.; Kang, I.S. Interactive feedback between ENSO and the Indian Ocean. *J. Clim.* **2006**, *19*, 1784–1801. [CrossRef]
23. Grunseich, G.; Subrahmanyam, B.; Murty, V.S.N.; Giese, B.S. Sea surface salinity variability during the Indian Ocean Dipole and ENSO events in the tropical Indian Ocean. *J. Geophys. Res. Oceans* **2011**, *116*. [CrossRef]
24. Richard, Y.; Trzaska, S.; Roucou, P.; Rouault, M. Modification of the southern African rainfall variability/ENSO relationship since the late 1960s. *Clim. Dyn.* **2000**, *16*, 883–895. [CrossRef]
25. Rouault, M.; Pohl, B.; Penven, P. Coastal oceanic climate change and variability from 1982 to 2009 around South Africa. *Afr. J. Mar. Sci.* **2010**, *32*, 237–246. [CrossRef]
26. Krug, M.; Tournadre, J. Satellite observations of an annual cycle in the Agulhas Current. *Geophys. Res. Lett.* **2012**, *39*. [CrossRef]
27. Rouault, M.; Penven, P.; Pohl, B. Warming in the Agulhas Current system since the 1980′s. *Geophys. Res. Lett.* **2009**, *36*. [CrossRef]
28. Biastoch, A.; Böning, C.W.; Schwarzkopf, F.U.; Lutjeharms, J.R.E. Increase in Agulhas leakage due to poleward shift of Southern Hemisphere westerlies. *Nature* **2009**, *462*, 495–498. [CrossRef] [PubMed]

![remote sensing]

MDPI

Article

Stability Assessment of the (A)ATSR Sea Surface Temperature Climate Dataset from the European Space Agency Climate Change Initiative

David I. Berry [1,*], Gary K. Corlett [2,3], Owen Embury [4,5] and Christopher J. Merchant [4,5]

[1]   National Oceanography Centre, University of Southampton Waterfront Campus, European Way, Southampton SO14 3ZH, UK
[2]   Department of Physics and Astronomy, University of Leicester, University Road, Leicester LE1 7RH, UK; gkc1@le.ac.uk
[3]   National Centre for Earth Observation, University of Leicester, University Road, Leicester LE1 7RH, UK
[4]   Department of Meteorology, University of Reading, Reading RG6 6AL, UK; o.embury@reading.ac.uk (O.E.); c.j.merchant@reading.ac.uk (C.J.M.)
[5]   National Centre for Earth Observation, University of Reading, Reading RG6 6AL, UK
*   Correspondence: dyb@noc.ac.uk; Tel.: +44-(0)23-8059-7740

Received: 24 November 2017; Accepted: 15 January 2018; Published: 18 January 2018

Abstract: Sea surface temperature is a key component of the climate record, with multiple independent records giving confidence in observed changes. As part of the European Space Agencies (ESA) Climate Change Initiative (CCI) the satellite archives have been reprocessed with the aim of creating a new dataset that is independent of the in situ observations, and stable with no artificial drift (<0.1 K decade^{-1} globally) or step changes. We present a method to assess the satellite sea surface temperature (SST) record for step changes using the Penalized Maximal t Test (PMT) applied to aggregate time series. We demonstrated the application of the method using data from version EXP1.8 of the ESA SST CCI dataset averaged on a 7 km grid and in situ observations from moored buoys, drifting buoys and Argo floats. The CCI dataset was shown to be stable after ~1994, with minimal divergence (~0.01 K decade^{-1}) between the CCI data and in situ observations. Two steps were identified due to the failure of a gyroscope on the ERS-2 satellite, and subsequent correction mechanisms applied. These had minimal impact on the stability due to having equal magnitudes but opposite signs. The statistical power and false alarm rate of the method were assessed.

Keywords: Along Track Scanning Radiometer (ATSR); sea surface temperature; stability; homogeneity; drifting buoys; Argo; Global Tropical Moored buoy Array (GTMBA); Penalized Maximal t Test

1. Introduction

Observations of the sea surface temperature (SST) forms one of the key components of the climate record (e.g., [1,2]), with in situ observations extending back over 150 years (e.g., [3–7]) and satellite-based estimates over 20 years (e.g., [8]). Whilst some confidence is gained from the agreement between the different datasets few of them are truly independent from each other. For example, all of the in situ based datasets (e.g., [3–7]) are derived from the International Comprehensive Ocean–Atmosphere Data Set (ICOADS; e.g., [9,10]). Blended datasets (e.g., [11]) tend to contain in situ observations from ICOADS, and one or more satellite based sources. Satellite-only estimates (e.g., [12]) have tended to be calibrated and validated using the same in situ sources as present in ICOADS.

In recognition of the importance of independent estimates the (A)ATSR Reprocessing for Climate (ARC) project [13] aimed to produce a satellite sea surface temperature record based on measurements from the (Advanced) Along Track Scanning Radiometer ((A)ATSR) series of sensors. Within ARC, independence from the in situ record was achieved through the use of radiative transfer modeling to determine the retrieval coefficients used to calculate the skin sea surface temperature from the observed brightness temperatures [13,14]. These were then converted to bulk temperatures and a standard time of 1030 am/pm local time using the models of Fairall et al. [15] and Kantha and Clayson [16].

More recently, in recognition of the importance of sustained observations, high quality independent climate data records and the contribution that Earth Observation (EO) data can make, the European Space Agency (ESA) launched the Climate Change Initiative (CCI) [17]. The primary aim of the CCI is to realize the full potential of EO data from archives held by both the ESA and other satellite agencies. As part of this initiative, the ESA SST CCI project [18] was set up with the goal of reprocessing the satellite based radiometric SST record from the Advance Very High Resolution Radiometer (AVHRR) and ATSR series of sensors to produce an independent satellite-based SST dataset with a target stability of 10 mK year^{-1} (0.1 K decade^{-1}) [18] and building on the previous effort of the ARC project.

Stability of observation is a key aspect of quality for any record intended for use in climate applications. Measurement of climatic change involves tracking of small differences over years and decades, with the signal being often similar in size to the error characteristics of the raw observations. By careful treatment of the observations the impact of errors can be minimized, allowing any long-term changes to be quantified. However, even with careful treatment, residual artifacts may exist, either in the mean values or the error characteristics. These artifacts may be step changes or long-term drift in the mean error or in the higher moments (such as variance). Detection of these residual artifacts, or inhomogeneity, relies on comparison with independent sources, which may themselves be inhomogeneous. Interpreting observed instability between sources then becomes an expert judgement taking account of the timescales, timing, and nature of the changes, and the availability of supplemental metadata.

Various algorithms exist for assessing time series for change points and homogeneity (e.g., see [19]) but these tend to have been developed using data from land stations in fixed locations. Within this paper we develop a method to assess the homogeneity of the ESA SST CCI dataset, using the Penalized Maximal t Test (PMT) [20] with satellite in situ differences aggregated over many different observations and platforms, and an ensemble approach to quantify the uncertainty in the timing and size of any change points. We then apply the method using three different in situ reference sources, using observations from the: Global Tropical Moored Buoy Array (GTMBA) (e.g., [21]); the global drifting buoy array [22]; and the array of Argo profiling floats [23]. Additionally we fit an auto-regressive trend model to identify any long-term drift in the satellite data relative to in situ data. Section 2 of this paper describes the different data sources used. Section 3 gives a brief summary of change-point analysis and describes how we have applied the PMT to our data. Results are given in Section 4, and a summary and conclusion are given in Section 5.

2. Data

Data from Experimental Version 1.8 of the ESA SST CCI project [24] have been used to develop and test the method presented in this paper. These data have been extracted from a multi-sensor match-up database (MMD) [25] containing collocated swath or level 2 pre-processed (L2P) data from the (A)ATSR series of satellites and in situ observations from a variety of platforms. The in situ observations and satellite retrievals are briefly described in this section.

2.1. In Situ Data

In situ data from the MMD was originally been extracted from the Met Office Hadley Centre Integrated Ocean Database (HadIOD) [26]. These were, in turn, extracted from Release 2.5 of the International Comprehensive Ocean and Atmosphere Data Set (ICOADS2.5; [9]) and the Hadley Centre EN4 dataset [27]. ICOADS2.5 and EN4 contain data from common sources, such as the

World Ocean Database (WOD) [28] Where an observation exists in both ICOADS2.5 and EN4, the observation from EN4 was retained in preference [26]. Whilst the match-up database contains observations from many different platforms, only those match-ups containing drifting buoy, GTMBA and near surface (<5 m) Argo data were used in this study. Temperature observations from ships, sub-surface measurements and extra-tropical moorings were been used due to either quality (ships [29], extra-tropical moorings [30]) or representativeness issues (sub-surface).

The drifting buoy temperature observations from HadIOD were extracted from ICOADS2.5 [9]. The data within ICOADS2.5 came from a number of overlapping sources, with observations duplicated between the different sources. These duplicates were removed by ICOADS2.5 processing [9]. Prior to ingestion into HadIOD, the observations also underwent quality control following Rayner et al. [4], with implausible values and gross errors in location (time and space) and temperature flagged. Additionally, the observations had a platform level quality check applied prior to ingestion into HadIOD, with the quality of the observations made by individual buoys tracked over time through comparison with a satellite-based analysis [31]. The GTMBA data underwent the same processing as the drifting buoy data and came from either ICOADS2.5 or EN4. As noted above, where a duplicate was found, the EN4 data were retained in preference.

The surface Argo data within HadIOD were extracted from EN4 [27]. As with ICOADS2.5, EN4 contains data from a number of sources, including from the Argo Global Data Assembly Centre (GDAC) [23], the World Ocean Database (WOD) [28], and the Global Temperature Salinity Profile Programme. As part of the EN4 processing, duplicate temperature profiles were identified and removed, with the Argo GDAC data kept in preference to the other sources. Prior to ingestion into EN4 the profiles undergo additional quality control checks, including *inter alia*: gray list checks, parameter range checks, profile checks (spikes, steps, etc.), bathymetry, and depth checks. Full details on the duplicate elimination and quality control can be found in Good et al. [27] and Ingleby and Huddleston [32].

The primary characteristics of the drifting buoy, GTMBA and near surface Argo data are summarized in Table 1. The uncertainty values listed are those as reported by Atkinson et al. [26]. Also listed are validation statistics for the initial version of the ESA SST CCI ATSR global analysis as reported by Merchant et al. [18]. The drifting buoy observations are the most numerous and closest to the surface but have higher uncertainties compared to the other sources. Alternative estimates of the uncertainty in the drifter data ranges from ~0.2 K [33] to ~0.5 K [34]. The manufacturer has little impact on the quality of drifting buoy data [30]. The GTMBA data have smaller uncertainties compared to the drifting buoy data but are limited to the tropics and at a slightly greater depth. The near surface Argo measurements have the smallest uncertainty due to measurement errors but are made at the greatest depth compared to the other sources, and with far fewer observations available. Overall, based on the validation results listed in Table 1, the expected error variance is broadly equal across the three different in situ platforms.

Table 1. Uncertainty characteristics and depth of the in situ observations following Atkinson et al. [26]. Also listed are the robust standard deviation of the differences between the ESA SST CCI product and listed platforms, the median difference and number of match-ups from phase 1 of the ESA SST CCI project [18].

Platform	Atkinson et al. [26]			Merchant et al. [18]		
	Uncertainty due to Noise/Random Errors (K)	Uncertainty due to Correlated Errors (K)	Nominal Depth	RSD (K)	Median Difference (K)	Number of Match-Ups
GTMBA	0.020	0.000	1 m	0.28	+0.09	25,492
Drifting buoys	0.260	0.290	0.2 m	0.22	+0.05	2,392,462
Argo	0.002	0.000	3–5 m	0.26	+0.04	8867

There have been few technological changes to the GTMBA, drifting buoy and Argo observations that will have significantly impacted the quality of the data over the study period. However, the density and distribution of the observing systems has changed significantly. There was a marked increase in the number of drifting buoy observations in the late 1990s and early 2000s. The number of available match-ups are shown in Figure 1. There has also been a spatial evolution in the location of the data. For example, prior to ~2000, drifting buoy observations of the SST tended to be poleward of the tropics, with very few observations within 10° of the equator. Figure 2 shows the percentage of drifting buoy match-ups available from the MMD in a given 5° latitude band calculated annually. The lack of tropical observations prior to 2000 was clearly seen.

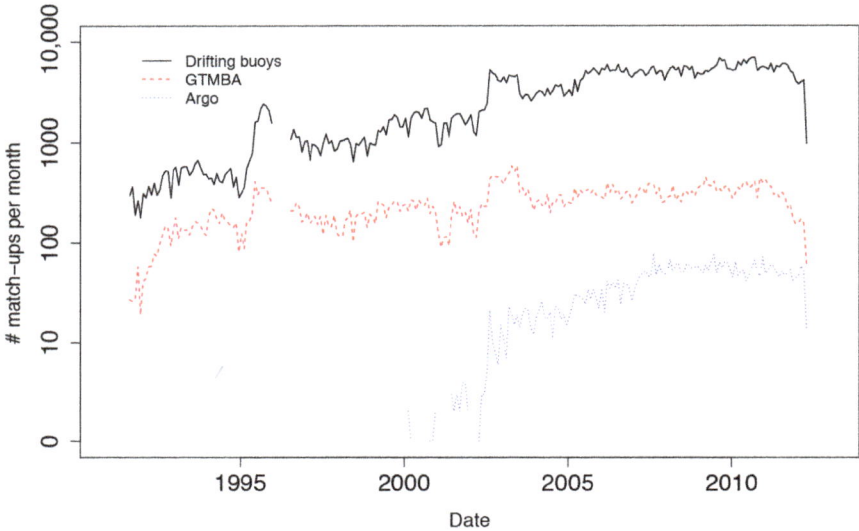

Figure 1. Number of match-ups per month between drifting buoy (black/solid line), GTMBA (red/dashed line) and Argo (blue/dotted line) observations and the ESA SST CCI data meeting the match-up criteria.

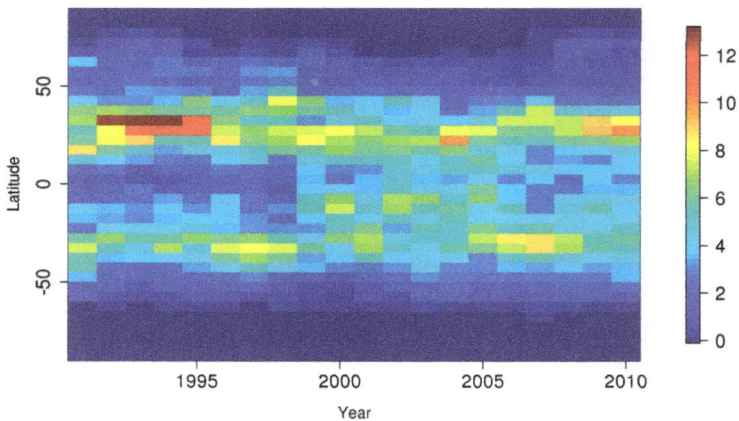

Figure 2. Percentage of drifting buoy match-ups available per 5° latitude band calculated annually.

2.2. Satellite Data

The match-up database from the ESA SST CCI project used within this study contained level 2 pre-processed (L2P) data from experimental version 1.8 of the project for the ATSR series of sensors. At the time of analysis, AVHRR data from v1.8 were unavailable. These will be available in subsequent versions. The algorithms used to generate the L2P data are summarized by Merchant et al. [18] and fully described in Merchant [24]. A general description of the (A)ATSR characteristics is given by Merchant et al. [35]. A brief summary is given in this section.

The L2P data contain estimates of the skin temperature (e.g., [36]) from a single swath with a resolution of 1 km, a swath width of 500 km, a 3-day repeat cycle and a local equator crossing time of either 1030 am/pm (ATSR and ATSR2) or 1000 am/pm (AATSR). For the ATSR sensors, the skin temperature estimates are based on the linear combination of 2 or 3 brightness temperature channels [24] following the ARC project [13,14]. The number of channels depends on whether the retrievals are for the daytime or nighttime and the satellite (See Table 2). The coefficients used in the linear combination of channels have been derived using radiative transfer modeling and atmospheric profiles from the ERA 40 reanalysis model. Cloud screening, trace gas atmospheric profiles and the impact of aerosols are accounted for as part of the retrieval process. Full details are given in Merchant [24]. Consistency between ATSR sensors was achieved using the overlap between the sensors and by referencing the brightness temperatures of all channels and sensors to be consistent with the 3.7 and 11 μm channels of the AATSR (e.g., [35]).

Table 2. Primary channels, satellite views and local equator crossing time for the different ATSR sensors and daytime/night time retrievals.

	ATSR1		ATSR2		AATSR	
	Day	Night	Day	Night	Day	Night
View	Dual	Dual	Dual	Dual	Dual	Dual
Channels (μm)	11, 12	11, 12	11, 12	3.7, 11, 12	11, 12	3.7, 11, 12
LECT	1030	2230	1030	2230	1000	2200
Period	August 1991–December 1995		August 1995–June 2003		July 2002–April 2012	

In addition to the skin temperature, the L2P files contain estimates of the sub-skin temperature and water temperature at different depths in the near surface ocean (0.2 m, 1 m and 5 m) adjusted to 1030 am/pm local time. The sub-skin temperatures were estimated using the Fairall et al. [15] model to adjust the skin temperatures to sub-skin temperatures. These were then adjusted to bulk temperatures using the model of Kantha and Clayson [16]. Both models were forced using the output from the ERA Interim Reanalysis model [37] and were used to adjust the temperatures to be representative of 1030 am/pm local time [24].

A number of events may impact on the quality of the (A)ATSR based retrievals. These are summarized in Figure 2 of Merchant et al. [35]. Shortly before the launch of the first ATSR sensor on board the ERS-1 satellite Mount Pinatubo erupted in June 1991, injecting a large amount of aerosol into the stratosphere. Whilst the (A)ATSR sensors have been designed to be robust to aerosol through the use of dual view sensors (e.g., [35]) there may be some residual effect present in the SST retrievals in the years shortly after 1991. In May 1992 the 3.7 micron channel failed on board the ATSR1 sensor. This can be seen in Table 2, with only two channel retrievals available for the ATSR1 sensor. Whilst data are available for 1996 from ATSR1, these are of lower quality and were excluded. Data from ATSR2 is available from August 1995 through to June 2003, with the exception of a six-month gap in the first half of 1996. During 2001 a gyroscope failed on the ERS-2 satellite, with data quality impacted between January 2001 and June/July 2001 when a zero gyro mode was implemented to improve the quality. Data from the AATSR sensor are available between July 2002 and April 2012. There are no known events thought to impact on the quality of the data in this period.

3. Method

The stability of the ESA SST CCI data was assessed using two different methods. The first tested for change points in the satellite data using a test based on the PMT algorithm [20]. The second tested for a residual trend in the time series of differences between the satellite data and in situ reference series. This section describes those tests. All tests were implemented using the R programming language [38].

3.1. Penalized Maximal t Test and Application to ESA SST CCI Data

3.1.1. Homogeneity Testing and the Penalized Maximal T Test

The majority of homogeneity tests for climate data are based on testing the relative homogeneity of a time series from one station with another nearby station that has a similar climatic signal. A discussion of the background and benchmarking of different homogeneity tests is discussed in Venema et al. [19]. Early tests, such as the Standard Normal Homogeneity Test (SNHT) [39], were based on classical statistical tests, comparing a target series to a reference series believed to be homogenous. More recently, tests have been developed to detect change points using inhomogeneous reference series and pairwise comparisons (e.g., [40]). In both techniques, using a reference series and those based on pairwise comparisons, test time series are created by either differencing or calculating the ratio between the target and reference time series or between pairs of time series across a local network of stations. A test statistic is then calculated for all possible change points and the most likely change point identified. If this exceeds some critical value it is then flagged as a change point. These tests are often applied recursively to find multiple change points.

Over the oceans, with the exception of the GTMBA [21], we have few high quality fixed stations spanning multiple years to use. The extra-tropical moorings tend to be in coastal regions where a small change in location can have a significant impact on the SST time series. Additionally, the quality of these observations is low compared to the GTMBA (e.g., see [26,30,41]). This makes application of tests based on multiple pairwise comparisons difficult with marine data. Instead we have opted to use the PMT [20] algorithm and aggregated time series of satellite and in situ data (see Section 3.1.2).

In the PMT algorithm [20], the test is applied to a test time series, X_t, with the null hypothesis given by:

$$H_0 : \{X_t\} \sim N\left(\mu, \sigma^2\right) \tag{1}$$

and the alternative hypothesis given by:

$$H_a : \begin{cases} \{X_t\} \sim N(\mu_1, \sigma^2), & t = 1, \ldots, k \\ \{X_t\} \sim N(\mu_2, \sigma^2), & t = k+1, \ldots, N \end{cases} \tag{2}$$

where X_t is the time series being tested for change points; $\{X_t\} \sim N(\mu, \sigma)$ indicates that the population $\{X_t\}$ follow a normal distribution with a mean μ and standard deviation σ. The test time series, X_t, is usually a time series of differences between the base time series, and a reference time series although ratios can be used. For the null hypothesis, the data from the test series (X_t) are assumed to come from a normal distribution with a mean difference (or ratio) of μ and standard deviation of σ. For the alternative hypothesis, the mean of the time series before the change point at time k is given by μ_1 and by μ_2 after time k.

Within the PMT algorithm, the null hypothesis was tested by calculating the test statistic for the two-sample t test with unknown but equal variance at every possible change point. The most likely position of a change point is then identified by the location where this test statistic, multiplied by a penalization factor, is a maximum. The penalization factor accounts for the increased false alarm rate observed when the sample sizes are unequal [20]. The test statistic for the two sample t test is given by:

$$T(k) = \frac{1}{\hat{\sigma}_k} \left[\frac{k(N-k)}{N} \right]^{0.5} (\overline{X_1} - \overline{X_2}) \tag{3}$$

where:

$$\overline{X_1} = \frac{1}{k} \sum_{1 \leq t \leq k} (X_t) \tag{4}$$

$$\overline{X_2} = \frac{1}{N-k} \sum_{(k+1) \leq t \leq N} (X_t) \tag{5}$$

$$\hat{\sigma}_k = \frac{1}{N-2} \left[\sum_{1 \leq t \leq k} (X_t - \overline{X_1})^2 + \sum_{(k+1) \leq t \leq N} (X_t - \overline{X_2})^2 \right] \tag{6}$$

and N is the number of time steps and k the position of the potential break point. The test statistic for the PMT is then given by [20]:

$$PT_{max} = \max_{1 \leq k \leq (N-1)} [P(k)T(k)] \tag{7}$$

where $P(k)$ is the penalization factor. When PT_{max} exceeds a critical value, the null hypothesis is rejected and a break point at position k identified. Both the penalization factor and critical values for PT_{max} have been determined empirically, and full details are given in Wang et al. [20]. The PMT algorithm has been updated [42] to pre-whiten the test time series prior to the calculation of step size, in order to account for any auto-correlation in the data. Within this study we used the original PMT algorithm [20] implemented in the RHTest software package [43]. This should not impact our conclusions as we were only interested in the detection of steps and not the subsequent adjustment.

3.1.2. Application of the PMT to ESA SST CCI Data

Three different change point analyses were using the PMT algorithm and in situ data. The analyses were identical other than the source platforms for the in situ observations. The platforms used were:

1. Observations from the GTMBA;
2. Observations from drifting buoys;
3. Near surface observations from the Argo profiling floats.

An additional assessment has been performed using synthetic data to test the sensitivity of the method to detect a step of 0.05 K at different points in time.

Pre-Processing and Selection of Match-Ups

The following criteria have been used to select match-ups:

1. Satellite quality level equal to 4 (acceptable) or 5 (highest) (see [24] for description of quality levels).
2. HadIOD quality control [31] passed for in situ observations.
3. Separation distance between in situ observation and satellite retrieval <100 km.
4. Maximum time separation between in situ observation and satellite retrieval $\leq$ 1 h.
5. Satellite in situ difference <5 standard deviations of all match-ups for a given platform.

Within the MMD match-ups are defined as the L2P pixel from the various satellite sensors containing the location of the reference in situ observation and where the overpass time is within $\pm$2 h of the reference observation [25]. The data extracted from the MMD contained one L2P match-up per reference observation for each (A)ATSR sensor plus the corresponding L2P data from the surrounding pixels on a 7 × 7 grid, with the grid centered on the match-up pixel. Pixels with quality level <4 are set to missing (criteria 1). As described in Section 2, the resolution of the pixels was ~1 km, and the L2P data was adjusted to 0.2 m depth and 1030 am/pm local time. In addition to the L2P data, the data extracted from the MMD included the in situ reference observation for each match-up plus observations nearby in time ($\pm$36 h) from the platform making the reference observation. Those failing the HadIOD QC were discarded (criteria 2). For Argo data, only the observation corresponding to the match-up was available due to the ~10 day repeat cycle for Argo floats.

Due to cloud screening and quality control, a significant proportion of the pixels in a 7 × 7 scene, including the central pixel, contained missing values. To mitigate the impact of missing data and increase the number of match-ups, the L2P data was averaged over the 7 × 7 scene to give an areal average, analogous to Level 3 data. To account for the time difference between the in situ observations and the time of the satellite data, adjusted to 1030 am/pm local time, the in situ observations, and location, either side of the satellite data and within 1 h (criteria 4) was interpolated to 1030 am/pm local time. The average location of the pixels containing valid data from the 7 × 7 scene and interpolated buoy locations were used to test criteria 3. This test was required due to a small number of erroneous drifting buoy locations in the HadIOD dataset. For Argo, a single observation was available and only those match-ups within 1 h of 1030 am/pm local time were used. It should be noted that the nearest pixel to the reference observation could have been selected in the case of missing data for the central pixel. However, during testing the choice of satellite data to use, nearest pixel or mean of 7 × 7 scene, made little difference to the results.

As a final step, following extraction and selection of the match-ups, the mean and standard deviation of the satellite in situ 0.2 m SST across all match-ups were calculated for the GTMBA, drifting buoy data and profiling floats separately. Any difference exceeding 5 standard deviations of the differences for the respective platform was then excluded from further processing (criteria 5).

Ensemble and Aggregation

As noted above, we have used aggregated time series in the application of the PMT. From Equation (3), it can be seen that by minimizing the standard deviation of the test series ($\hat{\sigma}_k$) we increased the sensitivity of the test. This term will include contributions from instrumental noise and errors due to both the satellite data and in situ observations. Climatic variations were minimized through use of collocated data. By averaging multiple match-ups per month the contribution of instrumental noise and errors to the monthly mean values was minimized. Based on propagation of error we would expect the standard deviation of the monthly mean values to be reduced. For independent normally distributed errors we expect the reduction in $\hat{\sigma}_k$ to be proportional to $\frac{1}{\sqrt{n}}$ where n is the number of match-ups per month.

As an example, we sub-sampled the selected match-ups for the drifting buoys and AATSR sensor for sample sizes between 1 and 100 match-ups per month and calculated the monthly mean difference. For each sample size 20 realizations were generated and the standard deviation of the monthly mean differences calculated. These were then averaged across all realizations of a given sample size and plotted as a function of sample size in Figure 3 (dots). The theoretical relationship was also shown, with a strong agreement between the observed and expected reduction in standard deviation with increasing sample size clearly seen. The rate of reduction was slightly less than expected for independent data, indicating that the samples were not strictly independent. This was to be expected as a given buoy may contribute to multiple match-ups in a given month. However, this impact was small.

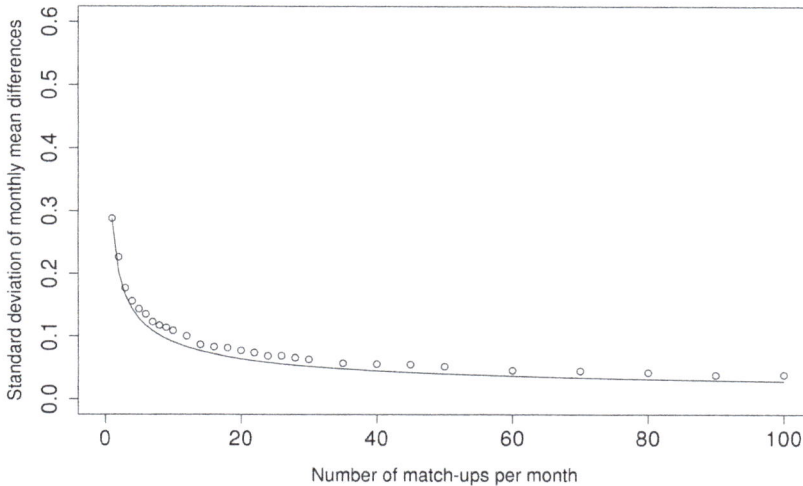

Figure 3. Standard deviation of monthly mean AATSR drifting buoy SST observations as a function of number of match-ups per month (dots). The theoretical relationship, $\frac{\sigma}{\sqrt{n}}$, is also shown (solid line).

To minimize the impact of errors, and increase the sensitivity of the PMT, aggregate time series were generated by sub-sampling the available match-ups and averaging to give monthly mean values. For each analysis, an ensemble of 1000 realizations was generated, with a fixed number of match-ups used to minimize the variance of the monthly standard deviation in time. The number of match-ups used depended on the data availability (see Section 4). The use of an ensemble allowed the impact of the sub-sampling on the analysis and the uncertainty in the date and size of any detected change points to be quantified.

3.1.3. False Alarm Rate

Within our application of the PMT we used a nominal false alarm rate (FAR) of 1% (see Section 4.1). Based on a FAR of 1%, for every 100 applications of the PMT to independent data, we expected one application to incorrectly identify a step change. Due to our use of ensembles that contained a proportion of the same match-ups between different ensemble members, our applications of the PMT were not independent. As a result, we expected a FAR that differs from the nominal FAR specified as part of the PMT. To quantify the impact of our method on the nominal FAR we calculated the observed FAR for each analysis presented by replacing the match-ups with synthetic data and reapplying the tests. Within the synthetic dataset, the in situ observations were replaced by zeros and the satellite data with normally distributed noise with a mean of zero and standard deviation corresponding to the observed standard deviation of the satellite in situ differences calculated each month and for each sensor/platform configuration.

3.2. Stability Assessment

The long-term stability of the satellite SST estimates relative to the in situ data was quantified by fitting a linear trend model to the satellite in situ differences. The (A)ATSR SST retrievals are known to contain small seasonally varying biases relative to the in situ data. Within a time series of the differences between the (A)ATSR and in situ temperatures this will appear as auto correlated errors. To take these into account when assessing the stability we used a lag 1 autoregressive (AR1) model, with the model given by (e.g., [44]):

$$X_i = \beta_0 + \beta_1 t_i + \epsilon_i \tag{8}$$

$$\epsilon_i = \rho\epsilon_{i-1} + e_i \qquad (9)$$

where X_i is the mean difference for time step i; β_0 the intercept for the trend model; β_1 the slope of the trend model; t_i the time variable for time step i; ϵ_i the auto regressive error at time step i; ρ the autocorrelation parameter at lag 1; and e_i the independent error or noise term for time step i. If the autocorrelation is not taken into account, the degrees of freedom used to calculate the confidence intervals will be overestimated and the uncertainty range will be too narrow.

4. Results

4.1. Step Change Analyses

Table 3 summarizes the configuration of the three different analyses and time period covered. The sample size lists the number of match-ups selected per ensemble member. The number selected was determined to balance the length of the time series available with the sensitivity of the analysis. The sample sizes for GTMBA and Argo were smaller than for the drifting buoy analysis due to the more limited availability of the data.

Table 3. Summary of the analyses performed using the PMT. The FAR column indicates the specified false alarm rate used in the tests.

Reference Platform (Period)	Sample Size	Ensemble Size	Reference SST	FAR
GTMBA (1991 onwards)	25	1000	Linear interpolation to 1030 AM/PM local time	1%
Drifting buoy (~1996 onwards)	100	1000	Linear interpolation to 1030 AM/PM local time	1%
Argo (~2004 onwards)	16	1000	Nearest observation to 1030 AM/PM local time (max separation 1 h)	1%

4.1.1. GTMBA

The first analysis repeated the homogeneity assessment made as part of the ARC project [35]. Within ARC, the PMT was applied to a single aggregate time series for a subset of the moorings from the GTMBA in the Pacific. Based on this analysis, the (A)ATSR record was found to be stable in the tropics after the effects of the Pinatubo eruption had dissipated (by around 1994/1995). Similar results were found within this analysis, but by using all available GTMBA moorings from the match-up database. Table 4 summarizes the results of the analysis, listing the number of change points detected per ensemble as a fraction of the ensemble size. A large number of ensemble members (~20%) had at least one change point detected compared to a nominal FAR of 1% and actual FAR of <1% based on the synthetic data. The majority of the ensemble members with a change point have a single change point detected (17%). A small fraction (1.6%) had two change points detected.

Table 4. Number of change points detected vs. percentage (%) of the 1000 ensemble members for the GTMBA analysis. Also listed is the number of change points detected when the observations are replaced with noise.

Number of Change Points	% of Ensemble Members with Detected Change Points	Observed False Alarm Rate (FAR) (%)
0	80.9	99.4
1	17.1	0.2
2	1.6	0.4
3	0.4	0.0

Figure 4 shows the results of the change point analysis as time series of the mean change point models averaged over those ensemble members with the same number of change points (panel a, black lines). Also shown are the monthly mean satellite in situ differences for the individual ensemble members (red lines). Figure 4b shows a histogram of the change point dates. The majority of change points detected occurred prior to 2000 and Figure 5 shows a scatter plot of the step size versus date for those ensemble members with a single change point occurring prior to the start of 2000. Three different clusters were seen, with a step size of ~0.08 K around the end of 1992 for the earlier cluster. Subsequent clusters decreased in magnitude of step size over time. Overall, the evidence suggests a warming in the satellite data during this period, with a median step of 0.054 K (95% confidence interval 0.022 to 0.089 K) occurring around February 1994 (October 1992 to December 1995). The confidence intervals were 2.5% and 97.5% quantiles estimated using the R function "quantile" and based on Hyndman and Fan [45]. The cluster of points in December 1995 coincided with the end of the ATSR1 data used, with the PMT algorithm and RHTest preferentially using the metadata to set the date of the change point.

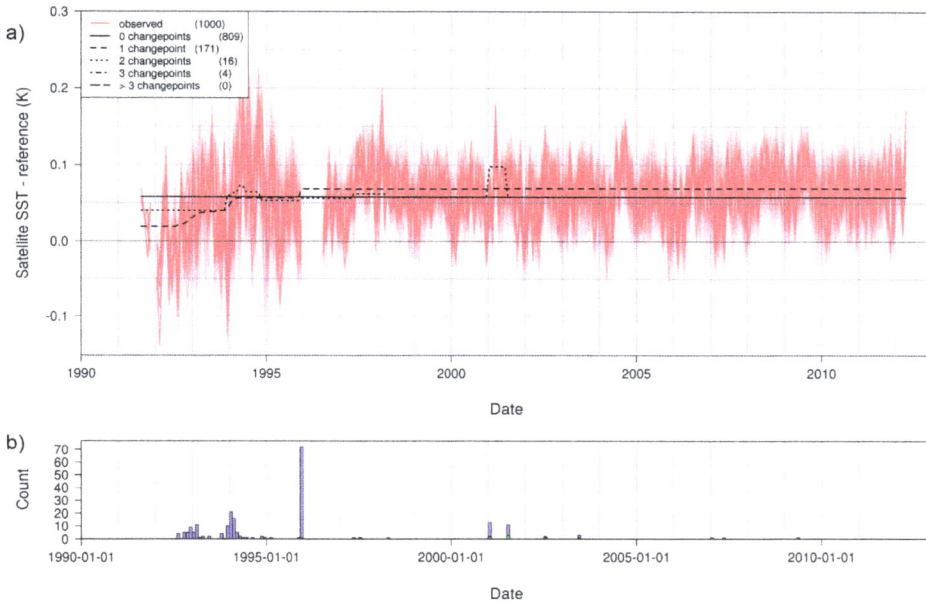

Figure 4. Results of the ensemble application of PMT using the GTMBA. (**a**) The monthly mean differences (satellite in situ) for the individual ensemble members are shown in red. Also shown are the means of the model fitted by PMT grouped by number of change points detected. (**b**) Histogram of the data of the change points detected across all ensemble members.

The evidence for the change points after 2000 shown in Figure 4 is weaker, with only 16 out of 1000 ensemble members showing the change points during January and July 2001. These change points are coincident with the gyroscope failure on the ERS-2 satellite and implementation of the zero-gyro mode in July 2001 to correct for the effects of this on the SST (e.g., Merchant et al., 2012). The size of these changes are +0.08 K and −0.07 K respectively. Without the supporting metadata neither of these change points would be significant based on the GTMBA data. However, similar change points have been detected in the drifting buoy analysis, but with opposite signs, suggesting that the impact of the gyro failure varies between the tropics and higher latitudes.

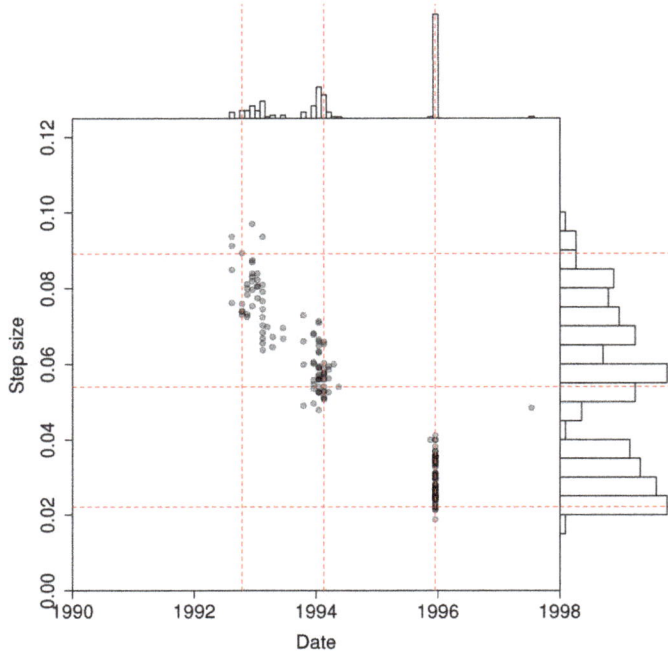

Figure 5. Scatter plot and histograms of the change point date versus step size for the GTMBA ensemble members where 1 change point was detected. Also shown are the median and 95% confidence intervals for the date and step size (red dashed lines).

4.1.2. Drifting Buoys

Table 5 summarizes the results of the change point analysis using the drifting buoy observations. One or more change points were detected in a large proportion of the ensemble members, with only 35% having no change point detected. As with the GTMBA, this is in contrast to a nominal FAR of 1% and actual FAR of 4.4% using the synthetic data. The majority of ensemble members with a change point have multiple change points detected. Figure 6 shows the timing of the change points, with the mean change point model averaged over ensemble members with the same number of change points shown in panel (a) (black lines) and histogram of change point dates in the panel (b). Also shown are the monthly mean differences for the individual ensemble members (panel (a); red lines). The change points occurring in January 2001 and July 2001 were very clear, both in the meantime series for the change point models and histogram of change point dates. There was also a clear warming in the satellite–buoy differences towards the end of the 1990s, but with more uncertainty over the timing of the change. For those ensemble members with a single change, the change occurs during the 1990s, with the record after ~2001 being relatively stable.

As noted above, the change points during 2001 (January and July) were associated with the failure of a gyroscope on the ERS-2 satellite and the period when only nadir (single) view SSTs were available. The impact of the use of single view SST retrieval during this period can be clearly seen in the uncertainty estimates provided in the L2P data (Figure 7). Figure 8 shows a scatter plot of the detected step sizes and dates across the ensemble (top) split into four groups. Individual scatter plots and histograms for the individual groups are also shown. The median values and 95% confidence intervals for the different groups are listed in Table 6 together with the size of each group. The first group (black circles, Figure 8a,b) is clustered in the mid to late 1990s, with a median step of 0.053 K and a median date of February 1999. This group is discussed in more detail below. The second group (red squares, Figure 8a,c) was coincident with the gyroscope failure, with a step of −0.060 K occurring during January 2001. The third group (green diamonds, Figure 8a,d) coincided with the commencement of the zero gyro mode in July 2007 and had a median step size of +0.054 K. The final group (blue triangles, Figure 8a,e) was relatively small and spread out over the AATSR record from middle 2002 onwards, with a cluster of change points detected around 2002–2003 and a small number later in the record. The median size change point for this group was −0.029 K with a date of July 2002. This final group coincided with start of the AATSR data, with the majority of change points flagged for 2002 across the ensemble only significant if supported by metadata.

Table 5. As Table 4 but for the drifting buoy analysis.

Number of Change Points	% of Ensemble Members with Detected Change Points	Observed False Alarm Rate (FAR) (%)
0	35.0	96.6
1	20.5	2.3
2	21.1	1.1
3	18.7	0.0
>3	4.7	0.0

Table 6. Identified step changes, size and dates for the drifting buoy analysis. The 95% confidence intervals have been estimated based on the 2.5% and 97.5% quantiles of the step sizes and dates.

Break Number (Count)	Step Size (K) (95% Confidence Interval)	Position (95% Confidence Interval)
1 (520)	0.053 (0.029–0.083)	15-02-1999 (15-12-1995 to 15-11-1999)
2 (424)	−0.060 (−0.114−−0.021)	15-01-2001 (15-07-2000 to 15-01-2001)
3 (356)	0.054 (0.015–0.102)	15-07-2001 (15-06-2001 to 15-07-2001)
4 (56)	−0.029 (−0.052−−0.017)	15-07-2002 (26-06-2002 to 02-01-2007)

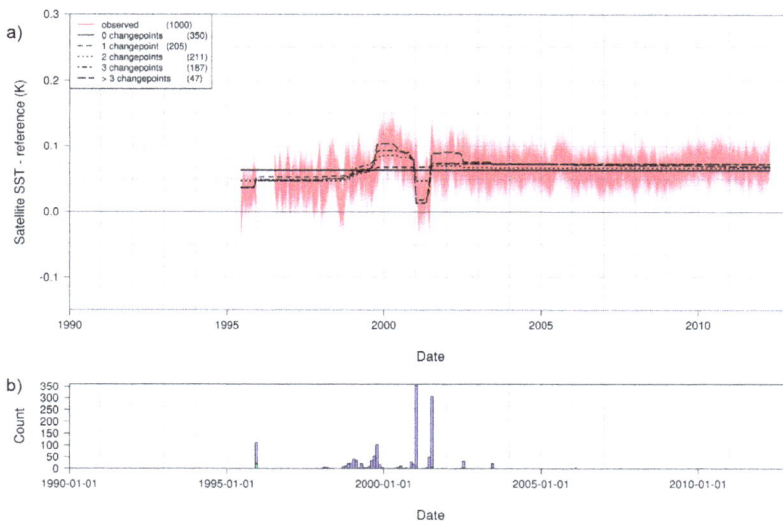

Figure 6. As Figure 3 but for drifting buoy observations. (**a**) The monthly mean differences (satellite in situ) for the individual ensemble members are shown in red. Also shown are the means of the model fitted by PMT grouped by number of change points detected. (**b**) Histogram of the data of the change points detected across all ensemble members.

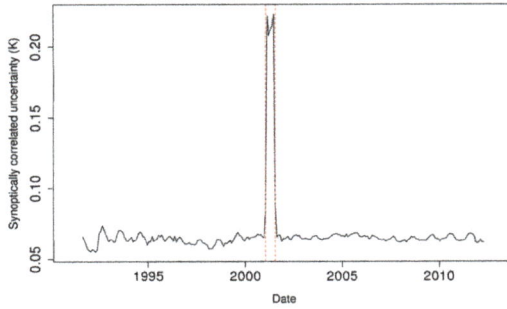

Figure 7. Monthly mean synoptically-correlated uncertainty estimate from the L2P data calculated across all available drifting buoy match-ups (black line). Also shown are the change points identified in 2001 (vertical red dashed lines).

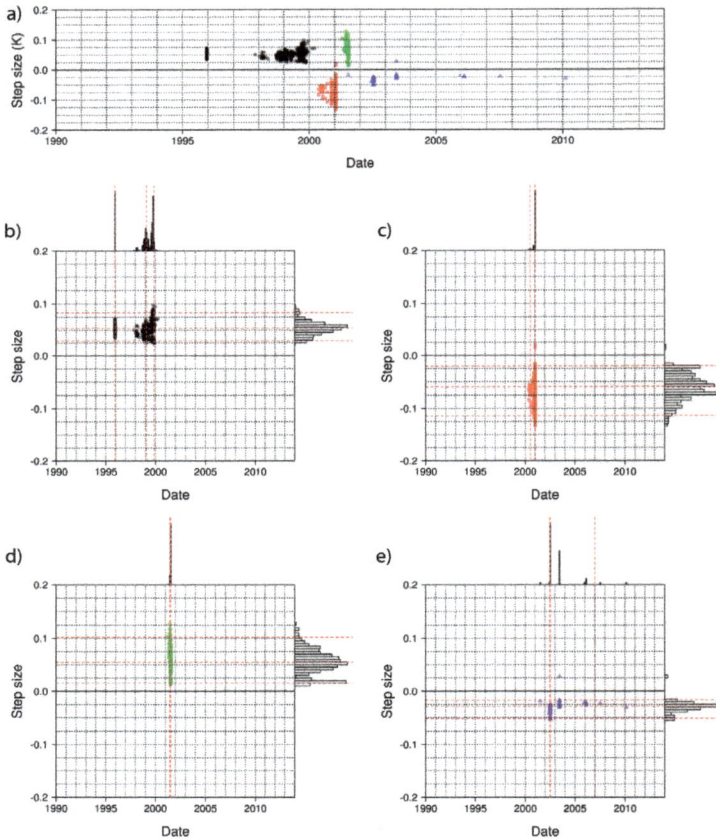

Figure 8. Scatter plot of the identified change point dates and step sizes for the drifting buoy analysis split into four groups. Panel (**a**) shows the steps identified over the full record—black represents break 1, red break 2, green break 3, and blue break 4. Panels (**b–e**) show scatter plots and histograms for the individual groups, the red dashed lines indicate the 2.5%, 50%, and 97.5% quantiles.

Whilst a large proportion of the ensemble had a change point detected pre-2000, this was a period where the drifting buoy network underwent a major evolution, with large changes to the spatial sampling by the drifting buoys (Figure 2). When these changes are coupled with regional differences previously reported for (A)ATSR in situ comparisons, with tropical differences tending to be positive and extra tropical differences cooler or negative (e.g., Figure 6 from [18]), we would expect to see some evidence in the time series of differences. As more observations and match-ups are made in the tropics we would expect to see a warming in the satellite–buoy differences, and this is what is seen in Figure 6 and the sign of the steps detected in the 1990s. As a result, it is likely that the pre-2000 change points were a feature of the evolution of the drifting buoy network and the changing spatial sampling by drifting buoys, rather than being attributed to the satellite data.

4.1.3. Argo

Figure 9 summarizes the results of the change point analysis using Argo data, with the monthly mean satellite in situ differences plotted for each ensemble member (red lines). In contrast to the drifting buoy network there were few match-ups available, e.g., Figure 1, and only 16 match-ups per month were used in this analysis. The impact of this reduced number of match-ups was evident in the variability of the monthly mean differences shown in Figure 9, and gaps in the record prior to the start of 2005. The lack of independence of the ensemble members was clearly seen, with less variability between the different ensemble members. This was also evident in the single change point detected compared to the ~10 that we would expect for an ensemble size of 1000 and 1% FAR. This suggests that there are no significant change points (relative to Argo) for the period 2005 onwards. The actual false alarm rate using synthetic data was slightly higher, with change points detected in 14 (1.4%) out of the 1000 ensemble members.

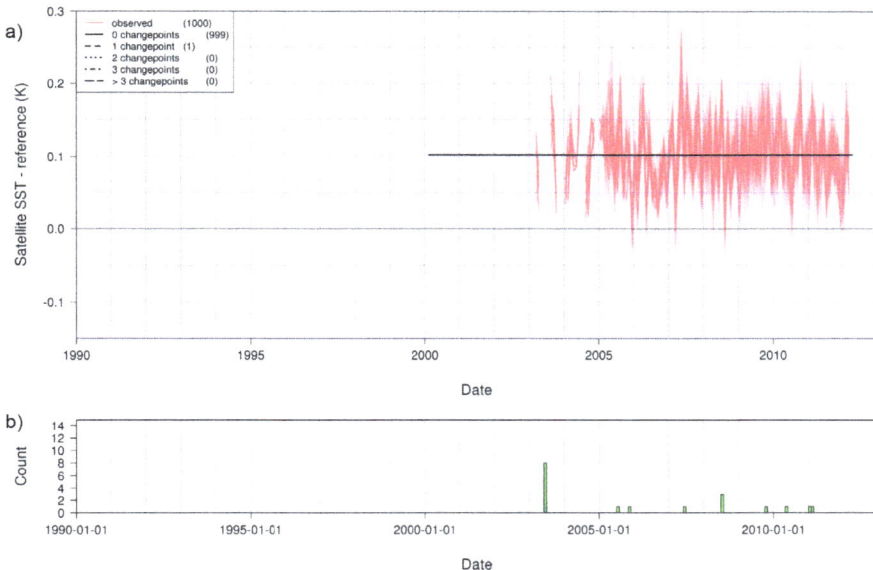

Figure 9. As for Figure 3 but for Argo observations. (**a**) The monthly mean differences (satellite in situ) for the individual ensemble members are shown in red. Also shown are the means of the model fitted by PMT grouped by number of change points detected. (**b**) Shows a histogram of the false alarms.

4.1.4. Sensitivity Tests

In addition to specifying our confidence in any detected change points and that they were not the result of chance (i.e., type I errors or FAR), it was also useful to specify our confidence of detecting a change point given our data and the probability of falsely accepting our null hypothesis. This concept is known as the statistical power of a test or the type II error rate, with the statistical power given by $1 - \beta$ where β is the type II error rate. The statistical power of the PMT has been quantified and published by Wang et al. [20]. Table 7 lists the ensemble median standard deviation of the monthly mean differences for the different analyses and number of months with valid data. Also listed are the mean hit rates of the PMT from Table 5 of Wang et al. [20], interpolated to match those in Table 7 and converted to statistical power (%).

Table 7. Median standard deviation of the monthly mean differences calculated across the different ensembles, ratio of the target step size (0.05) to standard deviation, number of months with valid data and estimated power interpolated from Table 5 of Wang et al. [20].

Analysis	Median Inter-Month Standard Deviation (σ)	$\frac{0.05}{\sigma}$	Number of Months	Power (%)
GTMBA	0.062	0.806	249	42.6%
Drifting buoys	0.039	1.282	203	69.6%
Argo	0.073	0.685	110	26.1%

From Table 7 it can be seen that the analysis based on the drifting buoy data had the greatest estimated statistical power, followed by the GTMBA and then the Argo analyses. To check the actual power of the analyses described we re-estimated the statistical power using the synthetic datasets described above. For each potential change point we inserted a step of 0.05 K into the satellite data in the synthetic datasets and repeated the change point analysis, sub-sampling, and applied the PMT 1000 times. We then calculated the power for each potential change point as the hit rate observed at that change point divided by 1000. The results expressed as % are shown in Figure 10. For the GTMBA array, the results indicated that there was a ~40–60% chance of detecting a step change using a single ensemble member over the majority of the period with data, but decreasing at either end of the time series. The impact of the sample size and reduction of noise with increasing sample size was seen, with a greater power of the drifting buoy analysis evident after ~2000. Based on this analysis, there was an ~80% chance of detecting a step of 0.05 K in a single ensemble member using the drifting buoys and sample size of 100. As with the GTMBA analysis, the power decreased towards either end of the time series. In contrast to the GTMBA and drifting buoy analysis, the Argo analysis showed very little power in detecting a step, due to the small number of match-ups available and larger inter-month standard deviation.

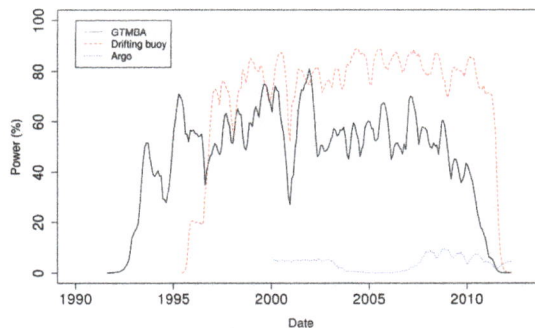

Figure 10. Estimated statistical power (%) for detection of 0.05 K step in the satellite data given the observing system characteristics and analysis configuration listed in Table 3.

These results were broadly comparable with those listed in Table 7. The results of the sensitivity tests were also broadly in line with the results for the different analysis presented above. Assuming that the detected change points were real, ~75% of the ensemble members for the drifting buoy analysis detected a change point, compared to a power of ~80%. For the GTMBA, change points were detected in ~20% of ensemble members, but towards the start of the time series. This was compared to a power of 40–50% in the middle of the time series, decreasing to 20% or less prior to 1995. Whilst the Argo statistical power was less than expected, the ensemble members used was not independent. This would lead to reduced power and an increased false alarm rate, as seen in this section and the previous section.

4.2. Stability Assessment

Figure 11 shows the monthly mean differences (black lines) for the different ensemble members from the GTMBA (panel (a)) and drifting buoy analysis (panel (c)) without adjustment for identified change points. Also shown are the results of fitting the AR1 trend model for each ensemble member (red lines). Figure 11 also shows the histogram of the fitted trend component for the GTMBA (panel (b)) and drifting buoys (panel (d)). No trend analysis has been performed using the Argo data due to the short time series available.

For both the GTMBA and the drifting buoy analysis the trends in the differenced time series were small. The median trend fitted to the GTMBA time series was 0.012 ± 0.015 K decade^{-1}, with the uncertainty equivalent to the median 95% confidence interval from the fitted AR1 models. The median trend component of the AR1 model fitted to the drifting buoy time series was 0.010 ± 0.014 K decade^{-1}. Table 8 lists the 2.5%, 50%, and 97.5% quantiles for both the trend estimate and confidence intervals across the ensemble. In both analyses, even with the detected change points, the stability of the satellite SST estimates relative to the in situ data was within the target 0.1 K decade^{-1}.

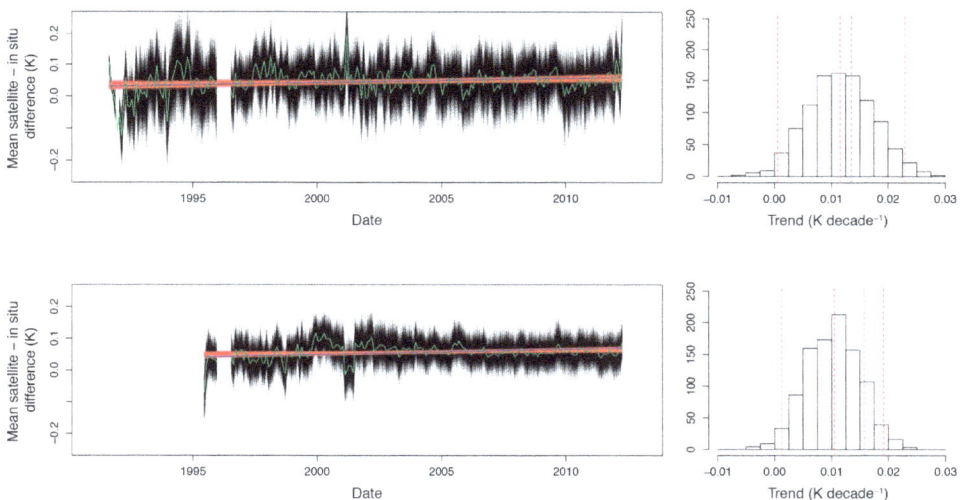

Figure 11. Panels (**a,c**): time series of the individual ensemble mean differences as a function of time (black lines) using observations from the (**a**) GTMBA and (**c**) drifting buoys, fitted AR1 trend models (red lines), mean differences averaged across all available match-ups (green lines) and AR1 model fitted to the mean of all available match-ups. Panels (**b,d**): histograms of the estimated trends across the ensembles for the (**b**) GTMBA and (**d**) drifting buoy data. The red dashed lines indicate the 2.5%, 50% and 97.5% quantiles. The blue dashed line indicates the trend fitted using all available match-ups.

Table 8. Trend and trend uncertainty estimates for the AR1 model applied to the GTMBA and drifting buoy ensembles.

Analysis	Trend (K decade^{-1})			Trend Uncertainty (K decade^{-1})		
	2.5%	50%	97.5%	2.5%	50%	97.5%
GTMBA	0.001	0.012	0.023	0.013	0.015	0.017
Drifting buoy	0.001	0.010	0.019	0.012	0.014	0.017

5. Conclusions

The ESA SST CCI project aims to generate an SST dataset that is independent of the in situ observations, free of inhomogeneity and with high temporal stability (drift <0.1 K decade^{-1}) in the global mean. Within this paper we have presented a method to test the homogeneity of the SST values from the CCI project in comparison to in situ observations, testing for change points in the SST retrievals. This work builds on that of others, extending the change-point analysis from stations at fixed locations to mobile station data through the use of aggregate time series and an ensemble approach. The use of aggregate time series increases the sensitivity of the tests applied by minimizing observational errors and noise through averaging. The use of ensembles allows us to quantify the sampling uncertainty due to our method and sub-sampling of the available match-up data to form our aggregate time series.

We have applied the method to three different sources of in situ data and SST retrievals based on the (A)ATSR series of sensors. The in situ sources were: (1) the GTMBA array, (2) the drifting buoy array (ATSR2 onwards), and (3) near surface Argo data (~2004 onwards). The results of the analysis using the GTMBA data were broadly comparable to that previously found in the ARC project [35], with evidence of a warming of the satellite SST retrievals compared to the GTMBA data between 1993 and 1995, and coincident with the decreasing levels of atmospheric aerosols following the Pinatubo eruption in 1991. In addition to recovery following Pinatubo, there is evidence of a pair of change points in the satellite data during 2001 in both the GTMBA and drifting buoy analysis. The first step detected occurs in January 2001 and is coincident with the failure of a gyroscope on the ERS-2 satellite. The second change point occurs in July 2001 and is coincident with beginning of the zero gyro mode SST retrievals in the ATSR 2 data. During this period (January–July 2001) only nadir view retrievals were available, with a substantial increase in the uncertainty in the satellite SST values. No change points were detected for the period including the Argo data.

The statistical power of the different analyses has been estimated, based on both the published power of the PMT [20] and an empirical assessment as part of the work presented in this paper. The drifting buoy analysis has the greatest statistical power, with a 70–80% chance of detecting a step of 0.05 K using a single ensemble member. For example, for an ensemble of 20 independent members containing a step of 0.05 K, we would expect 14 to 16 members to detect the step, and at most, one false positive. For a moderate ensemble size (~20–50) we can be virtually certain that a step of this magnitude would be detected. The power of the moored buoy analysis was in the range 40–50% for a step of 0.05 K. Again for a moderate ensemble size we can be virtually certain of detecting a change point in the data. The analysis using the Argo data showed much less statistical power and it is doubtful that we would be able to generate enough independent ensemble members to significantly increase the power. It should be noted that the success of the drifting buoy analysis and greater statistical power is primarily due to the large number of drifting buoys making SST observations. This can, in part, be credited to the use of design metrics, such as the equivalent buoy density and buoy need index of Zhang et al. [46], for the drifting buoy array.

In addition to the change point analysis, we have assessed the satellite in situ differences for drift by fitting an AR1 trend model to the monthly mean differences. We have applied the model to both the individual ensemble members and monthly mean differences using all available match-ups. Similar results are found using both methods, with small trends of order 0.01 K decade^{-1} present in the monthly mean differences. Uncertainties in the trend, both due to the model fit and sampling across the ensemble, are of a similar or slightly larger magnitude to the trend itself. The observed trends in the differences are significantly smaller than the target 0.1 K decade^{-1}.

Future assessments of the ESA SST CCI dataset(s) will be needed. For example, version 2 of the CCI dataset will contain SST retrievals from both the AVHRR and (A)ATSR sensors, and use updated algorithms. It is recommended that future assessments are made using the GTMBA match-ups in the tropics for the full period and drifting buoy data globally from ~2000 onwards. Whilst Argo has shown limited utility in the analyses presented, this is primarily due to the limited number of match-ups available for the (A)ATSR data. For AVHRR we would expect a greater number of match-ups due to the wider swath width, resulting in greater statistical power and an increased ability to detect small changes in the satellite data. This will be tested in future analyses.

Finally, the method developed in this paper is not restricted to the (A)ATSR sensors or to sea surface temperature. Provided that suitable sample sizes that are independent or near independent of each other can be generated from the available matchups and that the impact of changes to the in situ reference networks can be quantified the method can be applied. For example, the method could be applied to SST data from MODIS or from microwave sensors. Similarly, the data could be applied to wind speed or humidity retrievals where suitable reference data exists. Based on the results above for SST, and work by previous authors (e.g., [47]), a minimum sample size of ~100 matchups is recommended, but this will vary with the variable of interest and the quality of the reference data. For coastal regions, where sampling is more limited but there are more fixed stations, the use of the operational moored buoy network and methods such as the pairwise comparison test of Menne and Williams [40] may be more appropriate. However, this would require improvement to the operational buoy metadata record.

Acknowledgments: The work undertaken in this paper was funded by the European Space Agency (ESA ESRIN REF 3-13904/13/I-NB) Sea Surface Temperature Climate Change Initiative project. Analysis has been performed on the UK Science and Technology Facilities Councils (STFC) JASMIN supercomputing facility using version 3.4.0 of the R programming language. The RHTest v4 R package has been used to perform PMT tests and is available from http://etccdi.pacificclimate.org/software.shtml.

Author Contributions: David I. Berry wrote the paper, designed and developed the method to apply the homogeneity testing methods to mobile data, and performed the analysis. Gary K. Corlett provided the data from the match-up database used, provided guidance on how to use the match-ups, and contributed to the interpretation of the results. Owen Embury contributed to the interpretation of the results and provided guidance on events that may impact on the quality of the satellite data. Christopher J. Merchant conceived the assessment of the stability of the satellite SSTs using Argo and the other in situ platforms, building on the work of the ARC project, and contributed to the interpretation of the results.

Conflicts of Interest: The authors declare no conflict of interest.

Glossary

AATSR	Advanced Along Track Scanning Radiometer
AR1	Lag 1 Auto-Regressive Model
ARC	(A)ATSR Reprocessing For Climate Project
ATSR	Along Track Scanning Radiometer
AVHRR	Advanced Very High Resolution Radiometer
CCI	Climate Change Initiative
ECMWF	European Centre For Medium-Range Weather Forecasts
EN4	Hadley Centre EN4 Dataset
EO	Earth Observation
ERA	ECMWF Re-Analysis
ERS-1	European Remote Sensing Satellite 1
ERS-2	European Remote Sensing Satellite 2
ESA	European Space Agency
FAR	False Alarm Rate
GDAC	Global Data Assembly Centre

GTMBA	Global Tropical Moored Buoy Array
HadIOD	Hadley Centre Integrated Ocean Database
ICOADS	International Comprehensive Ocean-Atmosphere Data Set
L2P	Level 2 Pre-Processed
MMD	Multi-Sensor Match-Up Database
PMT	Penalized Maximal t Test
RSD	Robust Standard Deviation
SNHT	Standard Normal Homogeneity Test
SST	Sea Surface Temperature
WOD	World Ocean Database

References

1. Stocker, T.F.; Qin, D.; Plattner, G.-K.; Tignor, M.; Allen, S.K.; Boschung, J.; Nauels, A.; Xia, Y.; Bex, V.; Midgley, P.M. (Eds.) *IPCC Climate Change 2013: The Physical Science Basis. Contribution of Working Group I to the Fifth Assessment Report of the Intergovernmental Panel on Climate Change*; Cambridge University Press: Cambridge, UK; New York, NY, USA, 2013.
2. Blunden, J.; Arndt, D.S. State of the Climate in 2016. *Bull. Am. Meteorol. Soc.* **2017**, *98*, S93–S128. [CrossRef]
3. Huang, B.; Thorne, P.W.; Banzon, V.F.; Boyer, T.; Chepurin, G.; Lawrimore, J.H.; Menne, M.J.; Smith, T.M.; Vose, R.S.; Zhang, H.-M. Extended Reconstructed Sea Surface Temperature, Version 5 (ERSSTv5): Upgrades, Validations, and Intercomparisons. *J. Clim.* **2017**, *30*, 8179–8205. [CrossRef]
4. Rayner, N.A.; Brohan, P.; Parker, D.E.; Folland, C.K.; Kennedy, J.J.; Vanicek, M.; Ansell, T.J.; Tett, S.F.B. Improved analyses of changes and uncertainties in sea surface temperature measured in situ sice the mid-nineteenth century: The HadSST2 dataset. *J. Clim.* **2006**, *19*, 446–469. [CrossRef]
5. Kennedy, J.J.; Rayner, N.A.; Smith, R.O.; Parker, D.E.; Saunby, M. Reassessing biases and other uncertainties in sea surface temperature observations measured in situ since 1850: 2. Biases and homogenization. *J. Geophys. Res.* **2011**, *116*. [CrossRef]
6. Kennedy, J.J.; Rayner, N.A.; Smith, R.O.; Parker, D.E.; Saunby, M. Reassessing biases and other uncertainties in sea surface temperature observations measured in situ since 1850: 1. Measurement and sampling uncertainties. *J. Geophys. Res.* **2011**, *116*. [CrossRef]
7. Hirahara, S.; Ishii, M.; Fukuda, Y. Centennial-Scale Sea Surface Temperature Analysis and Its Uncertainty. *J. Clim.* **2014**, *27*, 57–75. [CrossRef]
8. Casey, K.S.; Brandon, T.B.; Cornillon, P.; Evans, R. The Past, Present, and Future of the AVHRR Pathfinder SST Program. In *Oceanography from Space: Revisited*; Barale, V., Gower, J.F.R., Alberotanza, L., Eds.; Springer: Berlin, Germany, 2010; pp. 273–287.
9. Woodruff, S.D.; Worley, S.J.; Lubker, S.J.; Ji, Z.; Freeman, J.E.; Berry, D.I.; Brohan, P.; Kent, E.C.; Reynolds, R.W.; Smith, S.R.; et al. ICOADS Release 2.5: Extensions and enhancements to the surface marine meteorological archive. *Int. J. Climatol.* **2011**, *31*, 951–967. [CrossRef]
10. Freeman, E.; Woodruff, S.D.; Worley, S.J.; Lubker, S.J.; Kent, E.C.; Angel, W.E.; Berry, D.I.; Brohan, P.; Eastman, R.; Gates, L.; et al. ICOADS Release 3.0: A major update to the historical marine climate record. *Int. J. Climatol.* **2017**, *37*, 2211–2232. [CrossRef]
11. Smith, T.M.; Reynolds, R.W. Extended reconstruction of global sea surface temperatures based on COADS data (1854–1997). *J. Clim.* **2003**, *16*, 1495–1510. [CrossRef]
12. Reynolds, R.W.; Smith, T.M.; Liu, C.; Chelton, D.B.; Casey, K.S.; Schlax, M.G. Daily high-resolution-blended analyses for sea surface temperature. *J. Clim.* **2007**, *20*, 5473–5496. [CrossRef]
13. Embury, O.; Merchant, C.J.; Filipiak, M.J. A reprocessing for climate of sea surface temperature from the along-track scanning radiometers: Basis in radiative transfer. *Remote Sens. Environ.* **2012**, *116*, 32–46. [CrossRef]
14. Embury, O.; Merchant, C.J. A reprocessing for climate of sea surface temperature from the along-track scanning radiometers: A new retrieval scheme. *Remote Sens. Environ.* **2012**, *116*, 47–61. [CrossRef]
15. Fairall, C.W.; Bradley, E.F.; Godfrey, J.S.; Wick, G.A.; Edson, J.B.; Young, G.S. Cool-skin and warm-layer effects on sea surface temperature. *J. Geophys. Res.* **1996**, *101*, 1295–1308. [CrossRef]

16. Kantha, L.H.; Clayson, C.A. An improved mixed-layer model for geophysical applications. *J. Geophys. Res. Ocean.* **1994**, *99*, 25235–25266. [CrossRef]
17. Hollmann, R.; Merchant, C.J.; Saunders, R.; Downy, C.; Buchwitz, M.; Cazenave, A.; Chuvieco, E.; Defourny, P.; de Leeuw, G.; Forsberg, R.; et al. The ESA climate change initiative Satellite Data Records for Essential Climate Variables. *Bull. Am. Meteorol. Soc.* **2013**, *94*, 1541–1552. [CrossRef]
18. Merchant, C.J.; Embury, O.; Roberts-Jones, J.; Fiedler, E.; Bulgin, C.E.; Corlett, G.K.; Good, S.; McLaren, A.; Rayner, N.; Morak-Bozzo, S.; et al. Sea surface temperature datasets for climate applications from Phase 1 of the European Space Agency Climate Change Initiative (SST CCI). *Geosci. Data J.* **2014**, *1*, 179–191. [CrossRef]
19. Venema, V.K.C.; Mestre, O.; Aguilar, E.; Auer, I.; Guijarro, J.A.; Domonkos, P.; Vertacnik, G.; Szentimrey, T.; Stepanek, P.; Zahradnicek, P.; et al. Benchmarking homogenization algorithms for monthly data. *Clim. PAST* **2012**, *8*, 89–115. [CrossRef]
20. Wang, X.L.; Wen, Q.H.; Wu, Y. Penalized maximal *t* test for detecting undocumented mean change in climate data series. *J. Appl. Meteorol. Climatol.* **2007**, *46*, 916–931. [CrossRef]
21. McPhaden, M.J.; Busalacchi, A.J.; Cheney, R.; Donguy, J.R.; Gage, K.S.; Halpern, D.; Ji, M.; Julian, P.; Meyers, G.; Mitchum, G.T.; et al. The tropical ocean global atmosphere observing system: A decade of progress. *J. Geophys. Res.* **1998**, *103*, 14169–14240. [CrossRef]
22. Lumpkin, R.; Centurioni, L.; Perez, R.C. Full access fulfilling observing system implementation requirements with the global drifter array. *J. Atmos. Ocean. Technol.* **2016**. [CrossRef]
23. Argo. Argo float data and metadata from Global Data Assembly Centre (Argo GDAC). *Seanoe* **2000**. [CrossRef]
24. Merchant, C.J. Algorithm Theoretical Basis Document (Phase II EXP 1.8). European Space Agency Contract Report SST_CCI-ATBD-UOR-202. Available online: http://www.esa-sst-cci.org/PUG/pdf/SST_CCI-ATBD-UOR-202_Issue-1-signed.pdf (accessed on 16 January 2018).
25. Corlett, G.K. MMD Content Specification. European Space Agency Contract Report SST_CCI-TN-UOL-001. Available online: http://www.esa-sst-cci.org/sites/default/files/Documents/public/SST_cciMMDContentSpecificationIssue1(20120504).pdf (accessed on 16 January 2018).
26. Atkinson, C.P.; Rayner, N.A.; Kennedy, J.J.; Good, S.A. An integrated database of ocean temperature and salinity observations. *J. Geophys. Res.* **2014**, *119*, 7139–7163. [CrossRef]
27. Good, S.A.; Martin, M.J.; Rayner, N.A. EN4: Quality controlled ocean temperature and salinity profiles and monthly objective analyses with uncertainty estimates. *J. Geophys. Res.* **2013**, *118*, 6704–6716. [CrossRef]
28. Boyer, T.P.; Antonov, J.I.; Baranova, O.K.; Coleman, C.; Garcia, H.E.; Grodsky, A.; Johnson, D.R.; Locarnini, R.A.; Mishonov, A.V.; Brien, T.D.O.; et al. *World Ocean Database 2013*; Levitus, S., Mishonov, A., Eds.; NOAA Printing Office: Silver Spring, MD, USA, 2013.
29. Kent, E.C.; Kennedy, J.J.; Smith, T.M.; Hirahara, S.; Huang, B.; Kaplan, A.; Parker, D.E.; Atkinson, C.P.; Berry, D.I.; Carella, G.; et al. A call for new approaches to quantifying biases in observations of sea surface temperature. *Bull. Am. Meteorol. Soc.* **2017**, *98*, 1601–1616. [CrossRef]
30. Castro, S.L.; Wick, G.A.; Emery, W.J. Evaluation of the relative performance of sea surface temperature measurements from different types of drifting and moored buoys using satellite-derived reference products. *J. Geophys. Res.* **2012**, *117*. [CrossRef]
31. Atkinson, C.P.; Rayner, N.A.; Roberts-Jones, J.; Smith, R.O. Assessing the quality of sea surface temperature observations from drifting buoys and ships on a platform-by-platform basis. *J. Geophys. Res.* **2013**, *118*, 3507–3529. [CrossRef]
32. Ingleby, B.; Huddleston, M. Quality control of ocean temperature and salinity profiles-Historical and real-time data. *J. Mar. Syst.* **2007**, *65*, 158–175. [CrossRef]
33. Lean, K.; Saunders, R.W. Validation of the ATSR Reprocessing for Climate (ARC) Dataset Using Data from Drifting Buoys and a Three-Way Error Analysis. *J. Clim.* **2013**, *26*, 4758–4772. [CrossRef]
34. Emery, W.J.; Baldwin, D.J.; Schlussel, P.; Reynolds, R.W. Accuracy of in situ sea surface temperatures used to calibrate infrared satellite measurements. *J. Geophys. Res.* **2001**, *106*, 2387–2405. [CrossRef]
35. Merchant, C.J.; Embury, O.; Rayner, N.A.; Berry, D.I.; Corlett, G.K.; Lean, K.; Veal, K.L.; Kent, E.C.; Llewellyn-Jones, D.T.; Remedios, J.J.; et al. A 20 year independent record of sea surface temperature for climate from Along-Track Scanning Radiometers. *J. Geophys. Res.* **2012**, *117*. [CrossRef]
36. Donlon, C.J.; Martin, M.; Stark, J.; Roberts-Jones, J.; Fiedler, E.; Wimmer, W. The Operational Sea Surface Temperature and Sea Ice Analysis (OSTIA) system. *Remote Sens. Environ.* **2012**, *116*, 140–158. [CrossRef]

37. Dee, D.; Uppala, S.; Simmons, A.; Berrisford, P.; Poli, P.; Kobayashi, S.; Andrae, U.; Balmaseda, M.; Balsamo, G.; Bauer, P. The ERA—Interim reanalysis: Configuration and performance of the data assimilation system. *Q. J. R. Meteorol. Soc.* **2011**. [CrossRef]

38. R Core Team. *R: A Language and Environment for Statistical Computing*; R Foundation for Statistical Computing: Vienna, Austria, 2017.

39. Alexandersson, H. A homogeneity test applied to precipitation data. *J. Climatol.* **1986**, *6*, 661–675. [CrossRef]

40. Menne, M.J.; Williams, C.N., Jr. Homogenization of Temperature Series via Pairwise Comparisons. *J. Clim.* **2009**, *22*, 1700–1717. [CrossRef]

41. Meindl, E.A.; Hamilton, G.D. Programs of the National-Data_buoy_center. *Bull. Am. Meteorol. Soc.* **1992**, *73*, 985–993. [CrossRef]

42. Wang, X.L. Accounting for autocorrelation in detecting mean shifts in climate data series using the penalized maximal t or F test. *J. Appl. Meteorol. Climatol.* **2008**. [CrossRef]

43. Wang, X.L.; Feng, Y. RHtestsV4 User Manual; Climate Research Division, Atmospheric Science and Technology Directorate, Science and Technology Branch, Environment Canada. Available online: http://etccdi.pacificclimate.org/software.shtml (accessed on 16 January 2018).

44. Neter, J.; Kutner, M.H.; Nachtsheim, C.J.; Wasserman, W. *Applied Linear Statistical Models*, 4th ed.; McGraw Hill: Boston, MA, USA, 1996.

45. Hyndman, R.J.; Fan, Y. Sample Quantiles in Statistical Packages. *Am. Stat.* **1996**. [CrossRef]

46. Zhang, H.M.; Reynolds, R.W.; Smith, T.M. Adequacy of the in situ observing system in the satellite era for climate SST. *J. Atmos. Ocean. Technol.* **2006**, *23*, 107–120. [CrossRef]

47. Kilpatrick, K.A.; Podestá, G.; Walsh, S.; Williams, E.; Halliwell, V.; Szczodrak, M.; Brown, O.B.; Minnett, P.J.; Evans, R. A decade of sea surface temperature from MODIS. *Remote Sens. Environ.* **2015**. [CrossRef]

remote sensing

MDPI

Article

Bayesian Cloud Detection for 37 Years of Advanced Very High Resolution Radiometer (AVHRR) Global Area Coverage (GAC) Data

Claire E. Bulgin [1,2,*], Jonathan P. D. Mittaz [1,3], Owen Embury [1,2], Steinar Eastwood [4] and Christopher J. Merchant [1,2]

[1] Department of Meteorology, University of Reading, Reading RG6 6AL, UK; j.mittaz@reading.ac.uk (J.P.D.M.); o.embury@reading.ac.uk (O.E.); c.j.merchant@reading.ac.uk (C.J.M.)
[2] National Centre for Earth Observation, Leicester LE1 7RH, UK
[3] National Physical Laboratory, Teddington TW11 0LW, UK
[4] Norwegian Meteorological Institute, Department of Research and Development, N-0313 Oslo, Norway; steinare@met.no
*   Correspondence: c.e.bulgin@reading.ac.uk; Tel.: +44-0118-378-6732

Received: 24 November 2017; Accepted: 9 January 2018; Published: 12 January 2018

Abstract: Cloud detection is a source of significant errors in retrieval of sea surface temperature (SST). We apply a Bayesian cloud detection scheme to 37 years of Advanced Very High Resolution Radiometer (AVHRR) Global Area Coverage (GAC) data, which is an important source of multi-decadal global SST information. The Bayesian scheme calculates a probability of clear-sky for each image pixel, conditional on the satellite observations and prior probability. We compare the cloud detection performance to the operational Clouds from AVHRR Extended algorithm (CLAVR-x), as a measure of improvement from reduced cloud-related errors. To do this we use sea surface temperature differences between satellite retrievals and in situ observations from drifting buoys and the Global Tropical Moored Buoy Array (GTMBA). The Bayesian scheme reduces the absolute difference between the mean and median SST biases and reduces the standard deviation of the SST differences by ~10% for both daytime and nighttime retrievals. These reductions are indicative of removing cloud contaminated outliers in the distribution, as these fall only on one side of the distribution forming a cold tail. At a probability threshold of 0.9 typically used to determine a binary cloud mask for SST retrieval, the Bayesian mask also reduces the robust standard deviation by ~5–10% during the day, in comparison with the operational cloud mask. This shows an improvement in the central distribution of SST differences for daytime retrievals.

Keywords: sea surface temperature; cloud detection; AVHRR; climate data record

1. Introduction

Remote sensing techniques are commonly used to retrieve geophysical properties of the Earth's atmosphere and surface. For the majority of applications using data at thermal infrared wavelengths, classification of observations into clear or cloudy skies is a fundamental preprocessing step. Cloud detection methods can be used either to isolate clouds for retrieval of cloud properties, e.g., cloud optical depth, cloud top height [1,2], or for identifying and discarding regions where cloud obscures the Earth's surface, essential for retrieval of surface parameters such as temperature, land cover classification and ocean colour [3–6].

Different approaches to image classification and cloud detection have been developed. One of the first approaches used to address the cloud detection problem (still commonly used) is to apply a series of threshold tests to the satellite imagery, based either on single channel properties, e.g., radiance, reflectance or brightness temperature, or using channel differences, normalised differences or textural measurements [7–9]. Typically, appropriate thresholds are determined using a selection of test data or case studies, and thresholds appropriate for one cloud type or atmospheric regime may not generalise well to other observation conditions.

This concept of applying a series of threshold-based tests to the observed data has been extended in the development of 'fuzzy logic' and neural networks for cloud detection. In these approaches binary thresholds are replaced with gradated boundaries defined using a cloud/clear weighting between 0–1. The outcomes of these tests are then combined, using a series of logical comparisons to give a final classification [10,11]. These logic models are developed using training data where the true classification of each observation is previously determined. Although neural networks may show good cloud detection skill under certain conditions, they are limited by the range of observations included in the training dataset.

A third approach is to use Bayes Theorem to determine a clear-sky probability given both the satellite observations and prior information on the background state [5,12]. We would expect this to be the optimum approach, as essentially cloud detection is a Bayesian problem, requiring an estimate of clear-sky probability given both satellite observations and prior knowledge of the surface conditions. In Bayesian approaches, this prior information is dynamic rather than static, increasing generality to a wide range of atmospheric conditions. A threshold can then be placed on this probability to determine a binary cloudmask appropriate to the application. In some retrieval algorithms a naive Bayesian approach is used assuming independence between classifiers (i.e., different channels or cloud tests) [13,14]. This can considerably reduce the size of the probability density space used to describe each classifier, and is often used to reduce the complexity of the problem or increase data processing speed. Arguably, however, the main benefit of a full Bayesian approach to cloud-screen multi-channel data comes from coherent assessment of joint probability distributions, as in reality individual classifiers are not independent.

Comparison studies have provided evidence that Bayesian techniques can more succesfully detect cloud under a range of different meteorological conditions than threshold-based methodologies [3,5]. Further, Bayesian approaches have been applied to retrievals of sea surface temperature [5,15] with their potential for use at high latitudes and over land also demonstrated [3,12]. Within the context of the European Space Agency's (ESA) Sea Surface Temperature Climate Change Initiative (SST CCI) and ATSR Reprocessing for Climate (ARC) projects, full Bayesian methods are applied to the dual-view Along Track Scanning Radiometer instruments [15,16] in the production of climate data records.

In this paper we consider the application of Bayesian methods to the Advanced Very High Resolution Radiometers (AVHRRs), assessing cloud detection skill against a heritage cloud mask provided in operational products. The paper is structured as follows: in Section 2 we describe the Bayesian cloud detection and its application to AVHRR, detailing the relevant instrument characteristics. In Section 3 we describe the application of single-sensor look-up tables to other instruments in the AVHRR data record, and in Section 4 we validate the cloud mask performance using SST comparison statistics and discuss the results. We conclude the paper in Section 5.

2. Bayesian Cloud Detection for AVHRR

Bayesian cloud detection methods as a pre-processing step for sea surface temperature (SST) retrieval have been used extensively with the Along-Track Scanning Radiometers (ATSRs) in the provision of SST data for climate applications [15,16]. The Bayesian methodology for cloud detection is described in detail in [5] and outlined only briefly here as this paper focuses on the applicability to data from the AVHRR sensors. The probability of clear-sky $P(c|\mathbf{y^o}, \mathbf{x^b})$ given both the observation vector $\mathbf{y^o}$ and prior knowledge of the background state $\mathbf{x^b}$ is:

$$P(c|\mathbf{y^o}, \mathbf{x^b}) = [1 + \frac{P(\bar{c})P(\mathbf{y^o}|\mathbf{x^b}, \bar{c})}{P(c)P(\mathbf{y^o}|\mathbf{x^b}, c)}]^{-1} \qquad (1)$$

where the observation vector consists of the satellite observations and the background state vector includes the prior sea surface temperature, total column water vapour and windspeed from the ERA-Interim numerical weather prediction (NWP) data [17]. $P(\bar{c})$ and $P(c)$ are the prior probabilities of cloudy and clear skies defined using the total cloud cover in the NWP data (where $\bar{c}$ and c denote cloud and clear conditions respectively). Limits are placed on these estimates that prevent the prior probability of cloud exceeding 95% or falling below 50%. These limits ensure that the cloud prior probability is both realistic and that the prior probability of cloud is not set to 100%. $P(\mathbf{y}^o|\mathbf{x_b})$ is the probability of the observations given the background state summed over clear and cloudy conditions, specified using an empricial look-up table for cloudy conditions (Section 2.4) and simulated using a radiative transfer model for clear-sky conditions (Section 2.3). Error covariance matrices used in the Bayesian calculation include terms for noise in the observations, uncertainty in the prior and forward model uncertainty.

2.1. Advanced Very High Resolution Radiometer Data

The Advanced Very High Resolution Radiometer (AVHRR) instrument comes in three different configurations. The AVHRR-1 instrument has the smallest compliment of channels at 0.6, 0.8, 3.7 and 10.8 µm. The AVHRR-2 instrument has an additional 12 µm channel and the AVHRR-3 instrument includes a 1.6 µm channel, which can be switched with the 3.7 µm channel on command. AVHRR data are available at two different resolutions. Full resolution data are 1.1 km at the sub-satellite point, and only routinely provided globally for the MetOp series of sensors. For earlier sensors, data are provided at a nominal resolution of 4 km at the sub-satellite point, and this data stream is referred to as 'global area coverage' (GAC) data. GAC data are derived by sampling four pixels across-track, missing one, and then sampling the next four, in every third scanline. The GAC data transmitted from the satellite are then an average of the signal across each four pixel block, and represent the geographical area of 3×5 full resolution pixels [18]. The first AVHRR instrument was TIROS-N, launched in 1978, and the AVHRR series provide an unbroken record of Earth observations over a 39 year time period, a data record length ideal for climate applications. Here we consider the data record from AVHRR-06 launched in 1979 through to MetOp-A at the end of 2016. Table 1 provides a list of the AVHRR instruments, their period of operation (note that for earlier sensors there may be data gaps within this record), available channels and equator overpass time.

Table 1. AVHRR instrument record indicating instrument name and type, period of operation, channel availability and daytime equator overpass time. Parenthesis around a given channel indicates that data from that channel are not routinely transmitted.

Instrument	Type	Period of Operation	Channel Availability (µm)	Overpass Time
NOAA-6	AVHRR-1	July 1979–March 1982	0.6, 0.8, 3.7, 10.8	Morning
NOAA-7	AVHRR-2	September 1981–February 1985	0.6, 0.8, 3.7, 10.8, 12.0	Afternoon
NOAA-8	AVHRR-1	May 1983–October 1985	0.6, 0.8, 3.7, 10.8	Morning
NOAA-9	AVHRR-2	February 1985–November 1988	0.6, 0.8, 3.7, 10.8, 12.0	Afternoon
NOAA-10	AVHRR-1	November 1986–September 1991	0.6, 0.8, 3.7, 10.8	Morning
NOAA-11	AVHRR-2	November 1988–December 1994	0.6, 0.8, 3.7, 10.8, 12.0	Afternoon
NOAA-12	AVHRR-2	September 1991–December 1998	0.6, 0.8, 3.7, 10.8, 12.0	Morning
NOAA-14	AVHRR-2	January 1995–October 2002	0.6, 0.8, 3.7, 10.8, 12.0	Afternoon
NOAA-15	AVHRR-3	October 1998–December 2010	0.6, 0.8, (1.6), 3.7, 10.8, 12.0	Morning
NOAA-16	AVHRR-3	January 2001–December 2010	0.6, 0.8, (1.6), 3.7, 10.8, 12.0	Afternoon
NOAA-17	AVHRR-3	June 2002–December 2010	0.6, 0.8, 1.6, 3.7, 10.8, 12.0	Morning
NOAA-18	AVHRR-3	May 2005–Present	0.6, 0.8, (1.6), 3.7, 10.8, 12.0	Afternoon
NOAA-19	AVHRR-3	February 2009–Present	0.6, 0.8, (1.6), 3.7, 10.8, 12.0	Afternoon
MetOp-A	AVHRR-3	October 2006–Present	0.6, 0.8, 1.6, 3.7, 10.8, 12.0	Morning

2.2. AVHRR Level 1 Calibration and Harmonisation

When considering long time series of data covering observations from multiple sensors, both the accuracy of individual sensor calibration and the biases between sensors are important. Instrument calibration can change over time due to sensor degradation, and individual sensors may show biases in relation to a reference sensor. In this section, we discuss how we address these issues for both the visible and infrared channels.

2.2.1. Visible Channels

For the AVHRR instruments there is no on-board calibration system to track any changes in the visible channel detector response but it is now known that significant instrument degradation over time has occurred e.g., [19]. From November 1996, the National Oceanographic and Atmospheric Administration (NOAA) has provided an operational monthly update to the visible channel coefficients based on the Libya-4 desert reference site [19]. However, data prior to November 1996 have not been corrected, and therefore the visible degradation can affect the cloud detection which could result in spurious trends in SST.

To remove these calibration trends and improve the visible channel data, there are several recalibration schemes available. Along with the operational updates discussed above, coefficients are available from the Community Satellite Processing Package (CSPP) software available from the Co-operative Institute for Meteorological Satellite Studies (CIMSS) (based on the Heidinger et al. algorithm [20]), from the NASA Langley Research Centre (LaRC) AVHRR Fundamental Climate Data Record (FCDR) project, from the NOAA Climate Data Record Program (e.g., [21]), from the University of Maryland (e.g., [22]) and others. The most recent of these are based on a number of different references including MODIS channels, desert sites, cloud tops and sites such as Dome-C in the Antarctic. In general, they have quoted accuracies of the order of 2–3%.

In the ESA SST CCI project we have used the CSPP coefficients (dated to late 2015) to calibrate the visible channels. These coefficients provide a time dependent model for the calibration coefficients based on time from launch. Beyond the update to the calibration coefficients we have also used a simple outlier rejection scheme to remove spurious counts from the space view data and have applied the same smoothing kernel to the space view data as is used for the IR calibration (see Section 2.2.2). These two extra processing steps will result in slightly different values than would be provided by the CSPP software directly but should improve stability.

2.2.2. Infrared Channels

The AVHRR has up to three IR channels and has an internal calibration target (ICT) which is used to determine variations in the instrument gain. To improve the calibration signal-to-noise we first use smoothed values of the calibration counts (space view and ICT view) and the ICT temperatures. Then to calibrate the IR channels we use the Walton et al. calibration algorithm [23] (the current operational calibration) using calibration coefficients taken from the CSPP. This is because the CSPP has coefficients for all the AVHRR sensors including coefficients for the AVHRR-1 instruments that do not have a 12 μm channel (NOAA-06, NOAA-08 and NOAA-10), which were not included in the original analysis [23]. To calibrate the IR channels, we first assume that the 3.7 μm channel is completely linear so the gain can be simply calculated from the calibration observations. The 10.8 μm and 12 μm channels are, however, known to be non-linear and the calibration first derives a linear radiance (including a bias correction term called the 'negative radiance of space') and then uses a quadratic based on this linear radiance as a non-linear correction term.

At this point, we note that all the calibration coefficients in the operational calibration [23] were based on an analysis of the pre-launch data which has subsequently been shown to have numerous problems ([24]) and are therefore suspect. Further, the difference between the thermal environment of the pre-launch testing (where uniform temperature gradients were a goal) and the thermal environment in-orbit (where directional heating from the Sun will cause strong thermal gradients) will impact the calibration. This variation in thermal gradients is particularly important when considering the long-term stability of the instrument because most of the AVHRRs are not in stable orbits but are in orbits that drift (see Figure 1), thereby varying the thermal state of the AVHRR on long timescales.

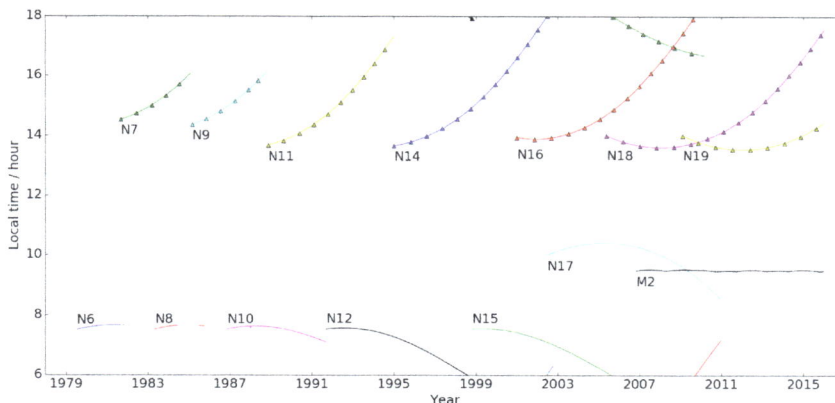

Figure 1. Daytime equator overpass times for AVHRR sensors as a function of instrument lifetime.

Biases in the operational IR AVHRR radiances [23] have been previously reported. For example, a strong (up to 0.5 K) scene temperature dependent bias has been seen in the AVHRRs flown on-board the MetOp series of satellites (e.g., [25,26]) and there is also evidence for a large time dependent bias due to changes in the AVHRR thermal state (e.g., [25,27]). To take both these sources of bias into account we have therefore implemented a two-step bias correction scheme applied to the operational calibration. We first determine the AVHRR bias relative to clear-sky radiances over ocean scenes (estimated using the ERA Interim atmospheric profiles together with RTTOV v11.3 as the radiative transfer model). Because there is a distinct lack of temperature sensors on the AVHRR (for the early sensors there are just the ICT PRT temperatures available) we have to estimate the AVHRR thermal state using the orbital average of the ICT temperature as a proxy. Then to model the bias we use a range of different model types such as simple polynomials and two state linear models (where there is a clear difference between trends as a function of orbital averaged ICT temperature). Different models are applied for different time periods all as a function of the orbital average ICT temperature. Full details of the time dependent biases seen in the AVHRR will be presented elsewhere.

Once the time dependent bias has been corrected for we then use the overlap period of the AVHRRs with the (A)ATSR sensors to determine an average scene temperature dependent bias. To ensure compatibility with the thermal state dependent bias discussed above we have again taken just the clear-sky matches between the AVHRR and (A)ATSR sensors. We have then used a double difference method (e.g., [28]) between the AVHRR and (A)TSR sensors which allows for large time separation in the matches (of order hours) as well as automatically correcting for spectral response function differences between sensors. A single linear fit of bias as a function of scene temperature was then made to all the (A)ATSR/AVHRR matchups which effectively links all the AVHRR sensors to the (A)ATSR sensors as a reference. Note that by fitting to all the overlapping sensors simultaneously we are also trying to capture the range of error in the calibration across all sensors and the standard deviation seen around the bias model is added to the total uncertainty of the IR calibration. For those sensors which do not overlap the (A)ATSR period we then double this uncertainty in an effort to take the lack of a reference pre-1995 (pre ATSR-2) into account.

Finally, an estimate of the radiance uncertainty for each pixel is derived from the counts noise seen in the space view coupled with the estimated instrument gain. To this we add the calibration uncertainty defined above and the total radiance uncertainty is converted into an noise equivalent differential temperature (NEdT, defined for a scene temperature of 300 K) for use with subsequent processing.

2.3. Simulating Clear-Sky Observations

For clear-sky conditions we use the fast forward radiative transfer model RTTOV v11.3, which has the ability to simulate top-of-atmosphere radiances at both visible and infrared wavelengths [29]. We have used infrared channel simulations extensively in previous applications of the Bayesian cloud screening to ATSR data [5,16] but here use similar information from the 0.6 and 0.8 μm visible wavelength channels in daytime cloud detection. RTTOV v11.3 radiative transfer is coupled with a Cox and Munk parameterisation of surface reflectance and glint [30], to determine suitability for cloud detection purposes. We compare simulated and observed reflectance in the 0.6 and 0.8 μm channels using an SST match-up database [31] including all AVHRR sensors used in the data record. Clear-sky matches were selected using Bayesian cloud detection with infrared channels only, including checks for bad data, navigation and calibration problems. Although the Bayesian cloud detection using only the 10.8 and 12 μm channels is not perfect, averaging the reflectances within radiative transfer model (RTM) reflectance bins minimises the impact of missed clouds on the resulting fit. Figure 2 shows the 0.6 and 0.8 μm data plotted in bins of 0.01 RTM reflectance including a linear regression for each channel. RTTOV v11.3 overestimates reflectance under brighter conditions compared to the observations in both channels. We therefore apply a simple linear correction to the RTTOV v11.3 simulations when processing the AVHRR data to align them more closely with the observations. A single correction is applied uniformly to all AVHRR sensors (similar results were found when considering each sensor individually but are not shown here), and Figure 3 shows a comparison for RTM minus observation for clear-sky matches with the correction applied. Linear regressions for both channels now lie almost directly along the 1:1 line, with a small deviation of the data at high reflectance. Most clear-sky observations have reflectance below 0.2 and in those regions these simulations are now accurate enough to use in the cloud detection scheme. Performance of the classifier under sunglint conditions, where the top of atmosphere reflectance will be higher, is specifically addressed in Section 4.

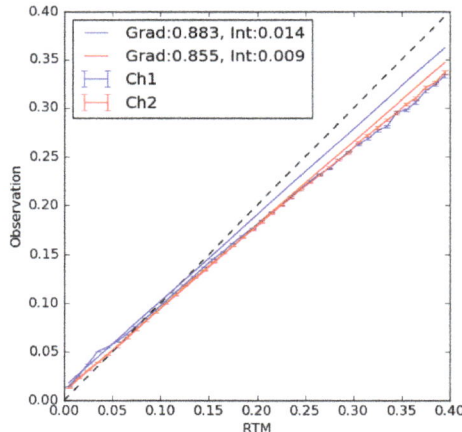

Figure 2. Observed reflectance as a function of RTTOV v11.3 radiative transfer model (RTM) reflectance for Channel 1, 0.6 μm (blue) and Channel 2, 0.8 μm (red). Joined data points show data binned at 0.01 in the RTM reflectance with associated error bars. Solid lines show a linear regression of the observed reflectance against the RTM in each channel. Dashed black line shows the 1:1 relationship. Legend shows the gradient (Grad) and intercept (Int) of the linear fits to the data.

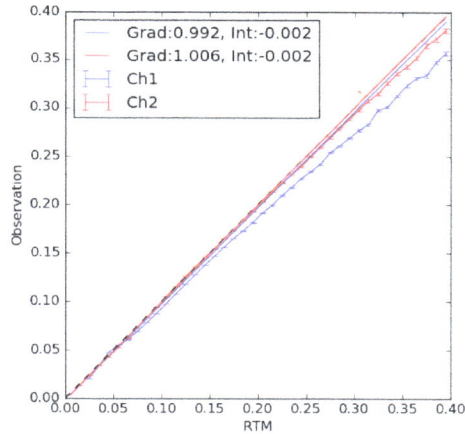

Figure 3. Observed reflectance as a function of the corrected RTTOV v11.3 radiative transfer model (RTM) reflectance for Channel 1, 0.6 μm (blue) and Channel 2, 0.8 μm (red). The linear correction used is that defined in Figure 1. Joined data points show data binned at 0.01 in the RTM reflectance with associated error bars. Solid lines show a linear regression of the observed reflectance against the RTM in each channel. Dashed black line shows the 1:1 relationship. Legend shows the gradient (Grad) and intercept (Int) of the linear fits to the data.

2.4. Empirical Cloudy PDFs

To calculate the probability of clear-sky, we need to determine the probability of the observations given the background state $P(\mathbf{y}^o|\mathbf{x_b})$ under both clear-sky and cloudy conditions. For cloudy conditions, simulating top of atmosphere radiances on-the-fly is computationally expensive. We therefore represent cloudy observations using pre-calculated empirical probability density functions (PDFs) derived from MetOp-A GAC data between 2006–2017. We generate the PDFs using a single iteration of our Bayesian cloud detection algorithm, bootstrapped using empircal cloudy PDFs for the Advanced Along Track Scanning Radiometer (AATSR) cloud detection. We use a brightness temperature shift at the point of indexing the look-up table to make the AVHRR observations 'look like' AATSR (the methodology is described in detail in Section 3). Further, as AATSR makes observations in two views (satellite zenith angles of 0°–22° in the nadir and 52°–55° degrees in the forward view), the AATSR PDFs are interpolated as a function of path length to represent the other viewing angles seen by MetOp-A. We also note that although the AVHRR GAC data does come with an operational cloud mask (CLAVR-x), we found that the performance of this mask was insufficient to prevent inclusion of large numbers of clear-sky observations in the empirical PDFs, rendering this unsuitable for bootstrapping the PDF generation.

Figure 1 shows daytime equator overpass times for all of the AVHRR sensors. Metop-A was selected as the baseline for creating the AVHRR cloudy PDFs due to the orbital stability. It is one of the few AVHRR instruments which does not show significant orbital drift over time, it has the maximum number of channels available for AVHRR instruments, and has a long data record spanning 10 years. Table 2 describes the AVHRR PDF construction, where separate look-up tables are used for spectral and textural features. There are three spectral PDFs generated for processing daytime data, and two for nighttime data (labelled PDF 1, PDF 2, PDF 3 in Table 2). The PDFs used are dependent on the sensor, for example during the day, AVHRR -1 instruments use daytime spectral PDF 1 and PDF 3, while AVHRR-2/3 instruments use daytime spectral PDF 1 and PDF 2. A single textural PDF is used regardless of the time of day.

Table 2. Empirical cloudy PDF specifications as applied for AVHRR cloud detection. Acronyms used are as follows: BT - brightness temperature, NWP - numerical weather prediction, SST - sea surface temperature, Diff - difference, SZA - solar zenith angle.

PDF	Sensor	Dimension	Lower Limit	Upper Limit	Bin Size	No. of Bins
Daytime Spectral PDF 1	1,2,3	0.6 μm refl	0.0	1.0	0.01	100
		0.8 μm refl	0.0	1.0	0.01	100
		SZA	0.0	95.0	2.5	38
		Path Length	1.0	2.4	0.35	4
Daytime Spectral PDF 2	2,3	10.8 μm BT–NWP SST	−20.0	10.0	1.0	30
		10.8–12 μm BT	−1.0	9.0	0.2	50
		NWP SST	260.0	310.0	1.0	50
		Path Length	1.0	2.4	0.35	4
		Day/Night	0.0	180.0	90.0	2
Daytime Spectral PDF 3	1	10.8 μm BT	260.0	305.0	1.0	45
		NWP SST	260.0	310.0	1.0	50
		Path Length	1.0	2.4	0.35	4
		Day/Night	0.0	180.0	90.0	2
Nighttime Spectral PDF 1	2,3	10.8 μm BT–NWP SST	−20.0	10.0	1.0	30
		10.8–12 μm BT Diff	−1.0	9.0	0.2	50
		3.7–10.8 μm BT Diff	−6.0	10.0	0.2	80
		NWP SST	260.0	310.0	2.5	20
		Path Length	1.0	2.4	0.35	4
		Day/Night	0.0	180.0	90.0	2
Nighttime Spectral PDF 2	1	10.8 μm BT–NWP SST	−20.0	10.0	1.0	30
		3.7–10.8 μm BT Diff	−6.0	10.0	0.2	80
		NWP SST	260.0	310.0	1.0	50
		Path Length	1.0	2.4	0.35	4
		Day/Night	0.0	180.0	90.0	2
Textural PDF	1,2,3	10.8 μm BT	260.0	305.0	1.0	45
		NWP SST	260.0	310.0	1.0	50
		Path Length	1.0	2.4	0.35	4

Spectral PDFs are generated under cloudy conditions only, while textural PDFs are calculated under both clear and cloudy conditions. In terms of individual PDFs, the visible channel spectral PDF 1 is used during the day and is applicable to all AVHRR instruments as it uses the 0.6 and 0.8 μm channels only. For the infrared daytime PDF, we use the 10.8 and 12 μm for AVHRR-2 and 3 instruments (daytime spectral PDF 2), and the 10.8 μm only for AVHRR-1 (daytime spectral PDF 3) as the 12 μm isn't present. At night, the spectral PDFs are extended to include the shortwave infrared 3.7 μm channel (nighttime spectral PDFs 1 and 2). Finally the textural PDF uses the local standard deviation of the 10.8 μm brightness temperature over a 3 × 3 pixel domain, centred on the pixel to be classified. This PDF is used consistently under day and nighttime conditions, for all AVHRR sensors.

Given the equator overpass time of the sensor however, MetOp-A does not make observations at the full range of solar zenith angles observed by other AVHRR sensors with different equator overpass times, and consequently the visible channel PDFs need to be extended in order to be applicable to the entire data record. To do this, we use data from NOAA-18, which has an afternoon equator overpass time. Although still currently operational, NOAA-18 shows a step-change in the SST data record post-2010. We therefore use only observations made between 2005–2009. MetOp-A routinely makes observations in the solar zenith angle range 25°–90° (with this range extending to smaller solar zenith angles at longer atmospheric path lengths). NOAA-18 routinely makes observations at solar zenith angles as low as 15° at nadir, and in the range not observed by MetOp-A we use NOAA-18 data to directly fill the corresponding bins in the look-up table. For lower solar zenith angles than those observed by either instrument, we use the average of the first three valid bins from the NOAA-18 data to define the PDF. This approach can be justified as conditions for solar zenith angles of ∼15° are very similar to when the sun is directly overhead. These data will be relevant for AVHRR sensors that show significant orbital drift throughout their lifetime, deviating substantially from the equator overpass times represented by MetOp-A and NOAA-18.

2.5. High Latitude Ice Masking

In regions where there is a chance that the sea surface freezes, cloud masking has to be extended to also include sea-ice masking. Clouds and sea-ice do have some similar spectral signatures, but for other features, the spectral signature for sea-ice can be closer to cloud-free sea. Therefore sea-ice masking should be included in the classification when processing SST at high latitudes. One solution could be to extend the two-way classifier described at the start of Section 2 to a three-way classifier on a global scale. Another option is to include a classifier adapted for sea-ice masking in regions where sea ice can occur as we do here.

The sea-ice masking has been set up using a three-way classifier trained for high latitude areas, and therefore only to be applied at high latitudes. The idea is that all pixels classified as cloud free by the two-way classifier at high latitudes undergo an additional classification by a three-way classifier that also includes sea-ice. This is used to potentially reclassify clear-sky pixels as 'not clear' only, by reducing the probability of clear-sky given by the Bayesian calculation. The clear-sky probability is set to the minimum of the clear-sky probability given by the two-way classification, and the clear-sky probability given by the three-way classification.

High latitudes can be defined as areas poleward of 50 N/S, or more specifically as areas within the climatological maximum sea-ice extent. We use the latter to define sea-ice affected regions using monthly data from National Snow and Ice Data Center (NSIDC) [32] . The sea-ice edge has been extended by 100 km into the open water definition to account for the daily sea-ice edge fluctuations.

The three-way classifier is defined in a similar way as the two-way Bayesian classifier (1), except that a third ice covered class is added. The probability of a pixel being clear-sky over ocean in a n-way classifier is given in (2) [12], where $class_n$ includes clear (c), cloud ($\bar{c}$) and ice (i):

$$P(c|\mathbf{y^o}, \mathbf{x^b}) = \frac{P(c)P(\mathbf{y^o}|\mathbf{x^b}, c)}{\sum_n P(class_n)P(\mathbf{y^o}|\mathbf{x^b}, class_n)} \qquad (2)$$

For this work, the three-way classifier developed in the EUMETSAT Oceans and Sea Ice Satellite Application Facility (OSI SAF) project and SST CCI Phase 1 [12] has been used. This classifier has been trained on AVHRR GAC data to work for all sensors in the series. This training is necessary to define the PDFs for the three classes (clear, cloud, ice) and uses an empirical approach. To generate the PDFs, the EUMETSAT Climate SAF cLoud, Albedo and surface RAdiation dataset from AVHRR data (CLARA-A2) dataset [33] and the EUMETSAT OSI SAF CDR on sea ice concentration [34] have been used. CLARA-A2 consists of both re-calibrated AVHRR channel data and cloud/ice masking products. The cloud/ice mask from CLARA-A2 and OSI SAF sea ice concentration are used to define occurrence of cloud and ice, and this is used together with the CLARA-A2 satellite data to define the PDFs for all the AVHRR sensors. For each AVHRR sensor, three years of GAC data are used to define the PDFs.

The CLARA-A2 dataset does not contain any AVHRR-1 data. To define PDFs for NOAA-6, 8 and 10 the AVHRR-2 sensor which is closest in terms of spectral response functions has been used as a proxy (NOAA-7). The PDFs used for the high latitude ice masking are listed in Table 3. Some PDFs are the same for all AVHRRs, while others are only used for AVHRR-2 and/or 3. This is identified with an 'a' or 'b' in the PDF name, and in the column specifying the sensor.

Table 3. Features used for three-way classification PDFs, dependent on AVHRR sensor and time of day. All PDFs are two dimensional with a feature and single dependency. The following acronyms are used: BT = brightness temperature, SZA = solar zenith angle, Atm = Atmospheric, Std = Standard Deviation, diff = difference, refl = reflectance.

PDF	Sensor	Feature	Dependencies	Range	Binsize
Daytime Spectral Feature 1	1,2,3	0.8/0.6 μm refl ratio	SZA	0–91	1
Daytime Spectral Feature 2a *	3	1.6/0.6 μm refl ratio	SZA	0–91	1
Daytime Spectral Feature 2b	1,2,3	3.7–10.8 μm BT diff	SZA	0–91	1
Daytime Spectral Feature 3	1,2,3	0.6 μm reflectance	SZA	0–91	1
Nighttime Spectral Feature 1	1	3.7–10.8 μm BT diff	Atm. Path Length	1–2.5	0.1
Nighttime Spectral Feature 2	2,3	10.8–12 μm BT diff	3.7–12 μm BT diff	−1.5–4	0.1
Nighttime Textural Feature 3a	2,3	Std(3.7–12 μm BT)	Atm. Path Length	1–2.5	0.1
Nighttime Textural Feature 3b	1	Std(3.7–10.8 μm BT)	Atm. Path Length	1–2.5	0.1

* This feature is only used when the 1.6 μm channel is available for AVHRR-3 instruments. Where this is unavailable, daytime spectral feature 2b is used.

3. Applying Metop-A Empirical PDFs to Other AVHRR instruments

In processing the AVHRR GAC archive, it is useful to be able to apply a single set of empirical PDFs to all instruments, providing consistency between sensors and minimising the use of disk space for storing ancillary data. In order to do this, we need to make observations from the sensor that is being processed closely match MetOp-A observations, by applying a brightness temperature shift at the point of indexing the look-up table. In order to calculate the shift required for each AVHRR instrument, we use RTTOV v11.3 to simulate clear-sky brightness temperatures for 2100 atmospheric profiles derived from ERA-40 [35,36]. These profiles are chosen to be representative of the variety of atmospheric conditions observed globally over the oceans, incorporating a range of surface temperatures, total column water vapour (TCWV) and ozone concentrations.

We run RTTOV v11.3 for each profile and each AVHRR instrument for two atmospheric path lengths: 1.0 and 1.8, corresponding to satellite viewing zenith angles of 0° and 56.25°, which covers the full range of angles observed by the AVHRR instruments (0°–55.37°) [18]. We calculate the difference between MetOp-A brightness temperatures and those from each of the other sensors in the 3.7, 10.8 and 12 μm channels (subject to channel availability). Figure 4 shows these differences for each sensor as a function of TCWV. The raw data are plotted for the nadir view only, and the solid and dashed lines show the cubic fit to the difference for atmospheric path lengths of 1.0 and 1.8 respectively. For any given observation path length we linearly interpolate between these cubic fits to get coefficients corresponding to the observation. The coefficients for atmospheric path lengths of 1.0 and 1.8 are provided in Tables 4 and 5.

The same process is used in generating the empirical cloudy PDFs bootstrapped using AATSR PDFs (Section 2.4), by processing these profiles for AATSR and computing the brightness temperature shifts using the AATSR minus MetOp-A difference. No shifts are applied for the reflectance data as a single correction is made for all sensors to modify the RTM output to more closely match the observations (Section 2.3).

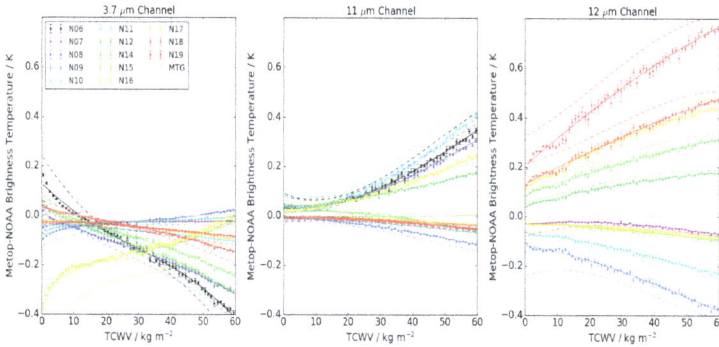

Figure 4. MetOp-A minus NOAA AVHRR brightness temperatures in the 3.7, 10.8 and 12 μm channels plotted as a function of total column water vapour (TCWV) using RTTOV v11.3 simulations under clear-sky conditions. Results are binned in 1 kg m^{-2} TCWV, with joined points showing actual data for an atmospheric path length of 1. Solid lines show a cubic fit to these data and dashed lines show a cubic fit for the differences for each AVHRR simulated using an atmospheric path length of 1.8. Data are not shown for the atmospheric path length of 1.8 for clarity.

Table 4. Cubic coefficients for AVHRR MetOp-A minus AVHRR sensor differences simulated using RTTOV v11.3 for an atmospheric path length of 1.0. Coeffiecients are a function of total column water vapour.

Sensor	Channel	a0	a1	a2	a3
NOAA-19	3.7 μm	−0.027	0.00016	-3.348×10^{-5}	2.56×10^{-7}
	10.8 μm	−0.001	3.566×10^{-5}	-2.275×10^{-5}	1.289×10^{-7}
	12 μm	0.205	0.0115	-2.9109×10^{-5}	-1.073×10^{-7}
NOAA-18	3.7 μm	0.0309	−0.00308	2.856×10^{-5}	-4.6126×10^{-7}
	10.8 μm	−0.00896	0.000132	-2.0128×10^{-5}	1.0472×10^{-7}
	12 μm	0.1309	0.00668	-2.45072×10^{-6}	-2.127005×10^{-7}
NOAA-17	3.7 μm	−0.01939	−0.00034	-2.4193×10^{-5}	1.5347×10^{-7}
	10.8 μm	−0.00015	−0.000258	-6.9899×10^{-6}	2.3774×10^{-8}
	12 μm	−0.03096	2.44656×10^{-5}	-1.9455×10^{-5}	8.2992×10^{-8}
NOAA-16	3.7 μm	−0.3313	0.0127	−0.000315	3.357×10^{-6}
	10.8 μm	0.02097	0.000595	7.068×10^{-5}	-2.9927×10^{-7}
	12 μm	0.11156	0.0079	-3.5232×10^{-5}	-8.4802×10^{-8}
NOAA-15	3.7 μm	−0.3154	0.01175	−0.000298	3.1463×10^{-6}
	10.8 μm	−0.00163	−0.0002	6.4334×10^{-6}	-3.1064×10^{-8}
	12 μm	−0.0305	−0.000288	-1.7848×10^{-5}	8.1289×10^{-8}
NOAA-14	3.7 μm	0.03976	−0.00571	7.01133×10^{-5}	-9.168×10^{-7}
	10.8 μm	0.013975	0.00105	-5.8323×10^{-5}	3.4936×10^{-7}
	12 μm	0.08675	0.0047	-1.2899×10^{-5}	-4.8781×10^{-8}
NOAA-12	3.7 μm	0.03904	−0.00704	8.4158×10^{-5}	-1.09379×10^{-6}
	10.8 μm	0.0355	0.00068	3.8612×10^{-5}	-1.79513×10^{-7}
	12 μm	0.04059	0.00383	-2.9693×10^{-5}	6.85518×10^{-8}
NOAA-11	3.7 μm	−0.0617	0.00287	-7.6207×10^{-5}	7.5237×10^{-7}
	10.8 μm	−0.0069	−0.0003	-1.6885×10^{-5}	9.2326×10^{-8}
	12 μm	−0.06774	−0.0011	-4.5607×10^{-5}	2.9325×10^{-7}
NOAA-10	3.7 μm	−0.03872	3.0642×10^{-5}	-3.3101×10^{-5}	2.557×10^{-7}
	10.8 μm	0.04135	−0.00228	0.000206	-1.0809×10^{-6}

Table 4. *Cont.*

Sensor	Channel	a0	a1	a2	a3
NOAA-09	3.7 μm	−0.08128	0.00386	-9.16637×10^{-5}	9.6075×10^{-7}
	10.8 μm	−0.00903	−0.00026	-3.6582×10^{-5}	1.9091×10^{-7}
	12 μm	−0.11753	−0.001987	-6.2965×10^{-5}	3.8112×10^{-7}
NOAA-08	3.7 μm	0.00687	−0.00601	5.8278×10^{-5}	-8.0757×10^{-7}
	10.8 μm	0.02989	−0.000924	0.000134	-6.8198×10^{-7}
NOAA-07	3.7 μm	−0.04225	0.00161	-4.7823×10^{-5}	4.5444×10^{-7}
	10.8 μm	−0.003	-9.51097×10^{-5}	-1.1605×10^{-5}	5.86227×10^{-8}
	12 μm	−0.03211	0.00133	-4.5503×10^{-5}	2.1078×10^{-7}
NOAA-06	3.7 μm	0.12815	−0.01287	0.000208	-2.3948×10^{-6}
	10.8 μm	0.03131	−0.00116	0.000152	-7.4657×10^{-7}

Table 5. Cubic coefficients for AVHRR MetOp-A minus AVHRR sensor differences simulated using RTTOV v11.3 for an atmospheric path length of 1.8. Coeffiecients are a function of total column water vapour.

Sensor	Channel	a0	a1	a2	a3
NOAA-19	3.7 μm	−0.0366	0.000166	-4.7607×10^{-5}	3.7265×10^{-7}
	10.8 μm	−0.0064	0.00057	-4.3107×10^{-5}	3.0404×10^{-7}
	12 μm	0.3245	0.00728	0.000145	-2.1723×10^{-6}
NOAA-18	3.7 μm	0.0584	−0.00481	5.6608×10^{-5}	-7.9549×10^{-7}
	10.8 μm	−0.02275	0.00082	-3.9996×10^{-5}	2.5976×10^{-7}
	12 μm	0.211	0.003698	0.000117	-1.6069×10^{-6}
NOAA-17	3.7 μm	−0.02343	−0.000648	-3.0342×10^{-5}	1.8785×10^{-7}
	10.8 μm	−0.00285	-1.2482×10^{-5}	-1.7495×10^{-5}	1.2649×10^{-7}
	12 μm	−0.0682	0.00217	-7.4383×10^{-5}	3.357×10^{-7}
NOAA-16	3.7 μm	−0.50446	0.01883	−0.00049	5.1165×10^{-6}
	10.8 μm	0.06066	−0.00156	0.00015	-9.8118×10^{-7}
	12 μm	0.15664	0.0076	1.71054×10^{-5}	-9.4979×10^{-7}
NOAA-15	3.7 μm	−0.47853	0.017397	−0.000462	4.7957×10^{-6}
	10.8 μm	−0.00207	−0.000342	1.1416×10^{-5}	-6.8027×10^{-8}
	12 μm	−0.0681	0.00192	-7.6334×10^{-5}	5.8077×10^{-7}
NOAA-14	3.7 μm	0.08511	−0.00922	0.00014	-1.6665×10^{-6}
	10.8 μm	0.017899	0.00216	−0.0001	7.3547×10^{-7}
	12 μm	0.13698	0.00294	5.9106×10^{-5}	-8.8745×10^{-7}
NOAA-12	3.7 μm	0.08578	−0.01131	0.000168	-1.98224×10^{-6}
	10.8 μm	0.06966	−0.00058	8.8986×10^{-5}	-6.336×10^{-7}
	12 μm	0.04942	0.00419	-2.0745×10^{-5}	-1.84225×10^{-7}
NOAA-11	3.7 μm	−0.09267	0.00428	−0.00012	1.152×10^{-6}
	10.8 μm	−0.01683	4.7522×10^{-5}	-3.4844×10^{-5}	2.6187×10^{-7}
	12 μm	−0.14462	0.00323	−0.000168	1.41525×10^{-6}
NOAA-10	3.7 μm	−0.04996	−0.000206	-4.1252×10^{-5}	3.1452×10^{-7}
	10.8 μm	0.11532	−0.00759	0.000401	-2.7889×10^{-6}
NOAA-09	3.7 μm	−0.1229	0.00572	−0.000143	1.4684×10^{-6}
	10.8 μm	−0.02512	0.0006	-7.4478×10^{-5}	5.3717×10^{-7}
	12 μm	−0.24302	0.00484	−0.00026	2.1553×10^{-6}
NOAA-08	3.7 μm	0.03502	−0.00963	0.000123	-1.5102×10^{-6}
	10.8 μm	0.08137	−0.0046	0.00028	-1.9549×10^{-6}
NOAA-07	3.7 μm	−0.0592	0.0022	-6.8678×10^{-5}	6.5164×10^{-7}
	10.8 μm	−0.00897	0.000148	-2.2912×10^{-5}	1.5846×10^{-7}
	12 μm	−0.08314	0.00453	−0.000123	8.1899×10^{-7}
NOAA-06	3.7 μm	0.24317	−0.02121	0.000403	-4.37867×10^{-6}
	10.8 μm	0.0903	−0.00547	0.000313	-2.1668×10^{-6}

4. Cloud Mask Validation and Performance Assessment

We validate the cloud mask performance using a match-up database (MD), of comparisons between satellite and in situ observations, covering all instruments in the sensor series [31]. We filter matches on the basis of in situ and satellite observations flagged as high quality, a maximum spatial separation of 10 km and a maximum time difference of four hours. The daytime SST uses the 10.8 and 12 µm channels for AVHRR-2/3 instruments and the 10.8 µm only for AVHRR-1. At night, the retrieval additionally uses the 3.7 µm channel. For a full description of the optimal estimation SST retrieval process please refer to [37].

We compare the Bayesian cloud detection algorithm to the operational cloud mask, CLAVR-x [13], which is a naive Bayesian algorithm. CLAVR-x uses six cloud features, and assumes that these probabilities can be considered as independent. We present the ratio (Bayesian/CLAVR-x) of the number of matches, SST standard deviation (SD) and SST robust standard deviation (RSD) using single pixel matches from the MD. To calculate the ratios we take the statistics of the of the retrieved minus in situ SST differences with each mask (Bayesian and CLAVR-x) applied in turn, and compare these. We threshold both the Bayesian and CLAVR-x probabilities of clear-sky at 0.9 (i.e., clear-sky probabilities must be greater than or equal to 0.9 for clear-sky classification) to generate the statistics. Tables 6 and 7 show comparison statistics against drifting buoy data for daytime and nighttime SST retrievals respectively.

The optimal outcome of improving the cloud detection algorithm would be to reduce the spread of the distribution of the satellite to in situ differences without significant loss of valid clear-sky data. For the daytime retrieval (Table 6), we typically see 5–10% fewer matches for the newer sensors (NOAA-12 onwards) and ~20% for the older sensors. We see a significant reduction in the standard deviation of the SST difference throughout the data record, and a smaller reduction in the robust standard deviation. This suggests that the Bayesian scheme is more successful at removing outliers (typically misclassified cloud) which result in large SST differences. At night (Table 7) we see a greater reduction in the number of matches, typically ~20% for newer sensors, and 30–40% for older sensors, suggesting that the Bayesian mask is more conservative at night than CLAVR-x when using this probability threshold. The standard deviation is reduced in all cases and robust standard deviations are either consistent with CLAVR-x or show small reductions.

Table 6. Ratio of two-channel (10.8 and 12 µm) daytime SST retrieval statistics using Bayesian and CLAVR-x cloud masks compared against drifting buoy data. Ratios are calculated using satellite minus in situ SST differences, first applying the Bayesian mask and then comparing against the same statistics with the CLAVR-x mask applied. Values less than one in each column indicate that the Bayesian mask has fewer matches, lower standard deviation or lower robust standard deviation.

Sensor	No. Matches	Std Dev	Robust Std Dev
NOAA-06 *	0.88	0.55	0.85
NOAA-07	0.85	0.86	0.94
NOAA-08 *	0.74	0.7	0.96
NOAA-09	0.87	0.83	0.9
NOAA-10 *	0.81	0.68	0.9
NOAA-11	0.87	0.83	0.93
NOAA-12	0.995	0.7	0.97
NOAA-14	0.94	0.91	0.95
NOAA-15	0.91	0.71	0.95
NOAA-16	0.97	0.91	0.98
NOAA-17	0.96	0.96	1.01
NOAA-18	0.82	0.82	0.9
NOAA-19	0.83	0.65	0.86
MetOp-A	1.01	0.67	0.99

* For these sensors a single channel retrieval (10.8 µm only) is used as the 12 µm channel is not present on AVHRR-1 sensors.

Table 7. Ratio of three-channel (3.7, 10.8 and 12 μm) nighttime SST retrieval statistics using Bayesian and CLAVR-x cloud masks compared against drifting buoy data. Ratios are calculated using satellite minus in situ SST differences, first applying the Bayesian mask and then comparing against the same statistics with the CLAVR-x mask applied. Values less than one in each column indicate that the Bayesian mask has fewer matches, lower standard deviation or lower robust standard deviation.

Sensor	No. Matches	Std Dev	Robust Std Dev
NOAA-06 *	0.71	0.55	0.87
NOAA-07	0.5	0.6	0.8
NOAA-08 *	0.6	0.95	1.001
NOAA-09	0.58	0.92	0.95
NOAA-10 *	0.56	0.61	0.78
NOAA-11	0.7	0.92	0.996
NOAA-12	0.73	0.91	0.97
NOAA-14	0.8	0.95	1.01
NOAA-15	0.74	0.91	1.01
NOAA-16	0.81	0.87	1.01
NOAA-17	0.78	0.91	0.98
NOAA-18	0.74	0.98	1.05
NOAA-19	0.82	0.74	0.98
MetOp-A	0.78	0.66	0.95

* For these sensors a two channel retrieval (3.7 and 10.8 μm) is used as the 12 μm channel is not present on AVHRR-1 sensors.

Tables 8 and 9 show the equivalent statistics, with comparison against the Global Tropical Moored Buoy Array (GTMBA) [38]. Only instruments from NOAA-09 onwards are included here, due to few in situ GTMBA measurements available for comparisons against earlier sensors. In comparisons against GTMBA, we see a smaller reduction in the number of matches, typically within 10–15% both day and night. The most significant reduction in the robust standard deviation is seen for daytime matches, while the standard deviation is consistently reduced by > 10% for both retrievals, with the exception of NOAA-16, 17 and MetOp-A during the day, which show reductions of a smaller magnitude.

Figure 5 shows the $\sqrt{SD^2 - RSD^2}$ metric for each sensor as a measure of uncertainty arising from outliers, for comparisons against drifting buoy and GTMBA data. For each in situ type, the day and nighttime statistics using each cloud mask are plotted. We see here that the metric for uncertainty arising from outliers is typically smaller when using the Bayesian algorithm, indicative of fewer cloud contaminated pixels which would increase the standard deviation. The uncertainty is also fairly consistent day and night, and throughout the sensor record. This is important in ensuring stability in the classification for the production of an SST climate data record. Uncertainties due to outliers are typically larger and more variable when cloud screening with CLAVR-x, particularly for the earlier sensors. The best performance for CLAVR-x is seen for NOAA-16, 17 and 18 during the day, where the metric is similar to that obtained when using the Bayesian algorithm. Overall, the uncertainty is smaller when comparing to GTMBA data than drifting buoys. This would be expected for two reasons: firstly, the GTMBA measurements are more accurate than those from drifting buoys as these instruments are calibrated. Secondly, GTMBA measurements are all made in the tropics under similar regimes, while the drifting buoy observations have a more global distribution. Although there may be a systematic bias in comparisons, these two factors would likely reduce the noise, thus reducing the standard deviation.

Despite the reduction in uncertainties due to outliers when using the Bayesian algorithm, the cold tail in the distribution of SST differences (largely responsible for the difference in the SD and RSD) is still more significant than in the ideal case. Cloud detection is performed after the radiance averaging in the GAC data production, and one possibility is that some cloudy pixels at the full resolution remain undetected following this averaging process. More generally, use of the textural PDF in the Bayesian masks aids the detection of cloud edges but may limit performance in regions of large SST gradients, for example across ocean fronts. The cloud detection methodology also has limited sensitivity to thin cirrus, but this is not critical for SST retrieval purposes.

Table 8. Ratio of two-channel (10.8 and 12 μm) daytime SST retrieval statistics using Bayesian and CLAVR-x cloud masks compared against GTMBA data. Ratios are calculated using satellite minus in situ SST differences, first applying the Bayesian mask and then comparing against the same statistics with the CLAVR-x mask applied. Values less than one in each column indicate that the Bayesian mask has fewer matches, lower standard deviation or lower robust standard deviation.

Sensor	No. Matches	Std Dev	Robust Std Dev
NOAA-09	0.86	0.82	0.89
NOAA-10 *	0.83	0.7	0.85
NOAA-11	0.91	0.82	0.92
NOAA-12	1.01	0.69	0.93
NOAA-14	0.94	0.85	0.92
NOAA-15	0.98	0.65	0.93
NOAA-16	0.99	0.92	0.97
NOAA-17	1.02	0.94	0.99
NOAA-18	0.85	0.81	0.88
NOAA-19	0.84	0.73	0.83
MetOp-A	1.07	0.93	1.01

* For this sensor a single channel retrieval (10.8 μm only) is used as the 12 μm channel is not present on AVHRR-1 sensors.

Table 9. Ratio of three-channel (3.7, 10.8 and 12 μm) nighttime SST retrieval statistics using Bayesian and CLAVR-x cloud masks compared against GTMBA data. Ratios are calculated using satellite minus in situ SST differences, first applying the Bayesian mask and then comparing against the same statistics with the CLAVR-x mask applied. Values less than one in each column indicate that the Bayesian mask has fewer matches, lower standard deviation or lower robust standard deviation.

Sensor	No. Matches	Std Dev	Robust Std Dev
NOAA-09	0.6	0.81	0.92
NOAA-10 *	0.72	0.58	0.69
NOAA-11	0.8	0.88	0.94
NOAA-12	0.88	0.87	0.95
NOAA-14	0.82	0.84	0.95
NOAA-15	0.93	0.89	1.002
NOAA-16	0.91	0.79	0.99
NOAA-17	0.91	0.82	0.94
NOAA-18	0.85	0.87	1.02
NOAA-19	0.92	0.85	1.03
MetOp-A	0.91	0.65	0.95

* For this sensor a two channel retrieval (3.7 and 10.8 μm) is used as the 12 μm channel is not present on AVHRR-1 sensors.

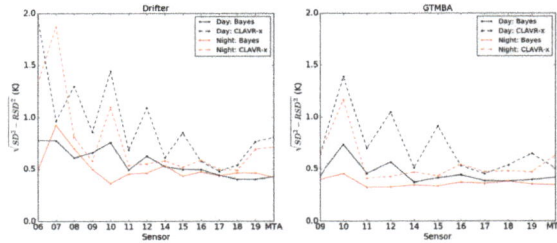

Figure 5. $\sqrt{SD^2 - RSD^2}$ as a function of AVHRR sensor, as a measure of uncertainty arising from outliers. SST difference statistics are derived with reference to drifting buoys (**left**) and GTMBA data (**right**). Each plot shows day and nighttime retrieval statistics for the Bayesian and CLAVR-x cloud masks.

We compare the absolute difference between the mean and median bias for each cloud mask with reference to drifting buoy data (Table 10). As the mean will be affected by outliers, a larger difference provides some indication of the effect of undetected cloud on the SST retrieval. For each sensor, the cloudmask with the smallest mean-median difference is highlighted in bold. During the day, for the Bayesian mask, the largest mean-median difference for any sensor is 0.07 K indicating fewer cloud contaminated pixels overall than for CLAVR-x, where differences are as large as 0.17 K. The Bayesian mask shows the smallest differences for all sensors except NOAA-07, 11 and 17. At night the differences tend to be larger as cloud detection is more difficult without the additional information available from the visible channels. The maximum difference is seen for NOAA-19: 0.13 K when using the Bayesian mask and 0.17 K for CLAVR-x.

Table 10. Absolute difference between mean and median bias for two-channel (10.8 and 12 μm) daytime and three-channel (3.7, 10.8 and 12 μm) nighttime SST retrieval statistics using Bayesian and CLAVR-x cloud masks compared against drifting buoy data.

Sensor	Daytime Mean-Med Diff		Nighttime Mean-Med Diff	
	Bayes	**CLAVR-x**	**Bayes**	**CLAVR-x**
NOAA-06 *	**0.017**	0.155	0.068	**0.059**
NOAA-07	0.076	**0.067**	**0.005**	0.157
NOAA-08 *	**0.002**	0.127	0.178	**0.146**
NOAA-09	**0.038**	0.045	**0.06**	0.068
NOAA-10 *	**0.006**	0.123	**0.021**	0.071
NOAA-11	0.041	**0.008**	**0.06**	0.057
NOAA-12	0.05	0.155	**0.049**	0.052
NOAA-14	**0.031**	0.062	0.085	**0.082**
NOAA-15	**0.044**	0.107	**0.072**	0.083
NOAA-16	**0.008**	0.032	**0.1**	0.117
NOAA-17	0.025	**0.024**	**0.088**	0.1
NOAA-18	**0.027**	0.077	0.108	**0.09**
NOAA-19	**0.049**	0.166	**0.127**	0.167
MetOp-A	**0.023**	0.083	**0.107**	0.15

* For these sensors the 12 μm channel is not used in the retrieval as it is not present for AVHRR-1 sensors.

Table 11 shows the equivalent results using comparisons agains GTMBA arrays. The mean-median differences are small using the Bayesian mask for the majority of sensors, with values typically ≤ 0.02 K with the exception of NOAA-09 and NOAA-11. For CLAVR-x, differences exceed 0.1 K for NOAA-10, 12, 15 and 19. As with the comparison against drifting buoys, the mean-median differences are larger at night, typically in the order of 0.1 K, but smaller using the Bayesian mask for all sensors apart from NOAA-10.

Table 11. Median bias, and absolute difference between mean and median bias for two-channel (10.8 and 12 µm) daytime and three-channel (3.7, 10.8 and 12 µm) nighttime SST retrieval statistics using Bayesian and CLAVR-x cloud masks compared against GTMBA data.

Sensor	Daytime Mean-Med Diff		Nighttime Mean-Med Diff	
	Bayes	CLAVR-x	Bayes	CLAVR-x
NOAA-09	0.11	**0.041**	**0.044**	0.126
NOAA-10 *	**0.004**	0.125	0.131	**0.055**
NOAA-11	0.099	**0.006**	**0.101**	0.13
NOAA-12	**0.028**	0.179	**0.072**	0.108
NOAA-14	**0.014**	0.07	**0.103**	0.143
NOAA-15	**0.026**	0.147	**0.085**	0.113
NOAA-16	**0.011**	0.025	**0.133**	0.18
NOAA-17	**0.001**	0.018	**0.131**	0.163
NOAA-18	**0.004**	0.064	**0.148**	0.159
NOAA-19	**0.01**	0.12	**0.132**	0.154
MetOp-A	0.017	**0.003**	**0.124**	0.169

* For this sensor the 12 µm channel is not used in the retrieval as it is not present for AVHRR-1 sensors.

In Figure 6 we compare the SD and RSD for the two cloud detection algorithms using comparisons against drifting buoys as a function of the number of clear-sky matches. We threshold the Bayesian and CLAVR-x clear-sky probabilities at intervals of 0.1 between 0–0.9, and intervals of 0.01 for probabilities between 0.9–1. We present results for NOAA-09 in the upper panels and NOAA-19 in the lower panels in order to compare an early and late sensor in the data record. Daytime retrievals are shown on the left and nighttime retrievals on the right.

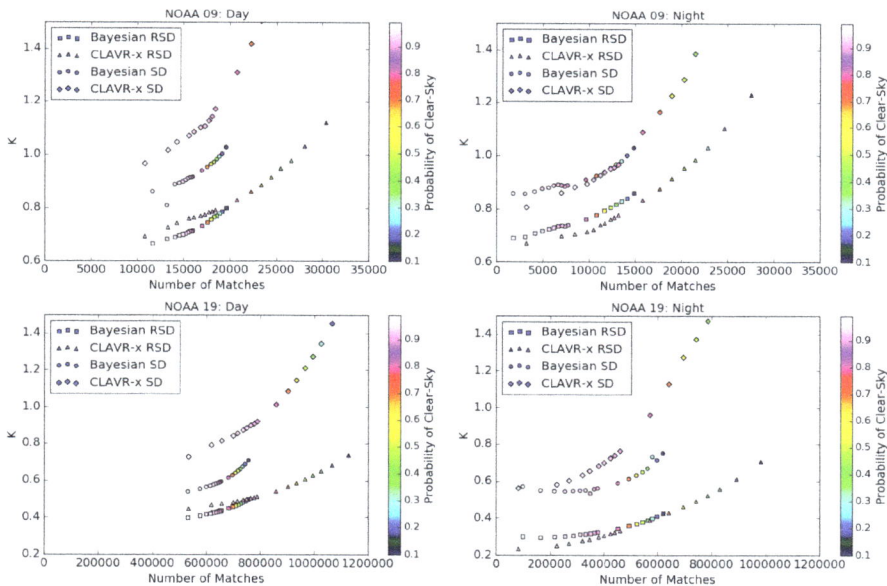

Figure 6. Standard deviation (SD) and robust standard deviation (RSD) of the satellite minus in situ SST differences against drifting buoys for NOAA-09 (**top**) and NOAA-19 (**bottom**). Results are shown for daytime (**left**) and nighttime (**right**) retrievals. The SD and RSD values are plotted as a function of the number of matches and the probability of clear-sky. The clear-sky probability is denoted by the colour of each symbol, and the shape refers to the the mask (Bayesian, CLAVR-x) and the statistic (SD, RSD) as shown in the legend.

We find for all retrievals, for the probability intervals plotted, that the CLAVR-x mask gives a much wider range in the number of clear-sky matches. This is because the Bayesian probability of clear-sky is generally much more bimodal than CLAVR-x, with the majority of pixels classified as very likely to be clear-sky ($\geq$0.9), or very likely to be cloud ($\leq$0.1). The CLAVR-x probability distribution includes significantly more pixels classified between 0.1–0.9, where the appropriate clear/cloud classification is less obvious. The points indicating probability thresholds of 0.1–0.9 are therefore more tightly grouped for the Bayesian mask.

For NOAA-09 and clear-sky probability thresholds above 0.9 during the day, the Bayesian RSD typically ranges between 0.67–0.71 K, with CLAVR-x RSD's between 0.69–0.78 K, also giving larger numbers of matches. At nighttime the RSD values are similar, with CLAVR-x giving comparable RSD values to the Bayesian mask for probabilities between 0.95–0.99, but with significantly more matches. The overall number of matches is fewer at night using both masks than during the day. We find that the Bayesian algorithm typically reduces the SD by $\sim$0.1 K during the day, with only moderate data losses at probabilities $\geq$0.9. At night, the standard deviations for the two algorithms follow a similar curve for total matches exceeding 12,500, but below this the CLAVR-x algorithm gives a lower SD than the Bayesian for a given number of matches.

A similar result is seen for NOAA-19 although the overall number of matches is significantly larger both due to the lifetime of the instrument and the increased number of in situ observations. The RSDs are lower overall, typically between 0.4–0.51 K during the day and 0.23–0.32 K at night. During the day, the Bayesian mask gives lower RSD values than CLAVR-x, and also at night for the probability threshold of 0.9. For probabilities $\geq$ 0.93, and for total matches fewer than five hundred thousand, the CLAVR-x outperforms the Bayesian mask at night. During the day, the Bayesian algorithm significantly reduces the SD, by a factor of $\sim$0.2 K for any given number of matches. At night, the standard deviations are lower for numbers of matches greater than two hundred thousand. The percentage reduction in the number of matches for a given probability at night is lower for NOAA-19 than NOAA-09. At a probability threshold of 0.9 we see a $\sim$45% for NOAA-09 and $\sim$15% for NOAA-19.

One region in which cloud detection is particularly challenging during the day is in the presence of sunglint. An example image from MetOp-A is shown in Figure 7, displaying the 0.6 and 0.8 μm reflectance in the left hand plots. The sunglint features can be seen on the right of these images as a band running vertically, most prominent in the top half of the image. Sunglint increases the brightness of the ocean surface, and where reflectance features are used to detect cloud, sunglint areas are typically partially or completely flagged as cloud. The plots on the right hand side of Figure 7 show the Bayesian and CLAVR-x probabilities of clear-sky over the same region. Note these are provided on different scales for each mask. We see that the CLAVR-x mask performs poorly in the sunglint region with clear-sky probabilities falling below 0.5 across some of the region. There is evidence of problems using some of the cloudy features in the CLAVR-x classifier here, as indicated by hard lines in the probability field in this region. Conversely, the Bayesian algorithm performs well. The colour scale used here is for clear-sky probabilities between 0.96–1.0. For most of the clear-sky regions in the image the probability exceeds 0.99, and we see only a very slight decrease in confidence to $\sim$0.98 in sunglint affected areas, without the loss of structure as seen in the CLAVR-x mask. Typically, we threshold the probability of clear-sky at 0.9 over the ocean to produce a binary clear/cloud mask, so this woud not result in a loss of clear-sky data.

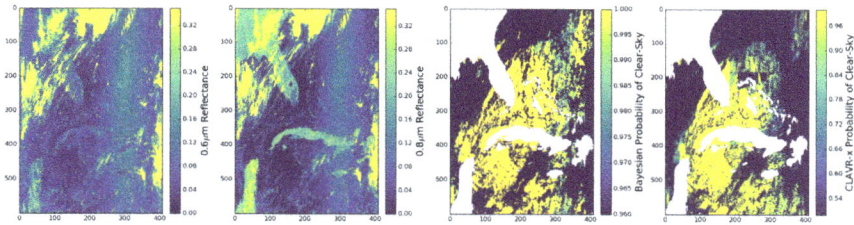

Figure 7. Bayesian and CLAVR-x clear-sky probabilities in a sunglint affected region. Data are from MetOp-A on the 1st of August 2012. From left to right, plots show the 0.6 μm reflectance, the 0.8 μm reflectance, Bayesian clear-sky probabilty and CLAVR-x clear-sky probability. For the clear-sky probability plots, white areas denote land.

5. Conclusions

We present here a Bayesian cloud detection scheme applied to the AVHRR data record from NOAA-06 through to MetOp-A. There are two novel concepts in the application of this scheme to AVHRR, beyond those published previously [5]: (1) Application of empirical cloudy look-up tables from a single sensor to the entire data record without the need to generate sensor specific auxiliary data files; (2) Use of the visible channel simulation from RTTOV v11.3 in daytime cloud detection. We compare the algorithm performance against the operational CLAVR-x cloud detection scheme, using matches with in situ data from drifting buoys and the GTMBA array. We find that the Bayesian cloud detection consistently reduces the uncertainty due to outliers indicative of removing undetected cloud from the distribution. For most sensors, using the Bayesian cloud detection scheme also reduces the robust standard deviation, most notably for daytime SST retrievals. At night, we find comparable RSD values for the Bayesian and CLAVR-x algorithms, but CLAVR-x gives significantly more matches. We present a case study over a sunglint region, highlighting the use of visible channel simulations for daytime cloud detection. Identification of clear-sky in sunglint is typically challenging for cloud detection schemes, but we find that the Bayesian scheme shows only small reductions in clear-sky probabilities. The Bayesian cloud detection performance is reasonably consistent across the data record, despite calibration issues affecting particularly earlier sensors in the data record, and differences in channel availability for AVHRR-1, 2 and 3 instruments for SST retrieval. We find that the Bayesian algorithm compares favourably against the well-established CLAVR-x operational cloud masking algorithm.

Acknowledgments: The work undertaken in this paper was funded by the European Space Agency (ESA ESRIN REF 3-1390413I-NB) Sea Surface Temperature Climate Change Initiative project.

Author Contributions: The lead author, Claire Bulgin, undertook the majority of this work, implementing the Bayesian cloud detection for AVHRR, determing the visible channel RTM corrections, generating the empirical cloudy PDFs and validating the cloud mask performance. Jon Mittaz was responsible for the brightness temperature harmonisation across the AVHRR data record. Owen Embury was involved in the development of the code base for applying the cloud detection to AVHRR within the SST CCI project. Steinar Eastwood was responsible for generating the ice detection PDFs applied to the data at high latitudes. Chris Merchant oversaw the science development team, providing critical scientific input in the development process.

Conflicts of Interest: The authors declare no conflict of interest.

References

1. Carbajal Henken, C.K.; Linstrot, R.; Preusker, R.; Fischer, J. FAME-C: Cloud property retrieval using synergistic AATSR and MERIS observations. *Atmos. Meas. Tech.* **2014**, *7*, 3873–3890.
2. Stengel, M.; Mieruch, S.; Jerg, M.; Karlson, K.-G.; Scheirer, R.; Maddux, B.; Meirink, J.F.; Poulsen, C.; Siddans, R.; Walther, A.; et al. The Clouds Climate Change Initiative: Assessment of state-of-the-art cloud property retrieval schemes applied to AVHRR heritage measurements. *Remote Sens. Environ.* **2015**, *162*, 363–379.
3. Bulgin, C.E.; Sembhi, H.; Ghent, D.; Remedios, J.J.; Merchant, C.J. Cloud clearing techniques over land for land surface temperature retrieval from the Advanced Along Track Scanning Radiometer. *Int. J. Remote Sens.* **2014**, *35*, 3594–3615.
4. Friedl, M.A.; McIver, D.K.; Hodges, J.C.F.; Zhang, X.Y.; Muchoney, D.; Strahler, A.H.; Woodcock, C.E.; Gopal, S.; Schneider, A.; Cooper, A.; et al. Global land cover mapping from MODIS: Algorithms and early results. *Remote Sens. Environ.* **2002**, *83*, 287–302.
5. Merchant, C.J.; Harris, A.R.; Maturi, E.; MacCallum, S. Probabilistic physically-based cloud screening of satellite infra-red imagery for operational sea surface temperature retrieval. *Q. J. R. Meteorol. Soc.* **2005**, *131*, 2735–2755.
6. Müller, D.; Krasemann, H.; Brewin, R.J.W.; Brockmann, C.; Deschamps, P.-Y.; Doerffer, R.; Fomferra, N.; Franz, B.A.; Grant, M.G.; Groom, S.B.; et al. The Ocean Colour Climate Change Initiative: II. Spatial and temporal homogeneity of satellite data retrieval due to systematic effects in atmospheric correction processors. *Remote Sens. Environ.* **2015**, *162*, 257–270.
7. Frey, R.A.; Ackerman, S.A.; Liu, Y.; Strabala, K.I.; Zhang, H.; Key, J.R.; Wang, X. Cloud Detection with MODIS. Part I: Improvements in the MODIS Cloud Mask for Collection 5. *J. Atmos. Ocean. Technol.* **2008**, *25*, 1057–1072.
8. Rossow, W.B.; Garder, L.C. Cloud Detection Using Satellite Measurements of Infrared and Visible Radiances for ISCCP. *J. Clim.* **1993**, *12*, 2341–2369.
9. Závody, A.M.; Mutlow, C.T.; Llewellyn-Jones, D.T. Cloud Clearing over the Ocean in the Processing of Data from the Along-Track Scanning Radiometer (ATSR). *J. Atmos. Ocean. Technol.* **2000**, *17*, 595–615.
10. Baum, B.A.; Tovinkere, V.; Titlow, J.; Welch, R.M. Automated Cloud Classification of Global AVHRR Data Using a Fuzzy Logic Approach. *J. Appl. Meteorol.* **1997**, *36*, 1519–1540.
11. Gomez-Chova, L.; Munoz-Mari, J.; Amoros-Lopez, J.; Izquierdo-Verdiguier, E.; Camps-Valls, G. Advances in synergy of AATSR-MERIS sensors for cloud detection. In Proceedings of the Geoscience and Remote Sensing Symposium, Melbourne, VIC, Australia, 21–26 July 2013; pp. 4391–4394.
12. Bulgin, C.E.; Eastwood, S.; Embury, O.; Merchant, C.J.; Donlon, C. Sea surface temperature climate change intiative: Alternative image classification algorithms for sea-ice affected oceans. *Remote Sens. Environ.* **2015**, *162*, 396–407.
13. Heidinger, A.K.; Evan, A.T.; Foster, M.J.; Walther, A. A Naive Bayesian Cloud-Detection Scheme Derived from CALIPSO and Applied within PATMOS-x. *J. Appl. Meteorol. Climatol.* **2012**, *51*, 1129–1144.
14. Islam, T.; Rico-Ramirez, M.A.; Srivastava, P.K.; Dai, Q.; Han, D.; Gupta, M.; Zhuo, L. CLOUDET: A Cloud Detection and Estimation Algorithm for Passive Microwave Imagers and Sounders Aided by Naive Bayes Classifier and Multilayer Perceptron. *IEEE J. Sel. Top. Appl. Earth Obs. Remote Sens.* **2015**, *8*, 4296–4301.
15. Merchant, C.J.; Embury, O.; Roberts-Jones, J.; Fiedler, E.; Bulgin, C.E.; Corlett, G.K.; Good, S.; McLaren, A.; Rayner, N.; Morak-Bozzo, S.; et al. Sea surface temperature datasets for climate applications from Phase 1 of the European Space Agency Climate Change Initiative (SST CCI). *Geosci. Data J.* **2014**, *1*, 179–191.
16. Embury, O.; Merchant, C.J. A reprocessing for climate of sea surface temperature from the Along-Track Scanning Radiometers: A new retrieval scheme. *Remote Sens. Environ.* **2012**, *116*, 47–61.
17. Dee, D.P.; Uppala, S.M.; Simmons, A.J.; Berrisford, P.; Poli, P.; Kobayashi, S.; Andrae, U.; Balmaseda, M.A.; Balsamo, G.; Bauer, P.; et al. The ERA-Interim reanalysis: Configuration and performance of the data assimilation system. *Q. J. R. Meteorol. Soc.* **2011**, *137*, 553–597.
18. EUMETSAT. *AVHRR Level 1b Product Guide*; EUMETSAT: Darmstadt, Germany, 2011; EUM/OPS-EPS/MAN/04/0029, v3A.
19. Rao, C.R.N.; Chen, J. Post-launch calibration of the visible and near-infrared channels of the Advanced Very High Resolution Radiometer on the NOAA-14 spacecraft. *Int. J. Remote Sens.* **1996**, *17*, 2743–2747.

20. Heidinger, A.K.; Straka, W.C., III; Molling, C.C.; Sullivan, J.T.; Wu, X. Deriving an inter-sensor consistent calibration for the AVHRR solar reflectance data record. *Int. J. Remote Sens.* **2010**, *24*, 6493–6517.

21. Bhatt, R.; Doelling, D.R.; Scarino, B.R.; Gopalan, A.; Haney, C.O.; Minnis, P.; Bedka, K.M. A Consistent AVHRR Visible Calibration Record Based on Multiple Methods Applicable for the NOAA Degrading Orbits. Part 1: Methodology. *J. Atmos. Ocean. Technlol.* **2016**, *33*, 2499–2515.

22. Vermote, E.; Kaufman, Y.J. Absolute calibration of AVHRR visible and near-infrared channels using ocean and cloud views. *Int. J. Remote Sens.* **1995**, *16*, 2317–2340.

23. Walton, C.C.; Sullivan, J.T.; Rao, C.R.N.; Weinreb, M. Corrections for detector nonlinearities and calibration inconsistencies of the infrared channels of the Advanced Very High Resolution Radiometer. *J. Geophys. Res.* **1998**, *103*, 3323–3337.

24. Mittaz, J.P.D.; Harris, A.R.; Sullivan, J.T. A Physical Method for the Calibration of the AVHRR-3 Thermal IR Channels 1: The Prelaunch Calibration Data. *J. Atmos. Ocean. Technol.* **2009**, *26*, 996–1019.

25. Mittaz, J.P.D.; Harris, A.R. A Physical Method for the Calibration of the AVHRR/3 Thermal IR Channels. Part II: An In-Orbit Comparison of the AVHRR Longwave Thermal IR Channels on board MetOp-A with IASI. *J. Atmos. Ocean. Technol.* **2011**, *28*, 1072–1087.

26. Wang, L.; Cao, C. On-Orbit Calibration Assessment of AVHRR Longwave Channels on MetOp-A Using IASI. *IEEE Trans. Geosci. Remote Sens.* **2008**, *46*, 4005–4013.

27. Mittaz, J.P.D.; Bali, M.; Harris, A.R. The calibration of broad band infrared sensors: Time variable biases and other issues. In Proceedings of the EUMETSAT Meteorological Satellite Conference, Vienna, Austria, 16–20 September 2013.

28. Wang, L.; Goldberg, M.; Wu, X.; Cao, C.; Iacovazzi, R.A., Jr.; Yu, F.; Li, Y. Consistency assessment of Atmsopheric Infrared Sounder and Infrared Atmospheric Sounding Interferometer radiances: Double differences verses simultaneous nadir overpasses. *J. Geophys. Res.* **2011**, *116*, doi:10.1029/2010JD014988.

29. Hocking, J.; Rayer, P.; Rundle, D.; Saunders, R.; Matricardi, M.; Geer, A.; Brunel, P.; Vidot, J. *RTTOV v11 Users Guide*; NWP SAF: Exeter, UK, NWOSAF-MO-UD-028. Available online: www.nwpsaf.eu/site/download/documentation/rtm/docs_rttov11/users_guide_11_v1.4.pdf?b188a8&b188a8 (accessed 8 January 2018).

30. Cox, C.E.; Munk, W. Measurements of the roughness of the sea surface form the sun's glitter. *J. Opt. Soc. Am.* **1954**, *44*, 838–850.

31. Block, T.; Embacher, S.; Merchant, C.J.; Donlon, C. High performance software framework for the calculation of satellite-to-satellite data matchups (MMS version 1.2). *Geosci. Model Dev. Discuss.* **2017**, doi:10.5194/gmd-2017-54.

32. National Snow and Ice Data Centre, Climatological Monthly Maximum Sea Ice Extent Dataset. Available online: www.nsidc.org/data/pm/ocean-masks (accessed 22 November 2017).

33. Karlsson, K.-G.; Anttila, K.; Trentmann, J.; Stengel, M.; Meirink, J.F.; Devasthale, A.; Hanschmann, T.; Kothe, S.; Jääkeläinen, E.; Sedlar, J.; et al. CLARA-A2: CM SAF Cloud, Albedo and surface RAdiation dataset from AVHRR data—Edition 2. *Satell. Appl. Facil. Clim. Monit.* **2017**. Available online: https://doi.org/10.5676/EUM_SAF_CM/CLARA_AVHRR/V002 (accessed 8 January 2018).

34. Tonboe, R.T.; Eastwood, S.; Lavergne, T.; Sorensen, A.M.; Rathmann, N.; Dybkjaer, G.; Pedersen, L.T.; Hoyer, J.L.; Kern, S. The EUMETSAT sea ice concentration climate data record. *Cryosphere* **2016**, *10*, 2275–2290.

35. Chevalier, F. *Sampled Database of 60-Level Atmospheric Profiles from ECMWF Analyses*; NWP SAF Report 4, ECMWF: Reading, UK, 2002. Available online: http://www.emwf.int/publications/library/do/references/show?id=83287 (accessed 8 January 2018).

36. Uppala, S.; Kalberg, P.W.; Simmons, A.J. The ERA-40 reanalysis. *Q. J. R. Meteorol. Soc.* **2005**, *131*, 2961–3012.

37. Merchant, C.J.; Le Borgne, P.; Marsouin, A.; Roquet, H. Optimal estimation of sea surface temperature from split-window observations. *Remote Sens. Environ.* **2008**, *112*, 2469–2484.

38. National Oceanic and Atmospheric Adminstration. Global Tropical Moored Buoy Array, Pacific Marine Environmental Laboratory. Available online: www.pmel.noaa.gov/gtmba/ (accessed 8 January 2018).

remote sensing

MDPI

Article

The Role of Advanced Microwave Scanning Radiometer 2 Channels within an Optimal Estimation Scheme for Sea Surface Temperature

Kevin Pearson [1],*, Christopher Merchant [1,2], Owen Embury [1,2] and Craig Donlon [3]

[1]   Department of Meteorology, University of Reading, P.O. Box 243, Reading RG6 6BB, UK;
      c.j.merchant@reading.ac.uk (C.M.); o.embury@reading.ac.uk (O.E.)
[2]   National Centre for Earth Observation, Department of Meteorology, University of Reading, P.O. Box 243,
      Reading RG6 6BB, UK
[3]   European Space Agency, ESTEC, Postbus 299, 2200 AG Noordwijk, The Netherlands; craig.donlon@esa.int
*   Correspondence: k.j.pearson@reading.ac.uk

Received: 10 November 2017; Accepted: 4 January 2018; Published: 11 January 2018

Abstract: We present an analysis of information content for sea surface temperature (SST) retrieval from the Advanced Microwave Scanning Radiometer 2 (AMSR2). We find that SST uncertainty of ~0.37 K can be achieved within an optimal estimation framework in the presence of wind, water vapour and cloud liquid water effects, given appropriate assumptions for instrumental uncertainty and prior knowledge, and using all channels. We test all possible combinations of AMSR2 channels and demonstrate the importance of including cloud liquid water in the retrieval vector. The channel combinations, with the minimum number of channels, that carry most SST information content are calculated, since in practice calibration error drives a trade-off between retrieved SST uncertainty and the number of channels used. The most informative set of five channels is 6.9 V, 6.9 H, 7.3 V, 10.7 V and 36.5 H and these are suitable for optimal estimation retrievals. We discuss the relevance of microwave SSTs and issues related to them compared to SSTs derived from infra-red observations.

Keywords: AMSR2; sea surface temperature; optimal estimation

1. Introduction

Sea surface temperature (SST) is a geophysical quantity of fundamental importance in the Earth system, since it is a controlling factor in air-sea fluxes [1,2] and therefore profoundly influences atmospheric and oceanographic thermodynamics [3], dynamics [4,5] and coupled interactions [6]. Near-real time estimation of global SST at adequate spatial resolution is crucial to weather forecasting by numerical weather prediction (NWP, [7]) and errors in knowledge of SST can materially degrade weather forecast skill [8,9]. SST is used as the measure of Earth's surface temperature over oceans [10–12] and is therefore a key metric of climatic variability and change whose global evolution can be estimated back to the mid-19th Century [12]. Historic observations of SST are relatively sparse prior to the satellite era [13], and centennial-scale reconstructions draw heavily on the relative completeness and detail of remotely sensed SST [14]. The series of Advanced Very High Resolution Radiometers (AVHRRs) have been operated since 1979 with channels supporting SST estimation, using differential-absorption-based techniques to account for the influence of the atmosphere on infra-red (IR) brightness temperatures [15–18]. Thus, reprocessing of multi-decadal satellite SST datasets has concentrated on IR sensors, namely, the AVHRRs [19] and Along Track Scanning Radiometers (ATSRs; [20]). Merchant et al. [21] more recently used both AVHRRs and ATSRs jointly to develop a blended, gap-filled analysis for climate applications, analogous to the SST analyses produced operationally for NWP [9,22], but with more attention to long-term stability

Microwave (MW) observations of SST were first attempted with the Scanning Multichannel Microwave Radiometer (SMMR) launched in 1978 and in 1999 the Tropical Rainfall Measuring Mission's (TRMM's) Microwave Imager began delivering SSTs of useful accuracy across the tropics. The record of globally SST-capable microwave radiometers is shorter, having commenced with the Advanced Microwave Scanning Radiometer-E (AMSR-E) in 2002. MW radiometry for SST has strengths and weakness relative to IR records. The primary advantage is coverage [23]: MW SSTs are available over the open ocean under non-precipitating cloud cover, while both precipitation and cloud cover strongly limits the sampling available in the IR. MW SST is not available near coasts, near sea-ice and in areas of persistent radio-frequency interference (RFI). The spatial resolution of MW SST is typically 50 km [24] compared to 1 km for IR, limiting the precision with which thermal ocean fronts can be located in MW imagery. The potential for confounding of SST signals by wind variability (via emissivity effects) is greater for MW SSTs than for IR SSTs. Nonetheless, since cloud cover is persistent in some seasons in climatologically significant regions, the coverage advantage of MW radiometry is such that the blending of MW and IR SSTs for climate data records should be considered.

AMSR2 is a microwave radiometer instrument flying on board the Japan Aerospace Exploration Agency's (JAXA) Global Change Observation Mission 1st-Water (GCOM-W1) satellite, launched in 2012. This forms part of the "A-train" [25] series of satellites that fly in the same orbit separated by a few minutes. It observes at 6.9, 7.3, 10.65, 18.7, 23.8, 36.5 and 89.0 GHz in both H and V polarizations. The 7.3 GHz channel is an addition compared to the predecessor AMSR-E instrument on Aqua and improves detection of radio frequency interference (RFI) from artificial sources.

This paper provides an information content analysis for the AMSR2 radiometer. Our aims are to establish the fundamental limits of retrieval uncertainty for AMSR2 SST retrieval in the framework of optimal estimation (OE), and to inform strategies about channel selection for developing a new MW SST product, ultimately intended for joint use with IR products in a climate data record. A previous study with similar objectives [26] neglected the importance of variable cloud liquid water in MW SST retrieval, and did not address itself to the prioritisation of channels, both addressed here.

In Section 2, we review some of the underlying physics relevant to MW SST retrieval, noting and contrasting the MW case from the IR case. Section 3 reviews some background theory relating to information content analysis and OE. These are applied to SST retrieval from the AMSR2 instrument in Sections 4 and 5.

2. Physical Considerations

Microwave thermal emission from the ocean surface occurs in the Rayleigh–Jeans tail of the Planck function. This is in contrast to the thermal IR, where the peak of the Planck function is in the 10.5–12.5 µm window that is often used for SST remote sensing. The ocean surface emissivity (ε) for the low-frequency AMSR2 channels is around ~0.5 compared to an emissivity of ~1 in the IR. The intensity of MW radiation at the top of atmosphere (TOA) is low, which is mitigated somewhat by the ability to use large (~m) antennae for microwave instruments. Despite this, the effective noise equivalent temperature difference (NEdT) is larger in the MW region than in the IR. The longer wavelengths involved also give rise to diffraction effects that limit the spatial resolution of AMSR2 to ~50 km. The MW emissivity of land and ice is significantly higher than the ocean. With contemporary instruments, this leads to side-lobe contamination of the ocean MW signal close to coasts and ice edges and prevents accurate SST retrievals in these areas. There is also a larger change in emissivity with polarisation over ocean compared to ice. This can be exploited for ice detection and classification [27].

A significant advantage of using MW measurements when attempting to achieve global coverage of SST is that microwaves can penetrate cloud, so they can observe the surface signal under cloudy conditions wherein IR instruments cannot. This is useful, in particular, in persistently cloudy regions such as winter high-latitudes. Here, the restriction of IR instruments to clear-sky conditions decreases the temporal frequency of the observations and thus increases sampling errors.

This study utilises simulations of AMSR2 brightness temperatures by the fast radiative transfer model "Radiative Transfer for TOVS" (RTTOV; whose acronym has evolved into a name). We use the v11.3 software package [28–31] to carry out the simulations in Sections 4 and 5. In the MW region, this uses the FAST EMissivity (FASTEM) code to calculate the surface emissivity which, for version 4, is described by Liu et al. [32]. In this study, we use the latest version, FASTEM-6. The MW emissivity model involves a complex calculation, which we summarise below.

There are several models for the emissivity and permittivity of seawater [33–39]. FASTEM-6 uses a method that starts from a formulation for the permittivity based on Ellison et al. [33]. This describes the complex permittivity with a double Debye model:

$$\epsilon = \epsilon_\infty + \frac{\epsilon_s - \epsilon_1}{1 + jv\tau_1} + \frac{\epsilon_1 - \epsilon_\infty}{1 + jv\tau_2} + j\frac{\alpha}{2\pi f \epsilon_0} \tag{1}$$

Here, ϵ_0 is the permittivity of free space and v the frequency of the electromagnetic wave. The other parameters have been derived by fitting to measurements: ϵ_∞ has a linear dependence on temperature; $\epsilon_s, \epsilon_1, \epsilon_1, \tau_1$ and τ_2 are represented by polynomial fits to temperature (T) and salinity (S); and α has a mixed polynomial and exponential dependence on temperature and salinity.

The modelled permittivity is used to calculate Fresnel reflectivities (R_p where p is v or h for vertical and horizontal polarization components respectively) from the standard Fresnel equations. These are subsequently modified to effective values that account for other factors such as foam and surface roughness. In general, these factors add a dependency of the final emissivity on the wind vector ($\mathbf{U}$). Surface roughness causes MW energy to be scattered both into and out of the direct line of sight of the surface by quasi-specular reflection events. FASTEM represents these with a two-scale model [32,40]. The small-scale waves have a size close to the wavelength of the emitted radiation. These small waves ride on the large-scale undulations of gravity waves. The correction to R_p for the small-scale features takes the form of a multiplicative factor $\exp(-y\cos^2\theta)$ where y is a polynomial fit to wind speed and frequency and θ is the zenith angle of the observation. The large-scale correction (L_p) takes the form of an additive term with polynomial fit to frequency, wind speed and $\sec\theta$. The wave orientation is accounted for by adding three cosine harmonics for the relative azimuth angle (ϕ) between the observation and wind vectors. The wind-speed factors here act as a proxy for what is in reality the mechanical stress on the ocean due to the wind. This drives the creation of small scale waves and thus changes the effective surface area.

Above wind speeds of a few metres per second, foam begins to form on the sea surface [38]. This is principally a mixture of water with air bubbles. FASTEM-6 calculates the fraction of the surface covered by foam (f) using the expression of Monahan et al. [41] where $f \sim |\mathbf{U}|^{2.55}$. (An alternative form $f \sim |\mathbf{U}|^{3.231}$ by Tang [42] is used in FASTEM-4.) The model then computes area-weighted mean values of foam emissivities ($\varepsilon_{p,f}$) and the modified sea water emissivities. The foam emissivities are calculated using a combination of the zenith angle polynomial fit of Kazumori et al. [43] with the linear frequency dependence from Stogryn [44]. The final form relating the effective emissivities (ε_p), Fresnel reflectivities and the correction factors is thus

$$\varepsilon_p = (1-f)\left(1 - R_p e^{-y\cos^2\theta} + L_p\right) + f\varepsilon_{p,f} + \sum_{m=1}^{3} a_{p,m}\cos(m\phi) \tag{2}$$

where the functional dependencies are $f(|\mathbf{U}|)$, $R_p(T,S)$, $y(|\mathbf{U}|,v)$, $L_p(|\mathbf{U}|,v,\theta)$, $\varepsilon_{p,f}(\theta,v)$.

2.1. Cosmic Microwave Background

The cosmic microwave background (CMB) is radiation from the recombination era of the early universe that has subsequently cooled due to the expansion of the universe and now forms a near isotropic source of background photons [45,46]. Its spectrum is characterised by an effective temperature of ~2.73 K [47]. We can make a simple estimate of the relative intensity of this source to emission from the Earth from the ratio of the black-body functions B_ν for the two sources:

$$\frac{F_{CMB}}{F_\oplus} = \frac{B_\nu(T_{CMB})}{\varepsilon B_\nu(T_\oplus)} = \frac{\exp\left(\frac{h\nu}{kT_\oplus}\right) - 1}{\varepsilon\left[\exp\left(\frac{h\nu}{kT_{CMB}}\right) - 1\right]} \tag{3}$$

$$\frac{F_{CMB}}{F_\oplus} \approx \begin{cases} 0.02 & \text{at 6.9 GHz} \\ 0.008 & \text{at 89 GHz} \end{cases} \tag{4}$$

for $T_\oplus = 290$ K and emissivity $\varepsilon = 0.5$. Although we have neglected surface roughness and atmospheric effects, this demonstrates that the contribution of the CMB to the observed TOA flux, although small, is not negligible and must be included in MW radiative transfer modelling.

2.2. Skin Depth

There is typically a cooling of order 0.2 K from a depth of ~1 mm at the top of the ocean (the sub-skin) to the interface where the atmosphere and ocean meet. At IR wavelengths, electromagnetic waves are absorbed in a distance of order 10 μm and sample the ocean at the top of the skin layer and are thus sensitive to "SST-skin". In contrast, microwaves have a frequency-dependent penetration depth measured in millimeters and so observations here are sensitive to SST-sub-skin. To compare or harmonise measurements made in the two wavelengths regions with those from in situ sources, retrievals must be corrected to the depth of in situ measurements, typically 10 cm to 1 m. This requires a model for the skin effect and the diurnal warming.

Robinson [48] gives an expression for the apparent temperature (T_{app}) seen by a radiometer assuming an exponential form for the temperature profile in the skin-layer. This temperature profile can written as

$$T(z) = T_{ss} + (T_0 - T_{ss}e^{-z/d_T}) \tag{5}$$

where T_0 is the surface (interface) temperature and T_{ss} is the sub-skin temperature. Using an e-folding distance d_μ for the absorption of radiation at the surface, results in

$$T_{app} = T_{ss}(1 - \gamma) + \gamma T_0 \tag{6}$$

where $\gamma = \frac{d_T}{d_\mu + d_T}$.

If the cooling across the skin layer is due to molecular conduction, we might expect the temperature profile through the skin layer to be linear. A similar derivation using a total skin thickness δ and such a linear assumption for $T(z)$ yields

$$T_{app} = T_0(1 - \gamma_1) + T_{ss}\gamma_1 \tag{7}$$

where $\gamma_1 = \frac{d_\mu}{\delta}(1 - e^{-\delta/d_\mu})$.

2.3. Salinity

Salinity has a negligible effect on emissivity in the IR region but can be significant at MW wavelengths. Figure 1 shows the change in brightness temperature with salinity for a given atmospheric profile. For the most SST-sensitive, low-frequency channels, the effect is relatively small across the typical range of global oceanic salinity (33–37 PSU). The effect is more significant, however, for the higher frequency channels and is temperature dependent. Including this effect in modelling would be more important in areas with a strong freshwater influence.

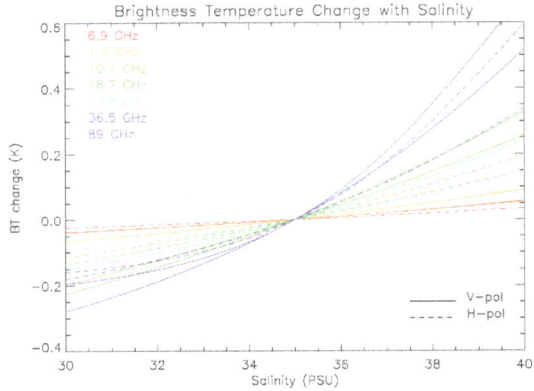

Figure 1. The change in the top-of-atmosphere brightness temperature from the value at 35 PSU as a function of salinity. The data were modelled by RTTOV for the AMSR2 instrument using the same atmospheric profile and with surface emissivities calculated by FASTEM. All channels are included ranging from 6.9 GHz (red) to 89 GHz (purple) with V-polarized channels indicated by solid lines and H-polarized channels by dashed lines.

2.4. Emissivity Dependence on Wind

As noted at the start of this section, the ocean emissivity in the MW region is affected by wind speed through the generation of foam and large- and small-scale waves. Accurate modelling of these processes is difficult particularly at low frequencies and is an ongoing area of research. Figure 2 shows the change in emissivity with wind speed for each of the channels for a SST of 297 K. The deviation from this azimuthal-mean emissivity value at a given wind speed is displayed against the separate wind-speed components in Figure 3. The lack of azimuthal symmetry means that it is possible, in principle, to derive some information about the separate wind components from MW observations. The small size of the deviation, however, implies that this is a weak constraint.

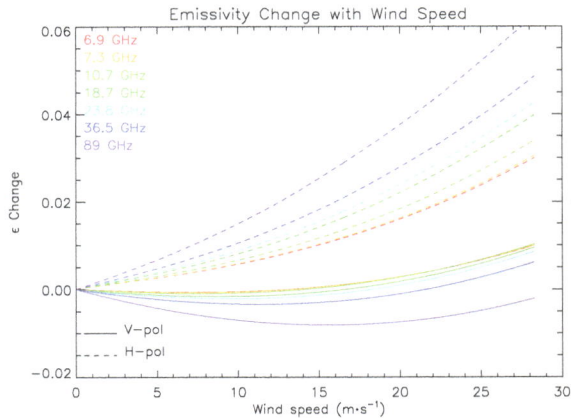

Figure 2. The change in emissivity as a function of wind speed using the FASTEM model for sea surface temperature (SST) = 297 K . All channels are included ranging from 6.9 GHz (red) to 89 GHz (purple) with V-polarized channels indicated by solid lines and H-polarized channels by dashed lines.

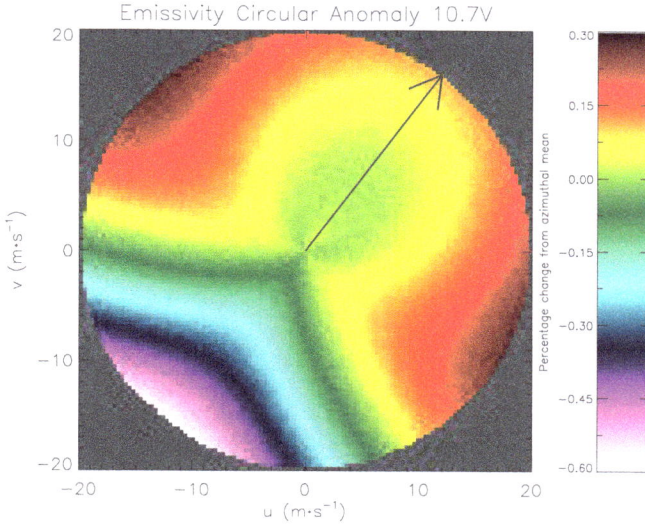

Figure 3. Deviation of emissivity from the azimuthal mean as a function of wind speed components. The lack of symmetry implies that the observation contains some information about the individual wind-speed components. The satellite azimuth angle has been chosen to be 37° here and is indicated by the arrow. A value of 0° would align the pattern along the v-axis.

2.5. Top-of-Atmosphere Radiance Dependence on Total Column Water Vapour

Water vapour acts as an additional source of absorption for radiation traveling through the atmosphere both at MW and IR wavelengths. There are interesting differences between the two regions, however. For illustrative purposes, consider radiative transfer for microwaves using a simple slab model of the atmosphere with absorptivity a (equal to its emissivity ε_a) and temperature T_a. Being in the Rayleigh–Jeans tail $B_\nu \propto T$ and, for convenience in this section, we absorb the constants of proportionality into the temperature units. The radiance of the upward emission by the atmosphere at temperature T_a is then

$$I_1 = \varepsilon_a\, T_a = aT_a \tag{8}$$

and, similarly, the downward emission by the atmosphere is

$$I_2 = aT_a \tag{9}$$

The radiance from the surface emission at temperature T_s

$$I_4 = \varepsilon T_s \tag{10}$$

and the amount that is transmitted through to the top of the atmosphere is

$$I_3 = (1-a)I_4 = (1-a)\varepsilon T_s \tag{11}$$

The total outward radiance is thus

$$I_{\text{TOA}} = I_1 + I_3 = aT_a + (1-a)\varepsilon T_s = \varepsilon T_s + a(T_a - \varepsilon T_s) \tag{12}$$

For a given column with fixed T_a and T_s, I_{TOA} can either increase or decrease with atmospheric absorption according to the sign of the final bracket. For $\varepsilon \approx 1$ (as in the IR part of the spectrum), I_{TOA}

will always decrease as the absorption in the atmosphere increases. In the MW region, however, where $\varepsilon \approx 0.5$, I_{TOA} can increase with increasing absorption.

In reality, the situation is obviously more complex. Not only is the atmosphere not isothermal, but, across the global ocean, there is a large-scale correlation between the total column water vapour (TCWV) and T_a. This sign of relationship, however, does occur and is counter to behaviour at IR wavelengths.

2.6. Top-of-Atmosphere Radiance Dependence on Total Cloud Liquid Water

At IR wavelengths, clouds are largely opaque, thus rendering observations of the surface impossible except perhaps in instances of thin cirrus. Microwaves penetrate non-precipitating clouds, although measured radiances are sensitive to the cloud liquid water content which must be included in any radiative transfer modelling. Figure 4 shows the change in modelled brightness temperature for the same conditions but with the cloud liquid water profile scaled to achieve different total cloud liquid water (TCLW) values. There is a significant effect on all of the channels as well as clear differences in the sensitivity between channels. Not only does this emphasise the importance of including these effects in any modelling but also suggests that TCLW can be retrieved to some degree.

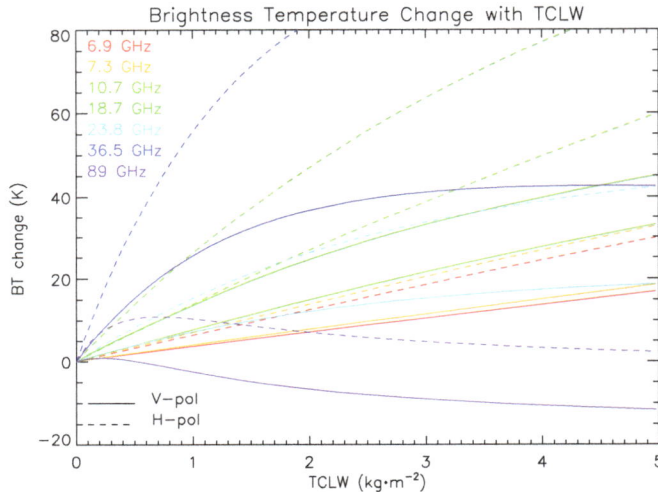

Figure 4. The change in measured brightness temperature as a function of total cloud liquid water. The data were modelled by RTTOV for the AMSR2 instrument using the same atmospheric profile but for scaled total cloud liquid water (TCLW). All channels are included ranging from 6.9 GHz (red) to 89 GHz (purple) with V-polarized channels indicated by solid lines and H-polarized channels by dashed lines.

3. Information Content and Optimal Estimation

OE provides a means to combine measured values from an instrument with initial a priori estimates of physical quantities of interest to provide a best estimate of the true value of the physical quantities. It does this by weighting the observations and a priori values via the appropriate covariance matrices of their uncertainties. The solution is always an optimised (minimised) function of the squares of residuals between observation and solution.

From Rodgers [49], the optimal estimate of the physical quantities in the state vector **x** is given by

$$\hat{x} = x_a + S_a\, K^T \left(K S_a\, K^T + S_\epsilon \right)^{-1} (y - y_a) \qquad (13)$$

This is the solution with maximum a posteriori probability given priori information and its uncertainty. In Equation (13), $\mathbf{y}$ is a vector containing the observations, $\mathbf{K}$ is the Jacobian matrix describing the sensitivity of each of the measurements to each physical quantity, $\mathbf{S}_a$ is the uncertainty covariance matrix of the a priori values for the physical quantities and $\mathbf{S}_\epsilon$ is the uncertainty covariance matrix for the measurements. The quantity $\mathbf{y}_a$ is the observation vector that would result from the a priori state $\mathbf{x}_a$. This must be calculated using a forward model and $\mathbf{y}_a(\mathbf{x}_a)$ is treated as linear in the region of $\mathbf{x}_a$. This equation can be interpreted as a form of multi-dimensional "weighted average" between the a priori values for the retrieved quantities and the values of the retrieved quantities that would give rise to the observations. Consider very small values of the a priori uncertainties. Here, the second term vanishes and the best estimate of the retrieval vector is the initial a priori values. Conversely, for large a priori uncertainties or very low measurement uncertainties, the best estimate is dominated by the observation vector. The degree to which observations and modelled values in the final bracket differ is translated from observation space into physical-quantity space by the preceding matrices. No assumption about the Gaussianity or otherwise of the uncertainty distributions is required in the derivation of this equation. In the particular case of Gaussian uncertainty distributions, the maximum a posteriori solution is also the solution with minimum error variance.

The expected uncertainty covariance matrix for the retrieved variables is

$$\mathbf{S} = \left(\mathbf{K}^T \mathbf{S}_a^{-1} \mathbf{K} + \mathbf{S}_\epsilon^{-1} \right)^{-1} \tag{14}$$

In principle, this approach allows all sources of information about a problem to be combined with the correct weighting no matter how weak their sensitivity to the variables we are interested in. In practice, imperfect forward modelling and the lack of exact knowledge of appropriate covariance matrices, limit the degree to which additional observations improve the accuracy of the retrieved quantities.

Without performing any retrievals, we can calculate the degrees of freedom for signal in a measurement system from

$$d_s = \mathrm{Tr}\left(\mathbf{K}\mathbf{S}_a\,\mathbf{K}^T \left[\mathbf{K}\mathbf{S}_a\,\mathbf{K}^T + \mathbf{S}_\epsilon \right]^{-1} \right) \tag{15}$$

d_s gives an estimate of the number of distinct quantities that may be inferred from the measurements. It is not, in general, an integer because usually retrieved variables are only partially constrained rather than precisely determined. A fuller description of optimal estimation as applied to retrieval of SST is given by Merchant et al. [50].

In the following sections, these techniques are applied to simulations using 2680 profiles over ocean taken from the EUMETSAT Satellite Application Facility on Numerical Weather Prediction (NWP SAF) 91-level dataset [51] sampled for specific humidity. The RTTOV simulation code is used as the forward model to generate $\mathbf{y}_a$ and $\mathbf{K}$ appropriate to the AMSR2 instrument. A constant salinity of 35 PSU is assumed for all the profiles.

Prigent et al. [26] carried out a similar analysis for a new mission concept, Microwat, simulating retrievals based on AMSR-E channel sensitivities. They retrieved SST and wind speed assuming initial uncertainties on these two quantities of 3.31 K and 1.33 m·s^{-1}, respectively. They also carried out an information content analysis including water vapour content uncertainties of 10% on model levels. To provide comparability, we conduct an analysis below based on this specification using, as did Prigent et al. [26], a retrieval vector containing the four variables SST (T_s), the natural logarithm of TCWV (W) and the two wind-speed components (u, v):

$$\mathbf{x}^T = \begin{pmatrix} T_s & W & u & v \end{pmatrix} \tag{16}$$

with an assumed-diagonal $\mathbf{S}_a$ populated with a priori uncertainties of 3.31 K in SST, 10% TCWV and 0.94 m·s^{-1} for each wind component. We also extend this approach using a retrieval vector with five variables:

$$\mathbf{x}^T = \begin{pmatrix} T_s & W & u & v & L \end{pmatrix} \tag{17}$$

that includes the logarithm of TCLW (L). With this formulation, we use a priori uncertainties of 1 K in SST, 10% in TCWV, 1.41 m·s^{-1} in each wind component and 10% in TCLW. Retrieving the logarithm of the integrated column values avoids retrieving unphysical negative estimates for quantities bounded at zero. The fractional uncertainties expressed on the quantities TCWV and TCLW transform into absolute uncertainties when expressed in log-space since, for a fractional uncertainty f on a quantity a, where

$$L = \ln a \tag{18}$$

and the absolute uncertatinty in L is

$$\sigma_L = \left| \frac{\partial L}{\partial a} \right| \sigma_a = \frac{1}{a}\sigma_a = f \tag{19}$$

$\mathbf{S}_e$ is also assumed to be diagonal with values filled by the NEdT for each AMSR2 channel. In ascending order of frequency, these are (0.34, 0.43, 0.7, 0.7, 0.6, 0.7, 1.2) K with both H- and V-components having the same value [52].

4. Information Content Analysis

The degrees of freedom for signal d_s, using all 14 channels, for each of the considered profiles, is shown in Figure 5. The mean value $\bar{d}_s = 2.86$ for the four-variable retrieval vector and $\bar{d}_s = 3.09$ when using the five-variable vector. These values are lower in both cases than the number of retrieved quantities and likely reflects the weak constraint that the observations place on the separate wind-speed components. There is also a noticeably wider spread of d_s values for the five-variable retrievals compared to the four-variable cases.

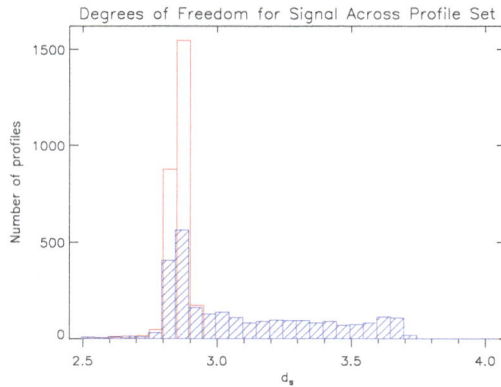

Figure 5. Degrees of freedom for signal across the profiles set used in the information content analysis using all 14 AMSR2 channels. Results assuming a four-variable retrieval vector are shown with open red bars and the five-variable retrieval vector with hatched, blue bars.

The estimated retrieval uncertainty matrix was calculated from Equation (14) for every profile for all possible channel combinations. For a given channel combination, we define the estimated average SST retrieval uncertainty (s) as the root mean squared expected uncertainty for SST across the profile set i.e.,

$$s = \sqrt{\frac{1}{n} \sum_{i=1}^{n} (S_{1,1})_i} \tag{20}$$

where n is the number of profiles (2680) that are indexed by i. Figure 6 shows s for the single-channel-only retrievals, illustrating which channels make the greatest individual contribution to reducing uncertainty in retrieved values of SST.

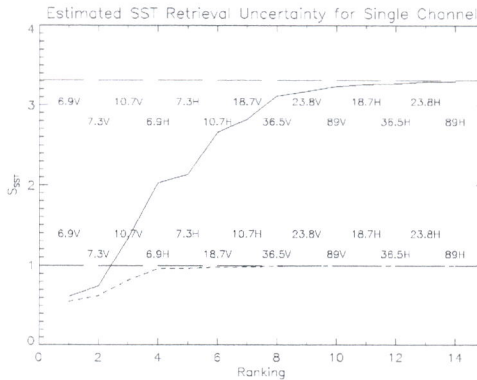

Figure 6. Estimated SST retrieval uncertainty for a single-channel retrieval from information content analysis. The upper set of channels and solid line come assume a four-variable retrieval vector and prior SST uncertainty of 3.31 K. The lower set and dashed line assunme a five-variable retrieval vector, in which cloud liquid water is additionally accounted for, and a priori SST uncertainty of 1.0 K. The prior uncertainty values in each case are shown in long-dashed lines.

Figure 7 shows the smallest value of s when a given number of channels is included in the observation vector along with the best channel to add. This is summarised in Tables 1 and 2.

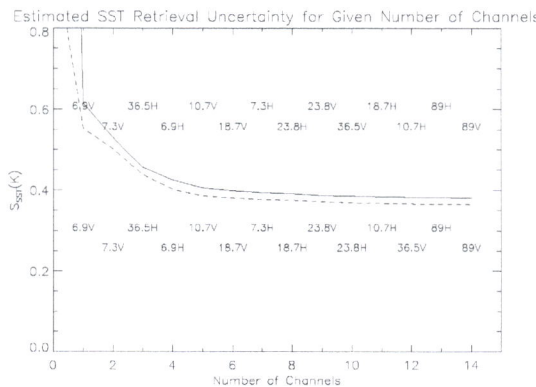

Figure 7. Estimated SST retrieval uncertainty against the number of AMSR2 channels used, for the best combination of the given number of channels, based on information content analysis. The new channel added to the set is indicated at each step and is chosen on the basis of minimizing the SST retrieval uncertainty. The upper set of channels and solid line come assume a four-variable retrieval vector and the lower set and dashed line assume a five-variable retrieval vector.

5. Simulated Retrieval

Simulated retrievals were carried out by randomly perturbing the NWP SAF profiles according to the S_a uncertainties for the two cases. A 10% variation was also applied to the total cloud liquid water (TCLW) profiles for the four-variable case even though this was not a retrieved variable. The water vapour and CLW values on each level of the profiles were uniformly scaled to give the perturbed TCWV and TCLW values. These perturbed profiles were treated as the unknown true values and corresponding simulated observations **y** were generated using RTTOV with random noise added consistent with S_ϵ. The unperturbed profiles were used both as the a priori state and linearisation point from which y_a and **K** were generated, again using values obtained from RTTOV.

The simulated retrieval error was calculated for every profile for all possible channel combinations. For a given channel combination, we define the simulated uncertainty (σ) as the standard deviation of the SST retrieval errors (e) across the profile set. Thus, for any retrieval

$$e = \hat{x}_1 - x_1 \tag{21}$$

and, for a given channel combination,

$$\sigma^2 = \frac{1}{n} \sum_{i=1}^{n} (e_i - \bar{e})^2 \tag{22}$$

Figure 8 shows the values of σ for single-channel-only retrievals, again illustrating which channels make the greatest individual contribution to a retrieval of SST. Figure 9 shows the smallest value of σ for a given number of channels included in the observation vector along with the best new channel to add to the existing set. These results are also summarised alongside the information content analysis in Tables 1 and 2.

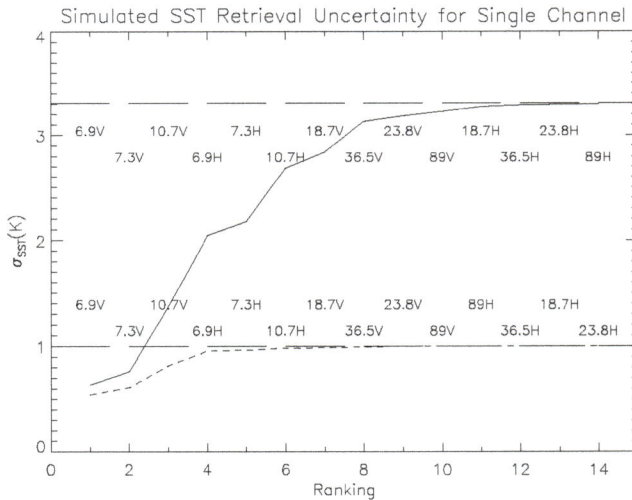

Figure 8. The standard deviation of the SST retrieval error for a single-channel retrieval from information content analysis. The upper set of channels and solid line come assume a four-variable retrieval vector and prior SST uncertainty of 3.31 K. The lower set and dashed line assume a five-variable retrieval vector, in which cloud liquid water is additionally accounted for, and priori SST uncertainty of 1.0 K. The prior uncertainty values in each case are shown with long-dashed lines.

Simulated SST Retrieval Uncertainty for Given Number of Channels

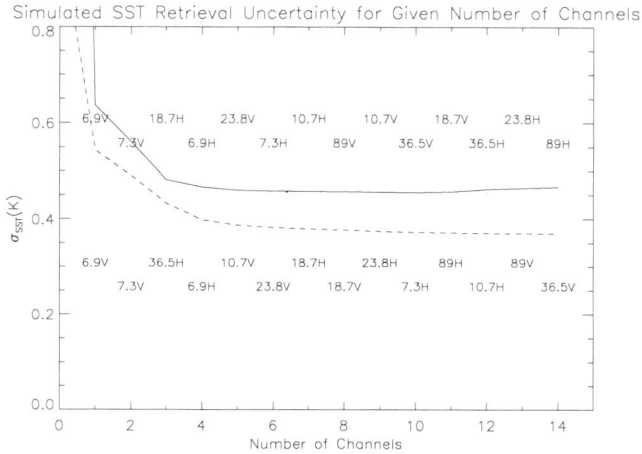

Figure 9. The standard deviation of the SST retrieval error for the best combination of the given number of channels from simulated retrievals. The additional channel added to the set is indicated at each step. The upper set of channels and solid line assume a four-variable retrieval vector and the lower set and dashed line assume a five-variable retrieval vector.

Table 1. The standard deviation of the sea surface temperature (SST) retrieval error and root-mean-squared predicted SST uncertainty from an information content analysis across all profiles for varying numbers of channels. The channel indicated on each row is the best one to add to the existing channel set for retrieving SST.

	4-Variable Retrieval				5-Variable Retrieval			
	Simulated Retrieval		Information Content		Simulated Retrieval		Information Content	
Number of Channels	Channel Added	σ_{SST} (K)	Channel Added	s_{SST} (K)	Channel Added	σ_{SST} (K)	Channel Added	s_{SST} (K)
1	6.9 V	0.636	6.9 V	0.615	6.9 V	0.543	6.9 V	0.553
2	7.3 V	0.561	7.3 V	0.530	7.3 V	0.491	7.3 V	0.502
3	18.7 H	0.481	36.5 H	0.457	36.5 H	0.432	36.5 H	0.438
4	6.9 H	0.466	6.9 H	0.425	6.9 H	0.398	6.9 H	0.402
5	23.8 V	0.460	10.7 V	0.405	10.7 V	0.387	10.7 V	0.386
6	7.3 H	0.458	18.7 V	0.398	23.8 V	0.382	18.7 V	0.381
7	10.7 H	0.457	7.3 H	0.394	18.7 H	0.379	7.3 H	0.377
8	89 V	0.457	23.8 H	0.391	18.7 V	0.377	23.8 H	0.375
9	10.7 V	0.456	23.8 V	0.387	23.8 H	0.374	23.8 V	0.372
10	36.5 V	0.455	36.5 V	0.385	7.3 H	0.372	18.7 V	0.369
11	18.7 V	0.457	18.7 H	0.384	89 H	0.372	36.5 V	0.368
12	36.5 H	0.462	10.7 H	0.383	10.7 H	0.371	10.7 H	0.366
13	23.8 H	0.464	89 H	0.382	89 V	0.370	89 H	0.366
14	89 H	0.466	89 V	0.381	36.5 V	0.370	89 V	0.365

Table 2. The standard deviation of the retrieval error and root-mean-squared predicted uncertainties from an information content analysis across all profiles using the best channel combinations for simulated SST retrieval i.e., 10 channels for the 4-variable retrievals and all 14 channels for the 5-variable retrievals.

	4-Variable Retrieval		5-Variable Retrieval	
	Simulated Retrieval	Information Content	Simulated Retrieval	Information Content
	σ	s	σ	s
SST (K)	0.455	0.403	0.370	0.365
ln(TCWV)	0.025	0.018	0.014	0.014
u (m·s^{-1})	0.745	0.679	1.06	0.972
v (m·s^{-1})	0.791	0.740	1.16	1.07
ln(TCLW)			0.086	0.084

6. Discussion

The OE framework provides a mechanism for combining all available information relating to an inverse problem with appropriate weighting. Since each channel brings some information, adding more channels to the observation vector results in progressively improving retrieval uncertainties if all sources of uncertainty are well-described by the error covariances used, and if the retrieved variables account for all significant variability in the observations. This is the behaviour that we see in the information content analyses summarised in Table 1 and Figure 7, where the predicted uncertainty monotonically decreases to the all-channel value at a declining rate as less informative channels are added.

The simulated uncertainty using the four-variable retrieval vector and $S_{a,4var}$ shows different behaviour, with the uncertainty increasing with added channels after the 10th. This arises because TCLW is missing from the retrieval vector. The OE method use a forward model run using the a priori values for the quantities in the state vector x_a to generate simulated observation. The differences between the simulated and observed values are ascribed to deviations between the a priori values in x_a and their true values. However, if the observed radiances additionally include variability due to TCLW (which is not in the 4-variable state vector), the scheme can only interpret any observational differences in terms of the other four state-vector variables. This misattribution is naturally largest for those channels that are most sensitive to TCLW where the "observed" values are most affected and which therefore result in the largest retrieval errors. These channels thus drop down the ranking of the best channel to add to the scheme. This effect is most obviously demonstrated in that adding the four least-favoured channels actually increases the SST retrieval error.

When TCLW is included in the retrieval vector, there is consistency between the behaviour of the estimated uncertainty (s) and simulated uncertainty (σ). Figure 5 bears out the above interpretation. Here, the degrees of freedom for signal of the four-variable retrieval has a lower mean value across the profile set, while the five-variable retrieval has a larger mean and a spread of values. In the five-variable retrieval case, the degrees of freedom for signal steadily increase with TCLW up to approximately 0.3 kg·m^{-2} (which includes 90% of the profiles) before plateauing. It then slowly declines again above about 1 kg·m^{-2} (4% of the profiles). The five-variable results indicate that a retrieval uncertainty for SST of $\sim$0.37 K may be achievable if TCLW is explicitly accounted for, whereas neglecting that aspect of variability would limit the achievable SST uncertainty to $\sim$0.45 K.

To check that the above difference is a result of including TCLW in the vector rather than merely an effect of the different a priori error covariance assumptions in the two case studies, we calculated results for a third configuration (not shown). This used the 5-variable retrieval vector with the error covariance assumptions used in the 4-variable case study. The error covariance assumption for L was 0.1^2 as used in the 5-variable case. When including all 14 channels, the values of $s = 0.383$ K and $\sigma = 0.395$ K are comparable to the 5-variable case. The value of σ also decreases monotically as channels are added to the scheme. This comparison proves that expanding the vector is more critical than the error covariance assumptions.

The analyses suggest a preferential ordering of channels for inclusion in the observation vector. We can interpret the channel ordering through Figures 10 and 11 for low- and high-TCLW profiles, respectively. In these figures, the axes represent the brightness temperature in pairs of channels in the order suggested by the five-variable information content analysis. The sensitivity of the two channels with respect to the retrieved quantities is scaled to a "typical" change in brightness temperature by multiplying by the a priori uncertainty on the quantities. In Figure 10, the panel (a) shows that the leading two channels (6.9 V and 7.3 V) are principally sensitive to SST and TCWV, with only small contributions from the other variables. In this case, the difference between the modelled and observed retrieval vectors is interpreted in proportion to the a priori uncertainties expressed as radiances. Panel (b) shows the next pair (7.3 V and 36.5 H) with very different responses for SST and TCWV. The 36.5 H channel is largely insensitive to SST in comparison to large changes due to TCWV, and it is consequently possible to remove the previous ambiguity and separate the two variables in the retrieval. It is not until the third pair (36.5 H and 6.9 H) that it begins to be possible to resolve wind speed effects and thus refine the small contributions they made to brightness temperature changes in the earlier channel combinations. The fact that the two wind-speed components are largely co-linear suggests that it is difficult to discriminate their individual contributions. This is the main reason that d_s is less than the number of state-vector variables.

The high TCLW profile shown in Figure 11 suggests significant ambiguity for brightness temperature changes between SST, TCWV and TCLW for the 6.9 and 7.3 V pair of channels. While the remaining panels show the effect of SST now being distinguishable from both TCWV and TCLW, these latter two variables remain largely co-linear. This figure also shows different sensitivities for the two wind-speed components largely indicative of the change in wind-speed sensitivity with wind speed. The u-component of the wind speed in this case is significantly smaller than the v-component. Consequently, the u-component sensitivity arrow is barely visible, whereas the v-component shows changes in some of the channel combinations comparable to TCLW.

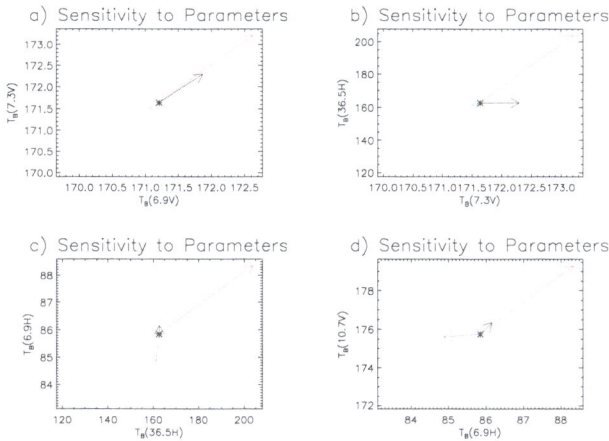

Figure 10. Sensitivity diagrams for an example atmospheric profile with low total cloud liquid water (SST = 296 K, TCWV = 36.2 kg·m^{-2}, u = 4.8 m·s^{-1}, v = -8.3 m·s^{-1}, TCLW = 0.017 kg·m^{-2}). Each arrow indicates, for two channels, the product of the brightness temperature sensitivity to a variable and the a priori uncertainty in that variable i.e., each arrow shows $\frac{\partial y}{\partial x_i}\sigma_i$ for the channels in **y** indicated on the figure axes. This is a measure of the magnitude of the response of the observations to the range of uncertainty in each retrieved variable. The variables are SST (black), ln(TCWV) (red), u (green), v (blue) and ln(TCLW) (cyan). The pairs of channels in each diagram are ordered (**a–d**) in accordance with the information content analysis for a five-variable retrieval vector in Table 1.

As alluded to in Section 2, modelling the emissivity in the MW region and particularly the wind speed dependency is a difficult task. In an effort to assess the effect of any shortcomings of the forward model in this respect, the information content and retrieval analysis were rerun doubling the sensitivity of brightness temperature to each of the wind components in **K**. The results are summarised in Table 3. From the information content analysis, the expected SST uncertainties for both retrieval vectors with all channels included change by around 0.01 K and although there is some slight reordering of the channels, the top five remain the best five to include. For the simulated retrievals with a five-variable retrieval vector, the 10.7 H channel has been promoted into the top five, but the best 14-channel retrieval changes by only 0.001 K. In the four-variable simulated retrieval case, the best retrieval error values is similarly small. Here, though, there are no changes to the channel order down to 7th place, perhaps reflecting that the absence of TCLW from the retrieval vector dominates the ordering.

As mentioned in relation to the increasing retrieval errors for the four-variable retrieval, including all channels in the retrieval is not necessarily the best approach in practice since there may be unrepresented physical processes (such as calibration errors) or poorly-estimated covariance matrices. Given the reasonable consistency of the channel ordering for the five-variable retrievals, we conclude that including the top five or six channels here is the optimum approach in practice when estimating SST using AMSR2.

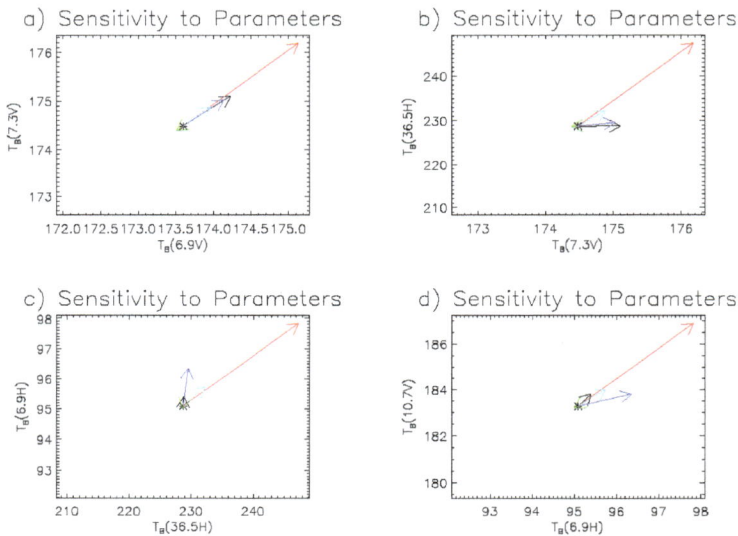

Figure 11. Sensitivity diagrams for an example atmospheric profile with high TCWV and TCLW (SST = 292 K, TCWV = 40.8 kg·m^{-2}, u = −2.7 m·s^{-1}, v = 14.8 m·s^{-1}, TCLW = 0.99 kg·m^{-2}). Each arrow indicates, for two channels, the product of the brightness temperature sensitivity to a variable and the a priori uncertainty in that variable i.e., each arrow shows $\frac{\partial y}{\partial x_i}\sigma_i$ for the channels in y indicated on the figure axes. This is a measure of the magnitude of the response of the observations to the range of uncertainty in each retrieved variable. The variables are SST (black), ln(TCWV) (red), u (green), v (blue) and ln(TCLW) (cyan). The pairs of channels in each diagram are ordered (a–d) in accordance with the information content analysis for a five-variable retrieval vector in Table 1.

Table 3. The standard deviation of the SST retrieval error and root-mean-squared predicted SST uncertainty from an information content analysis across all profiles for varying numbers of channels with the sensitivity to wind speed doubled in the Jacobian matrix.

Number of Channels	4-Variable Retrieval				5-Variable Retrieval			
	Simulated Retrieval		Information Content		Simulated Retrieval		Information Content	
	Channel Added	σ_{SST} (K)	Channel Added	σ_{SST} (K)	Channel Added	σ_{SST} (K)	Channel Added	σ_{SST} (K)
1	6.9 V	0.642	6.9 V	0.703	6.9 V	0.557	6.9 V	0.602
2	7.3 V	0.568	7.3 V	0.632	7.3 V	0.507	6.9 H	0.532
3	18.7 H	0.495	6.9 H	0.504	10.7 H	0.450	7.3 V	0.472
4	6.9 H	0.470	36.5 H	0.436	36.5 H	0.417	36.5 H	0.413
5	23.8 V	0.462	10.7 V	0.416	6.9 H	0.399	10.7 V	0.398
6	7.3 H	0.459	18.7 V	0.408	10.7 V	0.388	18.7 V	0.392
7	10.7 H	0.457	7.3 H	0.403	23.8 V	0.382	7.3 H	0.388
8	10.7 V	0.456	23.8 H	0.400	23.8 H	0.380	23.8 H	0.386
9	89 V	0.455	23.8 V	0.395	18.7 V	0.377	23.8 V	0.383
10	36.5 V	0.455	18.7 H	0.394	18.7 H	0.375	18.7 H	0.380
11	18.7 V	0.457	36.5 V	0.392	7.3 H	0.373	36.5 V	0.379
12	89 H	0.462	10.7 H	0.391	89 H	0.372	10.7 V	0.378
13	36.5 H	0.466	89 H	0.390	89 V	0.372	89 H	0.377
14	23.8 H	0.467	89 V	0.389	36.5 V	0.371	89 V	0.376

7. Conclusions

This information content analysis and simulated retrieval study for AMSR2 provides an ordered list of the best combination channels to use to retrieve SST in OE. In practice, the top five or six would be used in an OE scheme to minimise the accumulation of poorly understood errors (such as from calibration). The recommended channel set is 6.9 V, 6.9 H, 7.3 V, 10.7 V and 36.5 H. The 6.9 V and 7.3 V channels provide the greatest SST sensitivity to the retrieval and the contribution of TCWV is separated out with the addition of the 36.5 H channel. The 6.9 H and 10.7 V channels add in discrimination of the wind speed effects. These results will govern our approach in future work applying OE to real AMSR2 data for SST retrieval.

Acknowledgments: This work was carried out under the European Space Agency(ESA ESRIN REF 3-13904/13/I-NB) Sea Surface Temperature Climate Change Initiative project.

Author Contributions: K.P. and C.M. conceived and designed the experiments; K.P. performed the experiments; K.P. analyzed the data with contributions from C.M. and C.D.; O.E. contributed experimental software and analysis tools; K.P. wrote the paper with contributions from C.M. and C.D.

Conflicts of Interest: The authors declare no conflict of interest.

References

1. Clayson, C.A.; Chen, A. Sensitivity of a coupled single-column model in the Tropics to treatments of the interfacial parameterizations. *J. Clim.* **2002**, *30*, 1805–1831, doi:10.1175/1520-0442(2002)015<1805:SOACSC>2.0.CO;2.
2. Webster, P.J.; Clayson, C.A.; Curry, J.A. Clouds, radiation, and the diurnal cycle of sea surface temperature in the tropical Western Pacific. *J. Clim.* **1996**, *9*, 1712–1730.
3. Zhang, Y.; Vallis, G.K. Ocean heat uptake in eddying and non-eddying ocean circulation models in a warming climate. *J. Phys. Oceanogr.* **2013**, *43*, 2211–2229, doi:10.1175/JPO-D-12-078.1.
4. Boyle, E.A.; Keigwin, L. North Atlantic thermohaline circulation during the past 20,000 years linked to high-latitude surface temperature. *Nature* **1987**, *330*, 35–40.
5. Delworth, T.; Manabe, S.; Stouffer, R.J. Interdecadal variations in the thermohaline circulation in a coupled ocean-atmospere model. *J. Clim.* **1993**, *6*, 1993–2011.
6. Chelton, D.B.; Xie, S.-P. Coupled ocean-atmosphere interaction at oceanic mesoscales. *Oceanography* **2010**, *23*, 52–69.
7. Minobe, S.; Kuwano-Yoshida, A.; Komori, N.; Xie, S.-P.; Smal, R.J. Influence of the Gulf Stream on the troposphere. *Nature* **2008**, *452*, 206–209, doi:10.1038/nature06690.

8. Chelton, D.B. The impact of SST Specification on ECMWF Surface Wind Stress Fields in the Eastern Trtopical Pacific. *J. Clim.* **2005**, *18*, 530–550.

9. Donlon, C.J.; Martin, M.; Stark, J.; Roberts-Jones J.; Fiedler, E.; Wimmer, W. The Operational Sea Surface Temperature and Sea Ice Analysis (OSTIA) system. *Remote Sens. Environ.* **2012**, *116*, 140–158.

10. Kennedy, J.J.; Rayner, N.A.; Smith, R.O.; Saundby, M.; Parker, D.E. Reassessing biases and other uncertainties in sea surface temperatre observations measure in situ since 1850 part 1: Measurement and sampling errors. *J. Geohpys. Res.* **2011**, *116*, doi:10.1029/2010JD015218.

11. Kennedy, J.J.; Rayner, N.A.; Smith, R.O.; Saundby, M.; Parker, D.E. Reassessing biases and other uncertainties in sea surface temperate observations measure in situ since 1850 part 2: Biases and homogenisation. *J. Geohpys. Res.* **2011**, *116*, doi:10.1029/2010JD015220.

12. Rayner, N.A.; Parker, D.E.; Horton, E.B.; Folland, C.K.; Alexander, L.V.; Rowell, D.P.; Kent, E.C.; Kaplan, A. An improved in situ and satellite SST analysis for climate. *J. Geophys. Res.* **2003**, *108*, doi:10.1029/2002JD002670.

13. Kent, E.C.; Kennedy, J.J.; Berry, D.I.; Smith, R.O. Effects of instrumentation changes on sea surface temperature measured in situ. *WIREs Clim. Chang.* **2010**, *1*, 718–728, doi:10.1002/wcc.55.

14. Kaplan, A.; Cane, M.A.; Kushnir, Y.; Clement, A.C.; Blumenthal, M.B.; Rajagopalan, B. Analyses of global sea surface temperature 1856–1991. *J. Geophys. Res.* **1998**, *103*, 18567–18589, doi:10.1029/97JC01736.

15. Anding, D.; Kauth, R. Estimation of sea surface temperature from space. *Remote Sens. Environ.* **1970**, *1*, 272–220, doi:10.1016/S0034-4257(70)80002-5.

16. Deschamps, P.Y.; Phulpin, T. Atmospheric correction of infrared measurements of sea surface temperature using channels at 3.7, 11 and 12 *Boundary-Layer Meteorol.* **1980**, *18*, 131–143, doi:10.10007/BF00121320.

17. McClain, E.P. Global sea surface temperatures and cloud clearing for aerosol optical depth estimates. *Int. J. Remote Sens.* **1989**, *10*, 763–769.

18. Walton, C.C.; Pichel, W.G.; Sapper F.J.; May, D.A. The development and operational application of nonlinear algorithms for the measurements of sea surfrace temperatures with NOAA polar-orbiting environmental satellites. *J. Geophys. Res.* **1998**, *103*, 27999–28012, doi:10.1029/98JC02370.

19. Kilpatrick, K.A.; Podestá, G.P.; Evans, R. Overview of the NOAA/NASA advanced very high resolution radiometer Pathfinder algorithm for sea surface temperature and associated matchup database. *J. Gephys. Res.* **2001**, *106*, 9179–9197, doi:10.1029/1999JC000065.

20. Merchant, C.J.; Llewellyn-Jones, D.; Saunders, R.W.; Rayner, N.A.; Kent, E.C.; Old, C.P.; Berry, D.; Birks, A.R.; Blackmore, T.; Corlett, G.K.; et al. Deriving a sea surface temperature record suitable for climate change research from the along-track scanning radiometers. *Adv. Space Res.* **2007**, *41*, 1–11, doi:10.1016/j.asr.2007.07.041.

21. Merchant, C.J.; Embury, O.; Roberts-Jones, J.; Fiedler, E.; Bulgin, C.E.; Corlett, G.K.; Good, S.; McLaren, A.; Rayner, N.; Morak-Bozzo, S.; et al. Sea surface temperature datasets for climate applications from phase 1 of the European Space Agency Climate Change Initiative (SST CCI). *Geosci. Data J.* **2014**, *1*, 179–191, doi:10.1002/gdj3.20.

22. Reynolds, R.W.; Rayner, N.A.; Smith, T.M.; Stokes, D.C.; Wang, W. An improved in situ and satellite SST analysis for climate. *J. Clim.* **2002**, *15*, 1609–1625.

23. Wentz, F.J.; Gentemann, C.; Smith D.; Chelton, D. Satellite measurements of sea surface temperature through clouds. *Science* **2000**, *288*, 847–850, doi:10.1126/science.288.5467.847.

24. Chelton, D.B.; Wentz, F.J. Global Microwave Satellite Observations of Sea Surface Temperature for Numerical Weather prediction and Climate Research. *Bull. Am. Meteorol. Soc.* **2005**, *86*, 1097–1115.

25. Ecuyer, T.S.; Jiang, J.H. Touring the atmosphere aboard the A-Train. *Phys. Today* **2010**, *63*, 36–41, doi:10.1063/1.3463626.

26. Prigent, C.; Aires, F.; Bernardo, F.; Orlhac, J.C.; Goutoule, J.M.; Roquet, H.; Donlon, C. Analysis of the potential and limitations of microwave radiometry for the retrieval of sea surface temperature: Definition of MICROWAT, a new mission concept. *J. Geophys. Res. Oceans* **2013**, *118*, 3074–3086.

27. Cavalieri, D.J. A microwave technique for mapping thin sea ice. *J. Geophys. Res. Oceans* **1994**, *99*, 12561–12572, doi:10.1029/94JC00707.

28. Eyre, J.R.; Woolf, H.M. Transmission of atmospheric gases in the microwave region: A fast model. *Appl. Opt.* **1988** *27*, 3244–3249.

29. Hocking, J.; Rayer, P.; Rundle, D.; Saunders, R.; Matricardi, M.; Geer, G.; Brunel, P.; Vidot, J. *RTTOV v11 Users Guide*; Doc. No. NWPSAF-MO-UD-028; Version 1.4; EUMETSAT: Darmstadt, Germany, 2015.

30. Matricardi, M.; Chevallier, F.; Kelly, G.; Thepaut, J.-N. An improved general fast radiative transfer model for the assimilation of radiance observations. *Q. J. R. Meteorol. Soc.* **2004**, *30*, 153–173.

31. Saunders, R.W.; Matricardi, M.; Brunel, P. An improved fast radiative transfer model for assimilation of satellite radiance observations. *Q. J. R. Meteorol. Soc.* **1999**, *125*, 1407–1425, doi:10.1002/qj.1999.49712555615.

32. Liu, Q.; Weng, W.; English, S. An improved fast microwave water emissivity model. *IEEE Trans. Geosci. Remote Sens.* **2011**, *49*, 1238–1250.

33. Ellison, W.J.; Balana, A.; Delbos, G.; Lamkaouchi, K.; Guillou, C.; Prigent, C. New permittivity measurements of seawater. *Radio Sci.* **1998**, *33*, 639–648.

34. Ellison, W.J.; English, S.J.; Lamkaouchi, K.; Balana, A.; Obligis, E.; Deblonde, G.; Hewison, T.J.; Bauer, P.; Kelly, G.; Eymard, L. A comparison of ocean emissivity using the Advanced Microwave Sounding Unit, the Special Sensor Microwave Imager, The TRRM Microwave Imager, and airborne radiometer observations. *J. Geophys. Res.* **2003**, *108*, doi:10.1029/2002JD003213.

35. Guillou, C.; Ellison, W.J.; Eymard, L.; Lamkaouchi, K.; Prigent, C.; Delbos, G.; Balana, A.; Boukabara, S.A. Impact of new permittivity measurements on sea-surface emissivity modelling in microwaves. *Radio Sci.* **1998**, *33*, 649–667.

36. Klein, L.A.; Swift, C.T. An improved model for the dielectric constant of sea water at microwave frequencies. *IEEE Trans. Antennas Propag.* **1979**, *25*, 104–111.

37. Meissner, T.; Wentz, F.J. The complex dielectric constant of pure and sea water from microwave satellite observations. *IEEE Trans. Geosci. Remote Sens.* **2004**, *42*, 1836–1849.

38. Meissner, T.; Wentz, F.J. The emissivity of the ocean surface between 6 and 90 GHz over a large range of wind speeds and earth incidence angles. *IEEE Trans. Geosci. Remote Sens.* **2012**, *50*, 3004–3026.

39. Stogryn, A. Equations for calculating the dielectric constant of saline water. *IEEE Trans. Microw. Theory Tech.* **1971**, *19*, 763–766.

40. Liu, Q.; Simmer, C.; Ruprecht, E. Monte Carlo simulations of the microwave emissivity of the sea surface. *J. Geophys. Res.* **1998**, *103*, 24983–24989.

41. Monahan, E.C.; Spiel, D.E.; Davidson, K.L. A model of marine aerosol formation via whitecaps and wave disruption. In *Oceanic Whitecaps and Their Role in Air-Sea Exchange Process*; Monahan, E.C., MacNiocaill, G.D., Eds.; D. Reidel: Dordrecht, The Netherlands, 1986; pp. 167–174.

42. Tang, C. The effect of droplets in the air-sea transition zone on the sea brightness temperature. *J. Phys. Oceanogr.* **1974**, *4*, 579–593.

43. Kazumori, M.; Liu, Q.; Treadon, R.; Derber, J.C. Impact study of AMSR-E radiances in the NCEP global data assimilation system. *Mon. Weather Rev.* **2008**, *136*, 541–559.

44. Stogryn, A. The emissivity of sea foam at microwave frequencies. *J. Geophys. Res.* **1972**, *77*, 1658–1666.

45. Dicke, R.H.; Peebles, P.J.E.; Roll, P.G.; Wilkinson, D.T. Cosmic black-body radiation. *Astrophys. J.* **1965**, *142*, 414–419.

46. Penzias, A.A.; Wilson, R.W. A measurement of excess antenna temperature at 4080 Mc/s. *Astrophys. J.* **1965**, *142*, 419–421.

47. Fixsen, D.J. The temperature of the cosmic microwave background. *Astrophys. J.* **2009**, *707*, 916–920, doi:10.1088/0004-637X/707/2/916.

48. Robinson, I.S. *Measuring the Oceans from Space*; Springer: Berlin/Heidelberg, Germany, 2004.

49. Rodgers, C.D. *Inverse Methods for Atmospheric Sounding*; World Scientific Publishing: Singapore, 2000.

50. Merchant, C.J.; Le Borgne, P.; Marsouin, A.; Roquet, H. Optimal estimation of sea surface temperature from split-window observations. *Remote Sens. Environ.* **2008**, *112*, 2469–2484, doi:10.1016/j.rse.2007.11.011.

51. Chevallier, F.; Di Michele, S.; McNally, A.P. *Diverse Profiles from the ECMWF 91-Level Short-Range Forecasts*; Doc. No. NWPSAF-EC-TR-010; Version 1.0; EUMETSAT: Darmstadt, Germany, 2006.

52. Japan Aerospace Exploration Agency. *GCOM-W1 "Shizuku" Data Users Handbook*, 1st ed.; Japan Aerospace Exploration Agency, Earth Observation Research Center: Tsukuba, Japan, 2013.

remote sensing

MDPI

Article

Remote Sensing of Coral Bleaching Using Temperature and Light: Progress towards an Operational Algorithm

William Skirving [1,2,*], Susana Enríquez [3], John D. Hedley [4,5], Sophie Dove [6,7], C. Mark Eakin [1], Robert A. B. Mason [6,7], Jacqueline L. De La Cour [1,8], Gang Liu [1,8], Ove Hoegh-Guldberg [7,9], Alan E. Strong [1], Peter J. Mumby [7,10] and Roberto Iglesias-Prieto [3,11]

[1] Coral Reef Watch, National Oceanic and Atmospheric Administration, College Park, MD 20740, USA; Mark.Eakin@noaa.gov (C.M.E.); Jacqueline.Shapo@noaa.gov (J.L.D.L.C.); Gang.Liu@noaa.gov (G.L.); Alan.E.Strong@noaa.gov (A.E.S.)
[2] ReefSense Pty Ltd., Townsville, QLD 4814, Australia
[3] Unidad Académica de Sistemas Arrecifales Puerto Morelos, Instituto de Ciencias del Mar y Limnología, Universidad Nacional Autónoma de México, Cancun 77580, Mexico; susana.enriquezdominguez@gmail.com (S.E.); rzi3@psu.edu (R.I.-P.)
[4] Numerical Optics Ltd., Tiverton EX16 8AA, UK; j.d.hedley@gmail.com
[5] School of Biosciences, Exeter University, Exeter EX4 4PS, UK
[6] Coral Reef Ecosystems Laboratory, School of Biological Sciences, University of Queensland, St. Lucia, QLD 4072, Australia; sophie@uq.edu.au (S.D.); robert.mason1@uq.net.au (R.A.B.M.)
[7] ARC Centre of Excellence for Coral Reef Studies, University of Queensland, St. Lucia, QLD 4072, Australia; oveh@uq.edu.au (O.H.-G.); p.j.mumby@uq.edu.au (P.J.M.)
[8] Global Science & Technology, Inc., Greenbelt, MD 20740, USA
[9] Global Change Institute, University of Queensland, St. Lucia, QLD 4072, Australia
[10] Marine Spatial Ecology Lab, School of Biological Sciences, University of Queensland, St. Lucia, QLD 4072, Australia
[11] Department of Biology, The Pennsylvania State University, University Park, PA 16802, USA
*   Correspondence: William.skirving@noaa.gov; Tel.: +61-4-0489-3406

Received: 3 October 2017; Accepted: 19 December 2017; Published: 22 December 2017

Abstract: The National Oceanic and Atmospheric Administration's Coral Reef Watch program developed and operates several global satellite products to monitor bleaching-level heat stress. While these products have a proven ability to predict the onset of most mass coral bleaching events, they occasionally miss events; inaccurately predict the severity of some mass coral bleaching events; or report false alarms. These products are based solely on temperature and yet coral bleaching is known to result from both temperature and light stress. This study presents a novel methodology (still under development), which combines temperature and light into a single measure of stress to predict the onset and severity of mass coral bleaching. We describe here the biological basis of the Light Stress Damage (LSD) algorithm under development. Then by using empirical relationships derived in separate experiments conducted in mesocosm facilities in the Mexican Caribbean we parameterize the LSD algorithm and demonstrate that it is able to describe three past bleaching events from the Great Barrier Reef (GBR). For this limited example, the LSD algorithm was able to better predict differences in the severity of the three past GBR bleaching events, quantifying the contribution of light to reduce or exacerbate the impact of heat stress. The new Light Stress Damage algorithm we present here is potentially a significant step forward in the evolution of satellite-based bleaching products.

Keywords: coral bleaching; Light Stress Damage; LSD; DHW; remote sensing of coral bleaching; NOAA Coral Reef Watch; CRW; mass coral bleaching; light stress; F_v/F_m

1. Introduction

Corals live in an endosymbiotic relationship with unicellular algae forming what is referred to as the "holobiont". These dinoflagellate algae (genus *Symbiodinium*), collect light and perform photosynthesis, transferring energy to the coral. Coral bleaching refers to the dramatic loss of the *Symbiodinium* population that inhabits coral tissues, leaving the coral polyps transparent and making visible the underlying white calcium carbonate skeleton. This occurs when the symbiosis breaks down under any stressful condition that pushes the symbiosis beyond its limits of stability. The causes or "stressors" can include but are not restricted to: anomalous temperature (both hot and cold), anomalous increasing levels of light, anomalous levels of salinity (both high and low), reduction in water quality (e.g., heavy metals) and diseases [1–7]. Additionally, partial loss of coral pigmentation can occur during acclimation to high-light conditions [8–10] or too low nutrient availability [11]. Seasonal changes in the number of symbionts [12,13] and/or in symbiont pigmentation [14] can also lighten coral color during summer but these changes are unrelated to symbiosis instability. It is therefore important to distinguish between coral bleaching and coral holobiont homeostatic adjustments to the environment, since bleaching not only involves a dramatic change in coral pigmentation but is also an indication of a dysfunctional condition of the symbiotic relationship.

All the stressors listed above are known to cause bleaching on small to medium "local" scales (i.e., less than synoptic). However, heat stress is directly linked to synoptic-scale climate events and therefore is the only stressor demonstrated to have the capacity to cause mass coral bleaching (i.e., encompassing hundreds or more square kilometers and affecting many reefs at once). The first documented bleaching event was recorded in the 1870s [1] and the first recorded bleaching event attributed to heat stress was in 1911 [15]. Since the late 1970s the number, scale and intensity of coral bleaching events have grown significantly. While the first basin-scale (and possibly global) bleaching event occurred during the 1982–1983 El Niño [16,17], the first bleaching event demonstrated to have a fully global impact was not until 1998 [18]. Since the 1998 global coral bleaching event, heat stress events have occurred somewhere in the world each year and large-scale severe events are becoming more frequent, with subsequent global events in 2010 and 2014–2017 [19]. This trend has been linked to elevated ocean temperatures due to anthropogenic climate change [5,18].

To better understand the implications of climate change for coral communities, it is important to develop adequate tools to monitor these large-scale bleaching events and the stress that causes them. In doing so, we can compare the extent, duration and severity of different mass bleaching events in relation to physical disturbances and to the areas, coral communities and/or species particularly affected. This knowledge in combination with a better understanding of coral physiology and the cellular mechanisms that explain coral bleaching will help us to better model and predict the future risk to reefs from climate change, which will enable the development and encourage the adoption of better management tools and practices.

The National Oceanic and Atmospheric Administration's (NOAA) Coral Reef Watch program (CRW) developed and operates the world's only operational global satellite products designed specifically to monitor bleaching-level heat stress. Seawater temperature anomalies above the local maximum monthly mean (*MMM*); the average temperature of the climatologically-hottest month of the year [20] have been recognized to be the principal cause for mass coral bleaching events [5,21]. The HotSpot (*HS*) product, developed in 1997, is a sea surface temperature (*SST*) anomaly product comparing current *SST* to the *MMM*. In its current form, it provides a measure of the level of daily heat stress on corals [22,23]. This product was followed in 2000 with a near real-time satellite product called the Degree Heating Week (*DHW*), which provides a measure of the accumulated heat stress and has been shown to be an accurate predictor of coral bleaching [22–24].

These satellite-based coral bleaching products were designed to help coral reef managers identify and monitor oceanic heat stress and hence better predict mass coral bleaching. These products have been demonstrated to perform well when used to describe the onset of coral bleaching [25,26]. However, they occasionally predict bleaching for coral reefs where the phenomenon was not observed and vice versa. The products are based solely on *SST* and yet mass coral bleaching is known to be a result of the combined effect of temperature and light on the photosynthetic activity of *Symbiodinium* [27–30]. Indeed, mass bleaching has failed to materialize under persistently cloudy conditions despite the existence of sustained elevated sea temperatures that would otherwise elicit coral bleaching [31].

This paper describes a methodology that is planned to underpin a major evolution of the NOAA CRW satellite products. This new methodology (called Light Stress Damage, or LSD) enables satellite-derived *SST* data to be combined together with satellite-derived solar insolation data in a scientifically valid way. The LSD algorithm is based on physiological processes that incorporate the synergistic effects of light and elevated temperature to induce coral bleaching. Such stress causes a significant decline in holobiont photosynthetic performance due to the accumulation of light-induced damage to the photosynthetic apparatus (photodamage) [28–30] and can be used to measure the photodamage of *Symbiodinium*, even prior to visible bleaching. The LSD methodology is novel in that it allows the quantification of the severity of a particular heat stress event as a function of the variation in the magnitude of photodamage accumulation derived from the amplification of light stress under elevated temperature. Light stress is defined as the condition where the energy absorbed by the photosynthetic apparatus of a coral-algal symbiont exceeds its photoprotective abilities, causing photodamage accumulation. This excess energy also causes *Symbiodinium* to activate mechanisms to acclimate to high light [32,33]. Since light stress depends on the photoacclimatory condition of *Symbiodinium*, this algorithm also includes the effect of coral photoacclimation to counterbalance the enhancement of light stress during a heat stress event.

It is hoped that the LSD product will provide a more accurate measure of the onset and severity of a coral heat-stress event, allowing for improved bleaching predictions with dramatically reduced false positives. Improved characterization of the timing and levels of stress are expected to improve the predictions (in near real-time as well as hindcast) of the severity and mortality associated with mass bleaching events of scleractinian corals.

The development of the LSD algorithm took advantage of work being done at Universidad Nacional Autónoma de México (UNAM) in Puerto Morelos. The experiments carried out at UNAM were performed independently of the work described in this paper. Preliminary results from these experiments have been used in an opportunistic manner to guide the method by which light and temperature have been combined to form the LSD algorithm. The main purpose of this paper is to provide a description of this methodology and its nuances. An example of the application of the LSD algorithm is provided mainly to clarify the methodology but it also serves to demonstrate that the LSD algorithm has potential to provide improved bleaching predictions when compared to methods that use temperature only.

Coral Response to Variable Solar Irradiance

The amount of light available for photosynthesis (photosynthetically available radiation or *PAR*) rapidly increases from near zero just before dawn to a maximum at the local solar noon (on clear days) and then drops off to be near zero again just after dusk. The amount of light available for coral photosynthesis varies with cloudiness, sea surface roughness, water turbidity, depth and local shading.

Photosynthetic rates of *Symbiodinium* increase linearly with light at low irradiance levels and then gradually diminish until saturation (E_k) at a maximum rate (P_{max}) [34]. The slope of this linear increase (photosynthetic efficiency, α), E_k and P_{max} vary among coral species and photoacclimatory conditions. The amount of energy absorbed by the photosynthetic apparatus above the saturation level (E_k) cannot be incorporated into the photosynthetic processes and thus it has to be dissipated as heat by means of different photoprotective mechanisms. E_k is a key descriptor of the photoacclimatory condition of the photosynthetic apparatus, as it determines the amount of solar energy absorbed in excess by *Symbiodinium* and varies significantly among and within *Symbiodinium* species. Its variability is regulated by the plasticity of the species' photosynthetic response to light changes but it is also dependent on many other environmental changes that potentially affect photosystem II (PSII) excitation pressure and consequently the photoacclimatory response of the organism [35].

Photoprotection of the photosynthetic apparatus of all photosynthetic organisms requires continuous activity of cellular mechanisms for the repair of photodamage [36]. When the rate of damage to PSII exceeds the rate of repair, the photosynthetic apparatus accumulates photodamage, which is reflected in an incomplete diurnal recovery of the maximum PSII photochemical efficiency, measured as F_v/F_m, where F_v/F_m is the ratio of the difference between maximum fluorescence and minimum fluorescence (F_v) to the maximum fluorescence (F_m) as measured by a pulse amplitude modulated (PAM) fluorometer. This reduction in F_v/F_m can be used as a descriptor of photodamage accumulation at the level of the photosynthetic membrane [37,38]. F_v/F_m as employed here is a good descriptor of the balance between the rate of accumulation of heat- and light-damaged PSII and their rate of repair [39].

As PSII damage accumulates, it results in the leakage of reactive oxygen species into the coral host, with the consequent increase in coral tissue oxidative stress [40]. Once photodamage accumulation and/or photosynthesis inhibition crosses a physiological threshold for the stability of the holobiont, the symbiotic relationship can break down leading to coral bleaching. The breakdown can also occur when the algal symbiont no longer contributes nutritionally to the symbiosis, causing the coral to expel them from its tissues [41]. After the stressful conditions abate and if the coral host is able to cope with the physiological perturbation [42], *Symbiodinium* is able to re-populate the living coral tissue. However, if the conditions were sufficiently severe to cause irreversible cellular damage, polyps may not survive the heat stress event and eventually die. Stressed corals also are more susceptible to disease [43].

The highest value for F_v/F_m achieved by the photosynthetic apparatus of *Symbiodinium* is around 0.7 (70% efficiency) [10,39,44,45], substantially lower than the maximum value of 0.89 that can be achieved by higher plants [46]. Accumulated photodamage causes a drop in F_v/F_m. Once the accumulation of damaged PSII induces a dramatic reduction in *Symbiodinium* population (i.e., coral bleaching), the recovery rate of F_v/F_m will be dependent on PSII repair of the surviving algae and/or the repopulation of *Symbiodinium* [47,48].

Reduction in F_v/F_m is also associated with the high-light photoacclimatory response to reduce PSII pressure [49,50]. Whilst there are photoacclimatory mechanisms that involve photochemical quenching, only photoacclimation involving an upregulation of non-photochemical quenching, such as the accumulation of inactive PSII (a population of PSII that engage solely in mitigating excess excitation energy via heat dissipation), will reduce F_v/F_m [51]. The rate of inactive PSII accumulation and the consequent reduction in F_v/F_m is proportional to the light absorbed in excess by the photosynthetic membranes of the algae [52]. According to this physiological background, two main parameters need to be generated to quantify photodamage: the first describes the amount of excess solar energy absorbed by a symbiont that significantly contributes to increased rates of PSII photoinactivation, called Excess Excitation Energy (*EEE*, mol quanta m^{-2} day^{-1}); and the second is the amount of photodamage accumulated (i.e., inactive PSII). *EEE* varies with (1) light availability; (2) the holobiont's saturation irradiance threshold (E_k); and (3) holobiont capacity for photoprotection and repair. It then follows that *EEE* represents a convenient quantitative descriptor of the severity of light stress, including that induced by heat stress. This impact can be calculated as the diurnal change in F_v/F_m, which reflects the diurnal photodamage accumulation in *Symbiodinium*.

The premise that heat stress will enhance the impact of *EEE* and PSII photoinactivation in *Symbiodinium*, with the consequent negative effect on photosynthetic activity of the holobiont, is based on evidence that light stress and photosynthetic inhibition are key components of the adverse impact of heat stress on the physiology of symbiotic corals [28–30,44]. Photosynthesis is a temperature-dependent process, especially at the upper tolerance threshold, as photosynthesis can drop dramatically with rising temperature [53–55]. Such fast photosynthesis decline induces a quick *EEE* rise and subsequent PSII photodamage accumulation. Accordingly, heat stress will enhance the effect of *EEE* and PSII photoinactivation in *Symbiodinium*, with the consequent negative effect on the photosynthetic activity of the coral holobiont. After the stress is removed, processes allow F_v/F_m to recover, with F_v/F_m in stressed corals returning to the maximum value of 0.7 each year during the cool season [56].

To model this biological process, we have developed the Light Stress Damage (LSD) algorithm that combines *PAR* and *SST* to predict the impact of heat stress events on symbiotic corals. This algorithm allows quantification of energy available in excess for *Symbiodinium* as it is not possible to directly measure the energy in excess absorbed by *Symbiodinium* from satellite data. We quantified its negative impact on the symbiotic population using relative changes in F_v/F_m, due to variations in *PAR* and *SST* [37,38]. The LSD algorithm was derived from a combination of physiological processes and several empirical relationships determined by experimental manipulation of Caribbean corals. Thus, it describes key physiological relationships between F_v/F_m changes and the variation of *EEE* under optimal conditions and different levels of *PAR* and heat stress. The synergistic effects of light and temperature on F_v/F_m changes were combined into one index of light-induced damage equivalents (i.e., LSD index), with the aim of improving on the *DHW* measure of stress as an indicator of coral bleaching. As discussed below, constraints were placed on the form of the LSD algorithm so that it can be implemented on satellite data.

2. Methods

2.1. Definition of Relative F_v/F_m

The variation of the maximum photochemical efficiency (F_v/F_m) of each day was calculated as relative F_v/F_m (rel F_v/F_m) between two consecutive days:

$$\text{rel } F_v/F_m = [F_v/F_m]_{\text{today}}/[F_v/F_m]_{\text{yesterday}} \tag{1}$$

All modeled F_v/F_m values are subsequently derived from these relative values.

2.2. Photoacclimation

Photoacclimation in *Symbiodinium* occurs continuously and needs to be accounted for in the LSD algorithm. Anthony and Hoegh-Guldberg [57] found that photoacclimation in *Turbinaria mesenterina* occurred over a period of 5–10 days and provided figures for the daily acclimation of P_{max}. These assumed that acclimation rate is proportional to the difference between the current P_{max}, ($P_{\text{max}}(t)$) and the P_{max} that would be optimal for constant light at the current level (P_{maxS}),

$$dP_{\text{max}}/dt = \varepsilon(P_{\text{maxS}} - P_{\text{max}}(t)) \tag{2}$$

where t is measured in days. Anthony and Hoegh-Guldberg [57] quoted four resulting values for ε, two for acclimation upward (to higher light levels) and two for downward, ranging from 0.101 to 0.164. The associated standard errors suggest taking a single value for the upward and downward photoacclimation as the mean of 0.13 is reasonable, this corresponds to 50% acclimation in 5 days. However, that study suggested that upward and downward acclimation rates are not necessarily the same. This is an important topic for future research as the LSD algorithm and subsequent satellite products are still under development.

2.3. Definition of EEE

Excess Excitation Energy (*EEE*) is directly linked to the particular acclimatory and adaptive characteristics of an individual coral. This makes it impossible to use space-borne sensors for direct measurements. However, for any species or acclimatory condition, *EEE* = 0 implies that the organism is well acclimated to the environment and no net light-stress has occurred (no photodamage has been accumulated). On both a daily as well as a seasonal basis, *Symbiodinium* need to cope with a continuously changing light environment that combines days where they receive more sunlight than they can actually use for photosynthesis with days that allow net recovery. Thus, a useful proxy for daily *EEE* is the difference between daily *PAR* and the level of *PAR* to which a coral is acclimated. Hence, for the purpose of this paper, *EEE* will be defined as being the difference between *PAR* today and the coral's acclimated *PAR* level.

2.4. Definition of HotSpot

The HotSpot (*HS*) product, developed in 1997 by NOAA CRW, is a sea surface temperature (*SST*) anomaly product that measures the temperature (°C) above the local maximum monthly mean (*MMM*); the average temperature of the climatologically-hottest month of the year [20].

2.5. Experimental Quantification of the Synergistic Effect of Light and Temperature

To quantify the association between rel F_v/F_m and *EEE* and to incorporate the synergistic effect of temperature on the variation of *EEE*, we used results from a set of mesocosm experiments performed as part of another study at UNAM in Puerto Morelos for the species *Orbicella* (*Montastraea*) *annularis*. A description of the experimental design and methods can be found in [58]. The variation in F_v/F_m arising from these experiments is documented in [58] and the quantitative description of the combined effect of light and temperature used in the development of the LSD algorithm are the subject of a physiological paper, Enríquez (in prep.).

The results from these experiments provided this study with equations relating *EEE* to F_v/F_m for three separate temperatures (Figure 1) for coral colonies of *O. annularis*. These results will be described in detail by Enríquez (in prep.).

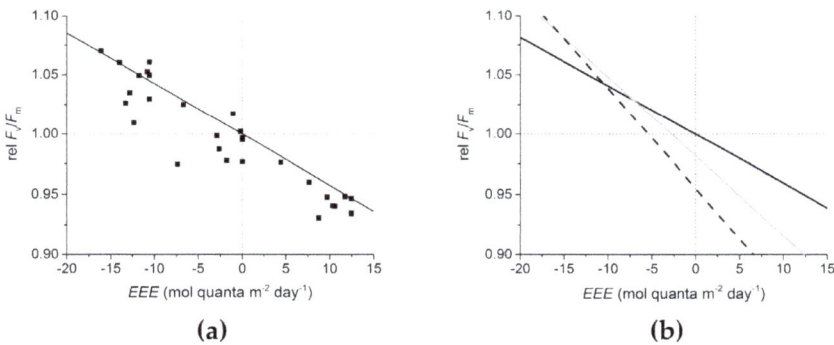

Figure 1. Experimental results for the variation of rel F_v/F_m as a function of the variation in *EEE* (mol quanta m^{-2} day^{-1}) for *Orbicella annularis* exposed to control conditions at (**a**) 28 °C and (**b**) the results for three temperatures, 28 °C (black), 30 °C (grey) and 32 °C (dashed). (Enríquez in prep.).

The results of these experiments are summarized in Figure 1. The plot of the data for the tank set at 28 °C (Figure 1a) is representative of non-stressful temperatures for *O. annularis* in Puerto Morelos. It is therefore representative of corals whose variation in F_v/F_m is driven by *EEE* in the absence of heat stress. It describes the relationship for 28 °C between rel F_v/F_m and *EEE* using the variation registered during 10 experimental days. Each data point is the average rel F_v/F_m (Equation (1)) on one evening over all corals for each light treatment.

For control conditions of no heat stress the relationship between rel F_v/F_m and *EEE* can be modeled by a linear regression (Figure 1a) giving:

$$\text{rel } F_v/F_m = 1.0 - 0.00426 \; EEE \tag{3}$$

The slope was derived from an unconstrained regression that produced a Y-intercept of 0.99, with $R^2 = 0.86$ and standard error of the slope of 0.0003. However, the intercept in Equation (3) was set at 1.0 since, by definition, rel $F_v/F_m = 1$ if $EEE = 0$.

As heat stress increases, the regression slopes become more negative and the Y-intercepts become progressively less than 1 (Figure 1b):

i.e., at 30 °C rel $F_v/F_m = 0.982 - 0.00663\ EEE$
at 32 °C rel $F_v/F_m = 0.955 - 0.00837\ EEE$

Although research of the association between rel F_v/F_m and EEE under heat stress is still in progress, in order to test the idea behind the LSD algorithm, we have used these preliminary results from Enríquez (in prep.) to model the change in coefficients of this relationship as a function of the experimental HotSpots (*HS*).

Using in situ data for the period 1992 to 2015 from the Puerto Morelos lagoon, the maximum of the monthly averages is 29.8 °C. The *MMM* methodology requires this value to be re-centered on 1988.3 [59], which lowers the value to 29.2 °C. Since the experiments were conducted with integer temperatures [58], the *MMM* used to derive *HS* should also be expressed as an integer. Hence, for the purpose of this paper, the *MMM* for Puerto Morelos is 29 °C and since the control was set at 28 °C, the resultant relationship between *EEE* and rel F_v/F_m (Equation (3)) is stable up until the *MMM* of 29 °C. The two experimental levels of thermal stress were set at 30 °C and 32 °C, which are equivalent to HotSpots of 1 and 3 respectively. A linear regression was applied to each of the coefficients of the three experimental regression outputs, allowing the Y-intercepts and slopes (from Figure 1b) to be expressed in terms of HotSpots. The slopes of each of these regressions were used along with the known Y-intercepts (from Equation (3)), giving:

$$\text{Y-intercept} = 1 - 0.01164\ HS \tag{4}$$

$$\text{Slope} = -0.00426 - 0.00130\ HS \tag{5}$$

2.6. LSD Algorithm Description

The idea behind the LSD algorithm was to track the effects of *EEE* on photosystem efficiency throughout a bleaching season, taking into account the amplifying effects of anomalous temperature (i.e., *HS* > 0). A stress threshold is established using a multi-year time series of F_v/F_m due to the effects of *EEE* with no temperature effects (using Equation (3); explained in more detail below). After temperature effects were included with *EEE*, the F_v/F_m values occasionally dropped below this threshold. When that occurred, the corals experienced abnormally low F_v/F_m values, most likely due to heat-induced stress. To gain a measure of the total stress for an event, for all F_v/F_m values less than the threshold, the area between the F_v/F_m curve and the threshold was integrated and the total is called the LSD index.

3. Results

3.1. LSD Algorithm Demonstration

This demonstration of the application of the LSD algorithm is designed to better describe the algorithm rather than providing a definitive proof of the applicability of the algorithm to the generic prediction of coral bleaching. It also serves to explain the differences between the temperature-only DHW product and the light/temperature LSD product and why the inclusion of light is potentially an important evolution in satellite monitoring of environmental stress for prediction of coral bleaching.

Demonstrating the LSD algorithm required a location that had a long, continuous dataset of quality *PAR* measurements, reliable *SST* data and a complete and thorough set of in-water surveys to ensure that all bleaching and non-bleaching events were known. No such sites were found in the Caribbean; however, a useful site was located in the southern Great Barrier Reef at the Keppel Islands,

Australia (Figure 2). The near-by Rockhampton airport has a climate-quality light station operated by the Australian Bureau of Meteorology (BoM) from which the *PAR* data were sourced. *SST*s were available from the NOAA Advanced Very High Resolution Radiometer (AVHRR) satellite sensors [60] and tested against in situ data from the Australian Institute of Marine Science (AIMS) and since there was no difference in the below example for either data set, we chose to use the satellite *SST*. Lastly, AIMS have annual (or more frequent) surveys of the Keppel Islands corals. These islands also have a marine park ranger who reports any signs of coral stress. For the purpose of this demonstration, the period 1999 to 2006 was used.

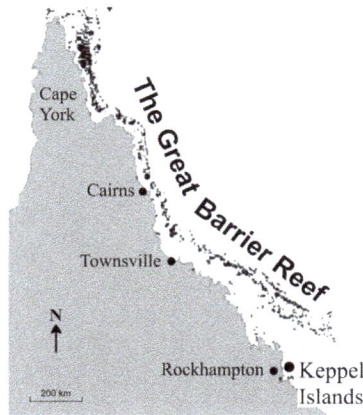

Figure 2. Map of Queensland showing the location of the Keppel Islands in the southern Great Barrier Reef and the township of Rockhampton.

The following presents an example of the application of the LSD algorithm. This example helps to demonstrate how the inclusion of light into the calculation of heat stress of corals can improve the determination of bleaching. Global implementation of the LSD algorithm will require testing its applicability in many more locations but since there is a scarcity of coral reef locations with sufficiently long time series of reasonable quality *PAR* measurements and *SST* measurements that are coincident with reliable bleaching reports, the validation of the general applicability of the LSD algorithm will be best done after implementation on satellite data (work currently in progress at NOAA CRW).

3.2. Definitions

A summary of all symbols and annotations used in the LSD algorithm example (below) can be found in Table 1.

Table 1. Summary of all symbols and annotations used in the LSD algorithm example.

Symbol	Description	Units
i	Day number	day
PAR_i	Daily integrated Photosynthetically Available Radiation on day i	mol quanta m^{-2} day^{-1}
acclim PAR_i	PAR to which the corals are currently acclimated on day i	mol quanta m^{-2} day^{-1}
EEE_i	Daily Excess Excitation Energy on day i	mol quanta m^{-2} day^{-1}
F_v/F_m	diurnal maximum PSII photochemical efficiency	-
rel F_v/F_m	change in F_v/F_m relative to yesterday's $F_v/F_m = (F_v/F_m)_i/(F_v/F_m)_{i-1}$	-
SST	Sea Surface Temperature	°C
MMM	Maximum Monthly Mean SST	°C
HS	HotSpot = $SST - MMM$	°C

3.3. Keppel Islands Example

To apply the LSD algorithm, first a seasonal F_v/F_m is calculated using *EEE* derived with *PAR* only (i.e., no temperature effect). This provides the LSD bleaching threshold. Next, the effects of temperature and light are included in the derivation of F_v/F_m by using a combination of *HS* and *PAR*. The results of this are then compared to the LSD bleaching threshold to derive the LSD index.

3.3.1. Step 1: Derive the Daily Value of rel F_v/F_m Due to *EEE* with No Temperature Effect

Daily total *PAR* was extracted from the Rockhampton BoM light station (Figure 3a). For this example, the LSD algorithm was set to 20 cm depth and an attenuation coefficient of $K_d = 0.202$ m^{-1}, the average reported attenuation for *PAR* at inshore reefs of the Whitsunday Islands, just north of the Keppel Islands [61]. Like light, temperature varies through the water column, it therefore makes sense to set the light values to be vertically at the same level in the water column as the temperature values in order to avoid mismatches between *EEE* and *SST* effects that vary with water depth. Since the satellite *SST* data used by CRW are derived from measurements made at the sea surface and calibrated against drifting buoys [62], which provide temperatures at 20 cm depth [63], it follows that the light should also be set to a water depth of 20 cm.

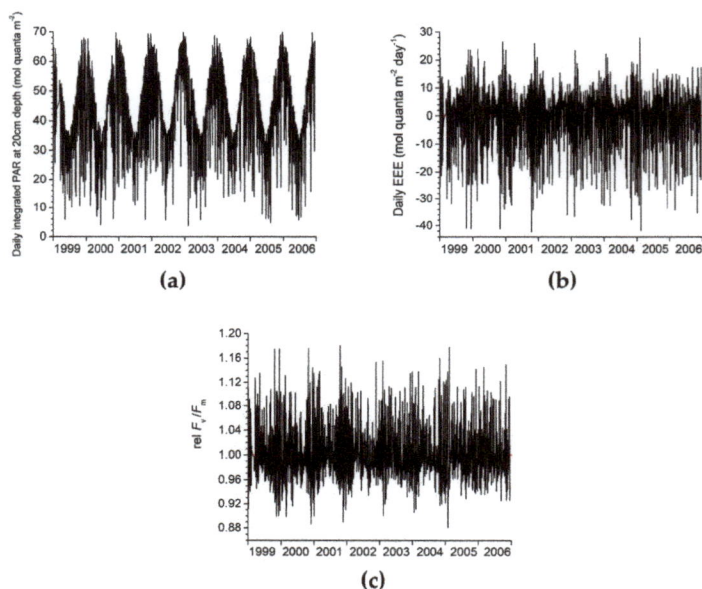

Figure 3. Rockhampton *PAR* at 20 cm depth (**a**), *EEE* (**b**) and change in relative F_v/F_m due to *EEE* with no heat stress (**c**) for the period 1999 to 2006.

The effect of light attenuation due to refraction and reflection at the sea surface was modeled prior to incorporation into the calculations. To cover both effects, NASA's Coupled Ocean and Atmosphere Radiative Transfer (COART, http://cloudsgate2.larc.nasa.gov/jin/coart.html) was used to simulate total daily *PAR* at the sea surface and at 1 cm below the surface. Since bleaching events often occur during low to no wind conditions [20] a wind speed of 2 ms^{-1} was chosen and a model resolution of 0.01 μm was used. The atmosphere in the model was set to "tropical" and the total downward flux of the combined direct and diffuse *PAR* was calculated at the surface and at a depth of 1 cm for each solar zenith angle from 0° to 85° at intervals of 5°. These were then totaled to simulate the total *PAR* for a day. The ratio of the daily total *PAR* at the surface vs the daily total at a depth of 1cm can then be used to estimate the attenuation at the air/sea interface to be 0.964 (i.e., slightly less radiation reaches a depth of 1cm than is incident at the sea surface. This is mostly due to reflection).

At each daily step, the LSD algorithm simulated light acclimation (acclim *PAR*) by determining the difference between the accumulated acclimation to PAR_{i-1} (i.e., acclim PAR_{i-1}) and the actual PAR_i, the difference was then reduced by a factor of 0.13 to simulate the rate of daily acclimation [57]. Acclim PAR_i was then used to calculate *EEE*.

Acclimated *PAR* at 20 cm depth (Figure 3a) was calculated to obtain the daily value of rel F_v/F_m due to *EEE* with no temperature effect (Figure 3b):

$$EEE_i = PAR_i - \text{acclim } PAR_{i-1} \tag{6}$$

$$\text{where acclim } PAR_i = \text{acclim } PAR_{i-1} + 0.13(PAR_i - \text{acclim } PAR_{i-1}) \tag{7}$$

These values were then converted to rel F_v/F_m using Equation (3) (Figure 3c).

3.3.2. Step 2: Derive the Daily Variation in F_v/F_m Due to *EEE* (with No Temperature Effect)

The daily rel F_v/F_m variation was used to calculate daily values of F_v/F_m throughout the year using Equation (8):

$$(F_v/F_m)_i = (F_v/F_m)_{i-1} \text{ (rel } F_v/F_m)_i \tag{8}$$

However, when Equation (8) was applied with no constraints, the resultant F_v/F_m values converged towards zero (Figure 4a). This is due to the asymmetry of *EEE* (Figure 3b). The bias towards negative anomalies resulted in the maximum for each consecutive year being progressively less than the previous year. To overcome this, at each winter solstice we assume that damage to the photosystems due to light stress was at a minimum and we reset F_v/F_m to 0.7, as being representative of the highest efficiency for *Symbiodinium*. This ensures that the seasonality described in [56] was reproduced in the model output.

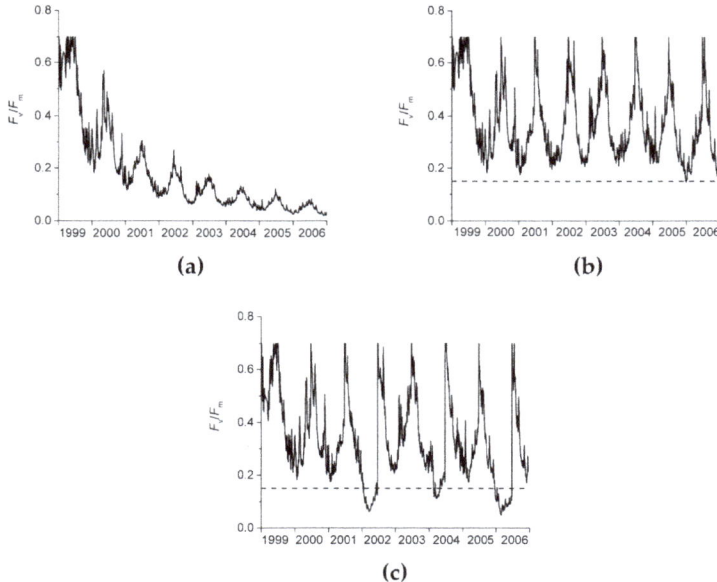

Figure 4. Seasonal variation of F_v/F_m over the period 1999 to 2006 with no annual reset and no heat stress (**a**), with annual reset and no heat stress (**b**) and with annual reset and heat stress (**c**). NB: the dashed line in (**b**,**c**) is the bleaching threshold of 0.15.

In keeping with the winter solstice reset, we also applied an upper limit to prevent F_v/F_m values from exceeding 0.7.

The reset value of 0.7 is somewhat arbitrary and may not be representative for *Symbiodinium* in all regions, however due to the makeup of the LSD algorithm presented here, changing this value does not alter the success of the algorithm, only its interpretation (i.e., the output graphs all look the same if another value is used but the range of the *Y*-axis alters). When applying the LSD algorithm to global satellite data it is likely necessary to use 0.7 in all regions so as to maintain a consistent interpretation of outputs from each region.

Starting with an F_v/F_m value of 0.7 at the winter solstice, Equation (9) was used to calculate values plotted in Figure 4b:

$$(F_v/F_m)_i = \text{minimum}((F_v/F_m)_{i-1} \text{ (rel } F_v/F_m)_i, 0.7) \tag{9}$$

The long-term minimum of F_v/F_m over the 8-year dataset was then used as the bleaching threshold, since it was expected that the corals have adapted to the local non-stressful ("normal") light conditions. Hence, light on its own should be able to drive the photosystem to the threshold of stress but not beyond.

This established the bleaching threshold for shallow corals in the Keppel Islands as 0.15 (Figure 4b).

Figure 4b demonstrates that the LSD algorithm reproduced the seasonality in F_v/F_m reported by [56] and that it is reproducing realistic values that vary according to field observations of F_v/F_m.

3.3.3. Step 3: Derive the Daily Value of rel F_v/F_m Due to EEE, Including Temperature Effects

Night-time *SST* values were derived from the NOAA AVHRR and plotted for the period of 1999 to 2006 in Figure 5a. HotSpot values were then calculated using *MMM*, where *MMM* = 27.5 °C, after the NOAA CRW methodology [20,63] and are plotted in Figure 5b.

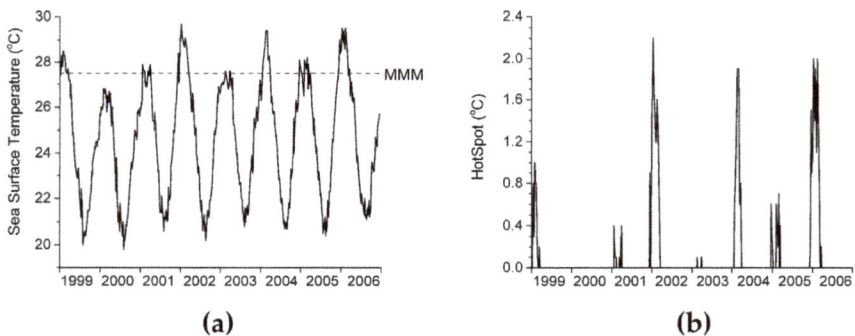

Figure 5. Plot of *SST* (a) and HotSpot anomaly (b) for the period 1999 to 2006.

For each day Equations (4) and (5) were used to calculate rel F_v/F_m for EEE including the effects of temperature.

i.e., (rel $F_v/F_m)_i = (1 - 0.01164 \; HS_i) + (-0.00426 - 0.00130 \; HS_i) \; EEE_i$

3.3.4. Step 4: Derive the Daily Variation in the Absolute Values of F_v/F_m Due to EEE, Including Temperature Effects

Daily F_v/F_m was then calculated using Equation (9), only this time, rel F_v/F_m includes the effects of temperature via step 3. As in Step 1, at each winter solstice, F_v/F_m was reset to the upper limit for

F_V/F_m of 0.7. The results are plotted in Figure 4c. Note that when adding the effect of temperature, F_V/F_m values occasionally drop below the stress threshold of 0.15, indicating bleaching-level stress.

3.3.5. Step 5: Calculate the Light Stress Damage Index

The LSD index is summed from the day on which F_V/F_m drops below the light stress threshold (0.15, determined in step 2, above) and continued to be summed until F_V/F_m rises above the stress threshold. The LSD index is therefore, a measure of the total accumulated stress during each stress event, derived by integrating between the F_V/F_m curve and the stress threshold (for all $F_V/F_m <$ *threshold*). This is similar to the methodology CRW uses for the calculation of *DHW* from *HS*.

The LSD index for this example is the integration between the 0.15 threshold and those F_V/F_m values less than 0.15:

$$LSD = \Sigma x_i, \text{ where } x_i = threshold - (F_V/F_m)_i; \; x_i \geq 0 \tag{10}$$

From Figure 4c it can be seen that there were three occasions when the F_V/F_m dropped below 0.15, in 2002, 2004 and 2006. However, the LSD index accumulation for 2004 is very small (Figure 6a).

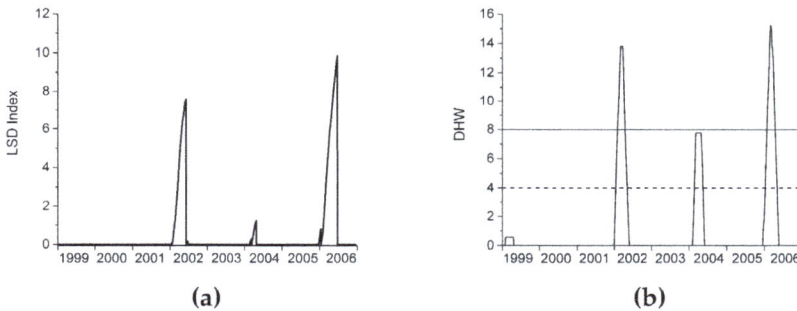

$$(a) \qquad\qquad\qquad (b)$$

Figure 6. Plot of Accumulated Light Stress Damage (**a**) and Degree Heating Weeks (**b**) for the Keppel Islands between 1999 and 2006. NB: 2002 and 2006 were bleaching years; the corals paled in 2004 with no bleaching; and all other years were stress-free. The dashed line in plot (**b**) at DHW = 4, is the threshold above which ecologically significant bleaching is expected. The solid horizontal line at DHW = 8, represents the threshold for widespread severe bleaching with some mortality.

For comparison purposes, *DHW* values were calculated using *HS* for the Keppel Islands for the period 1999 to 2006, using the methodology described in [20], see Figure 6b. Note that the DHW index is more pronounced than the LSD index for 2004.

4. Discussion

The LSD index (Figure 6a) suggest that there were two major stress events on the Keppel Islands during the period of 1999 to 2006. This was confirmed by annual surveys conducted by the Australian Institute of Marine Science (AIMS), in which only two bleaching events were recorded, one in early 2002 (100% bleached, 26% mortality) and the other in early 2006 (100% bleached, 35% mortality) (Australian Institute of Marine Science, unpublished data). The only other year in this record that came close to a bleaching event was in early 2004 when corals were recorded as having paled but not bleached and showed no mortality. The LSD index (Figure 6a) shows 2004 as a relatively small accumulation of light stress.

Figure 6b is a plot of the corresponding *DHW* values for the Keppel Islands over 1999 to 2006. Using the accepted key thresholds for the interpretation of the *DHW* product (DHW of 4 indicates ecologically significant bleaching and a DHW value of 8 or more indicates severe bleaching) [20,25]. These interpretations suggest that Figure 6b is indicating that the bleaching events in 2002 and 2006 were severe and of similar magnitude and that there was an ecologically significant bleaching event in 2004, which bordered on being classified as a severe bleaching event.

The 2004 heat event provides a good example where the use of temperature alone can lead to a false positive. A significant temperature anomaly existed in terms of the daily HotSpot (Figure 5b) and in terms of accumulated temperature anomalies in the *DHW* product (Figure 6b). However, surveys observed paling with no bleaching or mortality.

The LSD index more accurately predicted the 2002 and 2006 bleaching events (indicating that 2002 was significantly less severe than 2006) and a small amount of stress in 2004. Figure 7 shows three plots, one for each of the three temperature anomaly events experienced at the Keppel Islands in 2002, 2004 and 2006. The top grey line in each of the three plots shows the *HS* values, allowing visual identification of when temperature influenced the F_V/F_m values. The lower two black graphs in each of the plots are of F_V/F_m, with no heat stress effect (*EEE*-only, thin black line) and with the addition of the effect of heat stress (thick black line).

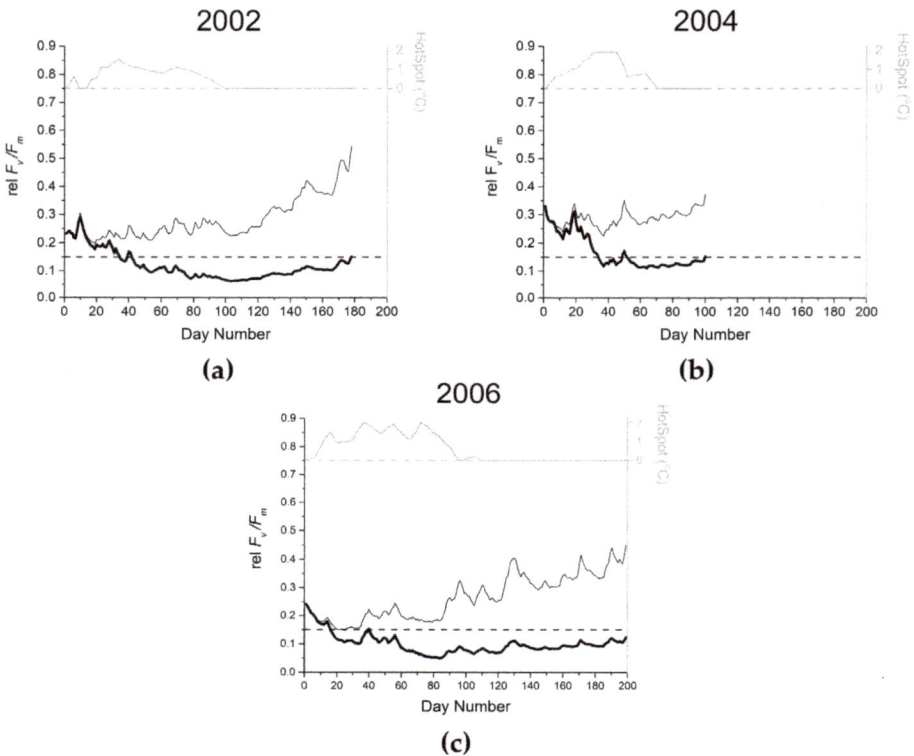

Figure 7. Plots of the three stress events in 2002 (**a**), 2004 (**b**) and 2006 (**c**). Plots are all on the same scale for easy comparison. Within each plot, the top (grey) line graph is the daily HotSpot values with the scale on the right of each plot. The two black line plots are F_V/F_m for the effect of *EEE*-only (thin black line) and for the combined effect of light and heat (thick black line).

The three plots in Figure 7 clearly demonstrate the influence that light has on the accumulation of heat stress (for the Keppel Island example) and why the outcomes for 2002, 2004 and 2006 were so different between the LSD index and DHW products. In general, the story is one of light buffering. When light conditions leading into the heat stress event are favorable (i.e., when *EEE* is consistently small), it is possible for these conditions to provide a "buffer" against the effects of the heat stress, allowing the corals to cope with larger amounts of heat stress before feeling its effects.

Figure 6b shows that the heat stress, as measured by DHW, was similar for 2002 (DHW = 13.8) and 2006 (DHW = 15.2). However, Figure 7a shows the *EEE*-only stress starting at around 0.25 and steadily rising over the period of the heat anomaly, serving to provide a modest buffer and hence, reducing the heat stress accumulation, resulting in an LSD value of 7.6. During 2006 however, Figure 7c shows the *EEE*-only stress starting at around 0.24 (similar to the 2004 level), however unlike 2004 where the *EEE*-only stress improved throughout the heat stress event, in 2006 it dipped right down to the stress threshold, providing no buffer for coral stress, allowing the full impact of heat stress to be felt by the corals. This resulted in an LSD value of 9.9, which is 30% larger than the 2004 value. The 2006 DHW value is only 10% larger than the 2004 value.

Since the AIMS data report both events having had 100% bleaching, the only measure of severity we can use is the level of mortality. 2006 had 35% mortality which is 35% larger than the mortality in 2004 of 26%. At least for this example, the LSD index seemed to do a marginally better job of characterizing these two bleaching events than did DHW.

It is the 2004 heat stress event that demonstrates the potential of the LSD algorithm. Figure 6b indicates that the DHW product recorded 7.8 in 2004, which was sufficient to have caused ecologically significant bleaching (above DHW of 4) and was bordering on being classed as a severe bleaching event (above DHW of 8) [20,25]. This is out of step with the AIMS bleaching records, which indicate that the corals paled with no bleaching. The LSD index recorded 1.2, which is only 15% of the 2002 LSD index. Had the 2004 DHW value been 15% of its 2002 value then it would have recorded a DHW value of 2, which equates to stress with no significant bleaching.

Figure 7b explains why the DHW product overstated the level of stress in 2004 while the LSD index provided a good measure of the level of stress.

In 2004, at the start of the heat event, F_v/F_m due to *EEE*-only was around 0.3 and although it fluctuated, F_v/F_m due to *EEE*-only remained quite high throughout the event. This countered the effect of the temperature anomaly and prevented F_v/F_m from dropping to the bleaching threshold until well into the heat stress event. As a result, the LSD only accumulated a small amount of stress. With light levels playing such a major role in offsetting the heat stress for this event, it is not surprising that the *DHW* product did not accurately describe this event.

These three stress events clearly demonstrate the important role of light in the lead up to and during a heat stress event. Favorable light conditions (i.e., consistently low *EEE* values to maintain a relatively high F_v/F_m value) in the lead up to a heat event can provide a significant buffer against temperature anomalies, allowing corals to be less susceptible to the effects of elevated temperatures, as they were during the 2004 Keppel Islands example. However, the converse can also occur as in the 2006 example, where only slightly higher temperature anomalies generated quite different outcomes in the severity and length of the stress event due to the lack of a light buffer because of unfavorable light conditions leading up to the event (i.e., consistently high *EEE* values, which pushed the F_v/F_m values close to the stress threshold at the beginning of the heat event).

The LSD was therefore able to take into account the impact of light on corals and their *Symbiodinium* and determine how much stress (i.e., combined effects of light and temperature) is needed during the temperature anomaly event to induce coral bleaching, significantly improving the ability to predict total stress when compared to a purely temperature derived stress index, such as the *DHW* product. Differences in the combined light and heat stress measured in the LSD index predicted both the bleaching and mortality much better than heat stress measured in the DHW.

As promising as the LSD algorithm is, there is still work to be done before the algorithm is complete. For instance, if the daily F_v/F_m value, within the LSD algorithm, is not reset to 0.7 at the winter solstice, F_v/F_m values converged towards zero after several years. This occurred in the calculation using *EEE* with no temperature effect, as well as when temperature was combined with *EEE*. While the annual reset worked for the production of a predictive algorithm, it was based on empirical observations rather than physiological mechanisms. Sensitivity testing identified two places where the addition of new terms to the LSD algorithm could prevent the need for this reset. The first was in the use of 0.13 as the daily acclimation rate [57]. This study suggested that acclimation rates may not be the same for increasing and decreasing light levels. We tested the application of asymmetric values and reducing the response to decreasing light by 5–10% stabilized the F_v/F_m over the eight-year period. This demonstrated that the multi-year stability of F_v/F_m was very sensitive to slight changes in the symmetry of the photoacclimation rate. However, imposing such a change did not stabilize the multi-year F_v/F_m simulation after a heat stress event. This suggested a term is needed to simulate the recovery of *Symbiodinium* after heat stress has ended. Both the physiological recovery of the photosynthetic apparatus through the up-regulation of the repair mechanisms and the recovery of *Symbiodinium* populations through asexual reproduction can contribute to this recovery. This can be simulated by maintaining the enhanced rate of recovery found in the slope of heat-stressed corals (Figure 1) for a period after heat stress has ended. It is reasonable to expect that this period of enhanced recovery may be maintained for up to 30 days [47,48]. Imposing such a change stabilized the multi-year F_v/F_m simulation after a heat stress event and was relatively insensitive to the length of time over which enhanced recovery was applied. It may be possible to quantify and include these two terms once the LSD algorithm is applied to remotely-sensed data (currently underway at NOAA) and a larger dataset becomes available for their derivation.

NOAA has a global *PAR* product that matches the temporal and spatial resolution of its 0.05 degree (approx. 5 km) Geo-Polar Blended *SST* product. Information about the GOES Surface and Insolation Products (GSIP) can be found at www.ospo.noaa.gov/Products/land/gsip/index_v3.html. Currently, there is 11 years of satellite-derived *PAR* data for the eastern Pacific and the Caribbean/Gulf of Mexico regions. Global coverage has been available since March 2014, however when the switch from MTSAT to Himawari-8 (Japanese geostationary satellites over the Australasian region) occurred on 5 December 2015, the GSIP product was not updated to account for Himawari-8 data. The inclusion of Himawari-8 data within the GSIP product is planned and will take place in the near future.

NOAA CRW have also been working closely with many collaborators (in particular the University of British Columbia) to collate a global dataset of quality controlled in-water coral bleaching observations against which to calibrate and validate algorithms such as the LSD and DHW.

When the *PAR* and *SST* satellite data sets are combined with the quality-controlled in-water bleaching observations, the path for continued development of the LSD algorithm will be in good shape.

5. Conclusions

The new methodology described herein will be used to underpin a new satellite product within the NOAA CRW coral bleaching decision support system.

The Keppel Islands example demonstrates that this methodology shows great promise to improve upon the already successful *DHW* heat stress product. However, more work is needed before the LSD index will be ready for use as an operational satellite product, including testing its applicability across multiple coral species and habitats around the world.

Changes in light attenuation with depth and through time (due to episodic events such as floods) will influence F_v/F_m and how it interacts with temperature variability. This can have important impacts on the LSD algorithm. These interactions will need to be investigated in the near future and are a planned step in the development of the LSD as a remotely-sensed product.

Time of acclimation to light for positive vs negative *EEE* is also important for the LSD algorithm and will need to be resolved before this algorithm can reach its full potential.

The Keppel Islands example helped to explain the LSD algorithm and why it works, rather than to provide a definitive test of the algorithm's ability to predict coral bleaching. Testing its applicability on other locations and species is fundamental to test the predictive capacity of the LSD algorithm for coral bleaching. However, considering the scarcity of quality in situ light data near coral reefs

Remote Sens. **2018**, *10*, 18

with documented bleaching records, this validation will have to wait until the algorithm has been fully implemented on satellite data, work currently underway at NOAA. Such work will need to consider both the light reaching the ocean surface and the reduction in light as it passes through seawater to reach the coral. For many applications, full implementation of the LSD algorithm for remote sensing may require not only temperature and surface insolation data but also the application of ocean color products that measure changes in light attenuation in the water column—products also in development at NOAA.

Acknowledgments: Development of NOAA Coral Reef Watch's Light Stress Damage product suite was supported by the NOAA Coral Reef Conservation Program (CRCP), the Australian Research Council, the University of Queensland and the Universidad Nacional Autónoma de México. The scientific results and conclusions, as well as any views or opinions expressed herein, are those of the author(s) and do not necessarily reflect the views of NOAA or the Department of Commerce.

Author Contributions: William Skirving, Susana Enríquez, Roberto Iglesias-Prieto, Sophie Dove and John Hedley were the main contributors to the concepts in this paper, with C. Mark Eakin, Robert A. B. Mason, Ove Hoegh-Guldberg, Alan E. Strong and Peter J. Mumby making smaller but significant contributions to the concepts in this paper. All experiments included in this paper were conducted by Susana Enríquez and involved Roberto Iglesias-Prieto, results from other experiments conducted by Robert A. B. Mason and involving Sophie Dove were not used in this work but did influence the concepts and algorithm development. Most of the algorithm development was performed by William Skirving, Susana Enríquez and Roberto Iglesias-Prieto, with significant input from John Hedley, C. Mark Eakin, Robert A. B. Mason and Sophie Dove. The manuscript was mostly written by William Skirving, Susana Enríquez and Roberto Iglesias-Prieto, with significant contributions being made by John Hedley, C. Mark Eakin, Sophie Dove, Robert A. B. Mason, Jacqueline L. De La Cour and Gang Liu.

Conflicts of Interest: The authors declare no conflict of interest. The founding sponsors had no role in the design of the study; in the collection, analyses, or interpretation of data; in the writing of the manuscript and in the decision to publish the results.

References

1. Glynn, P.W. Coral Reef Bleaching: Facts, Hypothesis and Implications. *Glob. Chang. Biol.* **1996**, *2*, 495–509.
2. Lesser, M.P. Elevated temperature and ultraviolet radiation cause oxidative stress and inhibit photosynthesis in symbiotic dinoflagellates. *Limnol. Oceanogr.* **1996**, *41*, 271–283. [CrossRef]
3. Brown, B.E. The significance of pollution in eliciting the 'bleaching' response in symbiotic cnidarians. *Int. J. Environ. Pollut.* **2000**, *13*, 392–415. [CrossRef]
4. Douglas, A.E. Coral Bleaching—How and why? *Mar. Pollut. Bull.* **2003**, *46*, 385–392. [PubMed]
5. Hoegh-Guldberg, O. Climate change, coral bleaching and the future of the World's coral reefs. *Mar. Freshw. Res.* **1999**, *50*, 839–866.
6. Hoegh-Guldberg, O.; Fine, M.; Skirving, W.; Johnstone, R.; Dove, S.; Strong, A. Coral bleaching following wintry weather. *Limnol. Oceanogr.* **2005**, *50*, 265–271. [CrossRef]
7. Dove, S.G.; Hoegh-Guldberg, O. The cell physiology of coral bleaching. In *Coral Reefs and Climate Change: Science and Management*; Phinney, J.T., Hoegh-Guldberg, O., Kleypas, J., Skirving, W., Strong, A., Eds.; American Geophysical Union: Washington, DC, USA, 2006; pp. 55–71.
8. Falkowski, P.G.; Jokiel, P.L.; Kinzie, R.A. Irradiance and Corals. In *Coral Reefs*; Dubinsky, Z., Ed.; Elsevier: Amsterdam, The Netherlands, 1990; pp. 89–107.
9. Iglesias-Prieto, R.; Trench, R.K. Acclimation and adaptation to irradiance in symbiotic dinoflagellates. I. Responses of the photosynthetic unit to changes in photon flux density. *Mar. Ecol. Prog. Ser.* **1994**, *113*, 163–175.
10. Hennige, S.J.; Suggett, D.J.; Warner, M.E.; McDougall, K.E.; Smith, D.J. Photobiology of *Symbiodinium* revisited: Bio-Physical and bio-optical signatures. *Coral Reefs* **2009**, *28*, 179–195.
11. Dubinsky, Z.; Stambler, N.; Ben-Zion, M.; McCloskey, L.; Muscatine, L.; Falkowski, P. The effect of external nutrient resources on the optical properties and photosynthetic efficiency of *Stylphora pistillata*. *Proc. R. Soc. Lond.* **1990**, *239*, 231–246.
12. Fagoonee, I.I.; Wilson, H.B.; Hassell, M.P.; Turner, J.R. The dynamics of zooxanthellae populations: A long-term study in the field. *Science* **1999**, *283*, 5403–843.

13. Fitt, W.K.; Mcfarland, M.E.; Warner, M.E.; Chilcoat, G.C. Seasonal patterns of tissue biomass and densities of symbiotic dinoflagellates and relation to coral bleaching. *Limnol. Oceanogr.* **2000**, *45*, 677–685. [CrossRef]
14. Brown, B.E.; Dunne, R.P.; Ambarsari, I.; Le Tissier, M.D.A.; Satapoomin, U. Seasonal fluctuations in environmental factors and variations in symbiotic algae and chlorophyll pigments in four Indo-Pacific coral species. *Mar. Ecol. Prog. Ser.* **1999**, *191*, 53–69. [CrossRef]
15. Berkelmans, R.; De'ath, G.; Kinimonth, S.; Skirving, W. Coral bleaching on the Great Barrier Reef: Correlation with sea surface temperature, a handle on 'patchiness' and comparison of the 1998 and 2002 events. *Coral Reefs* **2004**, *23*, 74–83. [CrossRef]
16. Glynn, P.W. Widespread coral mortality and the 1982–83 El Niño warming event. *Environ. Conserv.* **1984**, *11*, 133–146. [CrossRef]
17. Coffroth, M.A.; Lasker, H.R.; Oliver, J.E. Coral mortality outside of the eastern Pacific during 1982–1983: Relationship to El Niño. In *Global Ecological Consequences of the 1982–83 El Niño-Southern Oscillation*; Glynn, P.W., Ed.; Elsevier Oceanography Series; Elsevier: Amsterdam, The Netherlands, 1989; pp. 141–182.
18. Wilkinson, C. (Ed.) *Status of Coral Reefs of the World*; Australian Institute of Marine Science: Townsville, Queensland, Australia, 2004.
19. Eakin, M.C.; Liu, G.; Gomez, A.M.; De La Cour, J.L.; Heron, S.F.; Skirving, W.J.; Geiger, E.F.; Marsh, B.L.; Tirak, K.V.; Strong, A.E. Ding, Dong, The Witch is Dead (?)—Three Years of Global Coral Bleaching 2014–2017. *Reef Encount.* **2017**, *32*, 33–38.
20. Skirving, W.J.; Strong, A.E.; Liu, G.; Liu, C.; Arzayus, F.; Sapper, J.; Bayler, E. Extreme events and perturbations of coastal ecosystems: Sea surface temperature change and coral bleaching. In *Remote Sensing of Aquatic Coastal Ecosystem Processes*; Richardson, L.L., LeDrew, E.F., Eds.; Springer: Dordrecht, The Netherlands, 2006; pp. 11–25.
21. Brown, B.E. Coral bleaching: Causes and consequences. *Coral Reefs* **1997**, *16*, 129–138. [CrossRef]
22. Liu, G.; Heron, S.F.; Eakin, C.M.; Muller-Karger, F.E.; Vega-Rodriguez, M.; Guild, L.S.; De La Cour, J.L.; Geiger, E.F.; Skirving, W.J.; Burgess, T.F.R.; et al. Reef-scale Thermal Stress Monitoring of Coral Ecosystems: New 5-km Global Products from NOAA Coral Reef Watch. *Remote Sens.* **2014**, *6*, 11579–11606. [CrossRef]
23. Liu, G.; Skirving, W.J.; Geiger, E.F.; De La Cour, J.L.; Marsh, B.L.; Heron, S.F.; Tirak, K.V.; Strong, A.E.; Eakin, C.M. NOAA Coral Reef Watch's 5 km Satellite Coral Bleaching Heat Stress Monitoring Product Suite Version 3 and Four-Month Outlook Version 4. *Reef Encount.* **2017**, *32*, 39–45.
24. Eakin, C.M.; Morgan, J.A.; Heron, S.F.; Smith, T.B.; Liu, G.; Alvarez-Filip, L.; Baca, B.; Bartels, E.; Bastidas, C.; Bouchon, C.; et al. Caribbean Corals in Crisis: Record Thermal Stress, Bleaching and Mortality in 2005. *PLoS ONE* **2010**, *5*, e13969. [CrossRef] [PubMed]
25. Heron, S.F.; Johnston, L.; Liu, G.; Geiger, E.F.; Maynard, J.A.; De La Cour, J.L.; Johnson, S.; Okano, R.; Benavente, D.; Burgess, T.F.R.; et al. Validation of Reef-scale Thermal Stress Satellite Products for Coral Bleaching Monitoring. *Remote Sens.* **2016**, *8*, 59. [CrossRef]
26. Kayanne, H. Validation of degree heating weeks as a coral bleaching index in the northwestern Pacific. *Coral Reefs* **2017**, *36*, 63. [CrossRef]
27. Iglesias-Prieto, R.; Matta, J.L.; Robins, W.A.; Trench, R.K. Photosynthetic response to elevated temperature in the symbiotic dinoflagellate *Symbiodinium microadriaticum* in culture. *Proc. Natl. Acad. Sci. USA* **1992**, *89*, 10302–10305. [CrossRef] [PubMed]
28. Jones, R.J.; Hoegh-Guldberg, O.; Larkum, A.W.D.; Schreiber, U. Temperature-induced bleaching of corals begins with impairment of the CO_2 fixation mechanism in zooxanthellae. *Plant Cell Environ.* **1998**, *21*, 1219–1230. [CrossRef]
29. Warner, M.E.; Fitt, W.K.; Schmidt, G.W. Damage to photosystem II in symbiotic dinoflagellates: A determinant of coral bleaching. *Proc. Natl. Acad. Sci. USA* **1999**, *96*, 8007–8012. [CrossRef] [PubMed]
30. Takahashi, S.; Nakamura, T.; Sakamizu, M.; van Woesik, R.; Yamasaki, H. Repair machinery of symbiotic photosynthesis as the primary target of heat stress for reef-building corals. *Plant Cell Phys.* **2004**, *45*, 251–255. [CrossRef]
31. Mumby, P.J.; Chisholm, J.R.M.; Edwards, A.J.; Andrefouet, S.; Jaubert, J. Cloudy weather may have saved Society Island reef corals during the 1998 ENSO event. *Mar. Ecol. Prog. Ser.* **2001**, *222*, 209–216. [CrossRef]
32. Niyogi, K. Photoprotection revisited: Genetic and molecular approaches. *Annu. Rev. Plant Physiol. Plant Mol. Biol.* **1999**, *50*, 333–359. [CrossRef] [PubMed]
33. Niyogi, K. Safety valves for photosynthesis. *Curr. Opin. Plant Biol.* **2000**, *6*, 455–460. [CrossRef]

34. Anderson, J.M.; Chow, W.S.; Park, Y.I. The grand design of photosynthesis: Acclimation of the photosynthetic apparatus to environmental cues. *Photosynth. Res.* **1995**, *46*, 129–139. [CrossRef] [PubMed]

35. Chalker, B. Simulating light-saturation curves for photosynthesis and calcification by reef-building corals. *Mar. Biol.* **1981**, *63*, 135–141. [CrossRef]

36. Melis, A. Photosystem II damage and repair cycle in chloroplasts: What modulates the rate of photodamage "in vivo"? *Trends Plant Sci.* **1999**, *94*, 130–135. [CrossRef]

37. Vass, I.; Styring, S.; Hundal, T.; Koivuniemi, A.; Aro, E.; Andersson, B. Reversible and irreversible intermediates during photoinhibition of photosystem II: Stable reduced Q_A species promote chlorophyll triplet formation. *Proc. Natl. Acad. Sci. USA* **1992**, *90*, 1408–1412. [CrossRef]

38. Vásquez-Elizondo, R.M.; Enríquez, S. Coralline algal physiology is more adversely affected by elevated temperature than reduced pH. *Sci. Rep.* **2016**, *6*, 19030. [CrossRef] [PubMed]

39. Hill, R.; Brown, C.M.; DeZeeuw, K.; Campbell, D.A.; Ralph, P.J. Increased rate of D1 repair in coral symbionts during bleaching is insufficient to counter accelerated photo-inactivation. *Limnol. Oceanogr.* **2011**, *56*, 139–146. [CrossRef]

40. Weis, V.M. Cellular mechanisms of Cnidarian bleaching: Stress causes the collapse of symbiosis. *J. Exp. Biol.* **2008**, *211*, 3059–3066. [CrossRef] [PubMed]

41. Titlyanov, E.A.; Titlyanova, T.V.; Leletkin, V.A.; Tsukahara, J.; van Woesik, R.; Yamazato, K. Degradation of zooxanthellae and regulation of their density in hermatypic corals. *MEPS* **2006**, *139*, 167–178. [CrossRef]

42. Dunn, S.R.; Pernice, M.; Green, K.; Hoegh-Guldberg, O.; Dove, S.G. Thermal Stress Promotes Host Mitochondrial Degradation in Symbiotic Cnidarians: Are the Batteries of the Reef Going to Run Out? *PLoS ONE* **2012**. [CrossRef] [PubMed]

43. Burge, C.A.; Eakin, C.M.; Friedman, C.S.; Froelich, B.; Hershberger, P.K.; Hofmann, E.E.; Petes, L.E.; Prager, K.C.; Weil, E.; Willis, B.L.; et al. Climate Change Influences on Marine Infectious Diseases: Implications for Management and Society. *Ann. Rev. Mar. Sci.* **2014**, *6*, 249–277. [CrossRef] [PubMed]

44. Iglesias-Prieto, R. Temperature-dependent inactivation of photosystem II in symbiotic dinoflagellates. In Proceedings of the 8th International Coral Reef Symposium, Panama City, Panama, 24–29 June 1996; Macintyre, I.G., Ed.; Smithsonian Tropical Research Institute: Balboa, Panama, 1997; pp. 1313–1318.

45. Warner, M.E.; LaJeunesse, T.C.; Robinson, J.D.; Thur, R.M. The ecological distribution and comparative photobiology of symbiotic dinoflagellates from reef corals in Belize: Potential implications for coral bleaching. *Limnol. Oceanogr.* **2006**, *51*, 1887–1897. [CrossRef]

46. Schreiber, U.; Bilger, W.; Neubauer, C. Chlorophyll fluorescence as a nonintrusive indicator for rapid assessment of in vivo photosynthesis. In *Ecophysiology of Photosynthesis*; Schulze, E.D., Caldwell, M.M., Eds.; Springer: Berlin, Germany, 1995; pp. 49–70.

47. Rodríguez-Román, A.; Hernández-Pech, X.; Thomé, P.E.; Enríquez, S.; Iglesias-Prieto, R. Photosynthesis and light utilization in the Caribbean coral *Montastraea faveolata* recovering from a bleaching event. *Limnol. Oceanogr.* **2016**, *51*, 2702–2710. [CrossRef]

48. DeSalvo, M.K.; Sunagawa, S.; Fisher, P.; Voolstra, C.R.; Iglesias-Prieto, R.; Medina, M. Coral host transcriptomic states are correlated with *Symbiodinium* genotypes. *Mol. Ecol.* **2010**, *19*, 1174–1186. [CrossRef] [PubMed]

49. Maxwell, D.P.; Falk, S.; Huner, N.P.A. Photosystem II excitation pressure and development of resistance to photoinhibition. I. Light-harvesting complex II abundance and zeaxanthin content in *Chlorella vulgaris*. *Plant Phys.* **1995**, *107*, 687–694. [CrossRef]

50. Iglesias-Prieto, R.; Beltrán, V.H.; LaJeunesse, T.C.; Reyes-Bonilla, H.; Thomé, P.E. Different algal symbionts explain the vertical distribution of dominant reef corals in the eastern Pacific. *Proc. R. Soc. Lond.* **2004**, *271*, 1757–1763. [CrossRef] [PubMed]

51. Warner, M.E.; Fitt, W.K.; Schmidt, G.W. The effects of elevated temperature on the photosynthetic efficiency of zooxanthellae in hospite from four different species of reef coral: A novel approach. *Plant Cell Environ.* **1996**, *19*, 291–299. [CrossRef]

52. Matsubara, S.; Chow, W.S. Populations of photoinactivated photosystem II reaction centers characterized by chlorophyll a fluorescence lifetime in vivo. *Proc. Natl. Acad. Sci. USA* **2004**, *101*, 18234–18239. [CrossRef] [PubMed]

53. Kajiwara, K.; Nagai, A.; Ueno, S. Examination of the effect of temperature, light intensity and zooxanthellae concentration on calcification and photosynthesis of scleractinian coral *Acropora pulchra*. *J. Sch. Mar. Sci. Technol. Tokai Univ.* **1995**, *40*, 95–103.
54. Rodolfo-Metalpa, R.; Huot, Y.; Ferrier-Pagès, C. Photosynthetic response of the Mediterranean zooxanthellate coral *Cladocora caespitosa* to the natural range of light and temperature. *J. Exp. Biol.* **2008**, *211*, 1579–1586. [CrossRef] [PubMed]
55. Edmunds, P.J.; Brown, D.; Moriarty, V. Interactive effects of ocean acidification and temperature on two scleractinian corals from Moorea, French Polynesia. *Glob. Chang. Biol.* **2012**, *18*, 2173–2183. [CrossRef]
56. Warner, M.E.; Chilcoat, G.C.; McFarland, F.K.; Fitt, W.K. Seasonal fluctuations in the photosynthetic capacity of photosystem II in symbiotic dinoflagellates in the Caribbean reef-building coral *Montastraea*. *Mar. Biol.* **2002**, *141*, 31–38.
57. Anthony, K.R.N.; Hoegh-Guldberg, O. Kinetics of photoacclimation in corals. *Oecologia* **2003**, *134*, 23–31. [CrossRef] [PubMed]
58. Scheufen, T.; Kramer, W.E.; Iglesias-Prieto, R.; Enríquez, S. Seasonal variation modulates coral sensibility to heat-stress and explains annual changes in coral productivity. *Sci. Rep.* **2017**, *7*, 4937. [CrossRef] [PubMed]
59. Heron, S.F.; Liu, G.; Eakin, C.M.; Skirving, W.J.; Muller-Karger, F.E.; Vega-Rodriguez, M.; De La Cour, J.L.; Burgess, T.F.R.; Strong, A.E.; Geiger, E.F.; et al. *Climatology Development for NOAA Coral Reef Watch's 5-km Product Suite*; NOAA Technical Report NESDIS 145; NOAA/NESDIS: College Park, MD, USA, 2015.
60. Liu, G.; Rauenzahn, J.L.; Heron, S.F.; Eakin, C.M.; Skirving, W.J.; Christensen, T.R.L.; Strong, A.E.; Li, J. *NOAA Coral Reef Watch 50 km Satellite Sea Surface Temperature-Based Decision Support System for Coral Bleaching Management*; NOAA Technical Report NESDIS 143; NOAA/NESDIS: College Park, MD, USA, 2013.
61. Cooper, T.F.; Uthicke, S.; Humphrey, C.; Fabricius, K.E. Gradients in water column nutrients, sediment parameters, irradiance and coral reef development in the Whitsunday Region, central Great Barrier Reef. *Estuar. Coast. Shelf Sci.* **2007**, *27*, 503–519. [CrossRef]
62. Ignatov, A.; Zhou, X.; Petrenko, B.; Liang, X.; Kihai, Y.; Dash, P.; Stroup, J.; Sapper, J.; DiGiacomo, P. AVHRR GAC SST Reanalysis Version 1 (RAN1). *Remote Sens* **2016**, *8*, 315. [CrossRef]
63. Merchant, C.J.; Embury, O.; Roberts-Jones, J.; Fiedler, E.; Bulgin, C.E.; Corlett, G.K.; Good, S.; McLaren, A.; Rayner, N.; Morak-Bozzo, S.; et al. Sea surface temperature datasets for climate applications from Phase 1 of the European Space Agency Climate Change Initiative (SST CCI). *Geosci. Data J.* **2014**, *1*, 179–191. [CrossRef]

remote sensing

MDPI

Article

Reconstruction of Daily Sea Surface Temperature Based on Radial Basis Function Networks

Zhihong Liao [1,2,3], Qing Dong [1,*], Cunjin Xue [1], Jingwu Bi [1,2] and Guangtong Wan [1]

[1]   Institute of Remote Sensing and Digital Earth, Chinese Academy of Sciences, Beijing 100094, China;
    liaozh@radi.ac.cn (Z.L.); xuecj@radi.ac.cn (C.X.); jingwubi@163.com (J.B.); wanguangtong@126.com (G.W.)
[2]   University of Chinese Academy of Sciences, Beijing 100049, China
[3]   National Meteorological Information Center, China Meteorological Administration, Beijing 100081, China
*   Correspondence: qdong@radi.ac.cn; Tel.: +86-10-8217-8121

Received: 6 November 2017; Accepted: 18 November 2017; Published: 22 November 2017

Abstract: A radial basis function network (RBFN) method is proposed to reconstruct daily Sea surface temperatures (SSTs) with limited SST samples. For the purpose of evaluating the SSTs using this method, non-biased SST samples in the Pacific Ocean (10°N–30°N, 115°E–135°E) are selected when the tropical storm Hagibis arrived in June 2014, and these SST samples are obtained from the Reynolds optimum interpolation (OI) v2 daily 0.25° SST (OISST) products according to the distribution of AVHRR L2p SST and in-situ SST data. Furthermore, an improved nearest neighbor cluster (INNC) algorithm is designed to search for the optimal hidden knots for RBFNs from both the SST samples and the background fields. Then, the reconstructed SSTs from the RBFN method are compared with the results from the OI method. The statistical results show that the RBFN method has a better performance of reconstructing SST than the OI method in the study, and that the average RMSE is 0.48 °C for the RBFN method, which is quite smaller than the value of 0.69 °C for the OI method. Additionally, the RBFN methods with different basis functions and clustering algorithms are tested, and we discover that the INNC algorithm with multi-quadric function is quite suitable for the RBFN method to reconstruct SSTs when the SST samples are sparsely distributed.

Keywords: sea surface temperature (SST); radial basis function network (RBFN); improved nearest neighbor cluster (INNC) algorithm

1. Introduction

High-quality sea surface temperature (SST) data play an important role in many applications, including climate predictions [1], ocean data assimilation [2], and global change research [3]. However, raw SST data can vary significantly between different types of measurements [4], such as in-situ measurements (e.g., ships or buoys) and satellite sensors (e.g., infrared thermal radiometers or microwave radiometers on satellite platforms). In-situ SST measurements are accurate but sparse, while satellite-retrieved SST measurements have poor accuracy but provide a dense coverage globally. Therefore, one method to solve these problems is to combine different sources of SSTs to reconstruct new SST fields [4,5].

Even through many different sources of SSTs are combined, due to the limitations of satellite orbits and the field of view (FOV) of sensors, global ocean coverage remains difficult to achieve [6]. Additionally, high-quality SST samples usually have irregular distributions as some data are discarded, such as in-situ SST data from questionable drifters and satellite-retrieved SST data contaminated by clouds [6,7]. Especially for the SST data from infrared radiometers, the SSTs in some regions are largely missing since they can not get the information through clouds. Thus, it is critical to reconstruct SST fields from incomplete SST samples.

Several alternative methods, such as empirical orthogonal function (EOF) [8,9], data interpolating orthogonal function (DINEOF) [10,11], and optimum interpolation (OI) [7], have previously been used for SST reconstruction. Currently, the most popular method to reconstruct SST is with the OI algorithm, and many high-quality SST analysis products using the OI method have been designed with different kinds of temporal and spatial resolutions for various applications [5,6,12–17]. However, when the distribution of SST samples are very sparse and the local correlation scales are not large enough in the OI method, the SSTs are often replaced by the background fields and the SST accuracy in these regions is relatively poor [6].

A SST field can generally be described as a nonlinear pattern in space and time, which can be expressed as a set of complex functions. Radial basis function network (RBFN) is an artificial neural network that uses radial basis functions (RBFs) as activation functions in the field of mathematical modeling. It is a good function approximator when RBFNs have enough hidden neurons and samples [18–20]. RBFNs have been employed in many applications [21–23]. A RBFN-based response model for SST has recently been developed by Ryu et al. [24]. However, this model does not optimize the quantity and distribution of hidden knots, which are sensitive to the accuracy of the model and should be adjusted with different distributions of SST samples.

Thus, we introduce a RBFN method for reconstructing daily SST fields when the SST samples are limited and sparsely distributed, and design an improved nearest neighbor cluster (INNC) algorithm for optimizing the quantity and distribution of hidden knots for RBFNs in this study. The remainder of this paper is organized as follows. Section 2 provides an outline of the SST samples that are used in this study. Then, the RBFN method and the INNC algorithm are described in Section 3. Section 4 evaluates the performance of the RBFN method by using different basis functions and clustering methods for RBFNs, and analyze its SST accuracy with the OI method. Then, some characteristics of this method are discussed in Section 5. Finally, the conclusions are summarized in Section 6.

2. Data Description

Since the satellite-retrieved SST data from infrared radiometers are often contaminated by clouds, the SST samples that we can obtain is very limited, especially when typhoon or tropical storm comes with heavy clouds. This study considers data from the Pacific Ocean (10°N–30°N, 115°E–135°E) (see Figure 1a). The SST samples in the study are selected from both the in-situ SST and the AVHRR SST L2p data, but the coverage of SST samples is still very low when the tropical storm Hagibis arrived from 12 June to 18 June 2014 (see Figure 1b). The AVHRR SST L2p data have been provided by the group for high resolution sea surface temperature (GHRSST) (ftp://data.nodc.noaa.gov/pub/data.nodc/ghrsst/) and have been widely used in SST studies [4]. The in-situ SSTs are from the in-situ SST quality monitor (iQuam) system by using the highest quality flag value of 5 that is designed for high-accuracy applications (http://www.star.nesdis.noaa.gov/sod/sst/iquam/index.html) [25]. These SST samples combining the AVHRR L2p data and in-situ SST data have been averaged onto a 0.25° grid. To evaluate the SST errors only from RBFNs, the OISST product was used as a reference, and non-biased SST samples were selected from the OISST products according to the distributions of the SST samples. The OISST products are the level-4 data in GHRSST with a grid resolution of 0.25°, where in-situ data have been combined with AVHRR data through the analyses system of National Oceanic and Atmospheric Administration (NOAA)'s national climatic data center (NCDC) [6]. Thus, the distributions of SST samples are consistent with the AVHRR L2p data and in-situ SST data, while the values of SST samples were assigned by the OISST products. Accordingly, the global SST samples from 13 June 2014 are also used to test the performance of RBFN method on the global SST model. In addition, the reconstructed SSTs from this RBFN method are validated with the reference SSTs in the study.

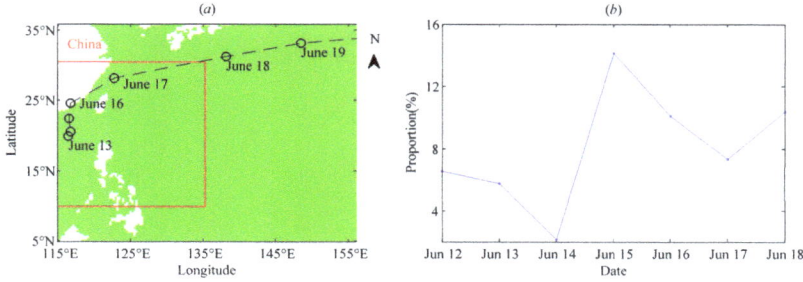

Figure 1. (**a**) The study area (10°N–30°N, 115°E–135°E) in red rectangle region and the track of the tropical storm Hagibis labeled by day. The white and green areas in (**a**) show land and ocean, respectively; (**b**) The coverage of sea surface temperature (SST) samples in the study region during the period from 12 June to 18 June 2014.

3. Methodology

3.1. RBFN Method

The daily SST can be considered as two parts: the background field and the SST increment [6,26]. The previous SST field is used as the background field. In particular, the values of climatology change are added to the background field. The reconstructed SST T_t can be defined as

$$
\begin{cases}
T_t = Y_t + T_t^b \\
T_t^b = T_{t-1} + (T_t^c - T_{t-1}^c)
\end{cases}
\tag{1}
$$

where Y_t is the SST increment, T_t^b is the background field, and T_{t-1} is the reconstructed SST for the previous time. For convenience, we use the OISST product as the a priori SST, and to minimize the effects of this, the procedure was run for 10 days in advance of the first date. The daily SST climatology, T_t^c, was obtained from 30 years (1982–2011) of OISST data [27]. $(T_t^c - T_{t-1}^c)$ is the SST climatology that changes between time t and $t - 1$. The SST increment Y_t is calculated from the RBFN proposed by Ryu et al. [24]. It should be noted that this RBFN is used to estimate SST increments, not SSTs as in Ryu's et al. study. The RBFN can be given by

$$
\begin{cases}
Y_t = f(X_t) + \varepsilon_t \\
f(X_t) = \beta_{t,0} + X_t^d \beta_{t,m}^T + \sum_{k=1}^{K_t} \phi(z_{t,k}) \beta_{t,k+5}
\end{cases}
\tag{2}
$$

where $f(X_t)$ is a RBFN function, $X_t = (X_{t1}, X_{t2})$ is a 2-dimensional matrix, and X_{t1}, X_{t2} are the longitude and latitude at time t, respectively. X_t^d denotes the set of all possible power with d degree. Here, $d = 2$ and the 2-degree polynomial can be expressed as $X_t^d = (1, X_{t1}, X_{t2}, X_{t1} X_{t2}, X_{t1}^2, X_{t2}^2)$. $z_{t,k} = \|X_t - \mu_{t,k}\|$ denotes the distance between X_t and hidden knots $u_{t,k}$ for $k = 1, ..., K_t$, and $\phi(z) = z$. $\beta_{t,0}, \beta_{t,m}, m = 1, ..., 5$; and $\beta_{t,k+5}$ denote regression coefficients for the intercept, polynomials, and basis functions, respectively. ε_t is an independent Gaussian random error with mean 0 and variance $\sigma_{y_t}^2$.

Using this RBFN method, the number of hidden knots K_t, the position, and error variance are unknown, and are strongly related to the precision of the reconstructed SST. The error variance is estimated by the squared deviation of the difference between $f(X_t)$ and SST increments Y_t. Since the number of hidden knots and the position can be affected by the distribution of SST samples, the INNC algorithm was designed to determine the hidden knots in the SST fields. Figure 2 shows the schematic diagram of this RBFN method for reconstructing SST.

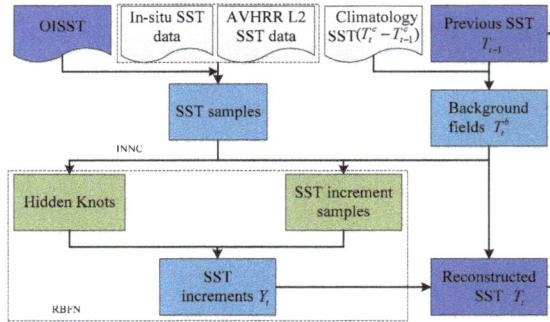

Figure 2. Schematic diagram of the RBFN method for reconstructing SST.

3.2. INNC Algorithm

The nearest neighbor cluster (NNC) algorithm is used to search the optimal hidden knots only from the SST samples for RBFNs [28,29], while the INNC algorithm can choose the hidden knots from both SST samples and background field, and consider the values of SST samples. The INNC algorithm is described as follows.

(1) Standardizing the original SST data, and make sure each variable of (x, y, z) in the SST matrix, with the mean of 0 and standard deviation of 1, where z is the value of the SST at the position of (x, y) in the SST matrix.

(2) Define a minimal distance D and set the first SST sample (x_1, y_1, z_1) as the first center c_1.

(3) For the second SST sample (x_2, y_2, z_2), the Euclidean distance s to the center c_1 is calculated. If $s > D$, then the position (x_2, y_2, z_2) is the next center c_2, otherwise the algorithm searches for the next SST sample (x_3, y_3, z_3).

(4) For the i-th SST sample (x_i, y_i, z_i), the Euclidean distance $\{s_k\}$ to each center $\{c_k\}$ is calculated, $k = 1, \ldots, K$. K is the number of center. If the minimal distance $s_m > D$, then the position (x_i, y_i, z_i) is the next center c_{K+1}, otherwise the algorithm searches for the next SST sample, until the last one is found.

(5) The values from the background field T_t^b are used to fill the positions without SST samples, before repeating step (3) for each position to select the centers from the background field. This continues until all of the positions are processed in the SST matrix, and the hidden knots u_t are obtained by using the positions of centers $\{c_k\}$ in the SST matrix.

In terms of the INNC algorithm, the parameters K_t and u_t are determined when the minimal distance D is defined. The value of D ranges from 0.2 to 1.5 in this study, and the optimal D is selected with the minimal error variance $\sigma_{y_t}^2$ to obtain the optimal K_t and u_t for RBFN. Thus, the optimal K_t and u_t may vary with different quantities and distributions of SST samples. In addition, the SST values at the positions of the optimal hidden knots from the background fields are added to the SST samples, so that there is at least one SST sample in the SST field within the minimal distance of D.

3.3. Evaluating the Performance of the RBFN Method

As described above, a simplified multi-quadric function $\sqrt{z^2 + s_1^2}$ (s_1 equal 0 here) is chosen as a basis function, and the INNC algorithm is designed to search for the optimal hidden knots for RBFNs. In order to evaluate the performance of this RBFN method, a Gaussian function $\exp(-s_2 z^2)$ is selected for comparison and $s_2 = K_t / d_{\max}$ [30], because $d_{\max}$ is the maximum distance between the hidden knots, thus $d_{\max} = D$ in the study. K-means algorithm and Kohonen-map algorithm are also tested for selecting hidden knots for RBFNs. While the number of hidden knots K_t should be assigned for these two clustering algorithms, so the value of K_t ranges from 20 to 120 for K-means algorithm, and the Kohonen-maps with regular array of $m \times m$ hidden knots are tested and the value of m ranges from 5 to 15, then the optimal values of K_t and m are separately selected with the minimal error variance $\sigma_{y_t}^2$. More Details about K-means algorithm and Kohonen-map algorithm are described in [31–33].

Additionally, the OI method is used to compare with this RBFN method. Since the values of SST samples are assigned by the reference SST in the study, we select the noise-to-signal standard deviation ratio of AVHRR data for SST samples (the value is 0.5) and the average correlation scales for OI method (zonal and meridional correlation scales are 151 km and 155 km, respectively). Details of the OI method are described by Reynolds et al. [6].

To validate the accuracy of SST, the reconstructed SSTs are compared with the reference SSTs, and four commonly used error metrics are calculated: root mean square error (RMSE), mean absolute error (MAE), Pearson correlation coefficient (R), and signal-to-noise ratio (SNR). The SNR is the ratio of the standard deviation of the SST results to the standard deviation of the errors [34].

4. Results

In this section, the RBFN methods with different basis functions and clustering methods are tested, and the performance of the RBFN method for reconstructing daily SSTs is evaluated by comparing with the OI method.

4.1. Results from Different Basis Functions

Using the INNC algorithm, the optimal hidden knots can be chosen from SST samples and background fields. But the distributions of SST samples are very sparse in this study, the reconstructed SSTs from the RBFN method may be influenced by the type of the basis function.

Figures 3 and 4 display the SST increments that are estimated by using Equation (2) with Gaussian function and multi-quadric function on 13 June and 18 June, respectively. It is clear that the SST increments obtained from Gaussian function have a lot of anomaly circular regions, as shown in Figures 3a and 4a, where the SST increments are significantly different to the values from the regions behind them. While the SST increment fields using multi-quadric function in Figures 3b and 4b are quite close to their neighborhood regions. Besides, due to the characteristics of the Gaussian function, the effective areas of the hidden knots are limited, so the SST increments are strongly affected by the hidden knots that are selected from the background fields when there are few SST samples nearby, which may cause large errors on SST increments.

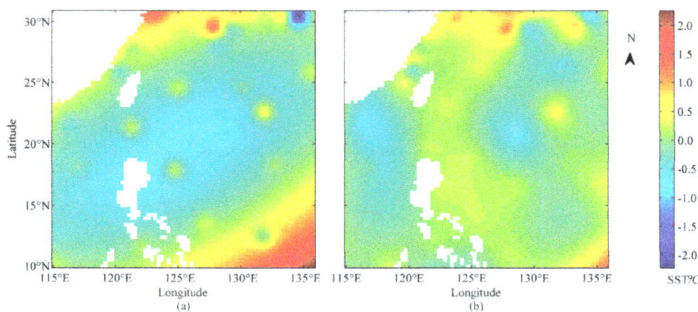

Figure 3. Reconstructed SST increments on 13 June 2014 obtained by using (**a**) Gaussian function and (**b**) multi-quadric function. White areas are lands in the images.

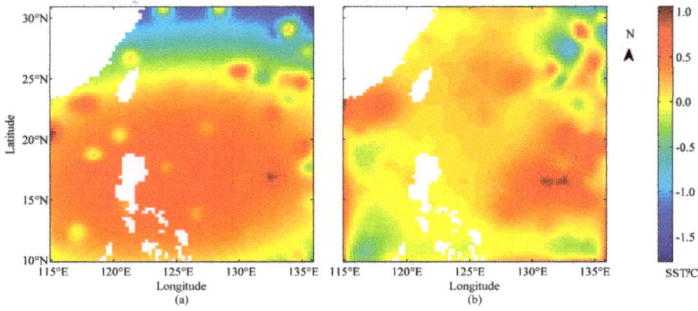

Figure 4. Reconstructed SST increments on 18 June 2014 obtained by using (**a**) Gaussian function and (**b**) multi-quadric function. White areas are lands in the images.

The statistical results from Tables 1 and 2 demonstrate that the multi-quadric function is significantly better than the Gaussian function as the basis function for the RBFN method, especially from the results on 18 June, the RMSE from the multi-quadric function is 0.47 °C, which is quite smaller than the value of 0.73 °C from the Gaussian function.

Table 1. Statistical results from different basis functions and clustering methods on 13 June 2014. The subscripts *G* and *M* stand for Gaussian function and multi-quadric function, respectively.

	RMSE (°C)	MAE (°C)	R	SNR
$INNC_G$	0.66	0.96	0.49	3.43
$INNC_M$	0.51	0.98	0.37	4.61
$K-means_M$	0.66	0.97	0.49	4.02
$Kohonen-map_M$	0.55	0.97	0.43	4.18

Table 2. Statistical results from different basis functions and clustering methods on 18 June 2014. The subscripts *G* and *M* stand for Gaussian function and multi-quadric function, respectively.

	RMSE (°C)	MAE (°C)	R	SNR
$INNC_G$	0.73	0.97	0.52	3.60
$INNC_M$	0.47	0.98	0.34	4.64
$K-means_M$	0.59	0.97	0.43	4.06
$Kohonen-map_M$	0.69	0.96	0.51	3.39

4.2. Results from Different Clustering Algorithms

A RBFN with too many or too few hidden knots is not conductive to simulating SST fields. Additionally, the distribution of hidden knots has a strong influence on the construction of a RBFN. Therefore, the quality of reconstructed SSTs from the RBFN method is highly related to the quantity and distribution of hidden knots, and it is important to select the suitable hidden knots for RBFNs.

Three clustering algorithms, including K-means algorithm, Kohonen-map algorithm, and INNC algorithm are compared to selecting the optimal hidden knots for RBFNs on 13 June and 18 June, respectively. Since the RBFN is quite sensitive to the hidden knots, the values of hidden knots are very important to the reconstructed SSTs. As a result, if a hidden knot is obtained from the background field, the value of which is close to 0 in the SST increment field in Equation (2), and the SST accuracy will decrease in this situation. Thus, it is conductive to select the hidden knots from the SST samples rather than from the background field.

Figures 5 and 6 show the distributions of hidden knots from these three algorithms. It is easy to discover that there are more hidden knots that are selected from the SST samples (green points) by the INNC algorithm than those by the other two algorithms, because the clustering centers of the K-means algorithm and the Kohonen-map algorithm are chosen by training the input samples, while the INNC algorithm directly selects the hidden knots from the SST samples and make sure that as many hidden knots as possible are obtained from SST samples within the minimal distance D. As the statistical results shown in Tables 1 and 2, the INNC algorithm in this study with the highest accuracy of reconstructed SSTs, is more suitable than the K-means algorithm and the Kohonen-map algorithm to select the optimal hidden knots for the RBFN method.

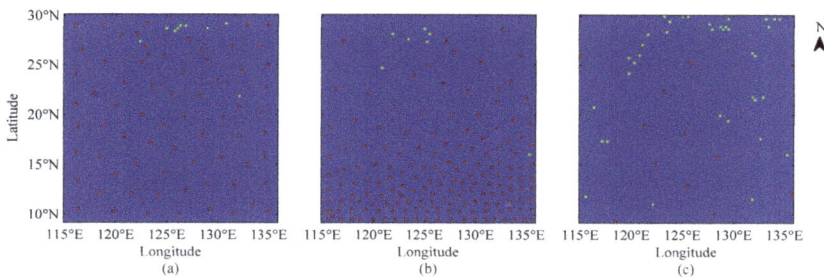

Figure 5. The distributions of hidden knots for radial basis function network (RBFN) on 13 June 2014 from (**a**) K-means algorithm; (**b**) Kohonen-map algorithm and (**c**) improved nearest neighbor cluster (INNC) algorithm respectively. The red points and green points are the locations of hidden knots for RBFNs separately selected from the background field and the SST samples.

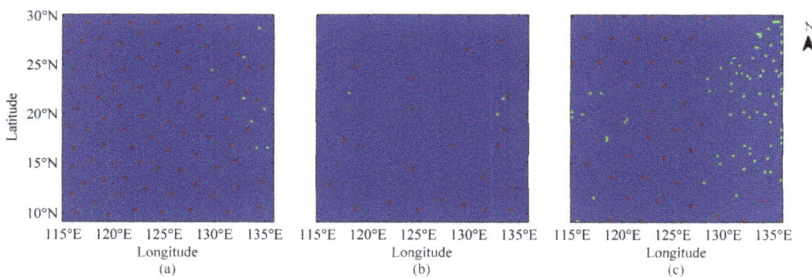

Figure 6. The distributions of hidden knots for RBFN on 18 June 2014 from (**a**) K-means algorithm; (**b**) Kohonen-map algorithm and (**c**) INNC algorithm respectively. The red points and green points are the locations of hidden knots for RBFNs separately selected from the background field and the SST samples.

4.3. Comparison with the OI Method

Due to the limitation of the correlation scales for the OI method, the SST increments will close to 0 when there are no SST samples behind them, while the SST increments from the RBFN method are obtained based on the tendency of the whole SST increment field. This is the key difference between the OI method and the RBFN method for reconstructing SST.

Figure 7 displays the time-series of each error metric from the OI and RBFN methods during the period from 12 June to 18 June 2014. The performance of each error metric from the RBFN method is much better than that from the OI method. The average values of these error metrics in Table 3 indicate that the RBFN method increases R and SNR from 0.96 and 3.82 to 0.98 and 4.94, respectively, and decreases RMSE and MAE from 0.69 °C and 0.46 °C to 0.48 °C and 0.35 °C, respectively. Although the RBFN method requires more computation time than the OI method, due to the selection of optimal hidden knots by using the INNC algorithm, the RBFN method has a better accuracy of reconstruction than the OI method.

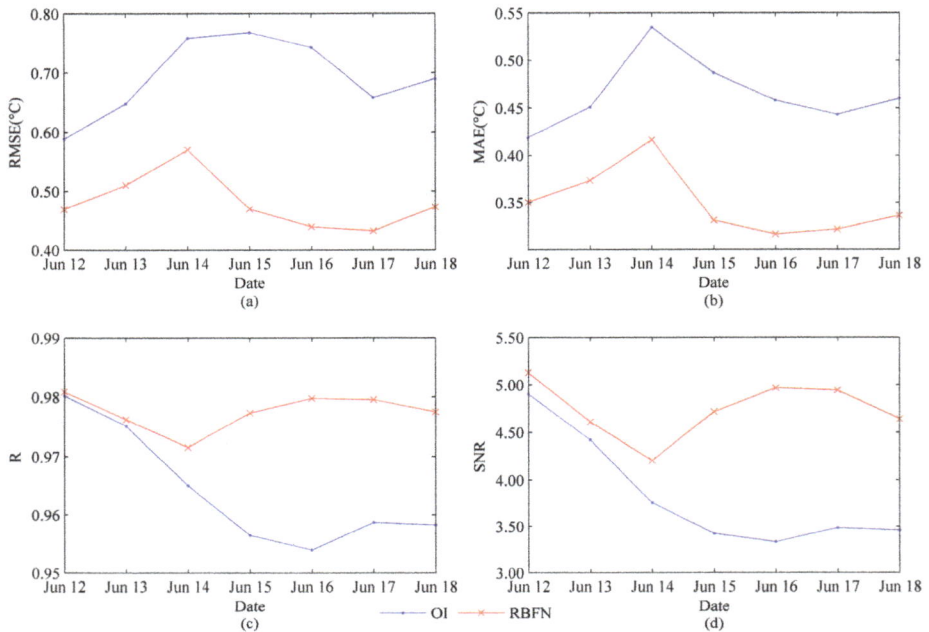

Figure 7. (a) root mean square error (RMSEs); **(b)** mean absolute error (MAEs); **(c)** Pearson correlation coefficient (Rs) and **(d)** signal-to-noise ratio (SNRs) of the reconstructed SSTs from the optimum interpolation (OI) and RBFN methods in the study region during the period from 12 June to 18 June 2014.

Table 3. Statistical results from the OI and RBFN methods.

	RMSE (°C)	MAE (°C)	R	SNR
OI	0.69	0.46	0.96	3.82
RBFN	0.48	0.35	0.98	4.94

In Figures 8 and 9, the reconstructed SSTs from 13 June and 18 June are separately displayed as examples. The results show that the reconstructed SSTs from the OI method (shown in Figures 8b and 9b) and the RBFN method (shown in Figures 8d and 9d) are quite similar to their reference SSTs (see Figures 8c and 9c). But, the SSTs in the region outlined by the square from the OI method have a significantly high SST anomaly, where these phenomenon are not obvious in the SSTs from the RBFN method and their reference data. This indicates that the RBFN method has a relatively better performance than the OI method to reconstruct SSTs when the original SST samples is quite sparse, as shown in Figures 8a and 9a.

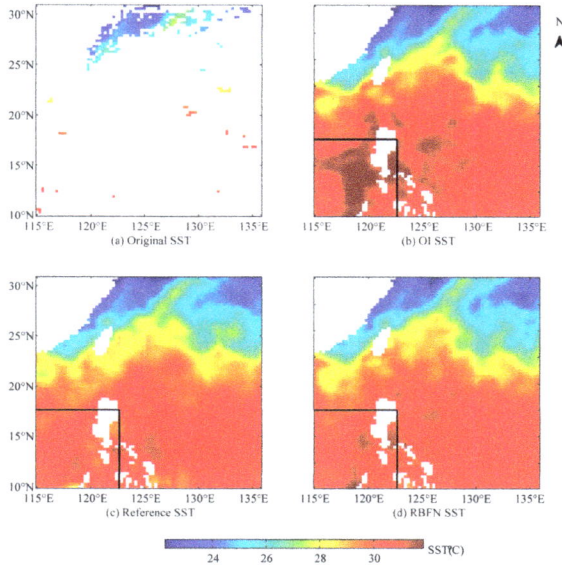

Figure 8. The distributions of (**a**) original SST samples; (**b**) OI SST; (**c**) reference SST; and, (**d**) RBFN SST on 13 June 2014. The region outlined by the square is used for comparison in detail. White areas in the images indicate no data or land.

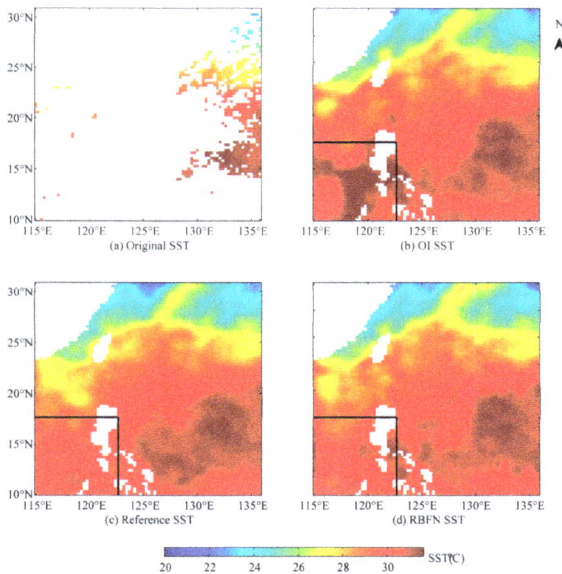

Figure 9. The distributions of (**a**) original SST samples; (**b**) OI SST; (**c**) reference SST; and, (**d**) RBFN SST on 18 June 2014. The region outlined by the square is used for comparison in detail. White areas in the images indicate no data or land.

However, the global distribution of original SST samples is much more complex than the local regions. As shown in Figure 10, the distribution of them is very irregular on 13 June 2014 in the global ocean, and the coverage of these SST data is not complete, especially in the equatorial regions, where the data is relatively limited. Therefore, in order to test the performance of RBFN method, we run the RBFN using the global SST samples and compare the results with the OI method. It should be noted that both the OI method and the RBFN method in this study choose the previous OISST product as the background field, and the global SST field from the RBFN method are combined by each 20° × 20° boxes of reconstructed SST samples that using the same scheme as the local experiment in Section 3.

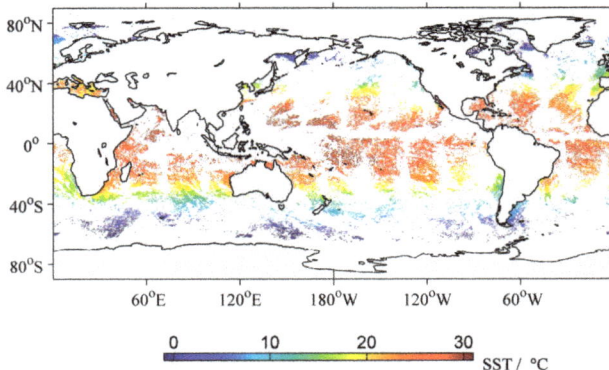

Figure 10. The global distribution of original SST samples on 13 June 2014. White areas in the images indicate no data or land, and the coastlines are highlighted with back contours.

Figure 11 displays the global distributions of SST biases separately from the OI method and the RBFN method on 13 June 2014 according to the corresponding reference SST. The results show that the variations of SST biases from these two methods are quite similar, but the SST biases from the RBFN method are relatively greater than those from the OI method when the SST samples in Figure 10 are not sparsely distributed. It indicates that this RBFN method might not be as good as the OI method when the coverage of SST samples is relatively high. Additionally, when compared with the global statistical results from the OI method in Table 4, the values of RMSE and MAD from the RBFN method are slightly larger, but its SNR value is relatively smaller. Thus the performance of RBFN method in the study is not very stable with different distributions of SST samples, and it still needs to be improved when applied in the global region.

Table 4. Global Statistical results from the OI and RBFN methods on 13 June 2014.

	RMSE (°C)	MAE (°C)	R	SNR
OI	0.18	0.08	0.99	65.38
RBFN	0.19	0.10	0.99	62.21

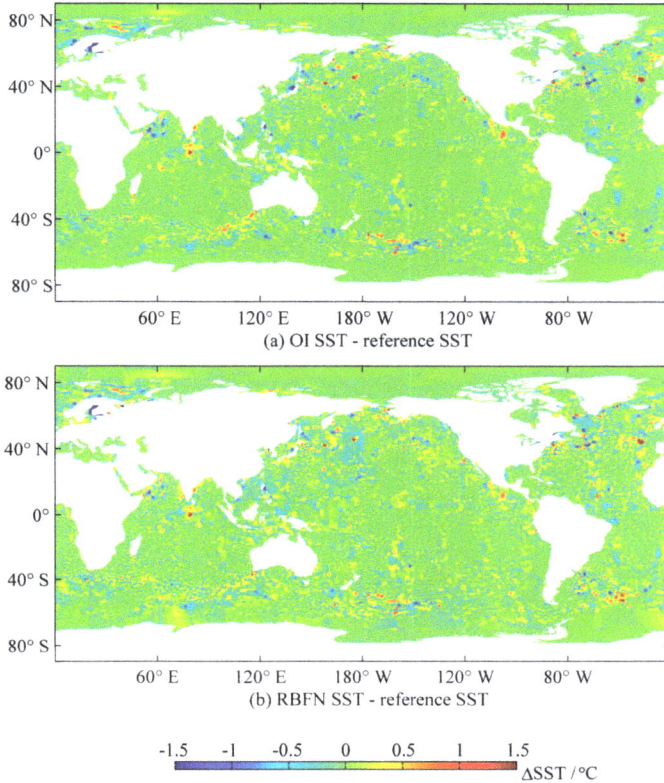

Figure 11. The global distributions of (**a**) OI SST—reference SST and (**b**) RBFN SST—reference SST on 13 June 2014. White areas in the images indicate no data or land.

5. Discussion

5.1. The INCC Algorithm for RBFNs

The hidden knots for RBFNs are commonly selected by clustering algorithms [28,29,35] or random sampling [24]. Since the hidden knots with different quantities and distributions should be tested in RBFNs, huge computational resources are required when random sampling is used to select the optimal hidden knots from SST fields. Commonly used clustering algorithms for RBFNs, such as K-means algorithm and Kohonen-map algorithm, are only used to determine the distribution of hidden knots when the number of hidden knots is given [32,35]. The K-means algorithm should be executed hundreds of times, which also takes a lot of time even though the algorithm has a fast convergence speed. For the Kohonen-map algorithm, an input vector is not only related to the nearest hidden knot, but also to its neighbor hidden knots, so the computation is also huge and the training step for the clustering centers consume a lot of time before the iteration ends. While using the INNC algorithm with a parameter D, both the quantity and the distribution of hidden knots could be directly determined without any iterative computation, and the values of D in this study were chosen from 0.2 to 1.5. Thus, the INNC algorithm is more efficient in selecting the optimal hidden knots for RBFNs.

Additionally, the INNC algorithm select the hidden knots from SST samples and background fields within the minimal distance D, and then the values of hidden knots from the background fields are added to the SST samples, which makes sure that the distributions of the hidden knots and the SST samples are close to uniform in the SST field. While the multi-quadric basis function for RBFNs considers large related domains in SST field, so the values from both the SST samples and the backgrounds fields will contribute to the reconstructed SSTs when there are few SST samples nearby.

5.2. SST Samples

The coverage of SST samples that we selected in the Pacific Ocean is very low because the tropical storm Hagibis come with heavy clouds, which makes poor quality of satellite-retrieved SSTs, and the SSTs we can used are very limited. Theoretically, the errors of reconstructed SSTs are from both background fields and SST increments. The SST increments are estimated by using current SST samples, and the quality of background fields is strongly influenced by the previous SST samples, thus the distributions of SST samples at both the previous time and the current time play a significant role in reconstructing SST. It can well explain the fact that the minimal RMSE did not happen on 15 June in Figure 7a, which has the highest coverage of SST samples, as shown in Figure 1b. Because the 14 June with the lowest SST coverage in the study has a low accuracy of reconstructed SST, and make a poor quality of the background field for 15 June, so the accuracy of reconstructed SST on 15 June is not the best even though it has the highest coverage of SST samples.

The quality of SST samples is not only relevant to the coverage of SST samples, but it also depends on their accuracy. Since the SST sample values that were used were assigned by the reference SSTs, which can be considered as non-biased SST data, the reconstructed SSTs are of relatively high quality. However, the satellite-retrieved SST is often contaminated by the atmosphere and the SST data from different sources may contain errors from their respective surveying systems, meaning that it is not easy to obtain non-biased SST data, and the accuracy will decrease if large errors exist in the SST samples.

The SST samples with high coverage are conductive to acquiring high-quality daily SSTs. While acquiring larger quantities of SST samples often needs the combination of different sources of SSTs, more SST errors may come from these different kinds of sensors, and this has a negative influence on the quality of the SST samples overall. For the purpose of obtaining high-accuracy SST data from the RBFN method, more SST data should be used in reconstructing the SST. Additionally, strict quality control is required for SSTs from various measurements, and bias corrections for satellite-retrieved SSTs are necessary to eliminate the errors in SST samples.

5.3. The Performance of the RBFN Method

Since the OI method estimates SSTs using only the information within a limited distance, the estimated SSTs in some areas rely on background fields, which create large errors in the SST fields. The RBFNs with enough hidden knots use all the SST samples to approximate the SST field, so the SSTs from the RBFN method contain more information on the whole SST field, not only the information about SST samples nearby. Thus, the accuracy of SST that is obtained using the RBFN method is better than that using the OI method when the distribution of SST samples is quite sparse.

As shown in Figure 7a, the RMSE on 15 June from the RBFN method is significantly smaller than that from the OI method when the quality of the background field is quite poor. This shows that when compared with the SSTs from the OI method, the quality of SSTs from the RBFN method is more stable when the SST errors are associated with the background field.

On the other hand, the OI method is effective in estimating SST using neighboring samples, which was conductive to obtaining a higher accuracy of SSTs when the coverage of the SST samples is very high. While the quantity and distribution of hidden knots in RBFNs are determined by the minimal error variance, which is a global variable of the SST field, and the distribution of hidden knots may not be suitable in some local regions, and the SST errors of RBFNs may occur even in some areas with large amount of SST samples nearby. This is may be the reason why the RBFN method is not as effective as the OI method in the global region. Because both the OI method and the RBFN method is designed to minimize the errors of the SST field, and if the SST samples is sparsely distributed,

Remote Sens. **2017**, *9*, 1204

the distribution of expected errors is relatively uniform in the SST field. In this case, a global parameter of RBFN is suitable for the whole SST field. But when the coverage of SST samples is high, the expected errors of RBFN in the regions with SST samples is much smaller than those without SST samples nearby, then the expected errors are distributed unevenly in the SST field, and a global parameter may not adapt for the RBFN to simulate the high quality of SST, which may make the errors much larger than those of the OI method, as shown in Figure 11.

In addition, although the INNC algorithm with D is very efficient in searching for hidden knots in SST fields, the optimal D of the INNC algorithm should be selected from the values between 0.2 and 1.2, which makes the RBFN method more computationally expensive than the OI method. Overall, the OI and RBFN methods have their own characteristics in terms of reconstructing SST, but the RBFN method is more effective than the OI method when the SST samples are sparsely distributed.

6. Conclusions

Reconstructing SST fields from a limited number of SST samples is important for data applications. In this study, a RBFN method is proposed and a INNC algorithm is designed to search for the optimal hidden knots for RBFNs. When compared with the other clustering algorithms as K-means algorithm and Kohonen-map algorithm, the INNC algorithm can obtain more hidden knots from the SST samples, which will contribute to acquiring a high quality of reconstructed SST. Using the multi-quadric function as the basis function is more effective than the Gaussian function to avoid the SST increment anomaly on the regions of hidden knots. Thus, the INNC algorithm with multi-quadric function is quite suitable for the RBFN method to reconstruct SSTs in the study.

To evaluate the accuracy of this RBFN method, SST samples with low coverage are reconstructed by using the RBFN and OI methods, respectively. The results show that when compared with the SSTs from the OI method, the quality of SST from the RBFN method is more stable when the SST errors are associated with the background field. Moreover, the RBFN method is less affected by missing values in SST samples, and the SSTs from the RBFN method have a higher accuracy than those from the OI method when SST samples are sparsely distributed. According to this characteristic of the RBFN method, it is quite suitable to be used for SST reconstruction that only combine the in situ data and the satellite retrieved SSTs from infrared radiometers.

When considering the efficiency of RBFNs, hidden knots with different quantities and distributions are determined simultaneously using various D values in the INCC algorithm, and the optimal D for RBFN is selected with the minimal error variance. The step-length of D is set as 0.02 in this study. We believe that if the step-length of D can be set smaller, then the accuracy may be improved.

However, the accuracy of reconstructed SSTs is strongly influenced by the quality of SST samples from both the present and current time, and the biases in SST samples will increase the errors in SSTs, thus the strict quality control and bias corrections to SST samples are required in practice. Though the RBFN method has a better performance than the OI method to reconstruct SST field with the limited SST samples, the advantage is not obvious when the coverage of SST samples is high (the details are described in the Supplemental Material), so this RBFN method is appropriate to be used in some local regions, not to the global region at the present stage, and we will improve this RBFN method in our future work.

Acknowledgments: The authors thank NESDIS/STAR, NOAA/NCDC and NSMC of China for providing the in-situ and SST products in the study. This work was supported by the National key research and development programme of China (2017YFA0603003), the National Natural Science Foundation of China (No. 41476154), the National key research and development programme of China (No. 2016YFA0600304) and the National Natural Science Foundation of China (No. 41371385 and No. 41671401).

Author Contributions: Zhihong Liao proposed the idea and the algorithm, wrote the introduction and the algorithm sections, and integrated the contributions from all authors as a whole. Qing Dong and Cunjin Xue designed the experiments and the structure of this manuscript, and collaborated with Zhihong Liao in writing the discussion and conclusion sections. Jingwu Bi and Guangtong Wan contributed the experiment, and collaborated with Zhihong Liao in writing the results section.

Conflicts of Interest: The authors declare no conflict of interest.

References

1. Saha, S.; Moorthi, S.; Wu, X.; Wang, J.; Nadiga, S.; Tripp, P.; Behringer, D.; Hou, Y.T.; Chuang, H.Y.; Iredell, M.; et al. The NCEP climate forecast system version 2. *J. Clim.* **2014**, *27*, 2185–2208. [CrossRef]
2. Shirvani, A.; Nazemosadat, S.; Kahya, E. Analyses of the Persian Gulf sea surface temperature: Prediction and detection of climate change signals. *Arabian J. Geosci.* **2015**, *8*, 2121–2130. [CrossRef]
3. Knutson, T.R.; McBride, J.L.; Chan, J.; Emanuel, K.; Holland, G.; Landsea, C.; Held, I.; Kossin, J.P.; Srivastava, A.K.; Sugi, M. Tropical cyclones and climate change. *Nat. Geosci.* **2010**, *3*, 157–163. [CrossRef]
4. Donlon, C.; Rayner, N.; Robinson, I.; Poulter, D.J.S.; Casey, K.S.; Vazquez-Cuervo, J.; Armstrong, E.; Bingham, A.; Arino, O.; Gentemann, C.; et al. The global ocean data assimilation experiment high-resolution sea surface temperature pilot project. *Bull. Am. Meteorol. Soc.* **2007**, *88*, 1197–1213. [CrossRef]
5. Martin, M.; Dash, P.; Ignatov, A.; Banzon, V.; Beggs, H.; Brasnett, B.; Cayula, J.F.; Cummings, J.; Donlon, C.; Gentemann, C.; et al. Group for High Resolution Sea Surface temperature (GHRSST) analysis fields inter-comparisons. Part 1: A GHRSST multi-product ensemble (GMPE). *Deep Sea Res. Part II* **2012**, *77*, 21–30. [CrossRef]
6. Reynolds, R.W.; Smith, T.M.; Liu, C.; Chelton, D.B.; Casey, K.S.; Schlax, M.G. Daily high-resolution-blended analyses for sea surface temperature. *J. Clim.* **2007**, *20*, 5473–5496. [CrossRef]
7. Reynolds, R.W. A real-time global sea surface temperature analysis. *J. Clim.* **1988**, *1*, 75–87. [CrossRef]
8. Smith, T.M.; Reynolds, R.W.; Livezey, R.E.; Stokes, D.C. Reconstruction of historical sea surface temperatures using empirical orthogonal functions. *J. Clim.* **1996**, *9*, 1403–1420. [CrossRef]
9. Smith, T.M.; Livezey, R.E.; Shen, S.S. An improved method for analyzing sparse and irregularly distributed SST data on a regular grid: The tropical Pacific Ocean. *J. Clim.* **1998**, *11*, 1717–1729. [CrossRef]
10. Ping, B.; Su, F.; Meng, Y. Reconstruction of satellite-derived sea surface temperature data based on an improved DINEOF algorithm. *IEEE J. Sel. Top. Appl. Earth Obs. Remote Sens.* **2015**, *8*, 4181–4188. [CrossRef]
11. Beckers, J.-M.; Rixen, M. EOF calculations and data filling from incomplete oceanographic datasets. *J. Atmos. Ocean. Technol.* **2003**, *20*, 1839–1856. [CrossRef]
12. Beggs, H.; Zhong, A.; Warren, G.; Alves, O.; Brassington, G.; Pugh, T. RAMSSA—An operational, high-resolution, Regional Australian Multi-Sensor Sea surface temperature Analysis over the Australian region. *Aust. Meteorol. Oceanogr. J.* **2011**, *61*, 1. [CrossRef]
13. Brasnett, B. The impact of satellite retrievals in a global sea-surface-temperature analysis. *Q. J. R. Meteorol. Soc.* **2008**, *134*, 1745–1760. [CrossRef]
14. Cummings, J.A. Operational multivariate ocean data assimilation. *Q. J. R. Meteorol. Soc.* **2005**, *131*, 3583–3604. [CrossRef]
15. Donlon, C.J.; Martin, M.; Stark, J.; Roberts-Jones, J.; Fiedler, E.; Wimmer, W. The operational sea surface temperature and sea ice analysis (OSTIA) system. *Remote Sens. Environ.* **2012**, *116*, 140–158. [CrossRef]
16. Gemmill, W.; Katz, B.; Li, X. *Daily Real-Time Global Sea Surface Temperature—High-Resolution Analysis: RTG_SST_HR*; NCEP No. 260; EMC Office Note: Maryland City, MD, USA, 2007.
17. Gentemann, C.L.; Wentz, F.J.; DeMaria, M. Near real time global optimum interpolated microwave SSTs: Applications to hurricane intensity forecasting. In Proceedings of the 27th Conference on Hurricanes and Tropical Meteorology, Monterey, CA, USA, 23–28 April 2006.
18. Park, J.; Sandberg, I.W. Universal approximation using radial-basis-function networks. *Neural Comput.* **1991**, *3*, 246–257. [CrossRef]
19. Holmes, C.; Mallick, B. Bayesian radial basis functions of variable dimension. *Neural Comput.* **1998**, *10*, 1217–1233. [CrossRef]
20. Powell, J. Radial basis function approximations to polynomials. In *Numerical Analysis 1987*; Longman Publishing Group: White Plains, NY, USA, 1987; pp. 223–241.
21. Liu, L.; Chua, L.; Ghista, D. Mesh-free radial basis function method for static, free vibration and buckling analysis of shear deformable composite laminates. *Compos. Struct.* **2007**, *78*, 58–69. [CrossRef]
22. Miazhynskaia, T.; Frühwirth-Schnatter, S.; Dorffner, G. Neural network models for conditional distribution under bayesian analysis. *Neural Comput.* **2008**, *20*, 504–522. [CrossRef] [PubMed]

23. Konishi, S.; Ando, T.; Imoto, S. Bayesian information criteria and smoothing parameter selection in radial basis function networks. *Biometrika* **2004**, *91*, 27–43. [CrossRef]

24. Ryu, D.; Liang, F.; Mallick, B.K. Sea surface temperature modeling using radial basis function networks with a dynamically weighted particle filter. *J. Am. Stat. Assoc.* **2013**, *108*, 111–123. [CrossRef]

25. Xu, F.; Ignatov, A. In Situ SST quality monitor (i Quam). *J. Atmos. Ocean. Technol.* **2014**, *31*, 164–180. [CrossRef]

26. Reynolds, R.W.; Smith, T.M. Improved global sea surface temperature analyses using optimum interpolation. *J. Clim.* **1994**, *7*, 929–948. [CrossRef]

27. Banzon, V.F.; Reynolds, R.W.; Stokes, D.; Xue, Y. A 1/4°-spatial-resolution daily sea surface temperature climatology based on a blended satellite and In Situ analysis. *J. Clim.* **2014**, *27*, 8221–8228. [CrossRef]

28. Diao, X.H. Study on regional financial risk early warning system based on uniform design method and Nearest Neighbo-Clustering RBFNN. In Proceedings of the 2011 IEEE International Conference on Intelligent Computing and Integrated Systems (ICISS), Guilin, China, 24–26 October 2011.

29. Zhu, M.; Zhang, D. Study on the algorithms of selecting the radial basis function center. *J. Anhui Univ.* **2000**, *3*, 72–78.

30. Lowe, D. Adaptive radial basis function nonlinearities, and the problem of generalisation. In Proceedings of the 1989 First IEE International Conference on Artificial Neural Networks, London, UK, 16–18 October 1989. No. 313.

31. Kanungo, T.; Mount, D.M.; Netanyahu, N.S.; Piatko, C.D.; Silverman, R.; Wu, A.Y. An efficient k-means clustering algorithm: Analysis and implementation. *IEEE Trans. Pattern Anal. Mach. Intell.* **2002**, *24*, 881–892. [CrossRef]

32. Latif, B.A.; Lecerf, R.; Mercier, G.; Hubert-Moy, L. Preprocessing of low-resolution time series contaminated by clouds and shadows. *IEEE Trans. Geosci. Remote Sens.* **2008**, *46*, 2083–2096. [CrossRef]

33. Kohonen, T. The self-organizing map. *Proc. IEEE* **1990**, *78*, 1464–1480. [CrossRef]

34. Zhang, H.; Qin, S.; Ma, J.; You, H. Using residual resampling and sensitivity analysis to improve particle filter data assimilation accuracy. *IEEE Geosci. Remote Sens. Lett.* **2013**, *10*, 1404–1408. [CrossRef]

35. Haykin, S. *Neural Networks and Learning Machines*, 3rd ed.; McMaster University: Hamilton, ON, Canada, 2009.

remote sensing

MDPI

Article

Submesoscale Sea Surface Temperature Variability from UAV and Satellite Measurements

Sandra L. Castro [1,*], William J. Emery [1], Gary A. Wick [2] and William Tandy Jr. [3]

[1]   Colorado Center for Astrodynamics Research, University of Colorado, 431 UCB, Boulder, CO 80309, USA; emery@colorado.edu

[2]   NOAA Earth System Research Laboratory, Physical Sciences Division, R/PSD2 325 Broadway, Boulder, CO 80305, USA; gary.a.wick@noaa.gov

[3]   Ball Aerospace, 1600 Commerce St., Boulder, CO 80301, USA; wtandy@ball.com

*   Correspondence: sandrac@colorado.edu; Tel.: +1-303-492-1241

Received: 24 September 2017; Accepted: 23 October 2017; Published: 25 October 2017

Abstract: Earlier studies of spatial variability in sea surface temperature (SST) using ship-based radiometric data suggested that variability at scales smaller than 1 km is significant and affects the perceived uncertainty of satellite-derived SSTs. Here, we compare data from the Ball Experimental Sea Surface Temperature (BESST) thermal infrared radiometer flown over the Arctic Ocean against coincident Moderate Resolution Imaging Spectroradiometer (MODIS) measurements to assess the spatial variability of skin SSTs within 1-km pixels. By taking the standard deviation, σ, of the BESST measurements within individual MODIS pixels, we show that significant spatial variability of the skin temperature exists. The distribution of the surface variability measured by BESST shows a peak value of $O(0.1)$ K, with 95% of the pixels showing $\sigma < 0.45$ K. Significantly, high-variability pixels are located at density fronts in the marginal ice zone, which are a primary source of submesoscale intermittency near the surface. SST wavenumber spectra indicate a spectral slope of -2, which is consistent with the presence of submesoscale processes at the ocean surface. Furthermore, the BESST wavenumber spectra not only match the energy distribution of MODIS SST spectra at the satellite-resolved wavelengths, they also span the spectral slope of -2 by ~3 decades, from wavelengths of 8 km to <0.08 km.

Keywords: spatial variability; sea surface temperature; submesoscale; wavenumber spectra

---

1. Introduction

Increased spatial resolution in observations of the upper ocean has revealed an abundance of processes on lateral scales of $O(1)$ km (e.g., Thomas et al. [1]). These processes, termed submesoscale, are vital for the transfer of energy from the mesoscale (10–100 km) to the small, three-dimensional processes at scales less than a kilometer (0.1–100 m). While processes at the large and small length scales have been studied extensively, the intermediate scales are less well understood, partly because limitations on instrumental sampling and computational resolution have hindered their research [1]. However, technological advances in the last decade mean that we are now able to achieve the required resolution in models and observations to capture this scale. Submesoscale variability is seen in high-resolution velocity fields from radar [2], sea surface temperature (SST) fields from high-resolution satellites [3], ice-tethered profiler measurements under ice in the Arctic Ocean [4], and hydrographic surveys from towed vehicles and gliders, just to name a few.

Increased scientific understanding has been enabled by the improvements in instrumentation and computational models, and has also established that submesoscale processes play an important role in the vertical transport and mixing of properties and tracers between the surface mixed layer and the thermocline. Submesoscale instabilities are shown to cause rapid changes (they operate on temporal scales spanning from hours to days) in the stratification and buoyancy transport of the mixed layer that cannot be explained by heating and cooling alone, and far exceed what can be achieved through mesoscale baroclinic instability [5–7].

An important question is the role that the submesoscale processes play in this transfer of energy from the meso to the small scales where the energy dissipation takes place [8]. The quasi two-dimensional mesoscale flow field is characterized by kinetic energy spectra with a slope of −3. Three-dimensional numerical simulations at progressively finer resolutions show that resolving submesoscale processes leads to the flattening of the kinetic energy spectra slope to −2 [9]. An aspect of relevance for this investigation is that submesoscales are known to be a source of spatial heterogeneity, or patchiness, at the ocean surface. This heterogeneity, or variance of the spatial distribution of properties and tracers (the SST is considered a passive tracer), is caused either by vertical advection at vertical gradients, or by horizontal stirring (lateral gradients at scales of $O(1)$ km that enhance lateral mixing), which promotes surface filamentation. Numerical experiments [1] show that, in the former case, submesoscale processes accelerate vertical stratification (sharper gradients, higher fluxes), introduce submesoscale concentration anomalies at the surface, and shift the spatial variance towards larger scales [10]. In the latter case, the vertical restratification of the surface layer slows down (weaker stratification and fluxes), stretching and stirring tracer filaments at the surface and increasing submesoscale spatial variance at smaller length scales [11].

Here, we present evidence that a substantial portion of the variance of satellite SST products with 1-km spatial resolution resides in the submesoscale range. It is common practice to treat the satellite-retrieved SST at a pixel as a point value, even though the measurement integrates the radiation coming from the surrounding area within the satellite footprint. There is an inherent uncertainty (estimation error) stemming from this representation, as spatial variability is always present in nature. The variance of the SST distribution within the pixel is referred to as the sub-pixel spatial variability. If the variance within the pixel is small, the point value is a good representation of the overall pixel, but as the variance increases, as in frontal regions and coastal and high-latitude regions, the estimation error also increases. This sub-pixel variability can, therefore, contribute to the estimated uncertainty of a satellite-derived SST retrieval when it is validated against an observation with a finer spatial resolution. Another factor contributing to the uncertainty in satellite SST products is the error arising from the discretization of the sampling, as satellite SST products are typically binned (gridded) in some form. A full description of the uncertainty budget of satellite SSTs can be found in Cornillon et al. [12] and the role of subpixel variability as it applies to Cornillon et al. [12] is explored in more detail in this paper.

Earlier studies of the spatial variability in SST using ship-based radiometric data suggested that the variability at scales smaller than 1 km is significant and affects the estimated uncertainty of satellite-derived skin SSTs, as derived from the in situ observations. In Castro et al. [13], we showed that, although satellite IR SSTs are more physically related to the ocean skin temperature, a satellite SST regression algorithm trained on subsurface temperatures had better accuracy (less variance) than when the regression coefficients were derived from coincidently measured skin temperatures. This was found by developing and testing parallel skin and subsurface SST regression algorithms using coincident temperatures from both research-grade thermometers and highly accurate radiometers deployed simultaneously from research ships. After comparing the SST variance from both types of algorithms (the subsurface-trained algorithm consistently outperformed the skin-trained algorithm), we concluded that the spatial variability of the skin layer was significantly larger than the variability below the ocean surface. The amount of noise (variability) that needed to be added to the retrieved subsurface temperatures in order to degrade their accuracy to the same level of the skin SSTs was on the order of 0.1–0.17 K. Variogram analysis suggested that differences in both measurement uncertainty and spatial variability with depth contributed equally to the levels of noise needed to produce equal accuracy between the two algorithms (~0.07–0.10 K for each source of uncertainty). A contribution of 0.1 K to the total uncertainty budget is not negligible if one considers that the SST accuracy needed to detect climate change requires that the uncertainties associated with the SST measurement be less than 0.3 K [14]. Contrary to popular belief, the results in Brink and Cowles [15] provide evidence that the radiometric satellite measurement, being a weighted average of the horizontal array of skin temperatures within the footprint and thus less variable than the point measurements from state-of-the-art in situ radiometers, is in better agreement with the smoother subsurface temperatures than to the shipborne radiometric skin temperatures.

Despite the emphasis on improving the characterization of the errors in satellite-derived SSTs (the suitability of a satellite-derived, climate-quality data record of SSTs relies on a stringent knowledge of the uncertainties), relatively little is known about satellite sub-pixel spatial variability, because limitations on instrumental sampling require measurements on all scales of interest. Calibration and/or validation of satellite-retrieved SSTs are usually done by comparing the retrieved values with in situ temperatures from IR radiometers deployed from ships or from thermistor chains on buoys. Unfortunately, these platforms cannot collect continuous in situ SSTs over an area as large as 1 km fast enough to resolve the spatial and temporal scales of the processes that control the variability of the skin layer of the ocean.

In this paper, we use SST data from a high-resolution (0.5 m) infrared (IR) radiometer, deployed on an unmanned aerial vehicle (UAV), to resolve the submesoscales on horizontal grid scales of $O(1)$ km. The collection of measurements of this resolution over the ocean are just becoming viable with the growing scientific use of UAVs, and only a few datasets of this type have been collected recently. We argue that submesoscale variability in the surface layer of the ocean is responsible for the sub-pixel variability in satellite retrievals of SSTs. The context of this investigation is the marginal ice zone (MIZ) in the offshore region of the northern Alaskan coast (the Beaufort Sea) during the summer melt season. There are observational indications that the MIZ has substantial submesoscale variability associated with the intense interweaving between warm open ocean water and cold water near the ice edge [15]. Previous studies (e.g., Timmermans et al. [4] and Toole et al. [16]) have shown the important role that submesoscale processes play in setting surface-layer properties and lateral density variability in the Arctic Ocean.

We begin Section 2 by defining the data used in this experiment, followed by a description of the technical challenges we encountered in the harsh Arctic environment that impacted the integrity of the field measurements. The solutions we implemented in order to obtain a viable signal are shared in Section 3. In Section 4, we quantify the sub-pixel variability in satellite SSTs with 1-km resolution, using the fast, repeated sampling from the UAV-deployed IR radiometer over the footprint of the satellite. The wavenumber spectral analysis in Section 4 is an attempt to corroborate whether the observed sub-pixel variability is the result of submesoscale variability. Finally, we discuss the implications of submesoscales for the Arctic mixed layer restratification and other phenomena, and provide a discussion of outstanding questions that justify the need for more measurement campaigns of this type, which can help resolve the space and time variability of the submesoscale, and the associated uncertainty in satellite remote sensing.

2. Data

The data used was collected during the 2013 Marginal Ice Zone Observations and Processes Experiment (MIZOPEX; https://ccar.colorado.edu/mizopex/), a multi-institutional airborne and in situ campaign, funded by the National Aeronautics and Space Administration (NASA) and the National Oceanic and Atmospheric Administration (NOAA) and led by Dr. James Maslanik at the University of Colorado, to survey sea and ice conditions in the MIZ during the summer melt season. In addition to the science objectives, MIZOPEX was conceived as a demonstration of the operational capabilities of UAVs in the polar environment, hence the use of UAVs as a main instrument deployment platform. The University of Colorado Ball Experimental Sea Surface Temperature (BESST) thermal infrared radiometer [17], referred to hereafter simply as BESST, was flown on a small commercial UAV, the Boeing–Insitu ScanEagle (https://en.wikipedia.org/wiki/Boeing_Insitu_ScanEagle), as part of the MIZOPEX suite of instruments for SST surveillance.

This paper examines the data collected during a six-hour flight on 4 August (day of year (DOY) 216) 2013, off Oliktok Point on the North Slope of Alaska. The ScanEagle was required to ferry out to international waters, along 150°W. Once in international waters, it began a "lawn mower" survey that spanned seven meridional transects spaced 0.05° apart, from 150°W to 149.65°W, and was confined to latitudes between 71.6°N and 72°N (Figure 1). The survey started at 13:00 LST (23:00 UTC) and ended at 16:50 LST (02:50 UTC on DOY 217). During this time, we obtained near-coincident SST measurements from NASA's Moderate Resolution Imaging Spectroradiometer (MODIS) aboard the Terra (19:55 UTC (9:55 LST) and 21:30 UTC (11:30 LST)) and Aqua (20:15 UTC (10:15 LST) and 21:50 UTC (11:50 LST)) satellites. The MODIS SST spatial resolution is 1.1 km, which is among the finest spatial resolution available for operational SST retrievals at the time. The Visible Infrared Imaging Radiometer Suite (VIIRS) with slightly higher resolution was flying on the Suomi National Polar-orbiting Partnership satellite at the time, but MODIS retrieval algorithms were more mature, and the available VIIRS data were subject to aggressive cloud screening. Other satellites, such as Landsat, have even higher resolution IR measurements, but lack the combination of channels required for global retrieval of SST.

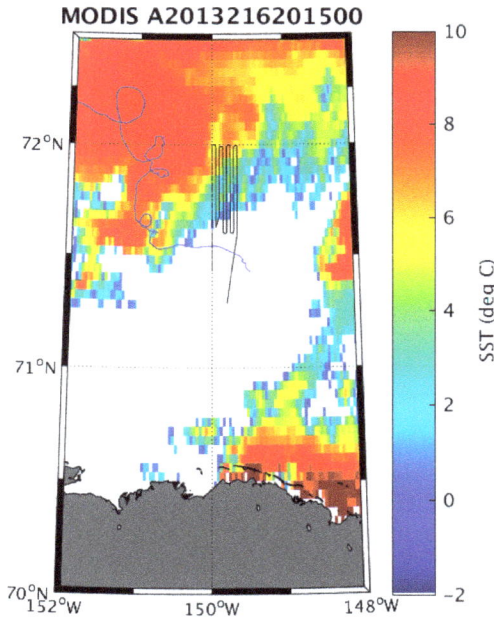

Figure 1. Ball Experimental Sea Surface Temperature (BESST) lawn mower survey during the Marginal Ice Zone Observations and Processes Experiment (MIZOPEX) (black). The aircraft entered the pattern heading northbound at the western edge of the pattern, and exited to the south at the eastern edge. The background image is from the Moderate Resolution Imaging Spectroradiometer (MODIS) AQUA sea surface temperature (SST) granule for 20:15 UTC on 4 August 2013, the day of the flight. The blue trace depicts the UpTempO buoy track while in the study area. This buoy was deployed the day after the unmanned aerial vehicle (UAV) survey, and was part of the MIZOPEX in situ instruments.

The BESST is a pushbroom IR imaging system especially designed for the aerial monitoring of SSTs from UAVs and small aircraft. As the aircraft moves forward, the radiometer generates thermal images or frames that are 320 × 256 pixels in size, but only the center 200 × 200 pixels are retained due to known optical distortions at the edge of the scan. The image spatial resolution depends on the aircraft altitude. With a field of view of 18° and 200 pixels per across-track scan line, the BESST swath width is about 1/3 of the flight altitude. During the MIZOPEX survey, the ScanEagle flew at an altitude of 300 m, resulting in a BESST ground spatial resolution of about 0.5 m and a swath width of 100 m. The system is configured to collect 130 sea-viewing frames in 53 s, followed by a 7 s onboard calibration period, resulting in about 21 frames of target data lost during this time. Thus, for a UAV mean speed over ground of 25 m·s⁻¹, the BESST samples ~1.3 km in the flight track direction, and skips ~0.2 km along track during calibration cycles. During the MIZOPEX survey, the ScanEagle coverage between calibration periods was ~1.4 km for the northward transects, and ~1.0 km for the southward transects, due to the relative winds. At this speed, the individual image frames were overlapping, resulting in some degree of oversampling. The flight altitude was below the cloud base. Collecting data at altitudes closer to the sea surface has the added benefit of minimizing the atmospheric error in the radiometric SST retrievals.

3. Data Processing

The BESST thermal sensing array consists of a microbolometer-based camera (a FLIR Photon Thermal Imaging Camera Core [17]), which is used for viewing both the target scene at the surface and the sky. The sky view is needed to correct for sky radiation reflected into the field of view of downward-looking radiometers. As part of the calibration system, the instrument includes two black bodies (BBs), one of which is allowed to drift with the internal ambient temperature, while the other is kept heated 12 K above ambient. A mirror changes the view of the microbolometer between view ports and the BBs. Thermistors continually monitor the temperatures of these elements for correction in processing the microbolometer data. This design, although effective in past deployments at milder latitudes [18], proved problematic for the cold ambient temperatures of the Arctic environment. After looking at the microbolometer's raw data, it was clear that the sensing-array measurements were noisy (large dispersions from pixel to pixel), and required extensive filtering to extract the SST signal from the background noise.

As part of the post processing, the measurements within the BESST SST frames were binned into 0.05 °C bins varying from −2 to 13 °C, and only temperature bins for that frame with 20 observations (the threshold determined based on histograms) or more were considered for further analysis. The exclusion of bins with low observation counts effectively removed portions of the edge of the scan with residual optical distortions, and damped the noise in the detectors. Binned SST data from "valid bins" were count-weighted averaged to obtain the mean retrieved temperature value for the entire frame. Thus, the thermal information contained in each of the BESST frames was condensed, via weighted spatial averaging, into single values representative of 100 m × 100 m images. However, since the frames were overlapping, these average values were spaced roughly every 10 m, yielding some degree of oversampling. Gaps due to calibration periods in the along-track direction were linearly interpolated, and the filled time series of mean-binned SSTs were smoothed using a boxcar moving average of width five (~50 m).

The geolocation of the BESST data relies on the geo-pointing referencing software from the Piccolo autopilot system, which is responsible for the autonomous navigation of the aircraft. Accurate latitude and longitude coordinates for each frame (geolocation) are obtained by matching the BESST acquisition time with the autopilot GPS time. To time-stamp the SST images, the BESST instrument uses a GPS receiver to acquire accurate time information. When a connection cannot be established between the BESST and the GPS satellites, the BESST defaults to the internal clock of the mini-computer that operates the system. Unfortunately, it is very difficult to establish communications with GPS satellites in the Arctic regions (north of 55°N), and the system defaulted to the central processing unit (CPU) internal clock during the MIZOPEX campaign most of the time. Computer internal clocks are renowned for their poor stability. To complicate matters, there is a mismatch between the frame acquisition rate (3 Hz) and the computer latency (less than 3 Hz), which results in a time delay of

333 ms between BESST frames and the CPU clock. This implies that there is a cumulative loss of time for every instance the system fails to make contact with the GPS satellite. This, plus the fact that BESST was initialized about an hour before the plane took off, made it impossible to retrieve the timing of the BESST frames by conventional means. To circumvent the lack of accurate information to time-tag the BESST frames and, by extension, to geolocate them, we had to use an unorthodox procedure of comparing the BESST SSTs against the MODIS SST themselves. This involved two processing steps that are usually carried out separately, but due to the technical difficulties described above, had to be performed simultaneously; that is, geolocating the BESST data while trying to match its measurements to the satellite SSTs (data collocation).

The first step towards the BESST–MODIS collocation procedure was to combine all of the satellite SSTs from the four MODIS L2 SST granules around the time of the experiment into a single, gridded image. This is justified since the different satellite overpasses occurred between 1–3 h before the start of the survey, and SSTs have a relatively slow and smooth rate of change. The combined approach takes advantage of the fragmented information contained in the individual granules (the spatial coverage in the MODIS granules varied significantly from one pass to the next, due to fast moving clouds over the survey domain and orbital geometry), and it can also minimize errors associated with the collocation procedure. The spatial resolution selected for the MODIS grid was 1.25 km by 1.25 km × sec (ϕ). The scaling factor in the meridional direction (sec (ϕ), where ϕ is the latitude of the grid center) takes into account the lateral distortions that occur near the poles when mapping the latitude and longitude coordinates of the satellite orbital path onto a Cartesian coordinate system. The gridded image, shown in Figure 2, was obtained by stitching the multi-temporal granules using a Maximum Value Composite (MVC) technique, i.e., only the highest value within a pixel is retained for that grid location. The collocation of the SST observations from BESST with MODIS was done by averaging the BESST measurements within the MODIS pixels in the image composite, based on the distance between the location of the aircraft and the center of the grid.

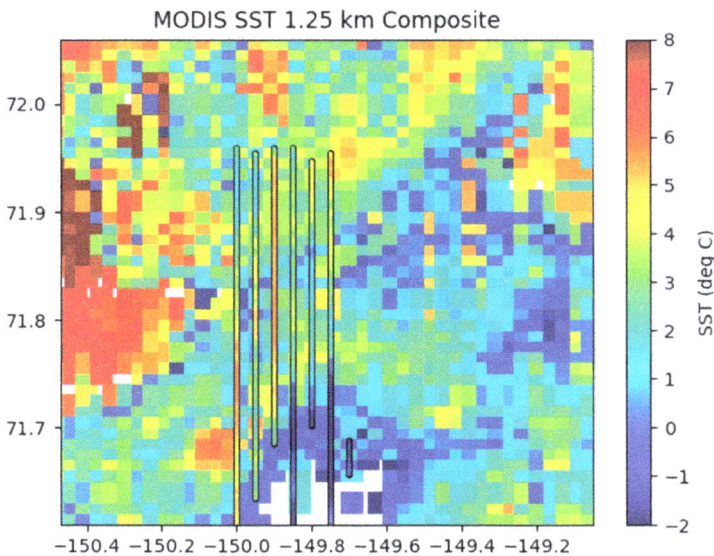

Figure 2. MODIS SST Maximum Value Composite with superimposed BESST back-shifted tracks, color-coded by the corresponding BESST SSTs. The western track was sampled first, with the aircraft heading northbound.

To simultaneously geolocate and collocate the BESST data, we first assigned the latitude, longitude, and time records from the Piccolo, starting from when the aircraft was still parked on the runway, to BESST, even though the latter underwent checkup procedures for about an hour prior to the Piccolo start time. The lag time between the two data records was found through trial and error by successively matching the measurements from the two radiometers, then shifting the BESST data records by a variable number of time steps and re-evaluating the matchups with MODIS until the discriminative features in both of the time series of matches overlapped when plotted together. The metric used to determine simultaneity between the time series was the correlation coefficient of the set of matches. A peak correlation of 0.43, which corresponded to a lag of 50 min, was deemed appropriate to geolocate the BESST point measurements.

The location information just derived was further used to separate the BESST data into flight segments along meridional transects. Figure 3 shows the back-shifted BESST time series of SSTs for a 50-min lag with corresponding MODIS matchups. The BESST SSTs are color coded by the predicted location of the flight tracks along meridional transects. Despite a global correlation of 0.43, it can be seen from Figure 3 that the agreement between the BESST and MODIS SSTs is good for the last three transects (correlation for the last three legs is 0.77). Correlations for the matchups along −149.85°, −149.80°, and −149.75° longitude are 0.73, 0.75, and 0.78, respectively. The good agreement between the BESST and the MODIS background temperatures for the last three transects is further exemplified in Figure 2, which also shows the geolocated BESST track, superimposed on the MODIS SST composite, color coded by BESST SSTs.

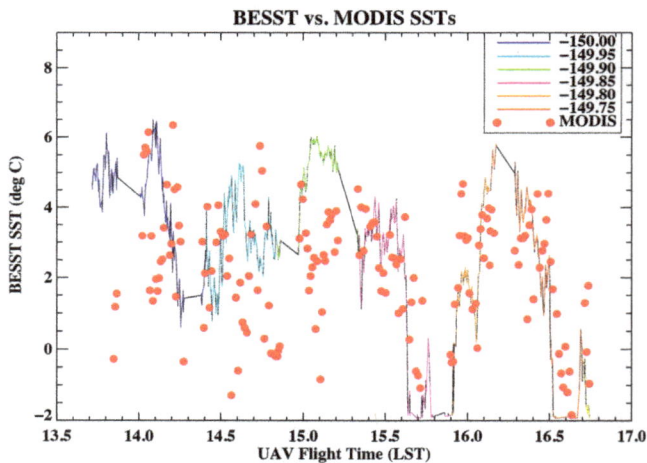

Figure 3. BESST SST time series, shifted back in time by 50 min, with superimposed MODIS SST matchups (red dots) from the maximum value composite.

The poorer correlation between MODIS and BESST for the first three transects could have been for any number of reasons, but the BESST observations being significantly warmer than the MODIS SSTs could be an indication that the MODIS measurements over these transects were contaminated by clouds, and hence appear colder than their actual value should have been. It is important to note that the BESST values were obtained below the cloud level. Moreover, due to the scarcity of satellite data coverage during the experiment, we opted to ignore the quality assessment of the MODIS data, as suggested by the data producer. The satellite sea ice concentration (SIC) product from the National Snow and Ice Data Center (not shown) indicated that the study area had a 60% ice concentration at the time of the flight, which was desirable for the overall campaign, as this was a study of the SST conditions in the MIZ. Most SST analyses do not report SST values when the SIC is 50% or higher, which means that there is no other satellite SST product available that can be used to corroborate the cold MODIS SSTs (the more stringent cloud clearing of the VIIRS SSTs produced no matches for this place and time). For the remainder of this paper, special attention will be given to the last three transects of the flight. It should be kept in mind that the use of MODIS thus far is to give an approximate location of the BESST radiometer, so that the observations can be separated and analyzed by transects. It should not affect the BESST centric analyses, as they do not rely on the MODIS SSTs directly.

4. Data Analysis

4.1. Spatial Variability

Since the goal of this paper is to see whether the BESST measurements provide accurate information about the sub-pixel variability within 1-km footprints, we look at the BESST SST variations within the MODIS grid. The metric of choice is the standard deviation, σ, of the BESST measurements that fall within an individual MODIS pixel. The standard deviation is used to estimate sub-pixel variability, because it gives a measure of the dispersion (variation) of the SSTs along 1.25-km flight line segments, as the aircraft flies over an area the size of the MODIS pixel. This analysis could have been done independently of the satellite just by binning the entire BESST record using a bin width, or horizontal grid point spacing that encompasses the scales of interest. The reason we concentrated on the subset of BESST measurements paired with MODIS is to be able to use the satellite SSTs as reference for interpreting features in the variability gathered from the BESST. The normalized frequency distribution (or probability mass function, PMF) for the sample standard deviations of BESST SST within MODIS pixels is shown in Figure 4. The frequency distribution of σ is right-skewed, with a peak value of $O(0.1)$ K, and a long right tail extending to a maximum of 0.8 K. The corresponding cumulative density function (CDF) shows that 95% of the pixels had σ < 0.45 K. It is interesting to note that the sub-pixel variability of this magnitude is comparable to the 0.4 K accuracy requirement for operational satellite SSTs, as cited by the Group for High Resolution Sea Surface Temperature (GHRSST) (e.g., Table 6.1 in the recommended GHRSST data specification document [19]). If the MODIS retrieval were to be validated against a point or very high spatial resolution measurement, temperature variability within the satellite pixel alone could cause differences equal to the required accuracy. The satellite minus validation difference would not accurately reflect the uncertainty in the satellite retrieval itself. In this manner, sub-pixel variability becomes a relevant component of the total uncertainty budget when satellite retrievals are validated against observations of finer resolution. The peak value of $O(0.1)$ K in the sub-pixel variability corresponds very well to the uncertainty contribution from spatial variability found by [15] for in situ skin SSTs measured with state-of-the art IR radiometers deployed from ships.

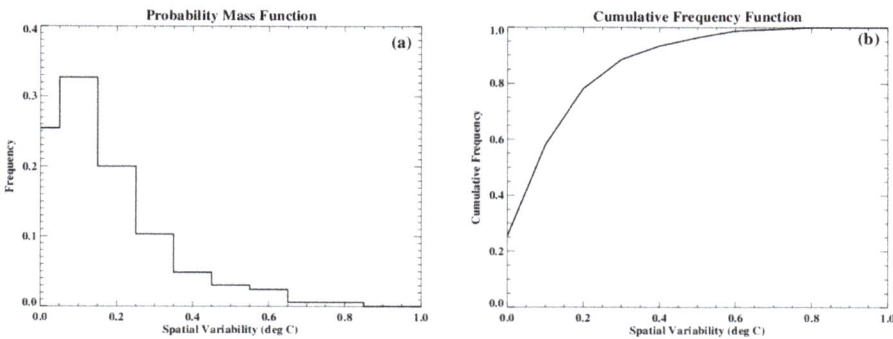

Figure 4. Probability mass function (**a**) and cumulative density function (CDF, (**b**) for the standard deviations of the BESST measurements within the MODIS Maximum Value Composite (MVC) SST pixels.

A long-tailed probability distribution can be interpreted as a statistical expression of intermittency or patchiness in the spatial and temporal distribution of submesoscale structures in the ocean surface [9]. This intermittency is manifested in the instantaneous patterns of surface temperature, density, and vorticity. Unfortunately, the 2D BESST images were reduced to linear transects of data to dampen the noise in the microbolometer, and thus, any visual indication of SST intermittency was significantly degraded. In order to look at where the high variability SST patches occurred, we looked instead at the time series of binned, MODIS-matched, BESST SSTs (Figure 5). From the collocation procedure described above, these are area-binned SSTs that result from averaging all of the BESST temperatures within 1.25-km cells, which are coincident with the MODIS grid. The triangles in Figure 5a correspond to the mean SSTs within the 1.25-km pixels of the MODIS MVC. Once again, the color of the triangles indicates the flight transect. The vertex of the triangles points to the flight direction of the UAV. Error bars correspond to 1-σ noise levels of the BESST measurements within the pixels. These error bars are a graphic representation of the spatial variability within the pixel. Colored error bars indicate grid cells where the standard deviation exceeded 0.4 K. Interestingly, the preferred location of pixels with high spatial variability was at the start and end of the meridional transects (Figure 5a).

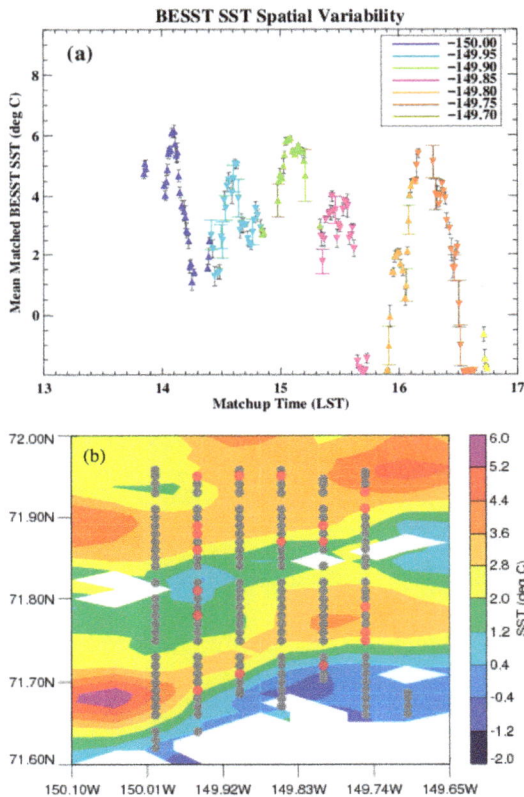

Figure 5. Time series (**a**) of time-lagged BESST SSTs matched to MODIS. The triangles represent the mean BESST temperature within the MODIS pixel. The color and vertex of the triangle indicates the flight transect and direction, respectively. Error bars correspond to 1-σ variability. Where σ ≥ 0.4 °C, the error bar is color coded by transect. The SST map shown in (**b**) is a contour representation of the MODIS composite (Figure 2) around the BESST survey (grey circles). Red circles illustrate the location of the BESST area-averaged measurements with σ ≥ 0.4.

To investigate whether the spatial distribution of high-variability pixels coincided with places of strong ocean dynamics, we looked at a contour map of the MODIS SST MVC shown in Figure 3, with the survey domain enlarged (Figure 5b). White areas are gaps in coverage due to clouds in the MODIS data. The dots show the location of the BESST matchups. Red dots indicate pixels where the standard deviation of the 1.25-km averaged BESST SSTs reflects the spatial variability in excess of the SST accuracy requirement of 0.4 K (i.e., where point-to-pixel differences can comprise a significant portion of inferred SST uncertainty estimates). The SST contour levels reveal a likely icy patch at the southern edge of the survey (water at freezing temperature), followed by alternating warm and cold filaments at mid-range and a warm area in the top third of the survey domain. There appears to be multiple surface vortices/eddies present in the warm patches of water. With the exception of the red dots at the northern end of the transects along $-149.95°$, $-149.90°$, and $-149.85°$ longitude, which appear to be related to the aircraft turning 90 degrees, all other high-variability pixels are located at or near regions with strong temperature gradients. If, as the evidence suggests, the area of interweaving warm and cold waters corresponds to the MIZ, then the transition zones at the ice edge and at the open water edge of the MIZ are where the high variability took place. The high-variability patches at the bottom end of the survey, in particular, align extremely well with the thermal front (there is an SST change of roughly 3.5 °C over 10 km) that marks the transition between the icy patch and the warm filament. This is consistent not only with long tail statistics [9], but also with horizontal density/temperature gradients (fronts) being a primary source of submesoscale intermittency near the surface (e.g., Thomas et al. [1]; Boccaletti et al. [8]; and Samelson and Paulson [20]). To look for more statistical evidence that the asymmetric distribution of the SST sub-pixel variability in the MIZ is the result of a submesoscale transition regime that develops near horizontal temperature gradients, we did some spectral analysis of the SST, which is shown next.

4.2. Spectral Analysis of SSTs

When dealing with spatial variability, one has to deal with the concept of "scale". In order to look at the spatial scales resolved by each of the IR radiometers, we next estimate the horizontal wavenumber spectra, for the BESST and the MODIS SSTs, using fast Fourier transforms. This methodology is appropriate to look at the scale content of oceanic variability, since the data were sampled at a regular interval. We emphasize that the spectra shown below are representative of the horizontal SST variance at the surface of the ocean (0 m depth), as both MODIS and BESST instruments have penetration depths on the order of 10–20 microns, which corresponds to the skin temperature of the ocean.

The MODIS SST horizontal wavenumber spectra were evaluated directly from the satellite swath data, as the level 2 product (geolocation is given in original satellite scan line/spot geometry) has higher spatial resolution than the MODIS maximum-value composite. Only the granules from the 20:15-Aqua and the 21:30-Terra overpasses were used in this analysis, as the other two lacked the complete data coverage and spatial continuity needed for a spectral analysis. The satellite wavenumber spectra are computed in the satellite along-track direction. Figure 6a,b show the portions of the MODIS scans over the survey area for the two granules considered. The Aqua granule covers an area approximately 250-km wide across-track by 400 km along-track, which is twice as much as the Terra granule. The dots indicate the center location of the 1-km pixels and the colors identify the selected scan spots, from successive scan lines, used to estimate the individual spectra. The sequence of successive scan spots was mostly complete over the survey domain, but in the few instances where there were gaps, these were interpolated by averaging the nearest neighbors to the selected scan spot.

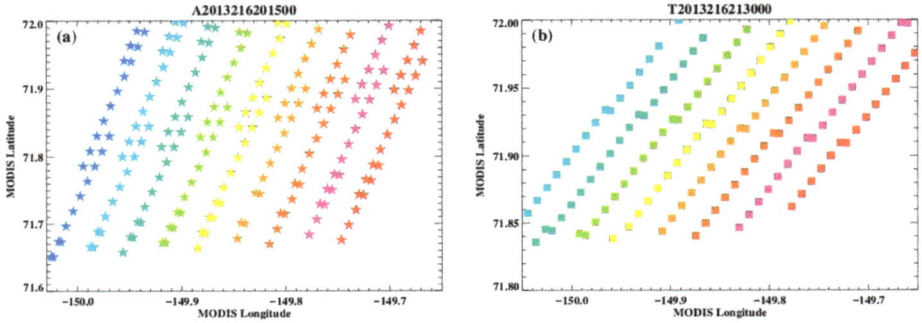

Figure 6. MODIS along-track scan spots for the Aqua (**a**) and Terra (**b**) granules that were used to compute the satellite horizontal wavenumber spectra.

The wavenumber spectra from the individual along-track scan spots were binned to a common wavenumber, ν, scale and band-averaged to obtain a blended wavenumber spectrum representative of their collective distribution in space. The selected wavenumber bandwidth was $\Delta\nu = 0.0417$ km^{-1} (or 24 km) for Aqua, and $\Delta\nu = 0.0714$ km^{-1} (or 14 km) for Terra. By averaging over wavenumbers in these bandwidths, the spectral densities for each of the satellites are assumed to be approximately constant over the 14–24-km length-scale range. The blended spectrum is a smooth spectral density estimate that is penalized by a loss in wavenumber resolution. In other words, we cannot resolve wavelengths finer than half the corresponding wavelength (i.e., ~7–8 km). Figure 7 shows the resulting blended spectra for the Aqua and Terra granules, in km^{-1}, in both semi-log and log–log scales. An Aqua–Terra ensemble-averaged wavenumber spectrum for MODIS is also shown in Figure 7.

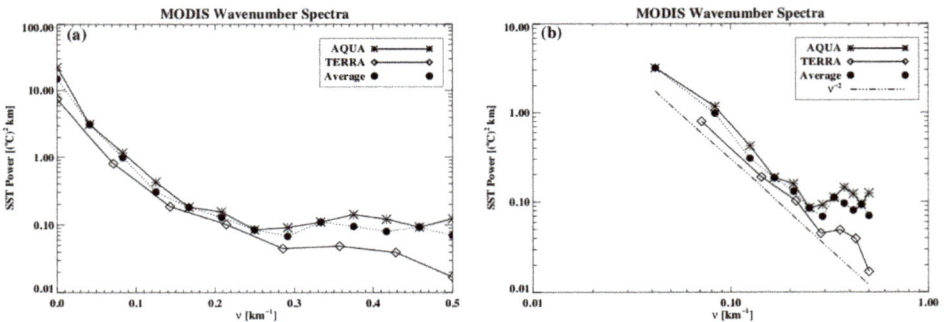

Figure 7. Ensemble-averaged, band-average horizontal wavenumber spectra for SST variance from MODIS, along the (**a**) semi-log and (**b**) log–log axes. MODIS spectra are very nearly proportional to the theoretical spectrum ν^{-2}.

The logarithm of the MODIS SST spectra against wavenumber, as displayed in Figure 7a, show very similar shapes for both the Aqua and Terra satellites, with a smooth exponential decay in power and no predominant peaks at the observed wavelengths. The log–log plot of the spectra (Figure 7b) shows that both the individual and the blended spectra are very nearly proportional to $\sim\nu^{-2}$ (ν is horizontal wavenumber) in the range $0.04 < \nu < 0.125$–0.25 km^{-1}, which corresponds to wavelengths $\lambda = \nu^{-1}$ between 25 and ~8 km. Near 4 km, the –2 power law breaks down, and a spectral peak appears close to $\lambda = 3$ km ($\nu = 0.33$ km^{-1}) in these three spectra, followed by a decrease toward the Nyquist wavelength (2 km). This log–log linear spectral shape with a slope of –2 is consistent with previous findings of horizontal wavenumber spectra of surface-layer temperature and density scaling as ν^{-2} over horizontal scales of $O(1)$–$O(100)$ km at mid-latitudes [20–22]. Thus, a slope of -2 in the temperature spectrum is consistent with submesoscale activity being the dominant source of the horizontal temperature variance, not only in the open oceans but, as our results suggest, in the MIZ as well. Since the total power (excluding the zero-term) will tend to equal the variance of the signal, the MODIS wavenumber spectra suggest that the source of the sub-pixel variability of satellite-derived SSTs within footprints of 1-km resolution resides in the submesoscale range. Dominant submesoscale features over the survey domain likely arise from temperature gradient production mechanisms such as frontogenesis and frontal instabilities.

The BESST wavenumber spectra were computed along individual meridional transects, and for the entire survey. A BESST wavenumber spectrum over the lawn-mower survey is possible under the assumption that the lawn-mower sampling was on a uniform grid. However, since the BESST frame separation varied in length around 10 ± 3 m, it was necessary to interpolate the BESST data onto a uniform 10-m length meridional grid before estimating the wavenumber spectrum along individual meridional transects. This grid resolution is consistent with the mean separation of the overlapping BESST frames. Figure 8 shows the horizontal wavenumber spectra, in km^{-1}, of the 10-m gridded BESST data along the meridional transects 149.85°W, 149.80°W, and 149.75°W, both in semi-log (left column) and log–log scales (right column), respectively. The wavenumber spectra for the transects that are not shown are very similar to those displayed in Figure 8. The collective MODIS Terra spectrum, in red, is also plotted for comparison. The black circles represent wavenumber spectra without any smoothing, whereas green traces depict spectra that were smoothed with a Tukey filter over 1-m length intervals. The density clustering of the circles illustrates clear determination of the BESST spectral shape for $\nu > {\sim}2.5$ km^{-1} ($\lambda < {\sim}400$ m), but under determination for $\nu < {\sim}1$ km^{-1} (i.e., for scales approaching the spatial resolution of the satellite). In all cases, the log–log plots of the spectra have a shape ${\sim}\nu^{-2}$ (this is illustrated in Figure 8 by the agreement between the smoothed spectrum in green and the theoretical spectrum ν^{-2} in dark blue) in the wavenumber range $0.04 < \nu < 40$ km^{-1} (25 km $> \lambda > 0.025$ km). A spectral slope of -2, once again, is a signature of submesoscale activity driving the surface temperature variability. The BESST spectrum extends the submesoscale range (10 km–100 m) at the finer wavelength resolutions an order of magnitude higher than most previously reported (from 100 m to ${\sim}25$ m). A flattening of the slope occurs at $\nu \sim 40$ km^{-1} ($\lambda \sim 25$ m), as the scales approach what can be resolved at the BESST resolution; thus, the change in slope might just reflect noise in the BESST data or aliasing of overlapping BESST measurements.

The spectral slopes of BESST and MODIS show excellent agreement despite the BESST spectra being less resolved at the low wavenumber end where both spectra overlap. Interestingly enough, the low wavenumber end of the semi-log spectrum along 149.80°W is indistinguishable from the MODIS Terra spectrum shown in red, with both spectra showing the same shape, featuring a spectral peak at 3 km (Figure 8c). This is surprising given the different nature of the instruments. The other two spectra (Figure 8a,e) show a broad peak around 1.5 km, outside the MODIS domain. In these two cases, the spectral levels between the two vary by roughly half a decade. A key aspect of this comparison is that the BESST high-resolution SST data spans the spectral slope of -2 by ${\sim}3$ decades in λ relative to MODIS, from 8 km to 25 m. This result, in a sense, validates the satellite findings. Since step functions also have Fourier transforms proportional to ν^{-2} [20], the MODIS behavior could possibly be interpreted as an artifact due to the presence of sharp fronts (the spectral slope of -2 in MODIS breaks down at 8 km, whereas the MODIS composite indicates the presence of frontal widths of ${\sim}10$ km). However, the ν^{-2} behavior from BESST persisting to sufficiently small scales where fronts do not resemble step functions corroborates that the spectral behavior observed by both instruments is not an artifact due to the presence of sharp fronts (step functions). Instead, it is an artifact from the secondary circulations associated with fronts where submesoscales are found to occur. In essence, the BESST instrument expands the ν^{-2} behavior resolved by the satellite well beyond the range of horizontal scales (by at least a decade in λ) required to capture the submesoscales ($\lambda = {\sim}500$ m according to Kunze et al. [23]).

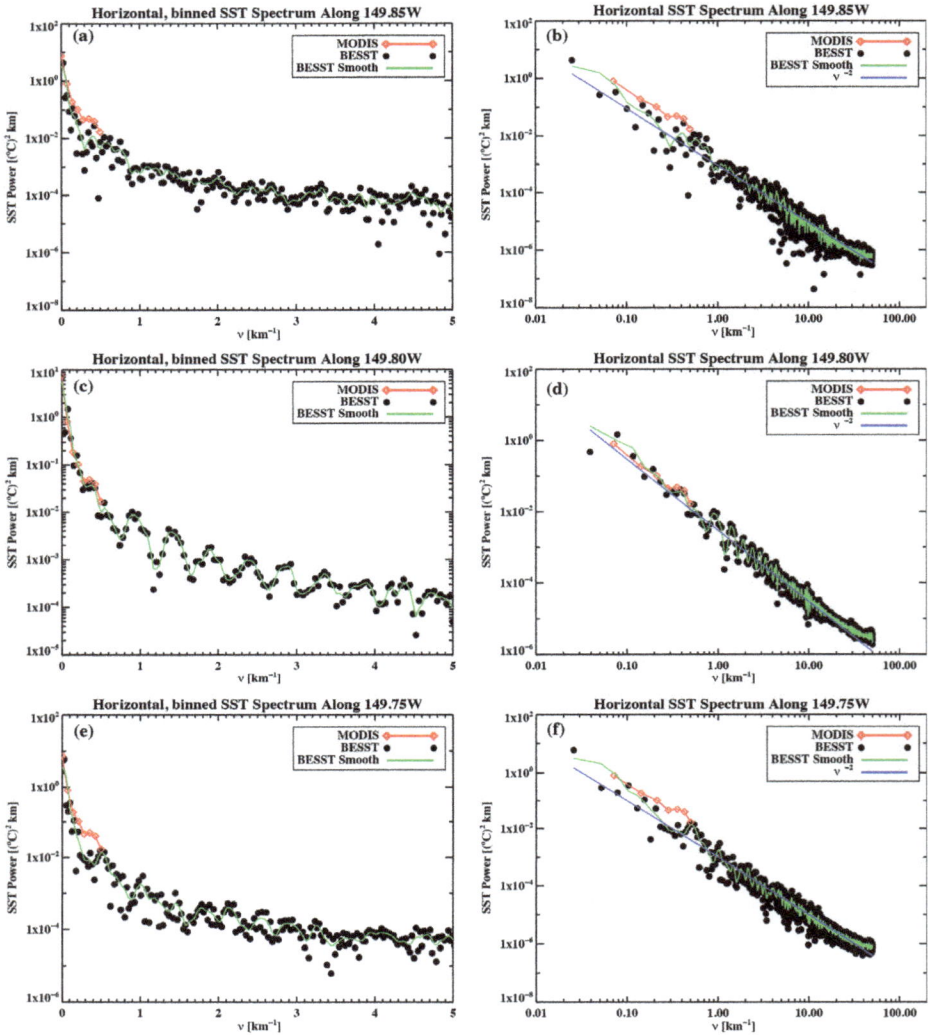

Figure 8. (a–f) Horizontal wavenumber spectra for SST variance from BESST, along the semi-log (left) and log–log axis (right), and along meridional transects: 149.85°W, 149.80°W, and 149.75°W.

For a blended BESST horizontal wavenumber spectrum representative of the whole survey area, the wavenumber spectra from the individual meridional transects were binned into non-overlapping, sequential wavenumber bands of width $\Delta v = 0.0286$ km^{-1} (or equivalently, $\Delta \lambda = 35$ km). In a manner consistent with the procedure used to derive the MODIS blended spectra, the ensemble-averaged BESST wavenumber spectrum for the survey domain was defined as the bin-wise average of the Fourier coefficients from the spectra from the six individual, uniformly gridded, lawn-mower transects. The ensemble waveband-average BESST wavenumber spectrum representative of the spatial domain is shown in Figure 9. The spectral distribution shown here is qualitatively very similar to the ones in Figure 8 for the individual transects, although they are also smoother due to the loss of wavenumber resolution resulting from the binning process. However, the low wavenumber end of the spectrum is better resolved, as it shows an offset of less than 0.5 decades in spectral power with respect to the MODIS spectra in the overlapping range. The log–log spectrum (black dots) in Figure 9b has a shape of $\sim \nu^{-2}$ in the wavenumber range $0.04 < \nu < 10$ km^{-1} (100 m–25 km horizontal wavelengths), spanning from the small to the submesoscales, as shown by the agreement with the light blue line representing a -2 spectral slope. The shape of the spectrum confirms, once again, that a substantial portion of the variance in the SST field over the 0.04- to 10-km^{-1} band is likely due to submesoscale processes present in the region at the time of the survey. The small peak present in the MODIS spectra at $\lambda = 3$ km is well resolved in the blended BESST spectrum, but appears shifted towards a slightly higher wavenumber ($\lambda = \sim 2$ km).

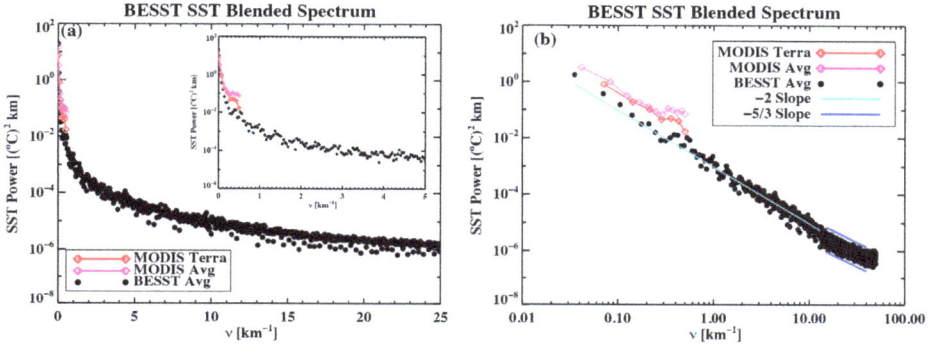

Figure 9. Blended BESST wavenumber spectrum over the entire survey domain. (**a**) Semi-log spectrum with the inset highlighting the range from 0–5 km^{-1}; (**b**) Corresponding log–log spectrum along with overlays of spectral slopes of -2 and $-5/3$ in the light and dark blue colors, respectively.

An interesting remark is that the break in the -2 slope for the blended spectrum happens at a longer wavelength than in the individual BESST spectra (100 m vs. 25 m). The bend in the spectrum observed here at $\lambda = \sim 100$ m is consistent with a transitioning regime from the submesoscale to small-scale flows. While the occurrence of this transition cannot be ascertained from the individual spectra in Figure 8, as the bend occurs near scales where the noise becomes predominant, noise has been significantly dampened in the blended spectrum, which is shown in Figure 9. Furthermore, the semi-log spectrum (Figure 9a) also shows a notable decrease in the spread of spectral power with wavenumber for $\nu > \sim 10$–13 km^{-1} (~ 80–100 km). The relevant aspect is that the spectral slope changes from -2 to $-5/3$, as shown by the dark blue lines bracketing the second regime in Figure 9b. The $\nu^{-5/3}$ scaling persists from $\sim 10 < \nu < 40$ km^{-1} (~ 100 m $> \lambda > 25$ m). A slope of $-5/3$ is consistent with the Kolmogorov spectrum for microscale turbulence [24]. Using Kolmogorov-like dimensional analyses, Obukhov [25] and Corrsin [26] predicted a temperature spectrum with a slope of $-5/3$ at the small-scale inertial range. Wavenumber spectra with two separate power law scalings have been reported [27,28]. If this is indeed the case, the horizontal SST variance transitions from a regime driven by submesoscale dynamics to an inertial regime driven by microscale turbulence at horizontal scales on the order of meters. However, conclusive evidence of a meaningful transition and a spectrum with dual power law scaling requires more in depth study.

5. Conclusions

The aim of this paper was to evaluate if, by flying an IR radiometer on a fast-moving platform such as an UAV, it would be possible to resolve the horizontal scales of SST variability in the MIZ. Moreover, we attempted to quantify how this spatial variability could impact uncertainty estimates of satellite-derived SSTs within 1-km footprints when validated with observations on finer spatial scales. The measurement campaign took place during a period of highly dynamic processes at the start of the melt season in the southern Beaufort Sea. Wavenumber spectral analyses from MODIS and the UAV-deployed BESST radiometer were used to describe horizontal SST variance over lengths of 0.025–25 km. Wavenumber spectra for both MODIS and BESST systematically showed a uniform spectral slope of -2 (Figures 7–9) for wavenumbers in the range $0.04 < \nu < 40$–50 km^{-1} (25 km $> \lambda >$ 25–100 m). The BESST blended spectrum for the region also suggests a second power law with $-5/3$ slope for $\lambda < 100$ m, which is consistent with the Kolmogorov–Obukhov–Corrsin power law scaling for the SST energy spectrum in the microscale turbulence regime. However, this requires further verification. The slopes of the BESST spectra not only aligned naturally with the satellite spectra at low wavenumbers, but also extended its range of horizontal wavenumbers (length scales) to higher ν (smaller λ), by about three orders of magnitude. A spectral shape of ν^{-2} is consistent with the submesoscale processes at the surface. So, it appears that the submesoscale features project onto a wide wavenumber spectrum, as seen when taken both individually and through their collective distribution in space [29]. Our interpretation is that the submesoscale processes increase the variance of the spatial distribution of tracers at the ocean surface, and hence are responsible for the horizontal scales contributing to the SST sub-pixel variability in the MIZ and in the open ocean. This is also supported by the long-tail probability distribution of the sub-pixel SST variance from BESST, which indicates strong patchiness (spatial variability) within 1-km footprints. Moreover, MODIS pixels with high spatial variability appeared to be associated with thermal fronts, where submesoscales are prone to occur. It is important to emphasize that the BESST observations were collected below cloud base, so cloud effects should not impact the observed variability.

Knowledge of the spectral slope alone is not sufficient to draw conclusions about the physical processes responsible for the observed SST variance in the MIZ, and certainly there is not sufficient data in this experiment to prove it unequivocally. However, given the pervasiveness of the ν^{-2}-dependence of the surface temperature variance on wavenumber for the submesoscale range, which is corroborated by extensive experimental evidence from throughout the open oceans [7–9,20–22], we speculate, with a certain degree of confidence, that submesoscale processes such as frontogenesis and submesoscale instabilities are responsible for the observed SST variability in the MIZ. This is somewhat expected if, as Saffman [29] pointed out, the energy spectrum density for random solutions of the equation:

$$\frac{\partial u}{\partial t} + u\frac{\partial u}{\partial x} = \frac{1}{Re}\frac{\partial^2 u}{\partial x^2} \tag{1}$$

should vary asymptotically similar to ν^{-2} in the range of wavenumbers between the macroscale and the microscale viscous cut-off. This was proven numerically by Hosokawa [30]. The differential equation above is the Burgers solution to the modified Navier–Stokes (NS) equation, $\partial u/\partial t = \chi(u) + f$, (also known as the Hopf equation [31]), which has an additional "fluctuation" term, f, which describes a random force that incorporates the "thermal agitation" caused by molecular motion. The thermal agitation term is usually neglected from NS, in part due to computational limitations in finding an analytical solution for the random force field [30]. What this suggests is that, by considering the natural thermal agitation (e.g., temperature variability) when describing turbulent motions in the range of scales that bridge the macro with the micros (e.g., the submesoscale regime), the energy spectrum of the fluid motion scales as ν^{-2}.

Submesoscales have temporal scales of hours to days (a couple of weeks at most). Therefore, the submesoscale variability is transient, and requires a continuous supply of variance to maintain it [23]. The examination of time series of temperatures with depth (Figure 10) from an UpTempO buoy [32], which was deployed in a nearby area (see Figure 1) as part of MIZOPEX on the same day of the survey, indicates that intense variability was present in the top 10 m of the ocean surface at the time of the UAV deployment, and persisted for 15 days after the survey. During this time, the temperature measurements from the buoy showed strong stratification in the top 10 m of the MIZ, and even a couple episodes of diurnal warming (DOY 219 and 228) in the top 2.5–5 m. After 19 August (DOY 231) 2013, the transient stratification was eradicated, and the surface layer became vertically homogenized. As shown in Figure 10, a well-mixed layer (isothermal temperature with depth) became apparent in the top 7.5 m after DOY 231, and in the top 10 m after DOY 248, which then persisted for the rest of the summer season. Similar Arctic mixed layer depths of 10–15 m were observed by Castro et al. [33] from other UpTempO buoys in the area for the summers of 2012 and 2013. Toole et al. [16] observed mixed layer depths in the Canada Basin varying between 10 and ~30–40 m over a few days during the winter; in the summer, the mixed layer depth averaged 16 m. Timmermans et al. [4] showed that mixed layers of this type are a common occurrence in the Arctic winter, and are the result of submesoscale restratification by horizontal density gradients. This circumstantial evidence, together with the spectral analyses described here, is further indicative of the importance of the submesoscales to the transient stratification that generates vertical gradients within the surface layer when atmospheric forcing is not strong enough to produce vertical mixing [4,5].

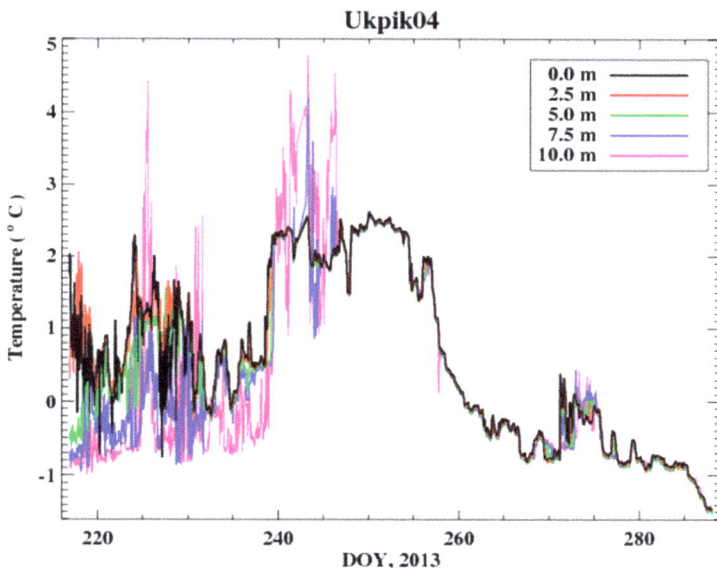

Figure 10. Temperatures in the top 10 m from the MIZOPEX UpTempO buoy (Ukpik04) deployed on 8 August 2013.

Boccaletti et al. [8] suggested that the horizontal density gradients at scales of $O(1)$ km that restratify the surface mixed layer on temporal scales of days are the result of baroclinic instabilities at frontal regions. These baroclinic instabilities have preferred length scales of ~$O(5)$ km. The spatial distribution of the BESST SST variance indicates that most of the sub-pixel variability is less than 0.1 K, but 5% of data can have a standard deviation well in excess of the accuracy requirements of satellite SSTs ($\sigma \geq 0.4$ K). An SST gradient spectrum with a slope of −2 is consistent with a front no wider than the smallest resolved length scales; in our case, this means that $\lambda < O(5–10)$ km. Not only are thermal fronts of this magnitude evident in the MODIS SST composite (Figure 5), but areas of high variability measured by BESST have preferential locations at or near these fronts.

It is noteworthy that MIZOPEX was not designed as a submesoscale measurement campaign, but rather as a marginal sea ice experiment. As such, there were no complementary measurements of density, winds, and surface fluxes to help us interpret the observed variability. Field campaigns aiming to resolve the submesoscale usually have multiple autonomous instruments with nested sampling covering entire areas and providing dozens of independent realizations of the submesoscale from different tracers. It was fortuitous that the BESST measurements captured the submesoscale. By using a UAV-deployed radiometer such as the BESST system, we have somewhat unintentionally revealed new features of the wavenumber spectrum of the SST variability between the mesoscale and the small scales that can be relevant for interpreting submesoscale processes of the upper ocean as well as for understanding the spatio-temporal variability of satellite remote sensing variables in general. Moreover, the horizontal scales involved in the restratification of the Arctic Ocean surface layer are not typically resolved in numerical models, which points to the need for additional high spatial and temporal resolution data that help demonstrate the role of submesoscale lateral processes in regulating the sub-pixel variability within high-resolution satellite data.

The growing literature on submesocale processes, as cited above, suggests that their impact is ubiquitous throughout the oceans. Additional high spatial resolution data from satellites, aircraft, and other new observing technologies will help to further verify this and study the processes in detail. New generation IR satellite sensors with spatial resolutions finer than 1 km will reveal increasing levels of pixel-to-pixel variability, and submesoscale processes represent an important potential contribution to that variability. More dedicated airborne and UAV observations such as those described here will be highly beneficial in quantifying the role and nature of these processes.

Acknowledgments: Sandra Castro, William Emery, and William Tandy Jr. were supported by the MIZOPEX project through NASA award NNX11AN57G. Gary Wick and Sandra Castro were also supported in part by a grant from the NASA Physical Oceanography Program, award NNH13AV34I (Eric Lindstrom). The authors also wish to thank James Maslanik for leading MIZOPEX, William Good for his additional advice on the function and performance of the BESST, and Mathew Tooth for early analysis of the BESST data. The UpTempO buoy SSTs from the Polar Science Center at the Applied Physics Laboratory in the University of Washington were obtained from the UpTempO Buoy Project web site at http://psc.apl.washington.edu/UpTempO/. Special thanks to NASA Cryospheric Sciences Program Manager Thomas Wagner. The authors also appreciate very constructive comments provided in a review by Peter Cornillon.

Author Contributions: Sandra Castro performed the data analysis and wrote the paper; William Emery deployed the BESST radiometer and had a vision for the application of the data; Gary Wick provided suggestions on the analysis and helped edit the paper; and William Tandy contributed instrument support and insight.

Conflicts of Interest: The authors declare no conflict of interest.

References

1. Thomas, L.N.; Tandon, A.; Mahadevan, A. Submesoscale Processes and Dynamics. In *Ocean Modeling in an Eddying Regime*; American Geophysical Union: Washington, DC, USA, 2008; pp. 17–23.
2. Shay, L.; Cook, T.; An, P. Submesoscale coastal ocean flows detected by very high frequency radar and autonomous underwater vehicles. *J. Atmos. Ocean. Technol.* **2003**, *20*, 1583–1599. [CrossRef]
3. Capet, X.; McWilliams, J.C.; Molemaker, M.; Shchepetkin, A. Mesoscale to submesoscale transition in the California Current system. Part II: Frontal processes. *J. Phys. Oceanogr.* **2008**, *38*, 44–63. [CrossRef]
4. Timmermans, M.-L.; Cole, S.; Toole, J. Horizontal density structure and restratification of the Arctic Ocean surface layer. *J. Phys. Oceanogr.* **2012**, *42*, 659–668. [CrossRef]
5. Hosegood, P.; Gregg, M.C.; Alford, M.H. Sub-mesoscale lateral density structure in the oceanic surface mixed layer. *Geophys. Res. Lett.* **2006**, *33*, 126–136. [CrossRef]
6. Lee, C.; D'Asaro, E.; Harcourt, R. Mixed Layer Restratification: Early Results from the AESOP Program. 2006. Available online: http://adsabs.harvard.edu/abs/2006AGUFMOS51E..04L (accessed on 6 September 2017).
7. Fox-Kemper, B.; Ferrari, R.; Hallberg, R. Parameterization of mixed layer eddies. Part I: Theory and diagnosis. *J. Phys. Oceanogr.* **2008**, *38*, 1145–1165. [CrossRef]
8. Boccaletti, G.; Ferrari, R.; Fox-Kemper, B. Mixed layer instabilities and restratification. *J. Phys. Oceanogr.* **2007**, *37*, 2228–2250. [CrossRef]
9. Capet, X.; McWilliams, J.C.; Molemaker, M.; Shchepetkin, A. Mesoscale to submesoscale transition in the California Current system. Part I: Flow structure, eddy flux, and observational tests. *J. Phys. Oceanogr.* **2008**, *38*, 29–43. [CrossRef]
10. Mahadevan, A.; Campbell, J. Biogeochemical patchiness at the sea surface. *Geophys. Res. Lett.* **2002**, *29*. [CrossRef]

11. Abraham, E. The generation of plankton patchiness by turbulent stirring. *Nature* **1998**, *391*, 577–580. [CrossRef]

12. Cornillon, P.; Castro, S.; Gentemann, C.; Jessup, A.; Kaplan, A.; Lindstrom, E.; Maturi, E. SST Error Budget—White Paper. 2010. Available online: http://works.bepress.com/peter-cornillon/1/ (accessed on 6 September 2017).

13. Castro, S.L.; Wick, G.A.; Minnett, P.J.; Jessup, A.T.; Emery, W.J. The impact of measurement uncertainty and spatial variability in the accuracy of skin and subsurface regression-based sea surface temperature algorithms. *Remote Sens. Environ.* **2010**, *114*, 2666–2678. [CrossRef]

14. Ohring, G.; Wielicki, B.; Spencer, R.; Emery, B.; Datla, R. Satellite instrument calibration for measuring global climate change: Report of a workshop. *Bull. Am. Meteorol. Sci. USA* **2005**, *86*, 1303–1313. [CrossRef]

15. Brink, K.H.; Cowles, T.J. The Coastal Transition Zone program. *J. Geophys. Res.* **1991**, *96*, 14637–14647. [CrossRef]

16. Toole, J.M.; Timmermans, M.L.; Perovich, D.K.; Krishfield, R.A.; Proshutinsky, A.; Richter-Menge, J.A. Influences of the ocean surface mixed layer and thermohaline stratification on Arctic sea ice in the central Canada basin. *J. Geophys. Res.* **2010**, *115*, C10018. [CrossRef]

17. Emery, W.J.; Good, W.S.; Tandy, W., Jr.; Izaguirre, M.A.; Minnett, P.J. A microbolometer airborne calibrated infrared radiometer: The Ball Experimental Sea Surface Temperature (BESST) radiometer. *IEEE Trans. Geosci. Remote Sens.* **2014**, *52*, 7775–7781. [CrossRef]

18. Good, W.; Warden, R.; Kaptchen, P.F.; Emery, W.J.; Giacomini, A. Absolute Airborne Thermal SST Measurements and Satellite Data Analysis from the Deepwater Horizon Oil Spill. In *Monitoring and Modeling the Deepwater Horizon Oil Spill: A Record-Breaking Enterprise*; Liu, Y., Macfadyen, A., Ji, Z.-G., Weisberg, R.H., Eds.; American Geophysical Union: Washington, DC, USA, 2011; pp. 51–61.

19. GHRSST Science Team. The Recommended GHRSST Data Specification (GDS) 2.0, Document Revision 5. Available from the GHRSST International Project Office (document reference GDS2.0r5.doc). 2012, p. 123. Available online: https://www.ghrsst.org/wp-content/uploads/2016/10/GDS20r5.pdf (accessed on 12 September 2017).

20. Samelson, R.M.; Paulson, C.A. Towed thermistor chain observations of fronts in the subtropical North Pacific. *J. Geophys. Res.* **1988**, *93*, 2237–2246. [CrossRef]

21. Cole, S.T.; Rudnick, D.L.; Colosi, J.A. Seasonal evolution of upper-ocean horizontal structure and the remnant mixed layer. *J. Geophys. Res.* **2010**, *115*, C04012. [CrossRef]

22. Hodges, B.A.; Rudnick, D.L. Horizontal variability in chlorophyll fluorescence and potential temperature. *Deep-Sea Res. I* **2006**, *53*, 1460–1482. [CrossRef]

23. Kunze, E.; Klymak, J.M.; Lien, R.C.; Ferrari, R.; Lee, C.M.; Sundermeyer, M.A.; Goodman, L. Submesoscale Water-Mass Spectra in the Sargasso Sea. *J. Phys. Oceanogr.* **2015**, *45*, 1325–1338. [CrossRef]

24. Pope, S.B. *Turbulent Flows*; Cambridge University Press: Cambridge, UK, 2000; p. 771.

25. Obukhov, A.M. The structure of the temperature field in a turbulent flow. *Izv. Akad. Nauk. SSSR Ser. Geogr. Geofiz.* **1949**, *13*, 58–69.

26. Corrsin, S. On the spectrum of isotropic temperature fluctuations in isotropic turbulence. *J. Appl. Phys.* **1951**, *22*, 469–473. [CrossRef]

27. Nastrom, G.D.; Gage, K.S. A climatology of atmospheric wavenumber spectra of wind and temperature observed by commercial aircraft. *J. Atmos. Sci.* **1985**, *42*, 950–960. [CrossRef]

28. McCaffrey, K.; Fox-Kemper, B.; Forget, G. Estimates of Ocean Macroturbulence: Structure Function and Spectral Slope from Argo Profiling Floats. *J. Phys. Oceanogr.* **2015**, *45*, 1773–1793. [CrossRef]

29. Saffman, P.G. On the Spectrum and Decay on Random Two-Dimensional Vorticity Distributions at Large Reynolds Numbers. *Stud. Appl. Math.* **1971**, *50*, 377–383. [CrossRef]

30. Hosokawa, I. Ensemble mechanics for the random-forced Navier-Stokes flow. *J. Stat. Phys.* **1976**, *15*, 87–104. [CrossRef]

31. Hopf, E. The partial differential equation $u_t + u u_x = u_{xx}$. *Commun. Pure Appl. Math.* **1950**, *3*, 201–230. [CrossRef]

32. Steele, M.; Rigor, I.; Ermold, W.; Ortmerer, M. UpTempO Buoys Deployed in 2013. NSF Arctic Data Center, 2015. Available online: http://psc.apl.washington.edu/UpTempO/Data.php (accessed on 14 September 2017).
33. Castro, S.L.; Wick, G.A.; Steele, M. Validation of satellite sea surface temperature analyses in the Beaufort Sea using UpTempO buoys. *Remote Sens. Environ.* **2016**, *187*, 458–475. [CrossRef]

remote sensing

MDPI

Article

Environmental Variability and Oceanographic Dynamics of the Central and Southern Coastal Zone of Sonora in the Gulf of California

Ricardo García-Morales [1], Juana López-Martínez [2,*], Jose Eduardo Valdez-Holguin [3], Hugo Herrera-Cervantes [4] and Luis Daniel Espinosa-Chaurand [1]

[1] CONACYT. Unidad Nayarit del Centro de Investigaciones Biológicas del Noroeste S.C. (UNCIBNOR+), Calle Dos No. 23. Cd. del Conocimiento., 63173 Tepic, Nayarit, Mexico; rgarcia@cibnor.mx (R.G.-M.); lespinosa@cibnor.mx (L.D.E.-C.)

[2] Centro de Investigaciones Biológicas del Noroeste S.C. Unidad Sonora, Campus Guaymas. Km. 2.35 Camino al Tular Estero de Bacochibampo, 85400 Heroica Guaymas, Sonora, Mexico

[3] Departamento de Investigaciones Científicas y Tecnológicas de la Universidad de Sonora, Luis Donaldo Colosio s/n, Colonia Centro, 83000 Hermosillo, Sonora, Mexico; jvaldez@guayacan.uson.mx

[4] Centro de Investigación Científica y de Educación Superior de Ensenada, Unidad La Paz. Calle Miraflores No. 334, 23050 La Paz, B.C.S., Mexico; hherrera@cicese.mx

* Correspondence: jlopez04@cibnor.mx; Tel.: +52-622-221-2237

Received: 28 June 2017; Accepted: 21 August 2017; Published: 6 September 2017

Abstract: This study analyzed monthly and inter-annual variability of mesoscale phenomena, including the El Niño Southern Oscillation (ENSO), the Pacific Decadal Oscillation (PDO) climate indexes and wind intensity considering their influence on sea surface temperature (SST) and chlorophyll a (Chl-a). These analyses were performed to determine the effects, if any, of climate indexes and oceanographic and environmental variability on the central and southern coastal ecosystem of Sonora in the Gulf of California (GC). Monthly satellite images of SST (°C) and Chl-a concentration were used with a 1-km resolution for oceanographic and environmental description, as well as monthly data of the climate indexes and wind intensity from 2002–2015. Significant differences ($p > 0.05$) were observed while analyzing the monthly variability results of mesoscale phenomena, SST and Chl-a, where the greatest percentage of anti-cyclonic gyres and filaments was correlated with a greater Chl-a concentration in the area of study, low temperatures and, thus, greater productivity. Moreover, the greatest percentage of intrusion was correlated with the increase in temperature and cyclonic gyres and a strong decrease of Chl-a concentration values, causing oligotrophic conditions in the ecosystem and a decrease in upwelling and filament occurrence. As for the analysis of the interannual variability of mesoscales phenomena, SST, Chl-a and winds, the variability between years was not significant ($p > 0.05$), so no correlation was observed between variabilities or phenomena. The results of the monthly analyses of climate indexes, environmental variables and wind intensity did not show significant differences for the ENSO and PDO indexes ($p > 0.05$). Nonetheless, an important correlation could be observed between the months of negative anomalies of the ENSO with high Chl-a concentration values and intense winds, as well as with low SST values. The months with positive ENSO anomalies were correlated with high SST values, low Chl-a concentration and moderate winds. Significant inter-annual differences were observed for climate indexes where the years with high SST values were related to the greatest positive anomaly of ENSO, of which 2002 and 2009 stood out, characterized as moderate Niño years, and 2015 as a strong El Niño year. The years with the negative ENSO anomaly were related to the years of lower SST values, of which 2007–2008 and 2010–2011 stood out, characterized as moderate Niñas. Thus, variability associated with mesoscale oceanographic phenomena and seasonal and inter-annual variations of climate indexes had a great influence on the environmental conditions of the coastal ecosystem of Sonora in the Gulf of California.

Remote Sens. **2017**, *9*, 925

Keywords: environmental variability; oceanographic dynamics; mesoscale phenomena; Gulf of California; SST; Chl-a; ENSO and PDO

---

1. Introduction

The Gulf of California (GC), one of the 24 marginal seas, one of the five largest gulfs of the Pacific Ocean [1] and considered one of the most productive seas of the planet [2,3], shows two periods separated by two short transition phases, a cold one with high biological productivity and a warm one with low productivity, attributed to mesoscale processes, such as thermocline, surface circulation induced by the wind, gyres, filaments and upwelling [4,5]. These mesoscale processes are the result of atmospheric forcing (wind) interaction in the eastern border of the Pacific, inducing a significant (barotropic) variability of eddy kinetic energy, which is associated with current forcing fluctuations in most regions [6,7].

The central region of the GC has intense seasonal upwelling, which takes place from January–April and from November–December [8,9]. The surface circulation pattern is influenced by seasonal winds with a flux toward the south in winter and toward the north in summer with important differences in temperature between both periods [10–14]. High temperatures and oligotrophic characteristics are generated during the summer with the arrival of the Mexican Coastal Current (MCC) intrusion to the area of study, generating a cyclonic circulation in this zone [13,15], while wind forcing in the winter generates filaments of cold water associated with high chlorophyll a (Chl-a) concentrations; likewise, cyclonic gyres (with counterclockwise circulation in the Northern Hemisphere) are responsible for the dispersion of high Chl-a concentration [16,17].

These oceanographic conditions of the GC vary in a wide interval of spatial scales (from a few to hundreds of km) and of temporal scales (monthly, seasonal, annual and inter-annual to decadal), exerting an influence on marine ecosystems and providing biologically-rich and productive habitats for a great diversity of ecologically- and commercially-important species [18–22]. These enriching processes of nutrient upwelling and mixing affect the trophic network, while the processes of particle concentration (gyres and currents) generate favorable conditions for spawning, survival and larval dispersion of phytoplankton organisms [14,23–26].

Many large-scale phenomena that act on varied time scales from the El Niño-Southern Oscillation (ENSO) occur in the Pacific Ocean with month-year cycles from decadal to multi-decadal frequency events, such as the Pacific Decadal Oscillation (PDO). Their effects are detectable from global to local ecosystems, and their aggregate contributions establish climate shifts that have shown fluctuations in atmospheric and oceanic conditions, such as sea temperature [27], wind fields associated with atmospheric pressure variations [28], ocean currents [29], coastal upwelling [30] and mixed layer depths [31]. Many of these climate variations occur via atmospheric-oceanic teleconnections [32], which extend from the troposphere to the ocean surface, including fluctuations in sea level pressure that are closely linked to changes in surface winds, sea surface temperature (SST), heat content, mixed layer and thermocline depth. Climate variability receives considerable attention because of large-scale influences in the North Pacific due to their impact on tropical and extra-tropical climate [33] and weather over North America [34].

The El Niño Southern Oscillation, perhaps the most studied large-scale phenomenon, comprises two phases. The warm El Niño phase, which has been studied extensively in the Pacific Ocean [12,32], is characterized by the weakening of trade winds, warming the sea surface layer in the tropical-subtropical eastern Pacific, a switch from low to high atmospheric pressure near Darwin, Australia, and the opposite effect near the Tahiti Islands, where it is termed the Southern Oscillation. The La Niña cold phase of ENSO is much less studied in the Pacific. It is characterized by an intensification of trade winds, cooling the sea surface layer in the tropical-sub-tropical Eastern Pacific, very low atmospheric pressure near Darwin, Australia, and very high pressure near the Tahiti Islands [35,36].

The Pacific Decadal Oscillation, another large-scale phenomenon, is a pattern of ocean variability over the entire Pacific, similar to ENSO in some respects, but with a much longer cycle [37,38]. It is also defined by two phases; the positive phase in the North Pacific occurs when SST anomalies are cold in the central North Pacific and warm along the Pacific coast and when sea level pressure is below average in the North Pacific [39], while the converse occurs during the negative phase. Both phases are calculated by the standardized difference between SSTs in the north-central Pacific and Gulf of Alaska. The PDO phases may be important in enhancing or dampening ENSO impacts [40].

The Pacific Ocean exerts a strong influence over the oceanographic conditions of the Gulf of California because of its connection to the eastern tropical Pacific [4]. The dynamic forcing from the Pacific over the gulf is one of the most important oceanographic features because it integrates relevant phenomena, such as salt and heat, global balances, thermohaline circulation and barotropic ocean circulation [12,41].

Studies have demonstrated the effects of these large-scale phenomena on the Gulf of California region, and satellite analyses have shown the influence of ENSO over SST variability. Based on satellite records of SST anomalies [11,42], it has been shown that temperature in the gulf registered up to 3 °C above normal during El Niño, while it was up to 3 °C below normal during La Niña.

Similarly, studies on the PDO have shown its influence on pressure, wind, temperature and precipitation patterns of the North Pacific [37,43,44]. Its temporal modulations are linked to several important biological and ecosystem variables in the ocean [45,46]. Nevertheless, other parameters, such as decadal fluctuations in salinity, nutrients and Chl-a in the eastern North Pacific, are often poorly correlated with the PDO [47].

The GC is complex in terms of climate variability in different time scales (ENSO), decadal to interdecadal and long-term trend besides the influence of climate change (CC) in the long-term trend [48] affecting marine communities [49,50]. It is located in the transition zone between the tropical and subtropical climate regimes and exposed to natural variability modes at a large scale, as ENSO, PDO and their spatial-temporal interactions [51]; in addition to the previous effects, this environmental variability causes changes in the limits of species distribution; mismatches between predators and prey; massive mortality events; and the increase of diseases, which are some of the biological effects attributed to CC [50,52]. Moreover, impacts in the GC ecosystem are caused by harmful algal blooms, deterioration in the mangrove area and morphology besides changes in hydrographic conditions in the coastal zone affecting aquaculture, mortality and changes in the distribution of benthic communities and important species for fisheries [53].

Because the GC remains free of clouds most of the year, this region is ideal for using high-resolution satellite-derived data to study surface variability and climate effects on some variables, such as SST and Chl-a. These effects have been well documented in the GC with more than three decades of satellite measurements (1980–2015). Soto-Mardones et al. [11], Lavín et al. [42] and Herrera-Cervantes et al. [13] examined high-resolution satellite-derived SST and chlorophyll data and their relationships to ENSO spatial signatures for different periods (1984–2004 and 1997–2006 respectively); they found that ENSO is the most important inter-annual variability signal. Kahru et al. [54], based on satellite data from the Ocean Color Temperature Scanner (OCTS), Sea-viewing Wide Field-of-view Sensor (SeaWiFS), the Moderate Resolution Imaging Spectroradiometer (MODIS-Aqua and MODIS-Terra), the Advanced Very-High-Resolution Radiometer (AVHRR) and the Vertically Generalized Production Model (VGPM) primary productivity model, found that the semiannual cycle of surface Chl-a concentration was higher during the spring and fall transition periods when the GC surface circulation was switching between cyclonic gyres in the summer and anti-cyclonic gyres in the winter. Satellite images, jointly with hydrographic surveys, suggest that the mesoscale variability in the GC is characterized by a complex pattern of filaments, meanders and semi-permanent eddy structures. These events carry organisms and properties from the entrance to the whole length of the GC [14], which cause important seasonal variability in the mean field of variables, such as SST and Chl-a, impacting the GC ecosystem, one of the most diverse marine biological communities in the world and a strategic region for marine fisheries in Mexico, ideal for the use of SST and Chl-a satellite images for its environmental characterization.

Therefore, the objective of this work was to perform a temporal-spatial characterization of the mesoscale processes and their relationships with the variations of SST and Chl-a concentration using high-resolution satellite-derived MODIS data. The influence of ENSO and PDO signals on SST, Chl-a and wind variations, as well as their influence on the coastal ecosystem of Sonora in the GC are included in the analysis.

2. Materials and Methods

2.1. Study Area

This study comprised the central and southern coastal regions of Sonora, Mexico, in the GC (from Bahía Kino to Yavaros, Sonora (Figure 1)), with several sub-basins, low tide amplitude, high productivity in winter due to coastal upwelling processes associated with winter wind patterns [8,9] and low productivity during the summer causing oligotrophic conditions [2,55]. Important differences in SST have been reported in the zone from winter to summer [11], as well as an important generation of cyclonic and anti-cyclonic gyres occupying the entire width of the GC [2,56].

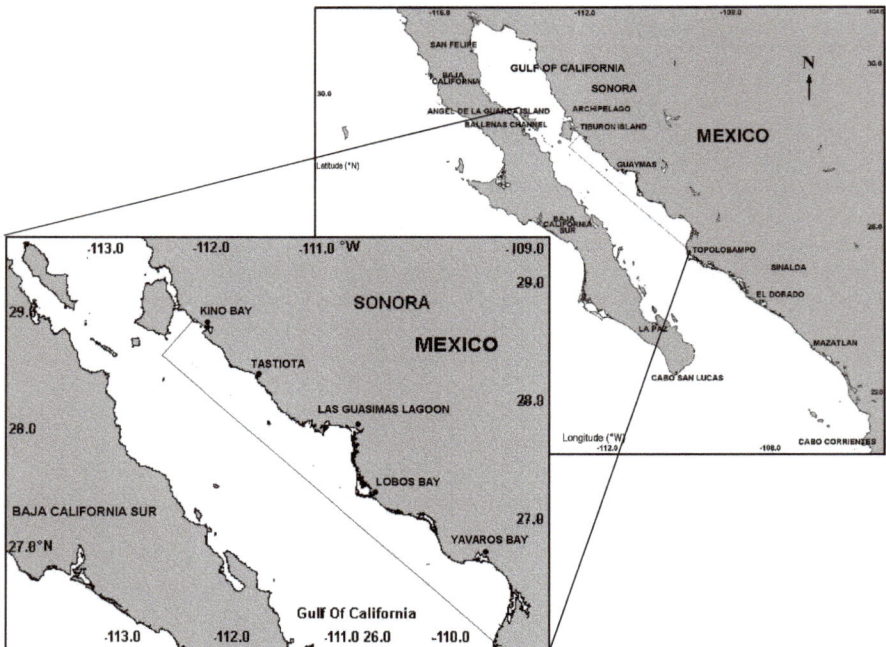

Figure 1. Central and southern coastal zone of Sonora in the Gulf of California and regionalization (continuous line) for environmental and oceanographic analyses.

2.2. Environmental Characterization and Oceanographic Dynamics

For spatial-temporal environmental characterization and oceanographic dynamics analyses, monthly averaged composite images of SST (°C) and Chl-a (mg/m^3) were used with a spatial resolution of 1 km, produced by Dr. Mati Kahru (Scripps Institution of Oceanography, http://www.wimsoft. com/CAL/). To produce monthly composite images, Kahru et al. (2012) processed and averaged daily images of SST and Chl-a Level-2 and unmapped datasets from multiple sensors: SeaWiFS, MODIS-Terra and MODIS-Aqua, MERIS (Medium Spectral Resolution Imaging Spectrometer) and VIIRS (Visible and Infrared Imager/Radiometer Suite) downloaded from the NASA's Ocean Color website (https://oceancolor.gsfc.nasa.gov/). Individual swaths were mapped to a common projection using an Albers Conic Equal Area projection to make each one comparable; because of the Earth's curvature, the viewing geometry is different for each pixel, and the ground projection of each satellite swath is also different. Cloud-free pixels from many satellite passes were merged to create composite images. As clouds move around during the day, it is possible to use holes in the clouds and composite data from multiple sensors and multiple overpasses during the day to get maximum coverage. Merging data from multiple sensors has improved spatial and temporal coverage compared to a single sensor and correspondence to in situ data because of regional optimization [57]. Sub-images of 1438 × 1397 pixels were cut for the gulf from the supplied images, and mean Chl-a and SST for each month were calculated using routines included in the Windows Image Manager Automation Module software (WIM/WAM) [58].

Satellite data were processed by Mati Karhu (www.wimsoft.com/Satellite_Projects.htm), who grouped them for the region of the California Current and at the same time included the region of the Gulf of California by using two types of measurements: sea surface temperature (SST) and superficial chlorophyll a concentration (Chl-a). In case they are required, other variables can be added, such as the diffuse attenuation of downwelling light at 490 nm (Kd490) and the remote sensing reflectances at various wavelengths (e.g., Rrs443, Rrs490, Rrs555, and so on).

For each year, these data are composed of different time intervals: day, 5 days, 15 days, month (www.wimsoft.com/CAL/). Likewise, if possible, similar data of different sensors are merged to increase coverage and reduce lost data due to clouds.

The set of high-resolution data (1 km) of SST and Chl-a with numerical values are in HDF4 and PNG format, which can be read by different image-processing programs where the attributes within the HDF files provide information on the same data. The complete resulting image is 3840 pixels wide and 3405 pixels high. Image annotations are reduced four times and include longitude/latitude grid, etc., so special care should be taken while interpreting pixel value. The corresponding latitudes and longitudes of each pixel can be obtained from the HDF4 file (spg.ucsd.edu/Satellite_Data/California_ Current/cal_aco_3840_Latitude_Longitude.hdf).

The SST and Chl-a products are simple averages of all valid data of all of the sensors available: MODIST and MODISA for SST; sensors such as MODIST, MODISA, MERIS and VIIRS for Chl-a. Not all sensors are always available. The most advanced data merging is described in Kahru et al. (2012, 2015, 2016). Both SST and Chl-a values in the HDF files use 1 byte per pixel with specific scaling.

Linear scaling is used for SST and logarithmic scaling for Chl-a. The scaling equations using the pixel value (PV) as the unsigned byte (from 0–255) are: SST (deg C) = 0.15 × PV − 3.0 and to Chl (mg m^{-3}) = 10^(0.015 × PV − 2.0), i.e., 10 to the power of 0.015 × PV − 2.0. Pixel values of 0 and 255 (and the corresponding scaled values) are considered invalid and must be excluded in any statistics (www.wimsoft.com/CAL/).

Additionally, a database was built for an area centered at Guaymas basin (27.5° N and 111° W) to observe the relationship between climate indexes, SST, Chl-a, mesoscale structures, wind and a monthly time series of wind data. These wind data derived from the NCEP Reanalysis Dataset of zonal and meridional wind website at https://www.esrl.noaa.gov/psd/data/timeseries/ from January 2002–December 2015.

To determine a more precise scheme of environmental fluctuations and their possible consequences on SST, Chl-a, and mesoscale phenomena in the gulf and central and southern coastal zone of Sonora, the monthly values of the following climate indexes were used: (1) Oceanic Niño Index (ONI) for ENSO, defined as the three-month running mean of SST anomalies in the Niño 3.4 region

(5°N–5°S, 120°–170°W); (2) Pacific Decadal Oscillation (PDO), the first principal component of the North Pacific SST anomaly field (20°N–70°N) with the subtracted global mean [59].

For the environmental and oceanographic analysis, a study zone was delimited, including from Bahía Kino to Bahía de Yavaros, Sonora, Mexico, given its importance in ecology and fisheries and to the diversity of oceanographic processes that occur there (Figure 1).

Each of the monthly images of SST and Chl-a were used to identify mesoscale phenomena (MCC water intrusion, filaments, upwelling and cyclonic and anti-cyclonic gyres), recording the monthly frequency of these phenomena observed (months in which the same phenomenon was observed) and the length (month in which they started, persisted and ended) in the study zone. Criteria established in the previous studies of Pegau et al. [56], López [60], Zamudio et al. [2] and García-Morales et al. [61] were followed for their identification. Upwelling and filaments were identified by a major concentration of pigments in their border compared to that in their surrounding; gyres, and their directions were determined by observing their origin and amplitude (originating mostly in the coastal zone with concentration values greater than the gyre border and the end of low concentrations; on many occasions, gyres did not show a closed circle, giving as a result the gyre direction). The observed length, monthly frequency and arrival of water induced by the MCC forcing toward the interior of the gulf were identified when monthly temperatures were greater than 26 °C [62,63] and when Chl-a concentrations were less than one mg/m^3 [17,64,65], taking them as warming and oligotrophic condition markers, respectively, in the area of study.

After identification, the observed cumulative monthly frequency of each mesoscale phenomenon was plotted for all of the years and months from 2002–2015; the length in months for each phenomenon observed was also determined, as well as SST and Chl-a variability of the coastline of Sonora and the adjacent oceanic zone from Bahía de Kino to Bahía Yavaros in the GC.

The percentage of observed mesoscale phenomena was calculated (MCC water intrusion, filaments, upwelling, cyclonic and anti-cyclonic gyres) and shown by month and year from 2002–2015; as a reference the total phenomena for each temporality (month and year) in such a period, as well as the correlation between these phenomena, SST, Chl-a, winds, PDO and ENSO climate indexes were taken to determine their influence and effects on the coastal ecosystems of Sonora, Mexico.

2.3. Statistical Analyses

A one-way analysis of variance (Kruskal-Wallis; $p < 0.05$) was applied to all data generated for percentages of the mesoscale phenomena observed per month (MCC water intrusion, filaments, upwelling, cyclonic and anti-cyclonic gyres), SST, Chl-a, winds, PDO and ENSO climate indexes per month and year from 2002–2015, in each case, all posterior to not complying with the normality Kolmogorov-Smirnov ($\alpha = 0.05$) test and/or Bartlett's ($\alpha = 0.05$) homogeneity of variance test. Significant differences were determined between months or years by phenomenon, SST, Chl-a, winds, PDO and ENSO by Tukey's ($\alpha = 0.05$) multiple comparison test. Square-root arcsine transformation was applied to data expressed as a percentage (MCC water intrusion, filaments, upwelling, cyclonic and anti-cyclonic gyres) for processing Zar, 1999. A Spearman analysis was performed to correlate the mesoscale phenomena observed (MCC water intrusion, filaments, upwelling, cyclonic and anti-cyclonic eddies), SST, Chl-a, winds, PDO and ENSO (R^2; $\alpha = 0.05$).

Time series analysis: The Fourier analysis was applied to both series (StatSoft, Inc. (2007). Statistica (data analysis software system), Version 8.0. www.statsoft.com), SST and Chl-a to obtain cyclical patterns and their lengths.

3. Results

3.1. Analyses of Cumulative Annual and Monthly Frequency of Mesoscale Phenomena Observed in the Central and Southern Coastal Region of Sonora in the Gulf of California from 2002–2015

A total of 336 monthly images of SST and Chl-a were analyzed, for which a total of 353 mesoscale phenomena were observed in the coastal region of Sonora and the adjacent deep zone to the GC with a cumulative monthly average of 29 ± 1.2, highlighting the months of February, March, April, May, June and July with greater frequency of mesoscale structures (37, 43, 42, 38, 29 and 29, respectively). The most frequent phenomena were cyclonic gyres, upwelling, and water intrusion with values of occurrence of 88, 90 and 84, respectively, followed by filaments and anti-cyclonic gyres. The upwelling seasonal pattern was clearly defined and extended from November–May with peaks in the months from January–April from 3.3–5.11 mg/m^3. On the other hand, cyclonic gyres showed two maximum peaks of occurrence from February–April and from June–July. Another very persistent phenomenon in the GC was MCC water intrusion, which showed its arrival to the gulf with major frequency starting from May and extended permanency up to September. The highest wind magnitudes occurred from March–July and a period of less intensity from October–February (Figure 2).

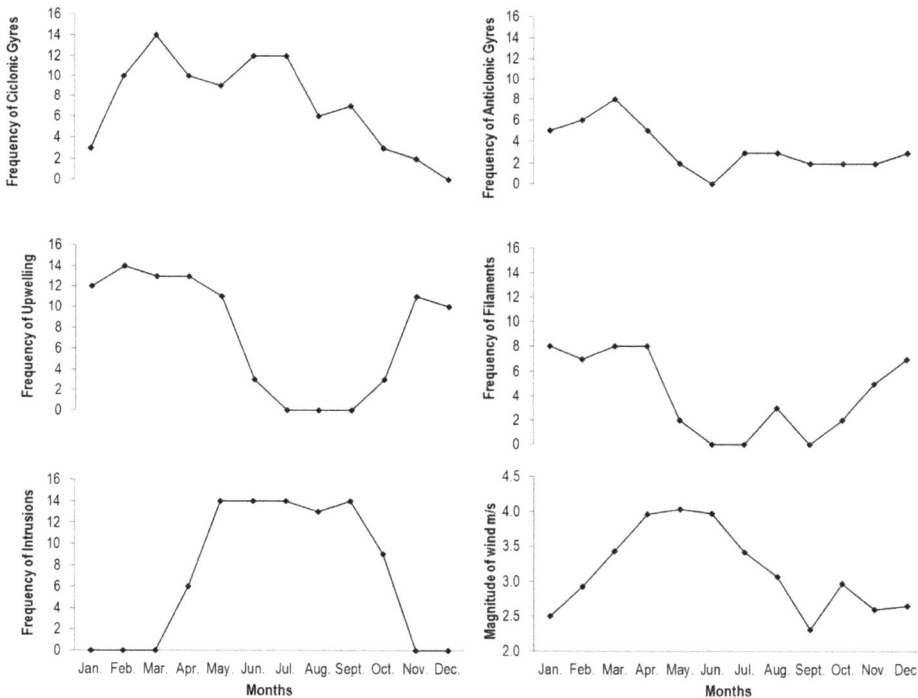

Figure 2. Cumulative monthly frequency observed of mesoscale phenomena and winds in the central and southern coastal region and adjacent deep zone in the Gulf of California.

Major structure frequencies were observed interannually in 2005, 2007, 2011, 2012 and 2014 with values of 28, 29, 31, 34 and 32, respectively, and a notable decrease in 2002, 2003, 2006, 2010 and 2015, of which 2015 was the year that showed the least number of phenomena with an annual total of 18. Likewise, an increasing trend was observed in the frequency of cyclonic gyres, filaments and intrusions and a decreasing trend in upwelling. As for anti-cyclonic gyres and upwelling, no trend was observed throughout the period of study, as well as no observation of anti-cyclonic gyres in 2006 due to the difficulty of determining them by the low concentration of Chl-a and intensification of the intrusion. The highest wind magnitudes occurred in 2006–2008 and 2011–2013. Variability was observed, although none of these trends were significant (Figure 3).

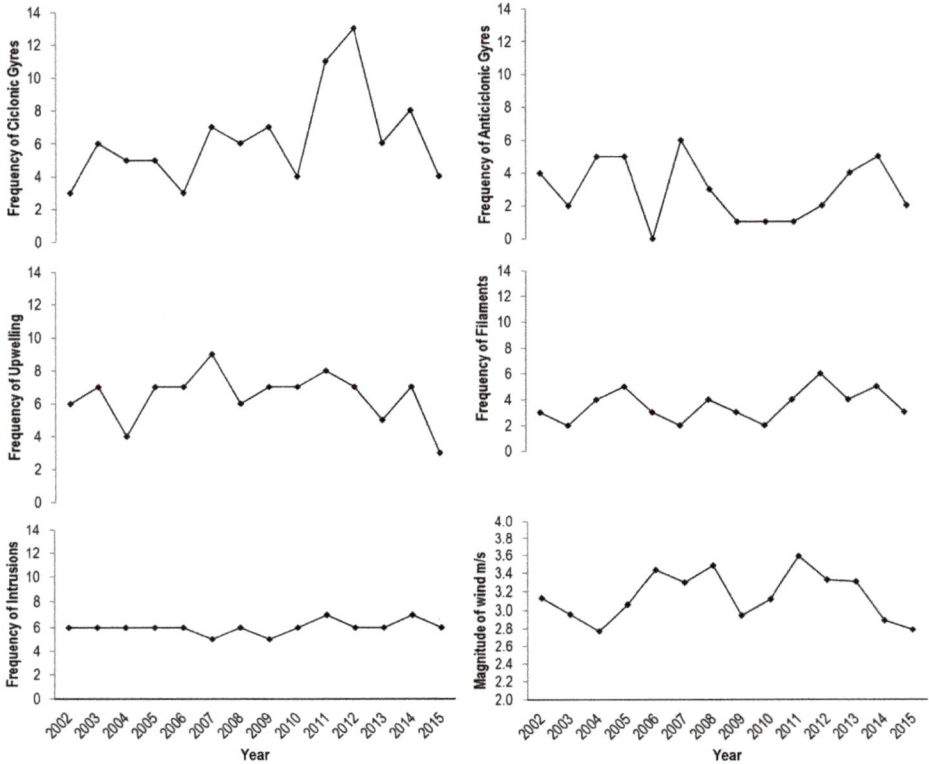

Figure 3. Cumulative annual frequency observed of mesoscale phenomena and winds in the central and southern coastal region of Sonora and adjacent deep zone in the Gulf of California.

3.2. Analyses of Mesoscale Phenomena Observed Monthly by Images of Chlorophyll a

Monthly images of Chl-a were taken indistinctly in 2004 and 2012 to exemplify the analysis of the mesoscale phenomena; the years were chosen at random after determining that no significant differences were observed in the phenomena between years. Figure 4 shows the analysis for 2004 with a total frequency of 24 phenomena observed, mainly cyclonic and anti-cyclonic gyres from February–September and upwelling from February–April and in November. The presence of intrusion was observed from April–September, but more intensity was detected in August, reaching inclusively the northern GC area. Filaments were also observed with higher concentration of Chl-a extending toward the deep zone from April–November (Figure 4).

The year 2012 (Figure 5) was characterized by the greatest monthly frequency of the mesoscale phenomena observed with a total of 34. It was also characterized by showing a greater presence of cyclonic and anti-cyclonic gyres from January–September when March and April stood out with trains up to four gyres and intense upwelling in the area of study during the first half of the year. Water intrusion from the Pacific also took place from April–October 2012.

Figure 6 shows an example of warm water intrusion from the tropical Pacific through the MCC with SST greater than 26 °C by monthly SST images in May 2004 and 2012.

The results of analyzing the number of monthly observations and the length of the phenomena determined that the cyclonic gyres had a minimum length from one month up to a maximum of nine where 2003, 2009, 2011 and 2012 were the years when the greatest length of this phenomenon was observed with 5, 7, 5 and 9 months, respectively; 2011 and 2012 were those showing the greatest observation frequency of this phenomenon, recording an increasing trend the first 11 years with a sudden decrease the last three (2013–2015) (Table 1 and Figure 5). As for anti-cyclonic gyres, a minimum length of one month was established for the phenomenon in most of the years with a maximum of three months in 2002, 2007 and 2014 with an observation frequency of 4, 6 and 5, respectively, and a decreasing trend in the period of study. It is worth mentioning that no observation was recorded on this phenomenon in 2006.

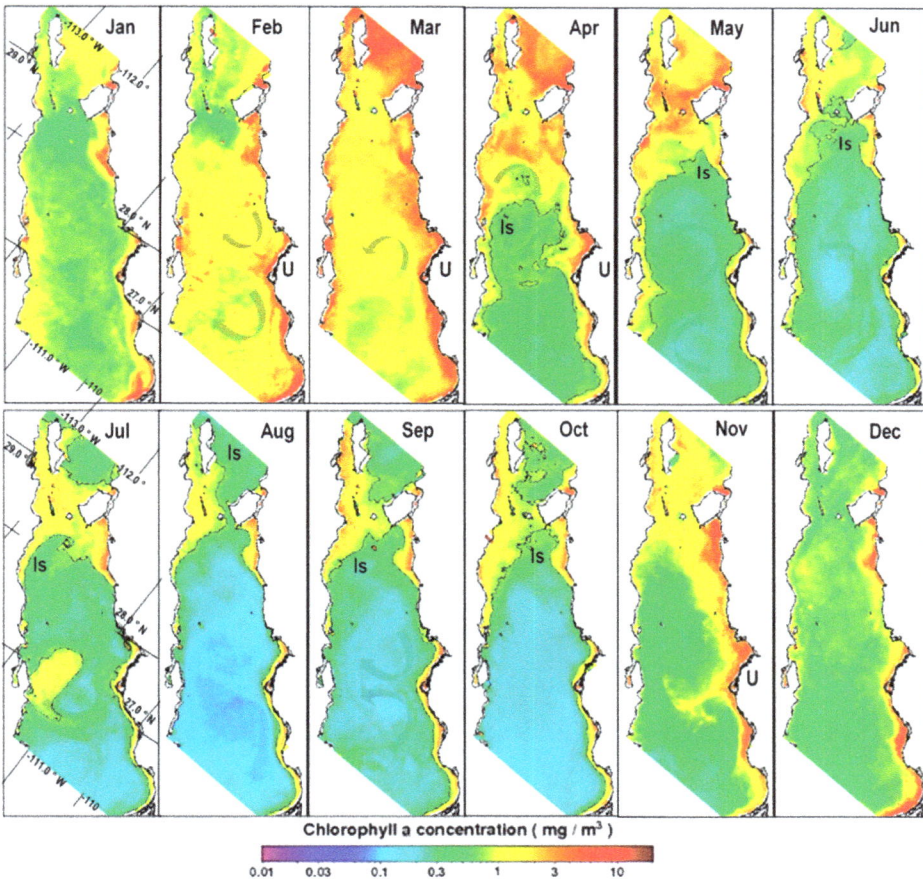

Figure 4. Satellite images of chlorophyll a mg/m^3 concentration in 2004. Isoline of Chl-a of 1 mg/m^3 (Is) and upwelling (U). Arrows in the clockwise direction indicate anti-cyclonic gyres, and those in the counterclockwise direction indicate cyclonic gyres.

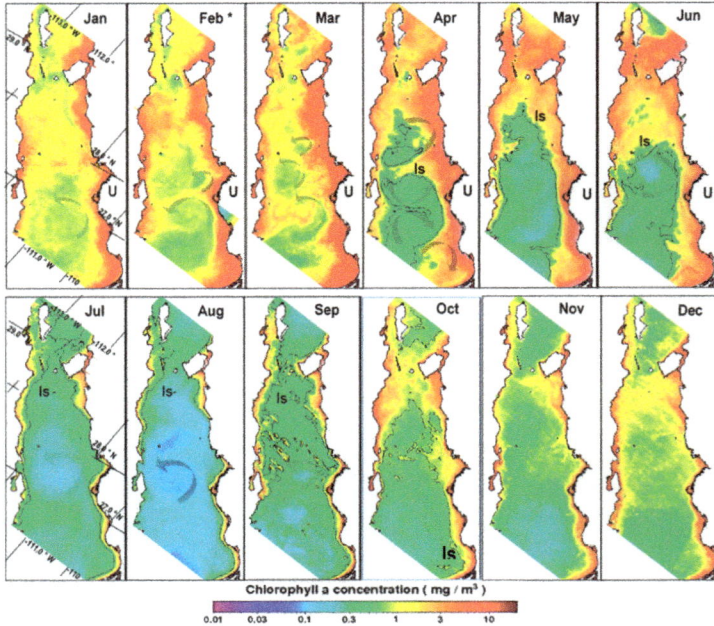

Figure 5. Satellite images of chlorophyll a mg/m^3 concentration in 2012. Isoline of Chl-a of one mg/m^3 (Is) and upwelling (U). Arrows in the clockwise direction indicate anti-cyclonic gyres, and those in the counterclockwise direction indicate cyclonic gyres.

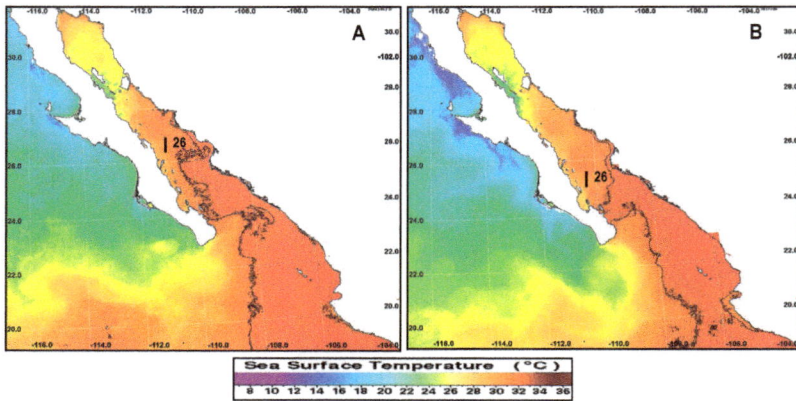

Figure 6. Satellite images of sea surface temperature °C of May 2004 (**A**) and 2012 (**B**) of the Gulf of California and temperature isolines of 26 °C (I 26) arriving to the central and southern coastal region of Sonora and the adjacent deep zone in the Gulf of California.

Table 1. The number of monthly observations of phenomena per year, length in months of phenomenon per year, its mode of observation per month and its maximum length in the central and southern coastal region of Sonora and adjacent deep zone in the Gulf of California, Mexico, from 2002–2015.

Phenomena	Year													
Cyclonic Gyres	2002	2003	2004	2005	2006	2007	2008	2009	2010	2011	2012	2013	2014	2015
No. of monthly observations of the phenomenon per year	3	6	5	5	3	7	6	7	4	11	13	6	8	4
Length of the phenomenon in months per year	1 1	1 5	1 3 1	2 1	1 1 1	2 2 3	2 1 2	7	1 3	6 3 1	9 2 2	2 2	2 4	4
Mode of phenomena observations per month	1	1	1	1	1	1	1	1	1	2	1	1	1	1
Maximum duration of the phenomenon in months	1	5	3	2	1	3	2	7	3	5	9	2	4	4
Anticyclonic Gyres														
No. of monthly observations of the phenomenon/year	4	2	5	5		6	3	1	1	1	2	4	5	2
Length of the phenomenon in months per year	3 1	1 1	2 1 1	1 1 1 1		3 1 1 1	1 1 1	1	1	1	1	1 2	3 2	1 1
Mode of phenomena observations per month	1	1	1	1		1	1	1	1	1	1	1	1	1
Maximum duration of the phenomenon in months	3	1	1	1		3	1	1	1	1	1	2	3	1
Upwelling														
No. of monthly observations of the phenomenon/year	6	7	4	7	7	9	6	7	7	8	7	5	7	3
Length of the phenomenon in months per year	5 1	5 2	3 1	4 3	5 2	6 3	2 2 2	5 2	5 2	6 2	6 1	4 1	5 2	3
Mode of phenomena observations per month	1	1	1	1	1	1	1	1	1	1	1	1	1	1
Maximum duration of the phenomenon in months	5	5	4	4	5	6	2	5	5	6	6	4	5	3
Filaments														
No. of monthly observations of the phenomenon/year	3	2	4	5	3	2	4	3	2	4	6	4	5	3
Length of the phenomenon in months per year	1 1 1	2	1 1 2	2 1	3	1 1	1 1	2 1	2	3 1	4 1	1 2	2 1	3
Mode of phenomena observations per month	1	1	1	2	1	1	1	1	1	1	1	1	1	1
Maximum duration of the phenomenon in months	1	2	2	2	3	1	1	2	2	3	4	2	2	3
Intrusion														
No. of monthly observations of the phenomenon/year	6	6	6	6	6	5	6	5	6	7	6	6	7	6
Length of the phenomenon in months per year	6	6	6	6	6	5	6	5	6	7	6	6	7	4 2
Mode of phenomena observations per month	1	1	1	1	1	1	1	1	1	1	1	1	1	1
Maximum duration of the phenomenon in months	6	6	6	6	6	5	6	5	6	7	6	6	7	5

Numerical values by columns and rows show the total number of phenomena observed by month and year, the length of the phenomenon by month (when there was more than one observation of the phenomenon per month, its length was considered for one month); thus, in some cases, the sum of lengths per month does not agree with the total observation of the phenomena per month.

In the analysis of monthly upwelling, a minimum length of one month and a maximum of six were observed with two upwelling periods, one five-month and two-month periods in 2002, 2003, 2006, 2009–2010 and 2014, although periods of up to six months were observed in this phenomenon in 2007 and 2011 with a decreasing trend. Filaments with high Chl-a were observed in the area of study with a minimum concentration of one month and a maximum of up to four in 2012, as well as a period of predominant observation of two months for this phenomenon in the majority of the years of study with an increasing trend. As for MCC water intrusion, a minimum of five months was observed in 2007, 2009 and 2015 and a maximum of seven in 2011 and 2014, where a six-month period predominated in the rest of the years, keeping stable during all of the period of study (Table 1).

3.3. Monthly and Annual Analyses of Chlorophyll a, Sea Surface Temperature and Climate Indexes of the El Niño Southern Oscillation and Pacific Decadal Oscillation from 2002–2015

The average concentration of Chl-a showed a seasonal pattern with high values from 3.13–5.11 mg/m^3 from November–April, reaching a maximum average of 5.11 ± 1.68 mg/m^3 in March and low values with an average of 1.02 ± 0.25 mg/m^3 from May–October, contrary to the SST, which showed average values ≥ 26 °C from June–October with a maximum average of 31.67 ± 0.51 °C in August and low values from 18 ± 1.29–24.9 ± 0.84 °C from November–April. It is worth mentioning that when SST values were ≥ 26 °C in May, they showed a decline of Chl-a; and when these values were ≤ 26 °C, they showed increased concentration of Chl-a in November (Figure 7).

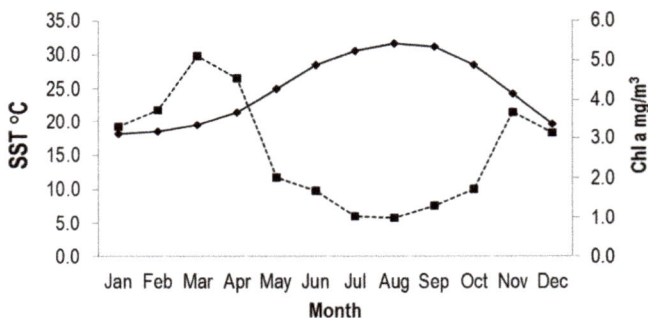

Figure 7. Variability of chlorophyll a (dotted line) and sea surface temperature (continuous line) in the central and southern coastal region of Sonora and adjacent deep zone in the Gulf of California.

An inverse lineal relationship ($R^2 = 0.73$) was observed between temperature and Chl-a concentration (Figure 8).

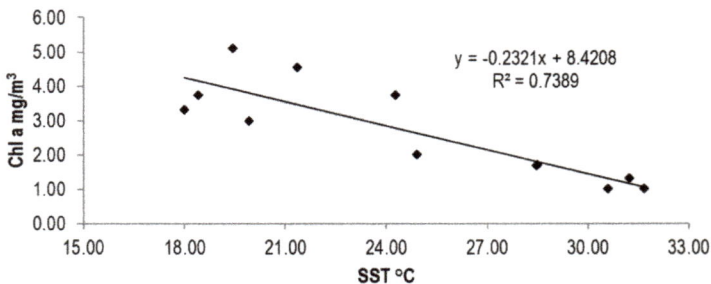

y = -0.2321x + 8.4208
R² = 0.7389

Figure 8. Relationship of sea surface temperature vs. chlorophyll a in the central and southern coastal region of Sonora and adjacent deep zone in the Gulf of California.

Higher concentrations of Chl-a were observed interannually from 2004–2012 with a temporal decrease in 2010. The highest one was recorded in 2009 with average values of 4.11 ± 3.58 mg/m^3; the minimum value observed was 1.79 ± 1.03 mg/m^3 in 2004 (Figure 9).

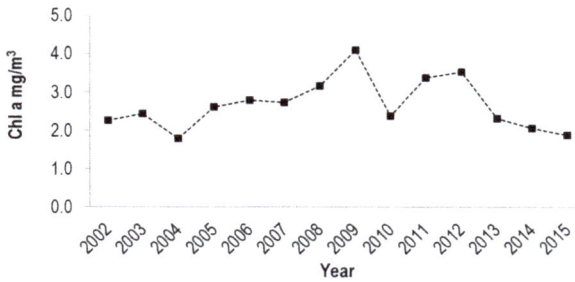

Figure 9. Average annual variability of chlorophyll a in the central and southern coastal region of Sonora, Mexico, and adjacent oceanic zone in the Gulf of California.

A minimum increasing trend in values of Chl-a was observed from 2002–2009, which started decreasing from 2013. As for average SST, fluctuations with values ranging from 23.89 ± 5.83–26.34 ± 4.39 °C were observed in 2008 and 2014, respectively, with a clear increasing trend (Figure 10).

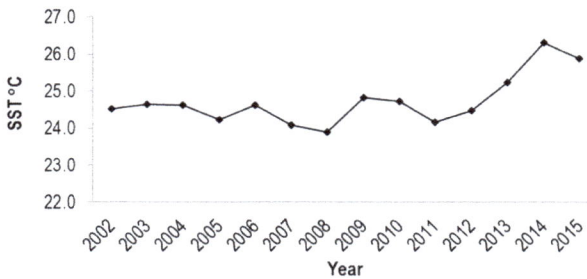

Figure 10. Average annual variability of sea surface temperature in the central and southern coastal region of Sonora, Mexico, and adjacent oceanic zone in the Gulf of California.

In the annual analysis of environmental variables and the percentage of the mesoscale phenomena, no statistical differences ($p > 0.05$) were observed in the years assessed, indicating lesser variability in yearly values for the percentage of phenomena reported (Table 2). Whereas the monthly analyses of the phenomena observed from 2002–2015 showed significant differences ($p < 0.05$) between months for temperature, chlorophyll, cyclonic gyres, upwelling and intrusions, no monthly statistical differences ($p > 0.05$) were observed in the anti-cyclonic gyres recorded by variability between months. For Chl-a, an increase in concentration was recorded from November–April with significantly higher concentrations in March (5.11 ± 1.68 mg/m^3) and April (4.55 ± 1.53 mg/m^3) ($p < 0.05$). Similar trends in upwelling and filaments were recorded with observations of significantly greater upwelling in November, January and February (55.00 ± 41.98, 42.86 ± 35.37 and $37.34 \pm 19.66\%$, respectively) and for filaments in December and January (35.00 ± 31.16 and $28.57 \pm 21.08\%$) ($p < 0.05$). For temperature and percentage of the intrusions observed, the increasing trend was recorded from June–September, showing August and September to be significantly hotter (31.67 ± 0.51 and 31.23 ± 0.39 °C, respectively, $p < 0.05$); for the intrusions observed, September showed a significantly greater percentage (60.37 ± 27.095; $p < 0.05$). According to the previous information, the increase in Chl-a concentration was related to similar increasing upwelling and filament trends in the same period; on the contrary lower concentrations agree with the periods where the percentages of upwelling and filaments decreased and with the temperature increase due to an increase of the MCC water intrusion percentage.

Table 3 shows the environmental variables and mesoscale phenomena with significant correlations (<0.05); the greatest correlations (R^2) were inversely observed between temperature and Chl-a in February and April (−0.78 and −0.74, respectively); between upwelling and filaments directly in October (0.74); and upwelling and intrusions inversely in June (−0.75), while inversely between cyclonic gyres in September (−0.76). According to the previous information, when SST decreased from February–April, greater Chl-a concentrations were observed; when water intrusion was observed from May–September, a decrease was observed in Chl-a, cyclonic and anti-cyclonic gyres, upwelling and filaments, and an increase in SST from September–October. It could be observed that when upwelling and filaments increased and in turn when upwelling increased, filaments also increased.

In the monthly analyses of the environmental variables, winds and data of the ENSO and PDO climate indexes, no statistical differences were observed for ENSO and PDO ($p > 0.05$) among the months assessed; however, a period of negative anomalies for ENSO could be appreciated from January–April with values from −0.05 ± 0.44 – −0.16 ± 0.71 and a period of positive anomalies from May–December with values from 0.01 ± 0.39–0.21 ± 0.96 with a negative anomaly of −0.21 ± 1.15 in August, which differs from the pattern previously observed. Whereas the annual analyses of climate indexes from 2002–2015 showed significant differences ($p < 0.05$) among the years analyzed, for ENSO and PDO, a five-year period of positive anomalies was observed from 2002–2006 with values from 0.16 ± 0.57–0.43 ± 0.26; it was followed by a restoration period of negative anomalies from 2007–2013 with values from −0.26 ± 0.10 – −0.70 ± 0.34 except for 2009 when a positive anomaly was recorded, different from the pattern that had been observed. The years that showed greater anomalies were 2002 and 2009, both characterized as moderate Niños, and 2015, characterized as a very strong Niño; the years with greater negative anomalies were 2007–2008 and 2010–2011, characterized as moderate Niñas. In terms of wind intensity, a period of greater intensity was observed from March–June with values from 3.4–4 m/s (Table 4).

The environmental variables, mesoscale phenomena, winds and climate indexes that showed significant correlations ($p < 0.05$) are shown in Table 5. Significant correlations (R^2) were observed inversely between ENSO, PDO, and Chl-a in January, February, May, June, July and October (−0.63, −0.70, −0.68, −0.67, −0.71, −0.59, −0.64, −0.71, −0.66, respectively), which indicates that when positive anomalies of PDO and negatives of ENSO are present, Chl-a concentrations decrease. Moreover, inverse correlations were observed between ENSO and winds in January, April, May, September, and October (−0.65, −0.53, −0.57, −0.55 and −0.53, respectively), which clearly showed that as the negative phase of ENSO increased, so did wind intensity. This same effect happened with the correlation of PDO and winds for the months of February, April, June and September (−0.53, −0.63, −0.56 and −0.55, respectively) when the positive phase of PDO decreased. In March, an inverse correlation was recorded between ENSO and filaments of −0.55 and between PDO and cyclonic gyres with an inverse correlation of −0.67, that is when negative ENSO anomalies decreased in the positive PDO phase, the frequency of these two mesoscale phenomena increased. On the other hand, when negative ENSO anomalies occurred by August, the frequency of anti-cyclonic gyres increased with a correlation of 0.65. The significantly positive correlations were recorded between PDO, ENSO and SST for January, February, March and December (0.63, 0.54, 0.62, 0.62, 0.61, 0.74, 0.72, 0.55 respectively), which indicated that temperature increased in the area of study in the months when positive PDO and negative ENSO anomalies were recorded. The positive PDO and ENSO correlations were recorded in January, February, April, May, September, November and December (0.68, 0.79, 0.59, 0.53, 0.60, 0.72 and 0.73, respectively). Thus, when PDO anomalies were positive, ENSO anomalies were negative.

Table 2. Monthly and annual analysis of temperature, chlorophyll a and percentage of mesoscale phenomena observed in the central and southern coastal regions of Sonora and adjacent deep zone in the Gulf of California, Mexico, from 2002–2015.

	Environmental Variables		Percentage of Mesoscale Phenomena				
	SST °C	Chl-a mg/m³	Cyclonic Gyres	Anti-Cyclonic Gyres	Upwelling	Filaments	Intrusions
Month							
January	18.00 ± 1.29 [bh]	3.31 ± 2.22 [ab]	10.71 ± 16.17 [ab]	17.86 ± 18.13 [a]	42.86 ± **35.87** [a]	**28.57 ± 21.08** [a]	0.00 ± 0.00 [c]
February	18.41 ± 1.83 [bgh]	3.74 ± 2.24 [ab]	27.03 ± 21.46 [ab]	16.22 ± 22.27 [a]	37.84 ± **19.66** [a]	18.92 ± 20.38 [ab]	0.00 ± 0.00 [c]
March	19.44 ± 1.46 [bfgh]	**5.11 ± 1.68** [a]	32.56 ± 22.63 [ab]	18.60 ± 21.56 [a]	30.23 ± 28.18 [ab]	18.60 ± 18.29 [ab]	0.00 ± 0.00 [c]
April	21.36 ± 1.28 [bdefgh]	**4.55 ± 1.53** [a]	23.81 ± 22.89 [ab]	11.90 ± 16.47 [a]	30.95 ± 16.77 [ab]	19.05 ± 23.18 [ab]	14.29 ± 21.15 [bc]
May	24.92 ± 0.84 [bcdef]	2.01 ± 1.07 [bcf]	23.68 ± 20.58 [ab]	5.26 ± 10.03 [a]	28.95 ± 20.26 [ac]	5.26 ± 7.26 [ab]	36.84 ± 20.08 [ab]
June	28.47 ± 0.65 [ae]	1.68 ± 1.68 [cf]	**41.38 ± 23.36** [a]	0.00 ± 0.00 [a]	10.34 ± 12.67 [bc]	0.00 ± 0.00 [b]	48.28 ± 27.90 [ab]
July	30.61 ± 0.65 [ac]	1.02 ± 0.25 [f]	**41.38 ± 23.99** [a]	10.34 ± 13.15 [a]	0.00 ± 0.00 [c]	0.00 ± 0.00 [b]	48.28 ± 28.21 [ab]
August	**31.67 ± 0.51** [a]	1.02 ± 0.37 [f]	24.00 ± 18.54 [ab]	12.00 ± 0.00 [a]	0.00 ± 18.62 [c]	12.00 ± 31.09 [ab]	52.00 ± 0.27 [ab]
September	**31.23 ± 0.39** [a]	1.31 ± 0.39 [cf]	30.43 ± 25.08 [ab]	8.70 ± 15.48 [a]	0.00 ± 0.00 [c]	0.00 ± 0.00 [b]	**60.87 ± 27.09** [a]
October	28.50 ± 0.76 [ad]	1.70 ± 0.65 [bcf]	15.70 ± 27.66 [ab]	10.53 ± 10.72 [a]	15.79 ± 18.62 [bc]	10.53 ± 14.47 [ab]	47.37 ± 47.86 [ab]
November	24.27 ± 1.26 [bcdefg]	3.73 ± 1.37 [ab]	10.00 ± 14.01 [ab]	10.00 ± 10.03 [a]	**55.00 ± 41.98** [a]	25.00 ± 20.69 [ab]	0.00 ± 0.00 [c]
December	19.92 ± 1.97 [bfgh]	3.00 ± 1.34 [ac]	00.00 ± 0.00 [b]	15.00 ± 16.98 [a]	50.00 ± 36.75 [ab]	**35.00 ± 31.16** [a]	0.00 ± 0.00 [c]
Year							
2002	24.52 ± 5.46 [a]	2.27 ± 1.62 [a]	13.64 ± 15.54 [a]	18.18 ± 17.53 [a]	27.27 ± 32.07 [a]	13.64 ± 17.21 [a]	27.27 ± 47.02 [a]
2003	24.65 ± 5.04 [a]	2.45 ± 1.90 [a]	26.09 ± 23.67 [a]	8.70 ± 10.76 [a]	30.43 ± 45.59 [a]	8.70 ± 15.05 [a]	26.09 ± 32.11 [a]
2004	24.61 ± 5.53 [a]	1.79 ± 1.03 [a]	20.83 ± 21.32 [a]	20.83 ± 22.29 [a]	16.67 ± 22.98 [a]	16.67 ± 33.43 [a]	25.00 ± 31.38 [a]
2005	24.24 ± 5.19 [a]	2.61 ± 2.02 [a]	17.86 ± 22.34 [a]	17.86 ± 22.95 [a]	25.00 ± 22.91 [a]	17.86 ± 24.31 [a]	21.43 ± 43.25 [a]
2006	24.62 ± 5.76 [a]	2.80 ± 1.99 [a]	15.79 ± 22.61 [a]	0.00 ± 0.00 [a]	36.84 ± 37.69 [a]	15.79 ± 22.61 [a]	31.58 ± 43.30 [a]
2007	24.09 ± 5.22 [a]	2.75 ± 1.80 [a]	24.14 ± 21.65 [a]	20.69 ± 23.69 [a]	31.03 ± 28.45 [a]	6.90 ± 15.54 [a]	17.24 ± 22.98 [a]
2008	23.89 ± 5.83 [a]	3.17 ± 1.99 [a]	24.00 ± 23.81 [a]	12.00 ± 16.57 [a]	24.00 ± 32.82 [a]	16.00 ± 14.97 [a]	24.00 ± 38.92 [a]
2009	24.83 ± 5.44 [a]	4.11 ± 3.58 [a]	30.43 ± 23.96 [a]	4.35 ± 7.22 [a]	30.43 ± 31.67 [a]	13.04 ± 19.82 [a]	21.74 ± 32.92 [a]
2010	24.71 ± 4.90 [a]	2.39 ± 1.34 [a]	20.00 ± 33.43 [a]	5.00 ± 9.62 [a]	35.00 ± 41.72 [a]	10.00 ± 16.60 [a]	30.00 ± 33.43 [a]
2011	24.16 ± 6.18 [a]	3.39 ± 1.78 [a]	35.48 ± 26.42 [a]	3.23 ± 4.81 [a]	25.81 ± 36.32 [a]	12.90 ± 16.76 [a]	22.58 ± 42.12 [a]
2012	24.48 ± 5.70 [a]	3.53 ± 1.94 [a]	38.24 ± 21.12 [a]	5.88 ± 8.25 [a]	20.59 ± 17.48 [a]	17.65 ± 18.45 [a]	17.65 ± 22.32 [a]
2013	25.23 ± 4.81 [a]	2.32 ± 1.71 [a]	24.00 ± 22.60 [a]	16.00 ± 18.28 [a]	20.00 ± 30.64 [a]	16.00 ± 17.53 [a]	24.00 ± 38.57 [a]
2014	26.34 ± 4.39 [a]	2.07 ± 1.05 [a]	25.00 ± 22.71 [a]	15.63 ± 18.23 [a]	21.88 ± 17.81 [a]	15.63 ± 18.23 [a]	21.88 ± 31.67 [a]
2015	25.90 ± 4.73 [a]	1.89 ± 1.08 [a]	22.22 ± 20.55 [a]	11.11 ± 11.49 [a]	16.67 ± 13.97 [a]	16.67 ± 13.97 [a]	33.33 ± 46.87 [a]

Median ± standard deviation. Different superindexes show statistical differences between months or years per environmental variable or mesoscale phenomenon ($p < 0.05$; i.e., $a \neq b \neq c$). Shaded values indicate the months that showed similar characteristics for each variable and mesoscale phenomenon. Numbers in bold represent the months that showed a greater difference than the other months in terms of their values and frequency of phenomena.

Table 3. Spearman's (R^2) correlation by month between temperature, chlorophyll a and mesoscale phenomena observed in the central and southern coastal regions of Sonora and adjacent deep zone in the Gulf of California, Mexico, from 2002–2015.

| | Environmental Variables | | Correlation Per Month between Mesoscale Phenomena and Environmental Variables | | | | | Effect |
| | | | Mesoscale Phenomena | | | | | |
	SST	Chl-a	Cyclonic Gyres	Anti-Cyclonic Gyres	Upwelling	Filaments	Intrusions	
January Cyclonic Gyres Anti-cyclonic Gyres					−0.56 −0.62			Upwelling ↑-Cyclonic and Anticyclonic Gyres ↓
February SST		−0.78						**SST ↓-Chl-a ↑**
March SST		−0.63						SST ↓-Chl-a ↑
April SST Filaments		−0.74					−0.53	**SST ↓ -Chl-a ↑**, Filaments ↑-Intrusions ↓
May Cyclonic Gyres Filaments					−0.67		−0.69 −0.65	Cyclonic Gyres ↑-Upwelling ↓, Intrusion ↑-Filaments and Cyclonic Gyres ↓
June Cyclonic Gyres Upwelling							−0.67 **−0.75**	Intrusions ↑-Cyclonic Gyres and **Upwelling** ↓
July Chl-a Cyclonic Gyres Anti-cyclonic Gyres			0.59				−0.64 −0.69	Cyclonic Gyres ↑-Chl-a ↑, Intrusions ↑-Cyclonic and Anticyclonic Gyres ↓
August Anti-cyclonic Gyres				−0.60		−0.51	−0.54	Anticyclonic Gyres and Filaments ↑-Chl-a ↓, Intrusions ↑-Anticyclonic Gyres ↓
September SST Cyclonic Gyres							0.54 −0.76	Intrusions ↑-SST ↑, **Intrusions ↑-Cyclonic Gyres** ↓
October SST Chl-a Upwelling			−0.55 −0.54		0.69	0.60 0.74	0.54 −0.56	Intrusions ↑-SST ↑, Cyclonic and Anticyclonic Gyres ↑-SST ↓, Upwelling and filaments ↑-Chl-a ↑, Intrusions ↑-**Chl-a** ↓, **Upwelling** ↑-**Filaments** ↑
December SST Chl-a					−0.58 0.65			Upwelling ↑-Chl-a ↑, SST ↑-Upwelling ↓

The numerical values by the intersection of rows and columns show Spearman's (R^2) correlation between the environmental variables of the mesoscale phenomena and among them with significant statistical differences between the two variables ($p < 0.05$); the highest correlations are shown in bold. Arrows pointing upward (↑) mean increase and downward (↓) decrease.

Table 4. Monthly and annual analyses of temperature, chlorophyll a, winds and climate indexes of the El Niño Southern Oscillation and Pacific Decadal Oscillation in the central and southern coastal regions of Sonora and adjacent deep zone in the Gulf of California, Mexico, from 2002–2015.

	Environmental Variables			Climate Index	
Month	**TSM °C**	**Chl-a mg/m³**	**Winds m/s**	**ENSO**	**PDO**
January	18.00 ± 1.29 [bh]	3.31 ± 2.22 [ab]	2.40 ± 0.57 [b]	−0.12 ± 0.84 [a]	0.22 ± 1.17 [a]
February	18.41 ± 1.83 [bgh]	3.74 ± 2.24 [ab]	2.90 ± 0.79 [bd]	−0.16 ± 0.71 [a]	0.16 ± 1.08 [a]
March	19.44 ± 1.46 [bfgh]	5.11 ± 1.68 [a]	**3.48 ± 0.59** [ad]	−0.11 ± 0.54 [a]	0.11 ± 1.07 [a]
April	21.36 ± 1.28 [bdefgh]	4.55 ± 1.53 [a]	**4.00 ± 0.49** [ac]	−0.05 ± 0.44 [a]	0.17 ± 0.96 [a]
May	24.92 ± 0.84 [bcdef]	2.01 ± 1.07 [bcf]	**4.04 ± 0.42** [a]	0.01 ± 0.39 [a]	0.23 ± 1.05 [a]
June	28.47 ± 0.65 [ae]	1.68 ± 1.68 [cf]	**4.05 ± 0.44** [a]	0.07 ± 0.42 [a]	0.06 ± 0.87 [a]
July	30.61 ± 0.65 [ac]	1.02 ± 0.25 [f]	**3.38 ± 0.40** [ad]	0.11 ± 0.51 [a]	−0.18 ± 1.15 [a]
August	31.67 ± 0.51 [a]	1.02 ± 0.37 [f]	0.65 ± 0.48 [bcd]	−0.21 ± 1.15 [a]	−0.14 ± 1.42 [a]
September	31.23 ± 0.39 [a]	1.31 ± 0.39 [cf]	2.34 ± 0.72 [b]	0.17 ± 0.80 [a]	−0.33 ± 1.21 [a]
October	28.50 ± 0.76 [ad]	1.70 ± 0.65 [bcd]	2.96 ± 0.51 [bd]	0.21 ± 0.96 [a]	−0.31 ± 1.08 [a]
November	24.27 ± 1.26 [bcdefg]	3.73 ± 1.37 [ab]	2.58 ± 0.59 [bd]	0.21 ± 1.02 [a]	−0.31 ± 1.14 [a]
December	19.92 ± 1.97 [bfgh]	3.00 ± 1.34 [ac]	2.60 ± 0.52 [bd]	0.14 ± 1.09 [a]	**0.06 ± 1.18** [a]
Year					
2002	24.52 ± 5.46 [a]	2.27 ± 1.62 [a]	3.13 ± 0.79 [a]	0.62 ± 0.52 [ac]	0.22 ± 0.86 [ae]
2003	24.65 ± 5.04 [a]	2.45 ± 1.90 [a]	2.96 ± 0.94 [a]	0.28 ± 0.31 [acd]	0.97 ± 0.59 [a]
2004	24.61 ± 5.53 [a]	1.79 ± 1.03 [a]	2.77 ± 0.67 [a]	0.43 ± 0.26 [acd]	0.35 ± 0.46 [ad]
2005	24.24 ± 5.19 [a]	2.61 ± 2.02 [a]	3.06 ± 0.86 [a]	0.14 ± 0.41 [ae]	0.38 ± 1.03 [ad]
2006	24.62 ± 5.76 [a]	2.80 ± 1.99 [a]	3.44 ± 0.59 [a]	0.16 ± 0.57 [ae]	0.19 ± 0.61 [ae]
2007	24.09 ± 5.22 [a]	2.75 ± 1.80 [a]	3.30 ± 0.62 [a]	**−0.40 ± 0.62** [bde]	−0.20 ± 0.63 [bcde]
2008	23.89 ± 5.83 [a]	3.17 ± 1.99 [a]	3.49 ± 0.91 [a]	**−0.68 ± 0.42** [be]	−1.29 ± 0.37 [b]
2009	24.83 ± 5.44 [a]	4.11 ± 3.58 [a]	2.95 ± 0.67 [a]	0.33 ± 0.70 [ac]	−0.61 ± 0.79 [bde]
2010	24.71 ± 4.90 [a]	2.39 ± 1.34 [a]	3.12 ± 0.77 [a]	**−0.33 ± 1.03** [bce]	−0.31 ± 0.95 [bcde]
2011	24.16 ± 6.18 [a]	3.39 ± 1.78 [a]	3.59 ± 0.94 [a]	**−0.70 ± 0.34** [be]	−1.23 ± 0.66 [be]
2012	24.48 ± 5.70 [a]	3.53 ± 1.94 [a]	3.33 ± 0.96 [a]	−0.12 ± 0.39 [bce]	−1.10 ± 0.58 [be]
2013	25.23 ± 4.81 [a]	2.32 ± 1.71 [a]	3.30 ± 0.76 [a]	−0.26 ± 0.10 [bde]	−0.52 ± 0.41 [bde]
2014	26.34 ± 4.39 [a]	2.07 ± 1.05 [a]	2.89 ± 0.84 [a]	0.01 ± 0.40 [bce]	1.13 ± 0.65 [ac]
2015	25.90 ± 4.73 [a]	1.89 ± 1.08 [a]	2.79 ± 0.77 [a]	1.26 ± 0.70 [a]	1.63 ± 0.49 [a]

Median ± standard deviation. Different superindexes show statistical differences among months or years per environmental variables, winds and climate indexes of the El Niño Southern Oscillation and Pacific Decadal Oscillation ($p < 0.05$; i.e., $a \neq b \neq c$). Shaded values in monthly analyses indicate the months that showed warm characteristics for ENSO and PDO climate indexes. Numbers in bold represent the months that showed cold characteristics for climate indexes and the months that showed higher wind intensity.

Table 5. Spearman's (R^2) correlation by month among temperature, chlorophyll a, mesoscale phenomena, winds and climate indexes observed in the central and southern coastal region of Sonora and adjacent oceanic zone in the Gulf of California, Mexico, from 2002–2015.

	Environmental Variables			Climate Index		Effect
January	SST	Chl-a	Winds	ENSO	PDO	PDO and ENSO ↑-SST ↑-Chl-a ↓,
SST				0.63	0.54	ENSO ↓-Winds ↑,
Chl-a				−0.63	−0.70	PDO ↑-ENSO ↑
ENSO			−0.65		0.68	
February	SST	Chl-a	Winds	ENSO	PDO	PDO and ENSO ↑-SST ↑-Chl-a ↓,
SST				0.62	0.62	PDO ↓-Winds ↑
Chl-a				−0.68	−0.67	PDO ↑-ENSO ↑
ENSO					0.79	
PDO			−0.53			
March	SST	Chl-a	Winds	ENSO	PDO	
SST				0.61	0.74	PDO and ENSO ↑-SST ↑,
Chl-a					−0.67	PDO ↓-Chl-a and Cyclonic Gyres ↑,
Cyclonic Gyres					−0.67	ENSO ↓-Filaments ↑
Filaments				−0.55		
April	SST	Chl-a	Winds	ENSO	PDO	
SST			−0.59			PDO and ENSO ↓-Winds ↑,
ENSO			−0.63		0.597	SST ↑-Winds ↓, PDO ↑-ENSO ↑
PDO			−0.53			
May	SST	Chl-a	Winds	ENSO	PDO	
SST			−0.63			SST ↑-Winds ↓, ENSO ↓-Winds ↑,
Chl-a				−0.71		ENSO ↑-Chl-a ↓ PDO ↑-ENSO ↑
ENSO			−0.57		0.53	
Jun	SST	Chl-a	Winds	ENSO	PDO	
Chl-a				−0.59	−0.64	PDO and ENSO ↑-Chl-a ↓,
PDO			−0.56			PDO ↓-Winds ↑
July	SST	Chl-a	Winds	ENSO	PDO	
Chl-a					−0.71	PDO ↑-Chl-a ↓
August	SST	Chl-a	Winds	ENSO	PDO	
Chl-a				−0.62		ENSO ↑-Chl-a ↓,
Anti-cyclonic Gyres				0.65		ENSO ↓-Anti-cyclonic Gyres ↑,
Intrusions				−0.53		ENSO ↓-Intrusions ↓
September	SST	Chl-a	Winds	ENSO	PDO	
ENSO			−0.55		0.60	PDO and ENSO ↑-Winds ↓,
PDO			−0.55			PDO ↑-ENSO ↑
October	SST	Chl-a	Winds	ENSO	PDO	
Chl-a					−0.66	PDO ↑-Chl-a ↓, ENSO ↑-Winds
Cyclonic Gyres				−0.63		and Cyclonic Gyres ↓,
Intrusions					0.58	PDO ↑-Intrusions ↑
ENSO			−0.53		0.77	
November	SST	Chl-a	Winds	ENSO	PDO	
ENSO					0.72	PDO ↓-ENSO ↑
December	SST	Chl-a	Winds	ENSO	PDO	
SST				0.72	0.55	PDO ↑- ENSO ↓-SST ↓,
ENSO					0.73	PDO ↑-ENSO ↑

The numerical values by intersection of rows and columns show Spearman's (R^2) correlation among the environmental variables of mesoscale phenomena and El Niño Southern Oscillation and Pacific Decadal Oscillation climate indexes and among them with significant statistical differences between the two variables ($p < 0.05$); the highest correlations are shown in bold. Arrows point upward (↑) mean increase and downward (↓) decrease.

The performed analysis showed an important annual component and minor variability associated with the semiannual cycle (4–6 months). Chl-a series also showed evidence of a higher variation period variation, related to interannual (three year) cycles, possibly generated by El Niño-La Niña events. However, those frequencies are not supported by the analysis, because the sampling does not expand 10-times the period of the event (Figure 11).

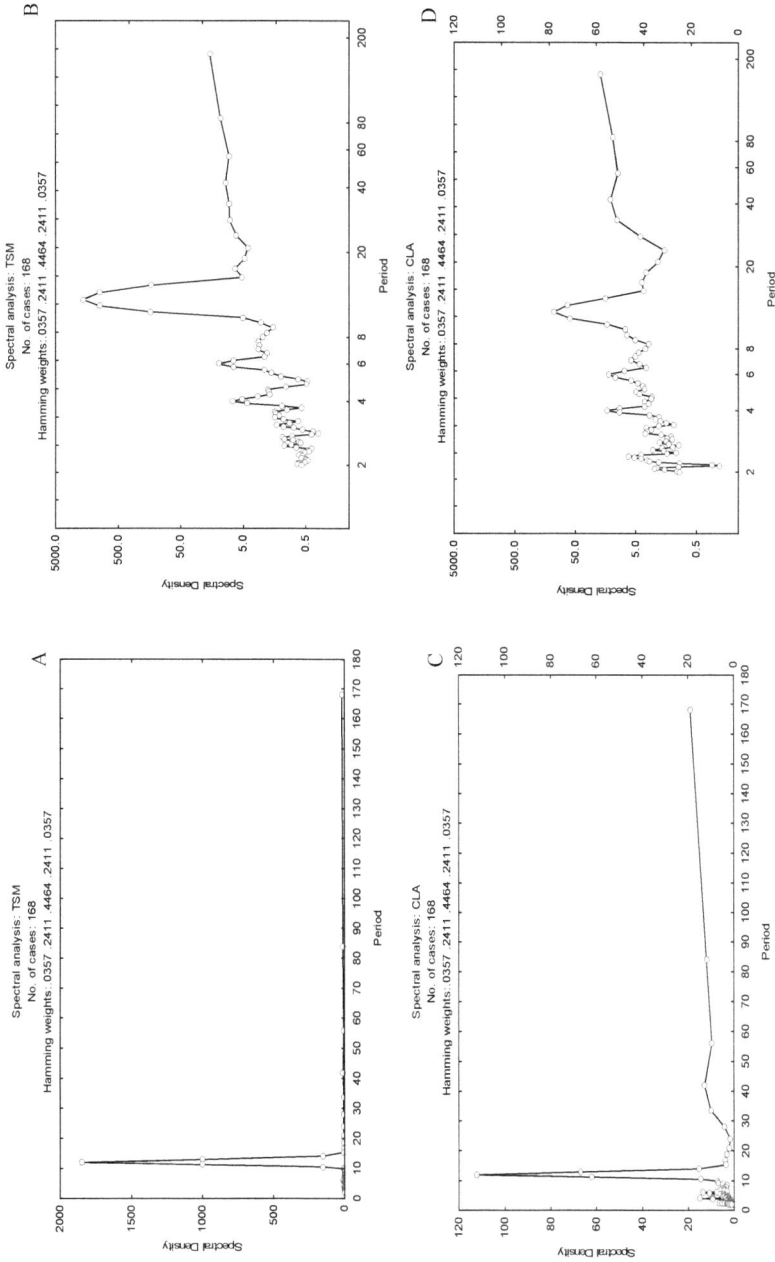

Figure 11. Spectral analysis and periodograms for sea surface temperature (SST) (**A**,**B**) and the concentration of chlorophyll a (Chl-a) (**C**,**D**).

4. Discussion

It is common to find mesoscale structures in the GC that have lead it to stand out as one of the most productive water bodies of the Pacific [5,12,62] because of the great number of oceanographic phenomena that occur, which agrees with the monthly analyses of the phenomena observed. They can be generated by the coupling processes of macroscale atmospheric phenomena and regional mesoscale events, which seem to occur each year [14].

The monthly frequency analyses showed a total of 353 mesoscale phenomena in the central and southern coastal regions of Sonora during the period of study where the months that stood out with greater observations per month of cyclonic gyres, upwelling and MCC water intrusion were from February–July. All of these phenomena have been the result of atmospheric forcing interaction in the Pacific where seasonal winds (with flux toward the southeast in winter and toward the northwest in summer) induced a significant barotropic variability in kinetic energy associated with gyres and upwelling, as well as in energy associated with fluctuations of the current force as mentioned by Badan-Dangón [15], Stammer and Wunsch [6] and Lavín et al. [14].

The results have shown a seasonal pattern of intense upwelling from November–May with a maximum observed from January–April, similar to that reported by Maluf and Lluch-Cota [8,9], who mentioned a period extending from November–April. In this same period, filaments with high Chl-a were observed, which according to Navarro-Olache et al. [16] Lavín et al. [54] and Zamudio [2] were generated by wind forcing in winter transporting cold water with a high concentration of Chl-a. They also mentioned that this same wind forcing forms cyclonic gyres responsible for high dispersion of Chl-a, as could be observed from January–April in our results. Likewise, these gyres play a fundamental role in the transport of suspended material, contributing to nutrient transport between the sub- and superficial layers, affecting the vertical and horizontal distribution of phytoplankton [66]. According to the previous information, the distribution and abundance of phytoplankton depend on the type of gyre and its time length, where anti-cyclonic gyres cause sinking of nutricline originating a low in primary local production. On the other hand, cyclonic gyres displace nutricline, and when this displacement is strong enough, it generates enrichment in the water column as for primary production exporting organic material to the ecosystem [67–70].

Likewise, according to the results of the monthly analyses and correlation, this pattern of upwelling and filaments of high Chl-a concentration was influenced by that of moderate and strong winds from November–April, caused by the cold phase of ENSO. It is characterized by wind intensification during these months, as well as a cooling factor of the superficial water layer in the tropical Pacific affecting the area of study [39,40]. Nonetheless, in the results of this study, the wind pattern was not correlated directly with the mesoscale phenomena, but the variations of these indexes affected wind intensity generating the type of mesoscale structures, such as upwelling, anti-cyclonic and cyclonic gyres, that are in charge of concentrating and dispersing high Chl-a concentration, respectively.

In the following months, climate indexes were directly correlated; in March, an increase in cyclonic gyres showed a decrease in the positive anomalies of PDO and an increase of filaments when ENSO showed a decrease in its negative anomalies; in August, an increase in anti-cyclonic gyres was observed when ENSO showed a sudden change from positive to negative anomalies, which also caused a decrease in intrusions. Finally, in October, a decrease of winds and cyclonic gyres was observed when the positive anomalies of ENSO increased. In the same manner, when the positive anomalies of PDO decreased in this month, intrusions increased.

The results showed a clear difference between cold and warm periods and variations between the same months in the central and southern coastal region of Sonora in the GC. The image analysis of Chl-a showed a similar variability to that of Espinosa-Carreón and Valdez-Holguín [17,71] who found a marked variability of Chl-a in the GC; they reported a high concentration during the cold period and a low one in the warm period, which agree with those shown in this study, where it could be clearly observed that as SST increased over 26 °C in the warm period, Chl-a showed a decrease;

and when SST decreased, Chl-a showed a recovery in the area of study. This monthly variability was due to gale winds and upwelling during the cold period, as well as to gentle winds and warm water intrusion with oligotrophic characteristic of the Pacific to the GC [54,65,72]. The variability in Chl-a concentration previously mentioned, caused by changes in wind intensity and CCM intrusion, was also modulated by changes in the ENSO and PDO phases where our results showed that when the positive PDO phase decreased, Chl-a concentration increased, and when the positive ENSO phase increased, such a concentration decreased.

The results of the SST analysis corroborated the presence of two extreme periods in the annual temperature cycle in the area of study previously reported by Soto-Mardones et al. [11] Lavin and Marinone [42] and Godínez-Sandoval [63]. They concluded that SST variability was due to changes in solar radiation incidence that changed evaporation rates and wind intensity, as well as to the strong influence of MCC water intrusion from the Pacific Ocean through the mouth of the GC by means of coastal Kelvin type waves that spread along the continental coastline of the gulf, which were the ones transmitting signals from the Pacific to the inner GC [13]. Additionally, such extreme periods of high temperatures that went from June–October were influenced by the positive ENSO phase and by the negative PDO phase. On the contrary the period of lower temperatures was modulated by the negative ENSO phase and the positive PDO phase from November–May.

On the contrary, one of the intriguing results in the statistical analyses was that no significant differences were observed among the years of study for the SST and Chl-a variables and mesoscale phenomena. Nonetheless, in the analysis of climate indexes of PDO and ENSO, significant statistical differences were observed when a five-year period of positive temperature anomalies was observed for both indexes; a period of negative anomalies of up to seven years was also observed for both indexes with extreme years, such as 2002 and 2009 with positive anomalies, which were characterized as moderate Niños; 2015 was characterized as a strong Niño followed by a restoration period named Niña from 2007–2008 and 2010–2011. As explained by Bernal et al. [73], they suggested that annual temperatures were modulated mainly by the influence of great scale phenomena in the Pacific, as ENSO and PDO.

Likewise, the mesoscale phenomena that occurred in the area of study and those at the mouth of the GC were modulated by the three time scales that occur in the Mexican Tropical Eastern Pacific: (1) seasonal current (California Current and Mexican Coastal Current); (2) gyres (3–5 months) at the mouth of the gulf; and (3) interannual anomalies as El Niño and La Niña. These two last ones apparently play a role in MCC intensification during El Niño, as well as an increase in the number of gyres generated [13,74–76], which affect oceanographic dynamics of the GC. Likewise, Chl-a concentrations were observed in our results where its gradual increase and a gradual decrease of SST could be appreciated during the negative PDO and ENSO anomalies (2003–2009) with a decrease in Chl-a and an increase in SST in 2004, 2010 and 2015 caused by extreme warm events characterized as Niños (ENSO-NOAA).

According to the previous information, the decrease of Chl-a indicates an oligotrophic area associated with a low presence of organisms integrating phytoplankton and zooplankton, thus causing changes in their distribution and abundance, such as species displacement toward more productive areas or habitats where planktonic food is more abundant [50–52,77]. This behavior of the organisms could be explained by rotation movements or mechanisms of cyclonic gyres, which transport nutrients vertically toward the eutrophic zone, fertilizing it, and the phytoplankton responds to this pulse causing a phytoplankton bloom according to Coria-Monter et al. [26].

Likewise, according to Maluf [8], the interaction of the water mass from the Pacific with that of the GC and its characteristic topography and important circulation processes allow the formation of oceanic fronts provoking a convergence of organisms in the frontal areas of this mass of water and in the phenomena provoked by the same intrusion. Moreover, gyres retain and transport organisms and enrichment processes as upwelling and filaments that favor the habitat of a great number of organisms [14,24–26].

Previously in other studies, Soto-Mardones et al. (1999) showed that on average, the sea surface temperature (SST) of the Gulf of California decreases from the mouth to the head and its variability increases by using fourteen years of infrared satellite images (1983–1996) to examine SST variability. The annual scale accounts for most of the variability of the SST, which oscillates in phase with small variations from north to south. Escalante et al. (2013) using satellite-derived data from 1997–2010, sea surface temperature, chlorophyll a and primary productivity variations showed that interannual signals of El Niño (EN) and La Niña (LN) were more evident at the entrance to the gulf.

Likewise coinciding with these two works and with a series of larger data, López-Martínez et al. (2001) showed that interannual variation is one of the dominant signals in the Gulf of California by making a spectral analysis of TSM from 1952–1993 with data from a tide-chart located in Guaymas, Sonora, in the central part of the gulf, which agrees with that obtained in this study.

All of the physical, biological and atmospheric processes mentioned previously take place at different temporal and spatial scales, as our results have shown in the number of monthly observations of phenomena per year, their length in months per year and maximum length. Likewise, climate indexes and winds exert an influence on the coastal pelagic ecosystem of Sonora, providing productive and biologically-rich habitats for a great diversity of commercially-important species. Moreover, these different types of mesoscale phenomena in the area of study and their length could exert a positive or negative effect on some species, such as the production of eggs and larvae and the growth and survival in the stages of the different species populations that inhabit the GC as mentioned by Hammann et al. [78], Daskalov [79] and Nevarez-Martinez [24,25,80].

5. Conclusions

During the period analyzed, a large number of mesoscale phenomena could be observed in the area of study by analyzing images of Chl-a and SST, which determined the seasonal and annual variations in these two variables, as well as the frequency among phenomena from 2002–2015.

The mesoscale structures that could be detected were anti-cyclonic gyres, filaments of Chl-a concentrations besides upwelling, cyclonic gyres and MCC water intrusion, which were those that stood out. Likewise, the frequency observed of each one of the phenomena and their length could be determined to establish their possible effect on the ecosystem.

The analyses of oceanographic dynamics performed for months and years in the coastal central and southern area of Sonora in the Gulf of California allowed determining, in great measure, the evolution of its mesoscale superficial dynamics, which permitted establishing the oceanographic framework affecting the coastal area.

As for the Chl-a concentration, a wider period with high concentrations could be identified compared to those traditionally proposed for this area due to the upwelling phenomena, the negative phase of ENSO and the positive phase of PDO that took place from January–May and from November–December with a maximum in winter and spring. The minimum values were recorded in summer because of the strong impact in the area of study due to the effects of MCC water intrusion; in previous months, highly favorable conditions prevailed for the primary sustenance of the trophic nets, and all of a sudden, these conditions strongly decreased by such an intrusion with a consequent increase in SST in the positive phase of ENSO and the negative phase of PDO.

It was also determined that in the annual analyses of environmental variables and the percentage of mesoscale phenomena recorded, no significant differences were observed between the years assessed. While the monthly analyses of the phenomena recorded from 2002–2015 showed significant differences between months, for temperature, Chl-a, cyclonic gyres, upwelling, filaments and intrusions, no statistical monthly differences were observed in anti-cyclonic gyres because of the variability between months. A high correlation was observed between some environmental variables and mesoscale phenomena.

All of the physical and biological processes mentioned in this work, as well as their length, have a bearing in the state of health of the pelagic ecosystem of the coastal zone of Sonora, and they exert a positive or negative effect on the requirements of some biologically- and commercially-important species. Therefore, because of their importance, this type of studies should be monitored periodically to have updated environmental information of the ecosystems and their possible implications on the organisms.

Acknowledgments: This research was financed by the projects of the Consejo Nacional de Ciencia y Tecnologia (CONACYT), PDCPN2013-01-00215355, SEMARNAT-CONACYT 249458 and SEP-CONACYT CB-2015-256477; Ricardo García Morales was a postdoctoral recipient of CONACYT 290754; the authors acknowledge CIBNOR staff at the Fisheries Laboratory at Guaymas Unit, specifically Eloísa Herrera Valdivia and Rufino Morales Azpeitia; we thanks Diana D. Fischer for editorial services in English. The authors acknowledge Dr. Mati Karhu for his valuable contribution to this manuscript.

Author Contributions: R.G.-M. wrote and edited this article, processed, analyzed and interpreted the satellite images. J.L.-M., J.E.V.-H. and H.H.-C. reviewed and edited this article. J.E.V.-H. is the Guest Editor of this Special Issue. L.D.E.-C. did the statistical analysis of the data obtained from the satellite images.

Conflicts of Interest: The authors declare no conflict of interest.

References

1. Salazar-Sparks, J. *Chile y la Comunidad del Pacífico*; Editorial Universitaria: Santiago, Chile, 1999; 253p, ISBN 956111528X, 9789561115286.
2. Zamudio, L.; Hogan, P.; Metzger, E.J. Summer generation of the Southern Gulf of California eddy train. *J. Geophys. Res.* **2008**, *113*. [CrossRef]
3. Lluch-Cota, S.E. Gulf of California. In *Marine Ecosystems of the North Pacific*; PICES Special Publication: Sidney, BC, Canada, 2004; Volume 1, pp. 1–7, 1280.
4. Lavin, M.F.; Beier, E.; Badan, A. Estructura hidrográfica y circulación del Golfo de California: Escalas estacional e interanual. In *Contribuciones a la Oceanografía Física en México: Monografía*; Lavín, F., Ed.; Unión Geofísica Mexicana: Ensenada, Baja California, México, 1997; pp. 141–171.
5. Álvarez-Molina, L.L.; Álvarez-Borrego, S.; Lara-Lara, J.R.; Marinone, S. Annual and semiannual variations of phytoplankton biomass and production in the central Gulf of California estimated from satellite data. *Cienc. Mar.* **2013**, *39*, 217–230. [CrossRef]
6. Stammer, D.; Wunsch, C. Temporal changes in eddy energy of the oceans. *Deep Sea Res.* **1999**, *46*, 77–108. [CrossRef]
7. Lluch-Cota, S.E.; Aragon-Noriega, E.A.; Arreguin-Sánchez, F.; Aurioles-Gamboa, D.; Bautista Romero, J.; Brusca, R.C.; Cervantes-Duarte, R.; Cortéz-Altamirano, R.; Del-Monte-Luna, P.; Esquivel-Herrera, A.; et al. The Gulf of California: Review of ecosystem status and sustainability challenges. *Prog. Oceanogr.* **2007**, *73*, 1–26. [CrossRef]
8. Maluf, L.Y. The Physical Oceanography. In *Island Biogeography in the Sea of Cortez*; Case, T.J., Cody, M.L., Eds.; University of California Press: Berkeley, CA, USA, 1983; pp. 26–45.
9. Lluch-Cota, S.E. Coastal upwelling in the eastern Gulf of California. *Oceanol. Acta* **2000**, *23*, 731–740. [CrossRef]
10. Ripa, P. Seasonal circulation in the Gulf of California. *Ann. Geophys.* **1990**, *8*, 559–564.
11. Soto-Mardones, L.; Marinone, S.G.; Parés-Sierra, A. Variabilidad espaciotemporal de la temperatura superficial del mar en el Golfo de California. *Cienc. Mar.* **1999**, *25*, 1–30. [CrossRef]
12. Marinone, S.G. A three-dimensional model of the mean and seasonal circulation of the Gulf of California. *J. Geophys. Res.* **2003**, *108*, 21–27. [CrossRef]
13. Herrera-Cervantes, H.; Lluch-Cota, D.B.; Gutiérrez-de-Velasco, G.; Lluch-Cota, S.E. The ENSO signature in sea-surface temperature in the Gulf of California. *J. Mar. Res.* **2007**, *65*, 589–605. [CrossRef]
14. Lavin, M.F.; Castro, R.; Beier, E.; Cabrera, C.; Godinez, V.M.; Amador-Buenrostro, A. Surface circulation in the Gulf of California in summer from surface drifters and satellite images (2004–2006). *J. Geophys. Res. Oceans* **2014**, *119*, 4278–4290. [CrossRef]
15. Badan-Dangon, A. Coastal circulation from the Galapagos to the Gulf of California. In *The Sea, Pan Regional*; Robinson, A.R., Brink, K.H., Eds.; John Wiley and Sons: Hoboken, NJ, USA, 1998; Volume 11, pp. 315–343.

16. Navarro-Olache, L.F.; Lavín, M.F.; Álvarez-Sánchez, L.G.; Zirino, A. Internal structure of SST features in the central Gulf of California. *Deep Sea Res. II* **2004**, *51*, 673–687. [CrossRef]

17. Espinosa-Carreón, L.; Valdez-Holguín, E. Variabilidad interanual de la clorofila en Golfo de California. *Ecol. Appl.* **2007**, *6*, 83–92. [CrossRef]

18. Hoffman, E.E.; Powell, T.M. Environmental variability effects on marine fisheries: Four case histories. *Ecol. Appl.* **1998**, *8*, 523–532.

19. López-Martínez, J.; Rodríguez-Romero, J.; Hernández-Saavedra, N.Y.; Herrera-Valdivia, E. Population parameters of the *Pacific flagfin* mojarra *Eucinostomus currani* (Perciformes: Gerreidae) captured by the shrimp trawling fishery in the Gulf of California. *Rev. Biol. Trop.* **2011**, *59*, 887–897. [CrossRef] [PubMed]

20. Rodríguez-Romero, J.; Barjau, E.; Galván, F.; Gutiérrez, F.; López, J. Estructura espacial y temporal de la comunidad de especies de peces arrecifales de la Isla San José, Golfo de California, México. *Rev. Biol. Trop.* **2012**, *60*, 2.

21. Contreras-Catala, F.; Sánchez-Velasco, L.; Lavín, M.F.; Godínez, V.M. Three-dimensional distribution of larval fish assemblages in an anticyclonic eddy in a semi-enclosed sea (Gulf of California). *J. Plankton Res.* **2012**, *34*, 548–562. [CrossRef]

22. García-Reyes, M.; Largier, J.L.; Sydeman, W.J. Synoptic-scale upwelling indices and predictions of phyto-and zooplankton populations. *Prog. Oceanogr.* **2014**, *120*, 177–188. [CrossRef]

23. Lluch-Cota, S.E.; Morales Zárate, M.V.; Lluch Cota, D.B. *Variabilidad del Clima y Pesquerías del Noroeste de México. En: Variabilidad Ambiental y Pesquerías de México*; López-Martínez, J., Ed.; Comisión Nacional de Acuacultura y Pesca: Mazatlán, Sinaloa, México, 2008; p. 216.

24. Sánchez-Velasco, L.; Lavín, M.F.; Jiménez-Rosenberg, S.P.A.; Godínez, V.M.; Santamaría del Angel, E.; Hernández-Becerril, D.U. Three-dimensional distribution of fish larvae in a cyclonic eddy in the Gulf of California during the summer. *Deep Sea Res. I Oceanogr. Res. Pap.* **2013**, *75*, 39–51. [CrossRef]

25. Sánchez-Velasco, L.; Lavín, M.F.; Jiménez-Rosenberg, S.P.A.; Godínez, V.M. Preferred larval fish habitat in a frontal zone of the northern Gulf of California during the early cyclonic phase of the seasonal circulation (June 2008). *J. Mar. Syst.* **2014**, *129*, 368–380. [CrossRef]

26. Coria-Monter, E.; Monreal-Gómez, M.A.; Salas de León, D.A.; Aldeco-Ramírez, J.; Merino-Ibarra, M. Differential distribution of diatoms and dinoflagellates in a cyclonic eddy confined in the Bay of La Paz, Gulf of California. *J. Geophys. Res.* **2014**, *6258–6268*. [CrossRef]

27. Schwing, F.B.; Murphree, T.; Green, P.M. The Northern Oscillation Index (NOI): A New Climate Index for the Northeast Pacific. *Prog. Oceanogr.* **2002**, *53*, 115–139. [CrossRef]

28. Parrish, R.H.; Schwing, F.B.; Mendelssohn, R. Mid-Latitude Wind Stress: The Energy Source for Climate Shifts in the North Pacific Ocean. *Fish. Oceanogr.* **2000**, *9*, 224–238. [CrossRef]

29. Wooster, W.S.; Hollowed, A.B. Decadal-Scale Variations in the Eastern Subarctic Pacific, Winter Ocean Conditions. *Can. Spec. Publ. Fish. Aquat. Sci.* **1995**, *121*, 81–85.

30. Bograd, S.J.; Chereskin, T.K.; Roemmich, D. Transport of Mass, Heat, Salt, and Nutrients in the Southern California Current System: Annual Cycle and Interannual Variability. *J. Geophys. Res.* **2001**, *106*, 9255–9275. [CrossRef]

31. Bernal, P.A.; Chelton, D.B. Variabilidad biológica de baja frecuencia y gran escala en la Corriente de California, 1949–1978 (Low Frequency and Large-Scale Biological Variability in the California Current, 1949–1978). In *Reports of the Expert Consultation to Examine Changes in Abundance and Species Composition of Neritic Fish Resources, Proceedings of the A Preparatory Meeting for the FAO World Conference on Fisheries Management and Development, San José, Costa Rica, 18–29 April 1983*; FAO Fisheries Report; Csirke, J., Sharp, G.D., Eds.; The Food and Agriculture Organization of the United Nations (FAO): Rome, Italy, 1984; Volume 1, p. 102.

32. Schwing, F.B.; Murphree, T.; deWitt, L.; Green, P.M. The Evolution of Oceanic and Atmospheric Anomalies in the Northeast Pacific during the El Nino and La Nina Events of 1995–2001. *Progr. Oceanogr.* **2002**, *54*, 459–491. [CrossRef]

33. Deser, C.; Alexander, M.A.; Timlin, M.S. Evidence for a Wind-Driven Intensification of the Kuroshio Current Extension from the 1970s to the 1980s. *J. Clim.* **1999**, *12*, 1697–1706. [CrossRef]

34. Barlow, M.; Nigam, S.; Berbery, E.H. ENSO, Pacific Decadal Variability, and US Summertime Precipitation, Drought, and Stream Flow. *J. Clim.* **2001**, *14*, 2105–2128. [CrossRef]

35. Chiew, F.H.; McMahon, T.A. Global ENSO-Streamflow Teleconnection, Streamflow Forecasting and Interannual Variability. *Hydrol. Sci. J.* **2002**, *47*, 505–522. [CrossRef]
36. Trenberth, K.E. The Definition of El Niño. *Bull. Am. Meteorol. Soc.* **1997**, *78*, 2771–2777. [CrossRef]
37. Mantua, N.J.; Hare, S.R.; Zhang, Y.; Wallace, J.M.; Francis, R.C. A Pacific Interdecadal Climate Oscillation with Impacts on Salmon Production. *Bull. Am. Meteorol. Soc.* **1997**, *78*, 1069–1079. [CrossRef]
38. Mantua, N.J.; Hare, S.R. The Pacific Decadal Oscillation. *J. Oceanogr.* **2002**, *58*, 35–44. [CrossRef]
39. Mantua, N.J. *The Pacific Decadal Oscillation and Climate Forecasting for North America*; Joint Institute for the Study of the Atmosphere and Oceans, University of Washington: Seattle, WA, USA, 1999.
40. Piechota, T.C.; Garbrecht, J.D.; Schneider, J.M. Climate Variability and Climate Change. In *Climate Variations, Climate Change, and Water Resources Engineering*; Garbrecht, J.D., Piechota, T.C., Eds.; ASCE: Reston, VA, USA, 2006; pp. 1–18.
41. Makarov, V.G.; Jiménez-Illescas, A.R. Barotropic Background Currents in the Gulf of California. *Cienc. Mar.* **2003**, *29*, 141–153. [CrossRef]
42. Lavín, M.F.; Palacios-Hernández, E.; Cabrera, C. Sea Surface Temperature Anomalies in the Gulf of California. *Geofís. Int.* **2003**, *42*, 363–375.
43. Latif, M.; Barnett, T.P. Interactions of the Tropical Oceans. *J. Clim.* **1995**, *8*, 952–964. [CrossRef]
44. Zhang, Y.; Wallace, J.M.; Battisti, D.S. ENSO-Like Interdecadal Variability: 1900–93. *J. Clim.* **1997**, *10*, 1004–1020. [CrossRef]
45. Hare, S.; Mantua, N. Empirical Evidence for North Pacific Regime Shifts in 1977 and 1989. *Progr. Oceanogr.* **2000**, *47*, 103–145. [CrossRef]
46. Martinez, E.; Antoine, D.; D'Ortenzio, F.; Gentili, B. Climate-Driven Basin-Scale Decadal Oscillations of Oceanic Phytoplankton. *Science* **2009**, *326*, 1253–1256. [CrossRef] [PubMed]
47. Di Lorenzo, E.; Schneider, N.; Cobb, K.M.; Chhak, K.; Franks, P.J.S.; Miller, A.J.; McWilliams, J.C.; Bograd, S.J.; Arango, H.; Curchister, E.; et al. North Pacific Gyre Oscillation Links Ocean Climate and Ecosystem Change. *Geophys. Res. Lett.* **2008**, *35*, L08607. [CrossRef]
48. Lluch-Cota, S.E.; Parés-Sierra, A.; Magaña-Rueda, V.O.; Arreguín Sánchez, F.; Bazzino, G.; Herrera-Cervantes, H.; Lluch Belda, D. Changing climate in the Gulf of California. *Prog. Oceanogr.* **2010**, *87*, 114–126. [CrossRef]
49. Walther, G.-R.; Post, E.; Convey, P.; Menzel, A.; Parmesan, C.; Beebee, T.J.; Fromentin, J.M.; Hoegh-Guldberg, O.; Bairlein, F. Ecological responses to recent climate change. *Nature* **2002**, *416*, 389–395. [CrossRef] [PubMed]
50. Helmuth, B.; Harley, C.D.G.; Halpin, P.M.; O'Donnell, M.; Hofmann, G.E.; Blanchette, C.A. Climate change and latitudinal patterns of intertidal thermal stress. *Science* **2002**, *298*, 1015–1017. [CrossRef] [PubMed]
51. Lluch-Cota, S.E.; Tripp-Valdez, M.; Lluch-Cota, D.B.; Lluch-Belda, D.; Verbesselt, J.; Herrera-Cervantes, H.; Bautista-Romero, J.J. Recent trends in sea surface temperature off Mexico. *Atmosfera* **2013**, *26*, 537–546. [CrossRef]
52. Edwards, M.; Richardson, A.J. Impact of climate change on marine pelagic phenology and trophic mismatch. *Nature* **2004**, *430*, 881–884. [CrossRef] [PubMed]
53. Páez-Osuna, F.; Sánchez-Cabeza, J.A.; Ruiz-Fernández, A.C.; Alonso-Rodríguez, A.C.R.; Cardoso-Mohedano, J.G.; Flores-Verdugo, F.J.; Carballo, J.L.; Cisneros-Mata, M.A.; Piñón-Gimate, A.; Álvarez-Borrego, S. Environmental status of the Gulf of California: A review of responses to climate change and climate variability. *Earth Sci. Rev.* **2016**, *162*, 253–268. [CrossRef]
54. Kahru, M.; Marinone, S.G.; Lluch-Cota, S.E.; Parés-Sierra, A.; Mitchell, G. Ocean color variability in the Gulf of California: Scales from the El Niño-La Niña cycle to tides. *Deep Sea Res. II* **2004**, *51*, 139–146. [CrossRef]
55. Lavín, M.F.; Beier, E.; Gómez-Valdés, J.; Godínez, V.M.; García, J. On the summer poleward coastal current off SW México. *Geophys. Res. Lett.* **2006**, *33*. [CrossRef]
56. Pegau, W.S.; Boss, E.; Martínez, A. Ocean color observations of eddies during the summer in the Gulf of California. *Geophys. Res. Lett.* **2002**, *29*, 6-1–6-3. [CrossRef]
57. Kahru, M.; Kudela, R.M.; Manzano-Sarabia, M.; Mitchell, B.G. Trends in the surface chlorophyll of the California Current: Merging data from multiple ocean color satellites. *Deep Sea Res.* **2012**, *77–80*, 89–98. [CrossRef]

58. Kahru, M. Windows Image Manager, WIM Software (Ver. 9.06) and User's Manual. 2016, p. 125. Available online: http://www.wimsoft.com/ (accessed on 28 June 2017).
59. Climate Prediction Center Internet Team NOAA/National Weather Service. National Centers for Environmental, Prediction, Climate, Prediction, Center. Available online: http://www.esrl.noaa.gov/psd/data/climateindices/list/ (accessed on 28 June 2017).
60. López, J.M. Variabilidad Anual e Interanual de la Clorofila-(SeaWiFS) y el Viento Superficial (QuikSCAT) en el Alto Golfo de California: Su Circulación y Asociación. Master's Thesis, Universidad Autónoma de Baja California, Ensenada, Mexico, 2005; p. 71.
61. García-Morales, R.; Shirasago-German, B.; Felix-Uraga, R.; Perez-Lezama, E.L. Conceptual Model of Pacific Sardine Distribution in the California Current. *Curr. Dev. Oceanogr.* **2012**, *5*, 27–47.
62. Lavín, M.F.; Marinone, S.G. An overview of the physical oceanography of the Gulf of California. In *Nonlinear Processes in Geophysical Fluid Dynamics*; Velasco Fuentes, O.U., Sheinbaum, J., Ochoa, J., Eds.; Kluwer Academic Publishers: Dordrecht, The Netherlands, 2003; pp. 173–204.
63. Godínez-Sandoval, V.M. Dinámica y Termodinámica en la Entrada Exterior al Golfo De California. Ph.D. Thesis, Facultad de Ciencias Marinas, UABC, México, 2011; p. 139.
64. Morel, A.; Berthon, J.F. Surface pigments, algal biomass profiles, and potential production of the euphotic layer: Relationships reinvestigated in view of remote-sensing applications. *Limnol. Oceanogr.* **1989**, *34*, 1545–1562. [CrossRef]
65. Santamaría del Ángel, E.; Álvarez-Borrego, S.; Muller-Karger, F. Gulf of California biogeographic regions based on coastal zone color scanner imagery. *J. Geophys. Res.* **1994**, *99*, 7411–7421. [CrossRef]
66. McGillicuddy, D.J., Jr.; Robinson, A.R.; Siegel, D.A.; Jannasch, H.W.; Johnsonk, R.; Dickey, T.D.; McNeil, J.; Michaels, A.F.; Knapk, A.H. Influence of mesoscale eddies on new production in the Sargasso Sea. *Nature* **1998**, *394*, 263–265. [CrossRef]
67. Duxbury, A.C.; Duxbury, A.B.; Sverdrup, K.A. *An Introduction to the World's Oceans*, 6th ed.; McGraw-Hill: New York, NY, USA, 2000; 528p.
68. Morrow, R.; Fang, F.; Fieux, M.; Molcard, R. Anatomy of three warm-core Leeuwin current eddies. *Deep Sea Res. II* **2003**, *50*, 2229–2243. [CrossRef]
69. Van Aken, H.M.; Van Veldhoven, A.K.; Veth, C.; De Ruijter, W.P.M.; Van Leeuwen, P.J.; Drijfhout, S.S.; Whittle, C.P.; Rouault, M. Observations of a young Agulhas ring, Astrid, during MARE in March 2000. *Deep Sea Res. II* **2003**, *50*, 167–195. [CrossRef]
70. McGillicuddy, D.J., Jr.; Anderson, L.A.; Bates, N.R.; Bibby, T.; Buesseler, K.O.; Carlson, C.A.; Davis, C.S.; Ewart, C.; Falkowski, P.G.; Goldtwaith, S.A.; et al. Eddy/wind interactions stimulate extraordinary mid-ocean plankton blooms. *Science* **2007**, *316*, 1021–1026. [CrossRef] [PubMed]
71. Robinson, C.J.; Gómez-Gutiérrez, J.; de León, D.A.S. Jumbo squid (*Dosidicus gigas*) landings in the Gulf of California related to remotely sensed SST and concentrations of chlorophyll a (1998–2012). *Fish. Res.* **2013**, *137*, 97–103. [CrossRef]
72. Godínez, V.M.; Beier, E.; Lavín, M.F.; Kurczyn, J.A. Circulation at the entrance of the Gulf of California from satellite altimeter and hydrographic observations. *J. Geophys. Res.* **2010**, *115*, C04007. [CrossRef]
73. Bernal, G.; Ripa, P.; Herguera, J.C. Variabilidad oceanográfica y climática en el bajo Golfo de California: Influencias del trópico y Pacífico norte. *Cienc. Mar.* **2001**, *27*, 595–617. [CrossRef]
74. Palacios, D.M.; Bograd, S.J. A census of Tehuantepec and Papagayo eddies in the Northeastern Tropical Pacific. *Geophys. Res. Lett.* **2005**, *32*, L23606. [CrossRef]
75. Kessler, W.S. The circulation of the Eastern Tropical Pacific: A review. *Prog. Oceanogr.* **2006**, *69*, 181–217. [CrossRef]
76. Zamudio, L.; Hurlburt, H.E.; Metzger, E.J.; Tilburg, C.E. Tropical wave-induced oceanic eddies at Cabo Corrientes and the Maria Islands, Mexico. *J. Geophys. Res.* **2007**, *112*, C05048. [CrossRef]
77. Mills, C.E. JellyFish blooms: Are populations increasing globally in response to changing ocean conditions. *Hydrobiologia* **2001**, *451*, 55–68. [CrossRef]
78. Hammann, M.G.; Nevárez-Martínez, M.O.; Green-Ruíz, Y. Spawning habitat of the Pacific sardine (*Sardinops sagax*) in the Gulf of California: Egg and larval distribution 1956–1957 and 1971–1991. *CalCOFI Rep.* **1998**, *39*, 169–179.

79. Daskalov, G. Relating fish recruitment to stock biomass and physical environment in the Black Sea using generalized additive models. *Fish. Res.* **1999**, *41*, 1–23. [CrossRef]
80. Nevárez-Martínez, M.O. Producción de Huevos de la Sardina Monterrey (Sardinops Sagax Caeruleus) en el Golfo de California: Una Evaluación y Crítica. Master's Thesis, CICESE, Ensenada, Mexico, 1990; p. 144.

remote sensing

MDPI

Article

Determining the Pixel-to-Pixel Uncertainty in Satellite-Derived SST Fields

Fan Wu [1,2,3,4], Peter Cornillon [2,*], Brahim Boussidi [2] and Lei Guan [1,4]

[1] Department of Marine Technology, College of Information Science and Engineering, Ocean University of China, 238 Songling Road, Qingdao 266100, China; wufan620@126.com (F.W.); leiguan@ouc.edu.cn (L.G.)

[2] Graduate School of Oceanography, University of Rhode Island, 215 South Ferry Road, Narragansett, RI 02882, USA; bboussidi@uri.edu

[3] Qian Xuesen Laboratory of Space Technology, China Academy of Space Technology, 104 Youyi Road, Beijing 100094, China

[4] Laboratory for Regional Oceanography and Numerical Modeling, Qingdao National Laboratory for Marine Science and Technology, 1 Wenhai Road, Qingdao 266237, China

* Correspondence: pcornillon@me.com; Tel.: +1-401-742-2911

Received: 18 July 2017; Accepted: 22 August 2017; Published: 23 August 2017

Abstract: The primary measure of the quality of sea surface temperature (SST) fields obtained from satellite-borne infrared sensors has been the bias and variance of matchups with co-located in-situ values. Because such matchups tend to be widely separated, these bias and variance estimates are not necessarily a good measure of small scale (several pixels) gradients in these fields because one of the primary contributors to the uncertainty in satellite retrievals is atmospheric contamination, which tends to have large spatial scales compared with the pixel separation of infrared sensors. Hence, there is not a good measure to use in selecting SST fields appropriate for the study of submesoscale processes and, in particular, of processes associated with near-surface fronts, both of which have recently seen a rapid increase in interest. In this study, two methods are examined to address this problem, one based on spectra of the SST data and the other on their variograms. To evaluate the methods, instrument noise was estimated in Level-2 Visible-Infrared Imager-Radiometer Suite (VIIRS) and Advanced Very High Resolution Radiometer (AVHRR) SST fields of the Sargasso Sea. The two methods provided very nearly identical results for AVHRR: along-scan values of approximately 0.18 K for both day and night and along-track values of 0.21 K for day and night. By contrast, the instrument noise estimated for VIIRS varied by method, scan geometry and day-night. Specifically, daytime, along-scan (along-track), spectral estimates were found to be approximately 0.05 K (0.08 K) and the corresponding nighttime values of 0.02 K (0.03 K). Daytime estimates based on the variogram were found to be 0.08 K (0.10 K) with the corresponding nighttime values of 0.04 K (0.06 K). Taken together, AVHRR instrument noise is significantly larger than VIIRS instrument noise, along-track noise is larger than along-scan noise and daytime levels are higher than nighttime levels. Given the similarity of results and the less stringent preprocessing requirements, the variogram is the preferred method, although there is a suggestion that this approach overestimates the noise for high quality data in dynamically quiet regions. Finally, simulations of the impact of noise on the determination of SST gradients show that on average the gradient magnitude for typical ocean gradients will be accurately estimated with VIIRS but substantially overestimated with AVHRR.

Keywords: spatial precision; sea surface temperature; VIIRS; AVHRR

1. Introduction

To date, a great deal of attention has been focused on the accuracy of satellite-derived sea surface temperature (SST) products (see, for example, [1–5], a very small fraction of the articles addressing this issue). By contrast, the spatial precision (defined below) of satellite-derived SST fields has only been addressed by Tandeo et al. [6] (Tan14 hereafter), and that peripherally in an analysis of the anisotropy of SST fields in the global ocean. Donlon et al. [7] provide an excellent overview of the general approach (based on the bias and variance of pixel SST values relative to co-located in situ values) used to determine satellite data product accuracy. A feature of this approach is that the satellite–in situ matchups are generally widely separated in space and time, because of cloud cover and the paucity of in situ data. This often results in a spatially slowly varying bias in the retrieved SST values, because a significant contribution to the uncertainty in satellite retrievals results from atmospheric contamination, the spatial scales of which are, in general, large compared with the pixel separation of infrared sensors. This means that the pixel-to-pixel uncertainty, the spatial precision, may be substantially smaller than the accuracy determined from in situ match-ups. Indeed, Tan14 estimates the spatial precision of Advanced Very High Resolution Radiometer (AVHRR) to be approximately 0.2 K, substantially smaller than the estimated accuracy of 0.5 K [1–3] and, as will be shown in the following, the difference between the spatial precision and the accuracy of Visible-Infrared Imager-Radiometer Suite (VIIRS) SST fields is even more pronounced, 0.08 K versus 0.4 K [4,5]. The lack of knowledge related to the spatial precision of satellite-derived SST fields makes selection of the appropriate dataset as well as the combination of datasets derived from different sources problematic at best for studies of processes at the one to ten pixel spatial scale.

We refer to the uncertainty of the retrieved SST relative to the actual SST as the values accuracy. By contrast, we refer to the uncertainty in SST following removal of a bias in the field associated with long-wavelength phenomena as the spatial precision of the field. The latter is important in studies related to the SST gradient, while the former to processes for which the specific value is important, such as those directly related to air–sea fluxes of a variety of properties. One might also refer to the temporal precision of the retrievals—the uncertainty of SST retrievals at a given location between consecutive satellite passes of the sensor from which the fields are being derived. However, the time scale separating consecutive retrievals for most satellite-borne infrared sensors is large relative to the time scale associated with atmospheric phenomena, hence the temporal precision will be close to the accuracy as described above.

In this study, we investigate the spatial precision of Level-2 (L2) [8] SST fields obtained from the VIIRS carried on the Suomi-National Polar-orbiting Partnership (Suomi-NPP) spacecraft launched in October 2012 and L2 SST fields obtained from the AVHRR carried on NOAA-15. Note that Tan14's estimate of spatial precision is based on comparisons with L3 fields, adding a processing step that is likely to add additional error to the fields. VIIRS fields were selected because of their high spatial precision as will be shown in Section 5.1. AVHRR fields were chosen because AVHRR instruments comprise the longest, global satellite-derived SST record, dating back to late 1981. L2 data were selected because they form the basis of all higher order products obtained from these sensors, hence provide a lower limit for the small-scale retrieval noise to be expected in their products. The contribution of instrument noise to the spatial precision for each of these datasets will be determined using two methods, one based on spectra, the other on variograms of the fields [6].

In Section 2, we describe the datasets, the study area and the period covered by the analysis. This is followed in Section 3 by a discussion of the preprocessing of the datasets and then of the two approaches used to estimate the "instrument" noise and from that the spatial precision under cloud-free conditions. The results of the analyses are in Section 4 and the related discussions are in Section 5.

However, first, we describe the error budget associated with satellite-derived SST fields.

The Error Budget of Satellite-Derived SST Fields

A number of factors contribute to the uncertainty in satellite-derived SST fields. These are described in a White Paper prepared by the National Aeronautics and Space Administration (NASA)-National

Oceanic and Atmospheric Administration (NOAA) SST Science Team [9] and summarized in Figure 1. Although the accuracy of an L2 skin temperature dataset is determined by the accumulation of the error elements shown in the upper gray box of Figure 1, which also shows the relationship between these errors and the level of processing, it is generally dominated by contributions from the atmosphere—the green block. As noted above, atmospheric retrieval errors tend to be long wavelength, with an e-folding distance of many pixels in the case of infrared retrievals. The spatial precision, on the other hand, is dominated by instrument noise and classification errors (e.g., cloud-contaminated pixels passing as clear pixels) for skin temperature L2 and L3U datasets. (In the case of "buoy" temperature L2 and L3U datasets, the error in extrapolating from the skin temperature, the quantity actually measured by the satellite, to the temperature at the depth of the buoy, generally 1 m below the surface, additional contributions to the spatial precision may result from the horizontal variability in the vertical temperature step, the orange block in the figure. Only L2 skin temperature SST fields are considered in this study, hence horizontal gradients in the surface to buoy depth temperature difference do not contribute to the uncertainty in retrievals discussed herein.) For L3C, L3S and L4 datasets, the collation and interpolation schemes used will likely contribute to a decrease in spatial precision—an increase in the pixel-to-pixel errors—but the degree to which this is the case has yet to be documented. Important in the analysis presented herein is the distinction between instrument noise (elements in the yellow block of Figure 1) and the noise associated with classification errors (one of the elements in the green block). Classification errors generally refer to the improper masking of cloud-contaminated pixels and this misclassification is thought to be dependent on cloud cover—the larger the fraction of the area contaminated by clouds, the larger the fraction of misclassified pixels. The contribution of misclassified pixels to the local error is also likely to depend on cloud type. Together, these observations suggest that the classification error may vary significantly geographically. For this reason, our focus is on instrument noise, which we assume to be less dependent on location, i.e., the estimates of instrument noise obtained in this work are thought to be good estimates in regions of low cloud cover and a lower bound in general.

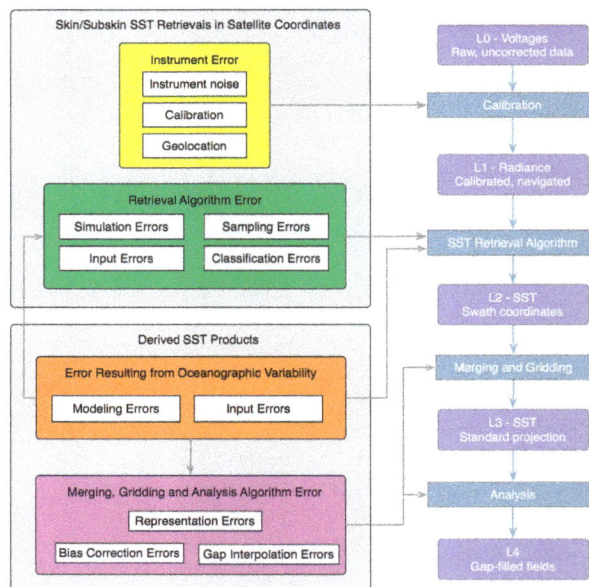

Figure 1. The error budget developed by the National Aeronautics and Space Administration (NASA)-National Oceanic and Atmospheric Administration (NOAA) sea surface temperature (SST) Science Team for satellite-derived SST fields.

2. Data

This study makes use of one dataset consisting of thermosalinograph (TEX) sections, one L2 SST dataset obtained from VIIRS radiances and one L2 SST dataset obtained from AVHRR radiances. These are discussed below along with the study area and period.

2.1. In Situ Temperature

The thermosalinograph, on which the in situ data are based, was mounted on the MV Oleander, a container ship making weekly round trips between Port Elizabeth, NJ, USA and Hamilton Harbor, Bermuda (Figure 2). Thermosalinograph temperature measurements were obtained from two thermistors, one from the seawater intake in the interior of the ship and the second directly at the intake, i.e., "external" to the hull. The exterior measure (referred to as TEX for "exterior" temperatures) is thought to be the most accurate [10] (Sch16 hereafter), hence only these are used in the work presented here. The SBE38 remote temperature sensor, on which the TEX data are based, has an accuracy of 0.0001 K, a resolution of 0.00025 K (although the TEX instrument noise is estimated to be 0.00069 K based on the variogram approach discussed in Section 3.3), and a response time of 0.5 s. The TEX sensor sampled every 10 s resulting in an approximate spatial resolution of 75 m at the typical 16 knots cruise speed of the Oleander. TEX data for the period September 2007 to fall 2013 were obtained from the Atlantic Oceanographic and Meteorological Laboratory. The quality control procedures used to screen these data are described in Sch16.

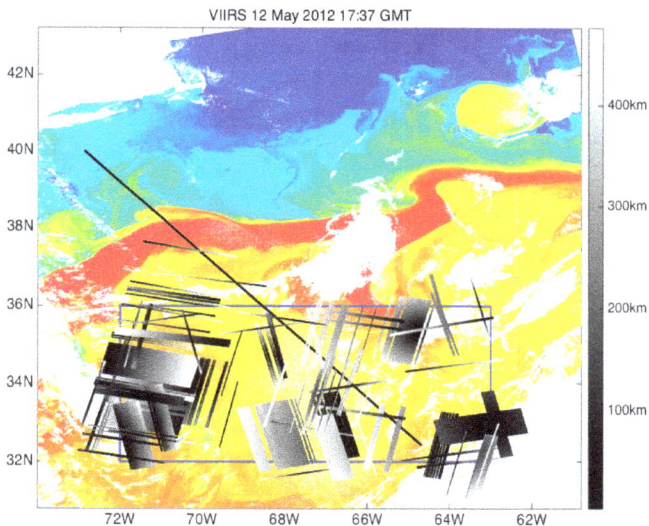

Figure 2. Visible-Infrared Imager-Radiometer Suite (VIIRS) SST image from 12 May 2012. The long black line (73.5°W, 40°N to 64.8°W, 32.6°N) indicates the nominal Oleander track. Blue frame (longitudes from 63°W to 72°W, and latitudes from 32°N to 36°N) denotes the region of the Sargasso Sea considered in this study. Shades of gray denote the location of sections extracted from VIIRS SST fields—discussed in Section 3.1.1. The gray scale indicates distance from nadir (discussed in detail in subsequent sections). Sections with a constant gray level are along-track sections; those with a gradient in gray are along-scan. Along-track (along-scan) sections with a negative slope and along-scan (along-track) sections with a positive slope are daytime (nighttime) sections. The SST field is simply provided as a background reference field and corresponds to only one of the images used.

2.2. Visible-Infrared Imager-Radiometer Suite (VIIRS)

The L2 VIIRS SST retrievals used here were derived from the VIIRS "Moderate Resolution Bands", which have a resolution of approximately 750 m at nadir. Because of the way in which the instrument samples, the resolution decreases very slowly (compared with other satellite-borne instruments, Figure 3) to approximately 1600 m at the scan edge, a ground distance of approximately 1500 km from nadir [11,12].

For this study, we used the VIIRS SST product obtained from NOAA's Comprehensive Large Array-data Stewardship System (the VIIRS Sea Surface Temperature Environmental Data Record (EDR) obtained from: http://www.nsof.class.noaa.gov/saa/products/search?datatype_family=VIIRS_EDR) produced with the Joint Polar Satellite System (JPSS). Only quality level 1 data, the "best" quality level, were used. Although screening at this level ideally removes all cloud contaminated pixels, some are still included in the analysis, leading to the misclassification error discussed above.

2.3. AVHRR Pathfinder SST

The AVHRR product used was derived with the Pathfinder retrieval algorithm developed at the University of Miami [13]. The algorithm was applied to the High Resolution Picture Transmission (HRPT) data stream obtained from the AVHRR on NOAA-15. Retrievals were performed at the University of Rhode Island. Only pixels with a quality level of 3 or higher were used. The nominal pixel spacing is 1.1 km although, as can be seen in Figure 3, it increases significantly from this value. This increase is what motivated use of pixels within 500 km of nadir as discussed below.

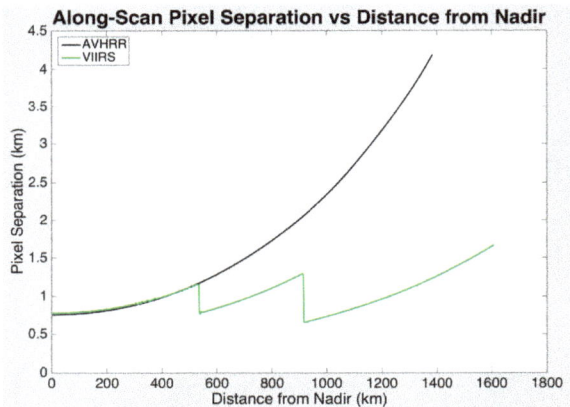

Figure 3. Spacing in the along-scan direction for Advanced Very High Resolution Radiometer (AVHRR) and VIIRS pixels in L2 fields as a function of distance from nadir.

2.4. The Study Area

MV Oleander traverses several distinct dynamical regimes: the shelf, the Slope Sea, the Gulf Stream, and the Sargasso Sea. In that the focus of this analysis is on the spatial resolving power of satellite-derived SST datasets, it is important to select a region in which the geophysical variability of the SST field does not overwhelm the uncertainty associated with the SST retrievals, be they driven by misclassification errors (the green block in Figure 1) or instrument/calibration issues (the yellow block). Specifically, this means selecting a dynamically "quiet" region in the ocean. The Sargasso Sea portion of the Oleander track between 32°N and 36°N meets this requirement. In order to increase the amount of satellite-derived data used in our analysis, we consider all data in the region between 63°W and 72°W, and 32°N and 36°N (Figure 2). This region was selected with the expectation that the dynamics are similar to those seen along Oleander track between 32°N and 36°N hence the spectra should be similar in both slope and energy. As shown in Sch16, spectra including the Gulf Stream are substantially more energetic than those for SST in the Sargasso Sea.

2.5. The Study Period

The analyses presented here are based on SST fields from the summer of 2012 only—June, July and August. Sch16 show that spring (March, April and May) and summer spectra tend to be about twice as energetic, over the spectral range examined, 1 to 100 km, as fall and winter spectra suggesting that the latter would be more appropriate for the evaluation proposed here, but the summer months are also substantially less cloud contaminated than the other seasons. Furthermore, the increased spectral energy is likely due in part to diurnal warming, the effect of which may be mitigated by selecting nighttime fields only as shown in Section 4.2. This raises a concern with regard to the TEX data because TEX sections are not synoptic, taking approximately 20 h to cross the study area. However, since the TEX samples between 5 and 6 m below the surface, diurnal warming is not thought to be a significant problem [14].

3. Methodology

3.1. The Spectral Approach

The spectral method, to determine retrieval noise at the pixel level, is based on an analysis of the large wavenumber tail of the power spectral density of SST temperature sections extracted from the SST fields. Spectra are based on the Discrete Fourier Transform (DFT) determined from the Fast Fourier Transform (FFT) (see Sch16 or Wang et al. [15], who also used the DFT to analyze TEX spectra). The FFT requires equally spaced, gap-free data, i.e., gaps, if they exist in the original series, must be filled and the data must be interpolated to equal spacing, if not already equally spaced, prior to applying the FFT algorithm. For satellite-derived fields, gaps result from cloud cover, intervening land values (not an issue for the region studied here) or missing scans while pixel spacing depends on the product. (The filling of gaps is discussed in Section 3.1.1.) In the case of L2 products the spacing of pixels in the along-scan direction varies with distance from nadir (Figure 3), as does the along-track spacing, although much less so (<0.5% change from nadir to the swath edge for both AVHRR and VIIRS). For the in situ data, intermittent system failures resulted in gaps although not to the extent of those in the satellite data and sample spacing depends on the ship speed, which varies.

Of importance to the analysis presented here is that interpolation, either to fill gaps or to regularize the spacing of samples on a section, impacts the resulting spectrum, with the impact generally increasing as the wavenumber increases, i.e., in the spatial range of most importance to the analysis here. Furthermore, the impact is a function both of the fraction of "good" values (defined as Q by Sch16), and the degree to which the "missing" data are clustered (referred to as cohesion and assigned the symbol C [16]). Sch16 found that " ... spectral slopes are increasingly biased low as Q decreases and C increases, and this effect becomes more pronounced as the true spectral slope increases". Based on this they only considered VIIRS spectra for $Q - C > 0.1$ and $Q > 0.5$ in their analysis. We found these thresholds to be too permissive for our purposes; the impact of interpolation on spectra in the 1 to 10 pixel range can overwhelm the underlying spectrum as will be shown below. We therefore chose more stringent constraints on Q, generally resulting in $Q > 0.9$. At this level, the cohesion of the data has a relatively small impact on the spectra for slopes in the range of those observed in the Sargasso Sea [10], so we did not impose an additional constraint on cohesion.

3.1.1. Selection of the Sections

Satellite-Derived Fields. The satellite-derived SST fields evaluated here are obtained from scanning radiometers, the characteristics of which may differ in the along-scan versus along-track directions. This is indeed the case for VIIRS due to the use of multiple detectors for each scan, which results in striping of the fields [17]. The decision was therefore made to separate the data into along-scan and along-track sections. The data were farther divided into day and night fields to allow analysis of the possible effect of diurnal warming on the spectral characteristics of the fields. This is of particular importance given the selection of the Sargasso Sea in summer months, a period when diurnal warming is significant [18].

Also with regard to the selection of sections from the L2 datasets is their distance from nadir. Both the area of each pixel (approximately the along-track spacing, 741 m for VIIRS and 1115 m for AVHRR, times the along-scan spacing) and the spacing of pixels along the scan increases away from nadir (Figure 3). Both of these factors impact along-scan spectra at small scales, while the increase in pixel size impacts along-track spectra, again at these scales. Although the pixel spacing of along-track sections is virtually independent of the distance from nadir, the size of the pixel is not, i.e., the SST values associated with pixels is averaged over increasingly larger areas away from nadir. This is similar to smoothing along-track with a moving average, which in turn depresses the power spectral density at small scales, this, independent of the preprocessing performed on the data and it affects along-track and along-scan spectra equally. Along-track interpolation (discussed below) to address the change in pixel spacing in the along-scan direction (Figure 3) also impacts the resulting spectra. In order to reduce the impact of both of these effects, only sections within 500 km of nadir are used for this analysis.

The final criterion used to select sections from the L2 fields relates to the gappiness of the data. For clarity, we combine this step with the interpolation to fill missing pixel values in the study area. The actual implementation of the algorithm is slightly different, to reduce processing time, but the result is the same. Missing values in the study area were replaced using a Barnes filter if 13 of the 24 pixels in a 5×5 pixel square surrounding the pixel of interest are cloud-free, otherwise the pixel remains flagged as missing. This corresponds to a decay scale associated with the averaging of 1.5 km for VIIRS and 2 km for AVHRR and follows the approach taken by Sch16. Following this gap filling, all complete (no missing values) 256 pixel, non-overlapping sections in the along-track direction meeting the distance from nadir criterion were selected as were all non-overlapping along-scan sections. Only a small fraction of sections used in the final analysis had more than 15 missing pixels in the original data (more than 6% of the pixels were filled on <10% sections). The impact of this on the final spectra was evaluated for the worst case scenario by using the Barnes filter to fill every point on a section (gap filling was still possible in that adjacent pixels were left as is, i.e., not set to missing values), not just the pixels with missing values. The result suggests that the gap filling performed only for pixels with missing values has little impact on the final spectra, because the number of missing values is in general small; less than 0.6% of all values contributing were replaced with the Barnes filter.

Oleander Sections. Only TEX sections that met the selection criteria of Sch16 were considered. Of these only sections with a maximum pixel separation of 150 m in the Sargasso Sea were selected. (Selection of temperature sections with maximum sample spacing in excess of 150 m resulted in a significant steepening of the spectral slope for wavelengths smaller than approximately 1 km. This is due to the nearest neighbor interpolation to 75 m spacing, which repeats samples for these large separations.) Barnes filtering with a decay scale of 0.2 km was used to fill these gaps and the resulting sections were nearest neighbor interpolated to a mean spacing of 74.9 m, the mean spacing averaged over all sections; the mean spacing varies from section-to-section with a minimum of 74.6 m and a maximum of 75.0 m [19].

Table 1 lists the number of satellite-derived sections by along-scan/along-track, day/night combination for the summer (June–August) of 2012 and the number of Oleander TEX sections for the summers of 2008–2013.

Table 1. Number of sections meeting the given selection criteria discussed in this section and in Sections 2.4 and 2.5.

	Day		Night	
	Along-Scan	Along-Track	Along-Scan	Along-Track
VIIRS	126	517	561	615
AVHRR	266	256	104	193
Oleander		42		

3.1.2. Interpolation to Equal Spacing

Satellite-Derived Fields. As previously noted the pixel separation in the along-scan direction changes with distance from nadir. Because the spectral energy determined with the standard FFT is a function of pixel spacing and the number of pixels in the section, combining data with different spatial resolutions tends to add noise to the spectra. To address this, along-scan sections were divided into three groups each for VIIRS and AVHRR based on mean pixel spacing. First, all adjacent temperature sections for a given satellite pass were grouped into subgroups and the mean separation of pixels for the subgroup was calculated (The subgroups ranged in size from 1 to O (100) sections depending on cloud cover.). Each subgroup was then assigned to the group indicated in Table 2 based on the mean pixel spacing of the subgroup. All of the temperature sections falling in a given group were then interpolated to the same pixel spacing, also shown in Table 2. This pixel spacing was determined from the mean pixel spacing determined from the contributing temperature sections for the given group. This, together with the relatively small size of the ranges, tended to eliminate problems associated with different spatial sampling and with an interference between the sampling frequency along the original section and that along the interpolated section. Nearest neighbor interpolation was used. Figure 4 shows the effective transfer function of three different interpolation algorithms available in Matlab: linear, nearest neighbor and cubic spline. To determine the most appropriate resampling strategy, SST values on the VIIRS sections were replaced with white noise and interpolated. Linear interpolation smooths the field the most resulting in a significant loss of energy at small wavelengths, the portion of the spectrum of most interest here. Cubic spline does better but still results in a loss of energy at small wavelengths. Nearest neighbor interpolation does not significantly alter the distribution of values but does alter the effective wavelength—by shifting the values in space. However, the effect on the spectrum is small since the values have been shifted to locations, which are on average relatively close to the original values—the use of the mean spacing of pixels (which varies from group-to-group) rather than a fixed spacing for all sections.

Table 2. Grouping of along-scan sections based on mean pixel spacing of the temperature section. The values indicated correspond to the lower limit on the range—the value to which temperature sections in the range is interpolated—the upper limit on the range.

	Group 1 (m)	Group 2 (m)	Group 3 (m)
VIIRS	770-805-820	860-885-910	940-995-980
AVHRR	760-765-810	820-865-920	940-947-980

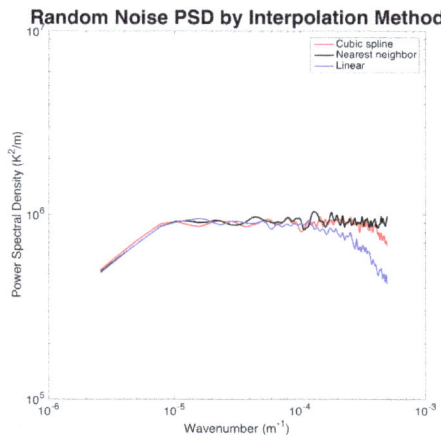

Figure 4. Spectral response of the interpolation methods applied to white noise.

3.1.3. Detrending

Typically, a windowing function is applied to time series (or temperature sections in this case) prior to obtaining the spectrum so as to reduce leakage between frequencies and the introduction of spectral energy due to step changes at the ends of the section. However, windowing functions tend to depress the amount of energy in the spectrum, which results in an underestimate of the instrument noise, so we elected not to window the data. Specifically, several different windowing functions, as well as simple detrending, were applied to simulations generated by adding white noise to randomly generated temperature sections with a linear power spectral density (in log-log space) typical of the spectra obtained from the SST sections but with random phase of the spectral elements between $-\pi$ and $+\pi$. Detrending provided the most accurate estimate of the imposed noise when compared to analysis of the data with the various windowing functions or to analysis of the data with no preprocessing.

3.1.4. FFT

Finally, the FFT function available in Matlab was used to obtain the spectra from the detrended temperature sections. For the along-scan direction, power spectral densities were ensemble averaged over each of the subgroups defined in Section 3.1.1. This resulted in approximately 100 subgroups for all groups of the AVHRR/VIIRS, day/night combinations, i.e., there was an average of eight subgroups for each of the defined groups. Similar averaging was performed for the subgroups of the along-track direction.

Oleander spectra were ensemble averaged over all of the selected sections.

3.2. Estimation of Instrument Noise

Instrument noise in the satellite-derived fields is estimated from the shape of the power spectral density on the short wavelength (large wavenumber) end of the retrieved spectra. To better understand the approach, consider the factors contributing to this portion of the spectrum. If adjacent values on a given temperature section are independent with no noise, then the shape of the spectrum is defined by the geophysical processes in the region. If the field has been smoothed or averaged over a significant region, there is little additional information in the value of one point relative to an adjacent one and the spectrum falls off more rapidly than the shape associated with geophysical processes. This is what we found for the spectra of the AVHRR SST fields associated with large scan angles (not shown here) as well as with the oversampled TEX sections with maximum spacing of samples in excess of 150 m resampled to a spacing of 75 m discussed in Section 3.1.1. To avoid the roll-off of the spectra at small wavelengths the data were not smoothed. If the field is not smoothed and, white noise is added to the values at individual pixels, the spectrum will tend to level off; the point at which it begins to do so being a function of the level of the added noise. Finally, if energy remains in the geophysical spectrum at wavenumbers larger than those at the end of the retrieved satellite-derived spectra, the spectra will also tend to level off near their end as a result of energy aliased from the larger wavenumbers. This is likely the reason the ensemble averaged Oleander TEX spectrum levels off (Figure 5). (It is not clear whether the slight fall off in the TEX spectrum beginning at approximately 1 km is a result of a fall-off in the geophysical signal or some form of averaging of the TEX data. However, this roll-off is very slight and ignored here.) In summary, the large wavenumber tail of the satellite-derived spectra is subject to the following:

- An increase in the magnitude of the slope of the spectrum due to averaging over the footprint of the sensor;
- A decrease in the slope due to geophysical noise aliased into the spectrum, especially at high wavenumbers; and
- A decrease in the slope due to instrument noise, the quantity of interest here.

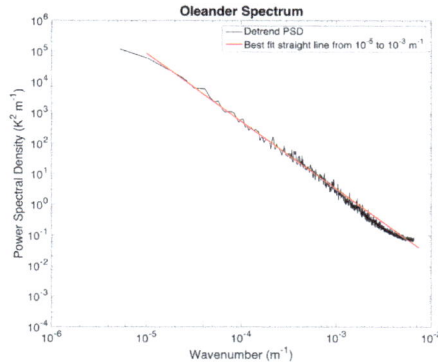

Figure 5. Power spectral density from Oleander TEX for all Oleander summer sections (June–August) of 2008 through 2013 with maximum sample separation less than 150 m. Temperature sections detrended prior to determining and ensemble averaging the spectra. Straight red line: least squares best fit straight line (slope = −2.12) of $\log_{10}$ (PSD) to $\log_{10}$ (wavenumber) between 10^{-5} and 10^{-3} m^{-1}.

In order to determine the instrument noise, i.e., to separate it from the other factors cited above, we defined a two-step process based on the following three assumptions:

1. $\log_{10}$ of the geophysical power spectral density in the study area falls off linearly with $\log_{10}$ of the wavenumber over the spectral range sampled by the satellite-borne sensors (1.5 to order 100 km).

2. The spectrum continues to roll-off with approximately the same slope, at wavenumbers larger than those associated with the Nyquist frequency of the satellite temperature sections. This and the previous assumption are borne out by the mean TEX spectrum shown in Figure 5 as well as from the analysis of the spectra from the two sensors.

3. The instrument noise for both sensors is white; i.e., it contributes equally at all wavenumbers associated with the given temperature sections. This is not quite the case for VIIRS hence one has to take a bit more caution with the results presented herein.

In the first step, the slope, intercept and noise level of a hypothetical spectrum yielding the best fit to the satellite spectrum is determined in a least squares sense. This is done by minimizing γ, the sum of the squared difference between the hypothetical spectrum and the satellite spectrum:

$$\gamma(slope, \; intercept, \; noise) \; = \; \sum_{i=1}^{N} \left(\left(10^{(slope*log_{10}k_i + intercept)} + noise\right) - PSD_i^{sat} \right)^2 \tag{1}$$

where *slope* and *intercept* define the straight line portion of the best fit spectrum in log-log space (assumption 1 above), *noise* is the noise level (assumption 3) also in spectral space, k_i is the wavenumber of the *i*th spectral component and PSD_i^{sat} the corresponding power spectral density of the satellite spectrum. In the second step, the constant *noise* level used to generate the spectrum in Equation (1) is related to white noise in the spatial domain. Specifically, 1000 noise-free temperature sections, with one tenth the sample spacing of that associated with the sensor of interest, are generated by inverse Fourier transforming spectra with the same slope and intercept found with Equation (1) but with the phase of each spectral component randomly selected between $-\pi$ and π. A 10-point moving average is then applied to each temperature section and the result is decimated by 10. Gaussian white noise of magnitude σ is then added to each point on each section, the sections are Fourier transformed, ensemble averaged and a new figure of merit is obtained:

$$\gamma(\sigma) = \sum_{i=1}^{N} \left(PSD_i^{Simulated}(\sigma) - PSD_i^{Best\ fit}(slope, intercept, noise) \right)^2 \qquad (2)$$

where $PSD_i^{Simulated}(\sigma)$ is the ensemble averaged power spectral density of the simulated temperature sections and $PSD_i^{Best\ fit}(slope, intercept, noise)$ is the best fit power spectral density to the satellite-derived power spectral density based on the slope, intercept and noise found with Equation (1). This is repeated over a range of white noise levels σ to find the level, which best corresponds to the noise level obtained with Equation (1). Of importance, is that generating temperature sections with 1/10 the spacing of the data associated with the sensors of interest, the energy at higher wavenumbers than those resolved by the instrument are aliased into the results thus allowing for a more accurate estimate of the instrument noise. In addition, averaging the oversampled temperature section simulates averaging performed over the footprint of the sensor. However, this does not take into account additional averaging, which takes place in the 2nd dimension of the sensor's footprint. This is not thought to contribute significantly to the determination of instrument noise outlined above.

3.3. The Variogram Approach

To determine instrument noise from variograms, a model, which includes instrument noise as one of its parameters, is fit to the empirical variogram. The model is intended to reflect the spatial characteristics of the underlying data, hence selection of an appropriate model for the data of interest is critical. Various models have been identified in the literature [20]. Tan14 used an exponential model of the form:

$$\gamma(\Delta_{x\ or\ y}) = \sigma^2 + \sigma^2 (1 - e^{(-\frac{\Delta_{x\ or\ y}}{L})}) \qquad (3)$$

where σ^2, referred to as the nugget, is the variance of the difference in the retrieval at a given location from that at a neighboring location as the separation between the two locations goes to zero, i.e., the instrument noise in this case, σ^2, referred to as the sill, is the variance associated with the variability for a spatial separation of L, the decorrelation scale. Note that the sill is a measure of the geophysical variance of the field, σ^2_{geo}, plus the "large" scale retrieval variance, which depends on the variance in the atmosphere, the variance of the surface emissivity, instrument noise, etc. So,

$$\sigma^2 \approx \sigma^2_{geo} + (\sigma^2_{retrieval} - \sigma^2) \qquad (4)$$

where $\sigma^2_{retrieval}$ is the total variance of the retrieval.

The formulation used by Tan14 works well for relatively homogeneous datasets for which the underlying variogram has an exponential form [6]. However, in the Sargasso Sea, the shape of the empirical variograms, for the L2 SST fields of interest, differ from subregion-to-subregion, not only in terms of parameters but also in terms of the model itself, with an exponential model fitting in some cases and a Gaussian model in others. In light of this, we have elected to use the "stable semivariogram" [20], a slightly modified single model, of the form:

$$\gamma(\Delta_{x\ or\ y}) = \sigma^2 + \sigma^2 (1 - e^{-(\frac{\Delta_{x\ or\ y}}{L})^w}) \qquad (5)$$

Note, in comparison to Equation (3), Equation (5) includes an extra parameter, w, which ranges from 1 for the exponential form to 2 for the Gaussian form. Although variograms can be developed in two dimensions for the model of interest, we chose to use variograms for the along-scan and along-track directions separately for much the same reasons presented in the discussion of the preliminary processing of the data,

As in Tan14 we use the formulation given by Cressie to estimate the variogram [21]:

$$\hat{\gamma}(\Delta x \text{ or } y) = \frac{\sum_{(s_i,s_j)}(SST(s_i) - SST(s_j))^2}{2n} \tag{6}$$

where $SST(s_i)$ is SST at location s_i, Δx or y is the spatial separation in kilometers of (s_i, s_j) pairs in the along-scan (x) and along-track (y) directions, and n is the number of such pairs, which varies with Δx or y and the number of cloud contaminated pixels.

For each of the combinations of interest (along-scan/along-track, day/night), a variogram was obtained (Equation (6)) for each of the interpolated, equally spaced temperature sections used in the spectral approach and described in Sections 3.1.1 and 3.1.2. Next, for each variogram, the values of σ_0, σ, L and w of Equation (5), which minimized the weighted squared difference between Equation (5) and the variogram, were obtained. The fit was performed over separations up to 20 km. (The nugget did not vary significantly for fits up to approximately 40 km. However, fitting to a larger range generally resulted in an increase in the nugget, which was thought to be unrealistic—the nugget wandered away from the variance at the smallest observed separation.) The weight assigned to each separation was equal to the number of pairs at that separation over the total number of separations contributing to the variogram, i.e., the weight assigned to a given separation decreased as the separation increased. The best-fit nuggets were then averaged for all temperature sections corresponding to a given sensor/day-night combination to obtain the estimate for instrument noise for that combination. Nuggets were also averaged by the subgroups identified in Section 3.1.1.

4. Results

The spatial precision of satellite-derived SST retrievals, the noise resulting from processes in the yellow and green boxes of Figure 1, which we refer to as instrument noise here, is shown in Table 3 for each of the along-scan/along-track, day/night combinations. The first row for each sensor (labeled *Spectra*) corresponds to the estimates obtained from the spectral method. Only subgroups consisting of five or more temperature sections and with a spectral slope steeper than -1 were used. The instrument noise for subgroups with shallower spectral slopes tended to dominate the geophysical signal increasing the uncertainty in the fit of Equation (1). The noise estimates provided in the table are the means of the estimates associated with each subgroup. The uncertainty is the square root of the variance of these means over the number of contributing subgroups. Variogram estimates follow in the next row (labeled *Variogram*) for each sensor. The means of the estimates are from the same subgroups used in the spectral approach and the uncertainties are calculated as for the spectral approach. The final row of the table (labeled *Upper Limit*) for each sensor is an "upper limit" on the instrument noise assuming that the pixel-to-pixel noise is white. This was obtained by noting that the variance of the difference of adjacent SST values, $\sigma^2(\Delta x_{min})$, is the sum of the variances of the noise of each of the two values, $2\sigma_i^2$, plus the contribution due to the geophysical variance between the two values, $\sigma_{geo}^2(\Delta x_{min})$:

$$\sigma^2(\Delta x_{min}) = 2\sigma_i^2 + \sigma_{geo}^2(\Delta x_{min}) \implies \sigma_i \leq \frac{\sigma(\Delta x_{min})}{\sqrt{2}} \tag{7}$$

If the noise is not white, for example, the actual level of noise may, in fact, be larger than the "upper limit".

4.1. AVHRR

Day-versus-night, along-scan instrument noise levels obtained for the AVHRR data are not statistically distinguishable. Nor are the along-track levels. The levels for the variogram estimates based on the same subgroups as the spectral estimates (2nd row) are also statistically similar. Furthermore, although somewhat larger, the variogram estimates are quite close to the spectral estimates, and all of the estimates are close to the upper limit (Equation (7)) for the given sensor/day-night/scan-track combination, suggesting that the instrument noise is white. It is possible that the pixel noise is correlated at small scales but, again, the mechanism for this is not obvious.

Table 3. Estimated instrument noise in satellite-derived SST fields. Numbers in parentheses are the number of subgroups from which the means are determined. The indicated uncertainty of the means is the square root of the variance of the contributing subgroups over the number of subgroups.

	Method	Day (K)		Night (K)	
		Along-Scan	Along-Track	Along-Scan	Along-Track
	Spectra	0.172 ± 0.001 (5)	0.209 ± 0.001 (7)	0.173 ± 0.003 (2)	0.209 ± 0.008 (4)
AVHRR	Variogram	0.185 ± 0.004 (5)	0.219 ± 0.006 (7)	0.183 ± 0.001 (2)	0.219 ± 0.006 (4)
	Upper Limit	0.189	0.218	0.194	0.208
	Spectra	0.046 ± 0.001 (4)	0.076 ± 0.002 (10)	0.021 ± 0.001 (24)	0.032 ± 0.002 (14)
VIIRS	Variogram	0.081 ± 0.013 (4)	0.097 ± 0.006 (10)	0.042 ± 0.004 (24)	0.056 ± 0.004 (13)
	Upper Limit	0.078	0.101	0.050	0.057

The along-scan AVHRR spectra are shown in Figure 6 for a daytime subgroup and a nighttime subgroup. The best-fit linear spectra with noise are also shown in Figure 6, obtained as discussed in Section 3.2. Figure 7 shows the corresponding along-track AVHRR spectra. In all four cases, noise is seen to impact the spectrum for wavelengths (wavenumbers) up (down) to approximately 25 km (0.04 km^{-1}). The approximately linear portion of the AVHRR spectrum corresponds to a small fraction (~10%) of the 129 spectral values, which is also apparent from these plots. This means that relatively small changes in the low wavenumber end of these spectra will have a more significant impact on the estimated background slope than for spectra less impacted by noise. However, the spectral method for determining instrument noise is relatively insensitive to this; significant changes in slope and intercept result in virtually identical values of instrument noise. For example, for the spectrum shown in the left panel of Figure 7, a slope, offset combination of $(-1.7570, -6.2730)$ yields the same level of instrument noise. This is because the instrument noise is one to two orders of magnitude larger that the assumed geophysical signal, the straight line portion of the spectrum, over a significant fraction of the spectrum (remember the fits are in regular, not log-log space) so changes in the slope do not result in a significant difference in the squared sum of the differences between the model and the observed spectrum. For spectra that level off substantially at large wavenumbers, the noise is effectively determined by the power spectral density level at these wavenumbers. This is readily seen in Figures 6 and 7; the high wavenumber end of the simulated spectra with noise are at a similar level for the along-scan sections and at a slightly higher level for the along-track sections. Care must be taken however when the level of instrument noise is similar, or smaller, in magnitude to the geophysical signal at these wavenumbers, as will become clear in the analysis of the VIIRS spectra.

AVHRR along-track instrument noise is approximately 20% larger than along-scan instrument noise. This is presumably due to the line-by-line calibration undertaken in the development of the L1b data product used as input to the L2 retrieval algorithm.

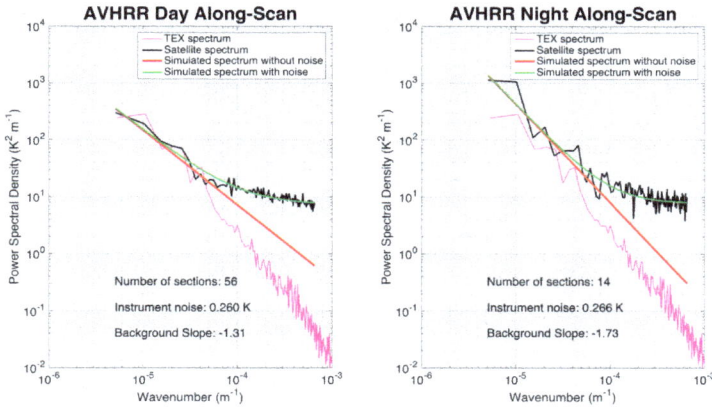

Figure 6. Mean AVHRR spectra for contiguous along-scan sections (black). Best-fit linear spectra with noise to the mean VIIRS spectra (green). Best-fit linear portion of the best-fit linear spectra with noise (red). Mean TEX spectrum shifted vertically to allow for comparison (magenta).

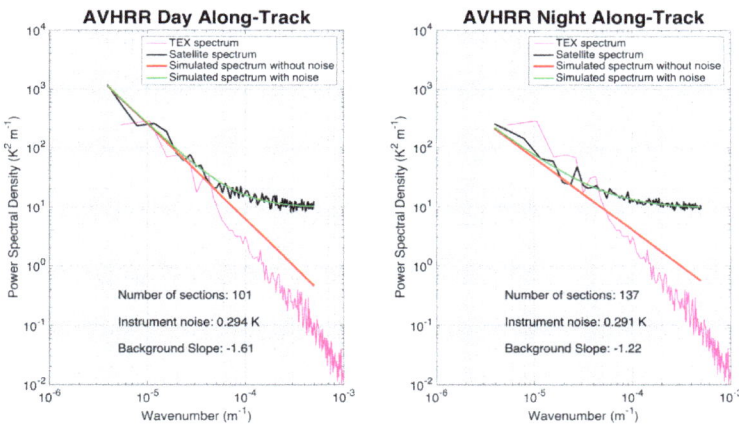

Figure 7. Mean AVHRR spectra similar to Figure 6 except for along-track sections. Daytime spectrum is for 21:08 GMT on 10 June 2012. Nighttime spectrum is for 09:34 GMT on 23 June 2012.

4.2. VIIRS

Mean VIIRS spectra similar to those shown for AVHRR in Figures 6 and 7 are shown in Figures 8 and 9, respectively. The spectra in these figures differ in several key ways from those associated with AVHRR. First, the level of instrument noise is, in all cases, *substantially* lower than that for AVHRR. Second, spectral peaks, especially in the daytime spectra, are evident at 1.5, 2.2, and 2.9 km as well as a broad peak at 12 km in the along-track spectra (Figure 9). There are 16 detectors for each of the VIIRS moderate resolution bands used for SST retrievals, hence, one scan of the instrument consists of 16 scan lines. The gain of these detectors may differ slightly and this difference is not regular, i.e., it changes along-scan and between scans. This is what gives rise to the observed peaks; the peaks at 1.5, 2.2, and 2.9 km correspond to a separation of one, two and three pixels and the peak at 12 km corresponds to the 16 pixel repeat scans of the instrument (750 m × 16 detectors = 12 km). Reassuringly, the along-scan spectra do not show these peaks. In addition, note that the noise from the different detectors contributes to a general elevation of the large wavenumber end of the spectrum—the simulated spectra with noise in Figure 9 tend to separate from the associated straight line spectrum at wavelengths smaller than approximately 8 km for along-track sections compared with approximately 5 km for along-scan sections. The point of separation is, of course, a function of the magnitude of the geophysical signal. In regions with a significantly larger geophysical signal, in the vicinity of the Gulf Stream for example, instrument noise will likely have no effect on the spectrum, with the possible exception of a few of the peaks.

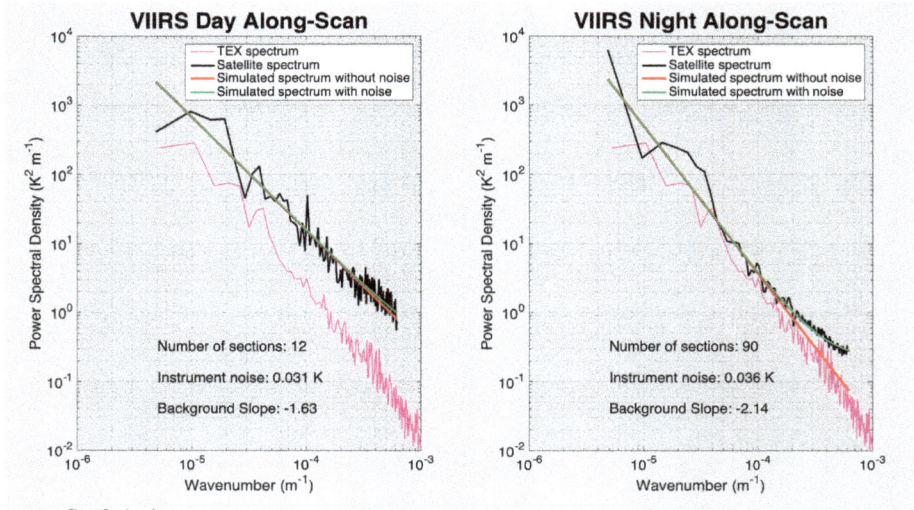

Figure 8. Mean VIIRS spectra similar to the AVHRR spectra in Figure 6.

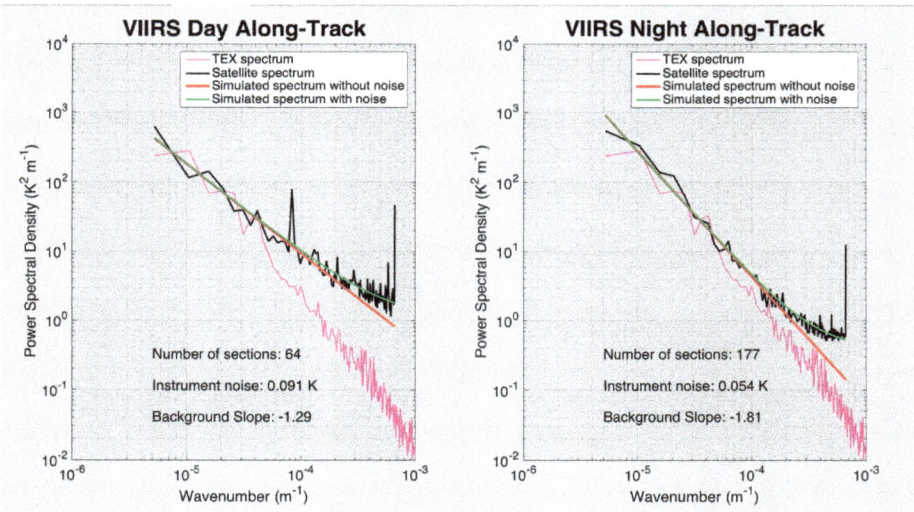

Figure 9. Mean VIIRS spectra similar to the AVHRR spectra in Figure 7.

The third significant difference between AVHRR and VIIRS spectra relates to the daytime spectra compared with the nighttime spectra. Specifically, there is a statistically significant difference between daytime and nighttime VIIRS spectra, with the daytime spectra being more energetic at wavelengths smaller than approximately 100 km. This is likely due to diurnal warming, which occurs frequently in the Sargasso Sea in summer months [14,18]. In addition, note that the slope of nighttime spectra for both along-scan and along-track sections is closer to that of the TEX spectrum than the daytime spectra. Surprisingly, the level of instrument noise is also larger at daytime than at nighttime as is evident both from the figures and from Table 3. This may result from the sensitivity of the banding to the energy in the SST field. Banding is difficult to correct for because it is not the entire scan line that has higher values than its neighbors, but rather, what appear to be randomly located segments of a given scan line. Furthermore, the magnitude of the difference in these regions appears to be related to the magnitude of the retrieved temperature.

Finally, the level of instrument noise estimated with the spectral approach is substantially smaller than (as much as one half) estimates based on the variogram. The reason for this is not clear. Although the spectral approach provides slightly better estimates of the noise added to simulated temperature sections than the approach based on the variogram, the estimates do not differ by the amounts seen in the actual data for VIIRS.

5. Discussion

5.1. Comparison of the AVHRR L2 Instrument Noise Estimates with the Results of Tandeo et al.

Tan14 estimated the nugget in the L3 Meteosat AVHRR data set produced by the O&SI SAF Project Team [6,22]. This product was assembled by remapping the full resolution nighttime AVHRR fields onto a regular $0.05° \times 0.05°$ global grid and averaging the results into 12 h fields. They found $\sigma_0 \approx 0.14$ K for the study area. This is larger than would be expected if instrument noise of the full resolution Meteosat AVHRR data is similar to that found for NOAA-15 AVHRR (on the order of 0.20 K) and if this noise is uncorrelated from pixel-to-pixel, the assumption made in the analyses presented herein. Specifically, we would expect the noise for the L3 product to be approximately 0.05 K since order 25 pixels are averaged for each $0.05° \times 0.05°$ SST estimate. It is possible that the level of instrument noise (elements in the yellow block of Figure 1) associated with the AVHRR on Meteosat is higher than that of NOAA-15. More likely however is that the difference results from misclassification errors associated with cloud flagging (the most significant element in the green block). Specifically, Tan14 processed all of the data for one year, 2008, i.e., they did not constrain their analysis to relatively cloud free fields as we did. Cloud-contaminated L2 pixels were, of course, excluded from the production of the L3 fields and Tan14 also excluded L3 pixels flagged as cloud-contaminated. However, the likelihood of misclassification, cloud-contaminated pixels not being flagged as such, increases as the fraction of cloud cover increases. Furthermore, classification errors tend to be small-scale errors, a small number of pixels here, a small number of pixels there, as opposed to large regions, which are misclassified. This means that such errors will likely contribute to noise at small spatial scales. A histogram of Tan14 nuggets (not shown here) shows a broad distribution ranging from σ_0 in the 0.05 K range to order 0.3 K with a peak around 0.14 K. If the nugget resulted primarily from instrument errors (those in the yellow block), one would expect a relatively narrow peak; the instrument noise is unlikely to vary substantially for the region. Thus, the broad σ range suggests that it is a combination of classification errors and instrument noise. Because our analysis required long sections of cloud-free pixels the data were likely much more clear, on average, than those of Tan14. Noise may be added through the combination of L2 fields to obtain the L3 product, which also contributes to the difference between our estimate of local noise and that of Tan14. Using nighttime only data, as Tan14 have done, will minimize, but not completely remove, this. Finally, we found that the model, which best fits the SST field in the Sargasso Sea, varies from an exponential form to a Gaussian form, hence our use of the standard model. Tan14 used the exponential form. This will likely also contribute to an overestimate of the instrument noise in regions in which a mixed form is more appropriate.

Lending credence to the values we find, approximately 0.2 K for at both day and night, are the estimates of uncertainty found by Keogh et al. [23] when they compare AVHRR SST values with the values obtained from a ship-borne radiometer. They examined eight sections in eight images, five nighttime and three daytime. The SST values were close in space, most probably consisting of adjacent pixels and were obtained within three hours of the satellite pass. The resulting standard deviations thus represent a lower limit on the spatial precision of the AVHRR SST fields at that location (the English Channel). They obtained a standard deviation of the differences for each section. The average of these standard deviations is 0.22 K.

5.2. Impact of Noise on Sobel Gradient

Of interest is how levels of noise, typical of the values found thus far, impact gradients and fronts. In order to address this, we simulated 10,000 3 × 3 pixel squares for a given gradient in *x*, added Gaussian white noise to each of the elements, applied the 3 × 3 Sobel gradient operator in *x* and *y* to these squares and then determined the mean gradient and the standard deviation of the gradient. This was done for gradients ranging from 0.001 to 0.1 K km^{-1}, values typical in the ocean, and for levels of instrument noise ranging from 0.01 K to 0.2 K. Figures 10 and 11 show the means and standard deviations of the *x*- and *y*-components of the gradient, respectively. The mean *x*- and *y*-components are unaffected by the noise; the mean *x*-component is the same as the initial value and the mean *y*-component is very nearly zero. The standard deviation of the components is very nearly independent of the imposed noise. For a noise level typical of VIIRS, 0.05 K, the vertical white lines in the figures, the uncertainty of each of the components is approximately 0.022 K and for a level typical of AVHRR, 0.2 K, the uncertainty in the components is 0.09 K. In general, the uncertainty in the given component is approximately one half of the level of imposed noise.

The impact on the gradient magnitude (Figure 12) is more dramatic. The mean of the estimated gradient is no longer equal to the magnitude of the imposed gradient. For example, for a relatively robust gradient of 0.05 K/km, the mean of the estimated gradient ranges from 0.05 to in excess of 0.1 K/km as the imposed noise increases from 0 to 0.2 K/km. Note that contours of the estimated gradient tend to become level for imposed noise levels less than approximately 0.07 K. This means that VIIRS estimates of the mean gradient magnitude will be centered on the actual value of the gradient, but that the gradient magnitude will be substantially overestimated in AVHRR fields. The uncertainty of the estimated gradient magnitude increases with the imposed noise, nearly doubling from the value associated with a zero imposed gradient to an imposed gradient of 0.1 K/km. These observations do not mean that a front with a gradient of this magnitude (0.05 K/km) is undetectable in a field with an AVHRR noise level but detection will be problematic. Simulations using front detection algorithms need to be undertaken to evaluate this. Although none of this is surprising, we are not aware of any studies involving the gradient magnitude of satellite-derived SST fields accounting for this—including many of our own.

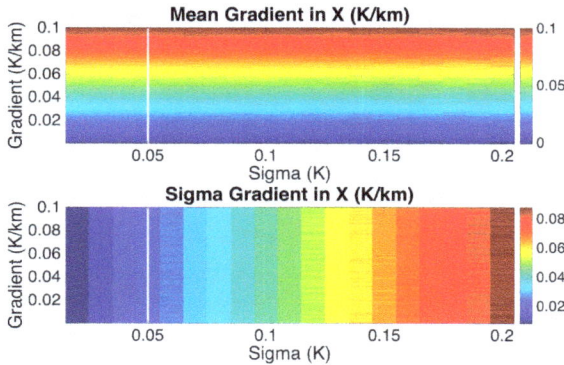

Figure 10. Simulated impact of Gaussian white noise of magnitude sigma imposed on a field with an *x*-gradient indicated on the vertical axis. The vertical white line is an imposed noise level typical of VIIRS values.

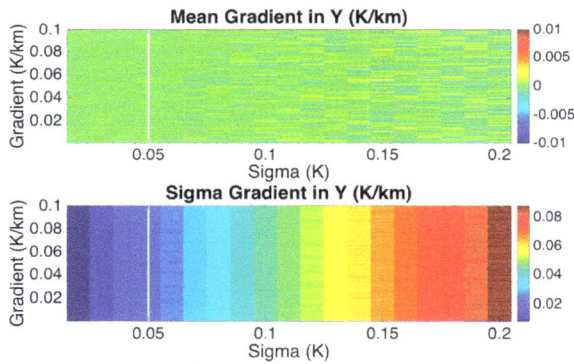

Figure 11. As in Figure 10 except for the *y*-component of the gradient.

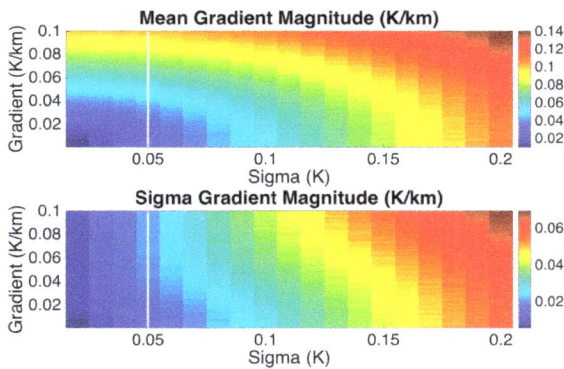

Figure 12. As for Figure 10 except for the gradient magnitude.

Remote Sens. **2017**, *9*, 877

6. Conclusions

The accuracy with which the local gradient of any digital field can be determined is a function of the spatial precision of the underlying data, where the spatial precision is defined as the square root of the variance of individual pixel values following removal of real trends in the data and removal of noise that is correlated over scales that are large compared with the scale used to calculate the gradient. In the case of fields obtained from satellite-borne sensors, this noise is attributed to characteristics of the sensor, "instrument noise", and to the retrieval process, "retrieval noise". Two approaches, a spectral-based approach and a variogram-based approach, were used to estimate the instrument portion of this noise in L2 AVHRR and VIIRS SST fields. To reduce the non-instrument portion of the local noise in the analysis, only cloud free sections were used, the assumption being that the dominant contribution to the non-instrument local noise is due to the misclassification of clouds. Because instrument noise was thought to differ between the along-scan and along-track directions and because the geophysical variance was thought to differ between day and night, the analysis was performed separately for the four along-scan/along-track and day/night combinations.

Both methods yielded similar results for AVHRR, with daytime and nighttime along-scan values of ~0.18 K and along-track values of 0.21 K. VIIRS instrument noise, on the other hand, was found to differ by method, scan geometry and day-versus-night–ranging from 0.021 K for the nighttime, along-scan spectral estimate to 0.097 K for the daytime, along-track variogram estimate. Day and night along-scan estimates based on the spectral approach are close to one half those based on the variogram. For both methods, the nighttime estimates are also roughly one half the corresponding daytime estimates. Finally, the along-track estimates are roughly 50% larger than the along-scan estimates for the spectral approach but only about 25% larger when based on the variogram. In all cases, the estimates were smaller than the "upper limit".

In summary, VIIRS instrument noise is substantially smaller than AVHRR instrument noise, with levels as low as 0.02 K in the along-scan direction at nighttime. In fact, VIIRS instrument noise under these conditions is near the level of the geophysical signal in the dynamically quietest regions in the ocean.

Acknowledgments: This research was supported by the Global Change Research Program of China (2015CB953901), the National Natural Science Foundation of China-Shandong Joint Fund for Marine Science Research Centers (U1606405), the National Natural Science Foundation of China (41376105, 41574014, 41774014), the Scientific and Technological Innovation Project of Qingdao National Laboratory for Marine Science and Technology (2016ASKJ16), the National Oceanographic and Atmospheric Administration (NA11NOS0120167), the National Aeronautics and Space Administration (NNX16AI24G) and the Frontier science and technology innovation project of the Science and Technology Commission of the Central Military Commission (085015). Salary support for F.W. was provided by the China Scholarship Council and Ocean University of China. Salary support for P.C. was provided by the state of Rhode Island and Providence Plantations.

Author Contributions: Fan Wu and Peter Cornillon conceived, designed and performed the experiments, and wrote the paper; Brahim Boussidi performed the experiments of variogram approach; and Lei Guan provided suggestion to the experiments and the analysis of the results.

Conflicts of Interest: The authors declare no conflict of interest.

Acronyms

AVHRR	Advanced Very High Resolution Radiometer
DFT	Discrete Fourier Transform
FFT	Fast Fourier Transform
GHRSST	Group on High Resolution Sea Surface Temperature
HRPT	High Resolution Picture Transmission
JPSS	Joint Polar Satellite System
NPP	National Polar-orbiting Partnership
PSD	Power Spectral Density
SST	Sea Surface Temperature
VIIRS	Visible-Infrared Imager-Radiometer Suite

References and Notes

1. Strong, A.E.; McClain, E.P. Improved ocean surface temperatures from space-comparisons with drifting buoys. *Bull. Am. Meteorol. Soc.* **1984**, *65*, 138–139. [CrossRef]
2. Guo, P.; Bo, Y. Validation of AVHRR/MODIS/AMSR-E satellite SST products in the west tropical Pacific. In Proceedings of the 2008 IEEE International Geoscience and Remote Sensing Symposium (IGARSS), Boston, MA, USA, 6–11 July 2008; Volume 4, pp. 942–945.
3. Park, K.A.; Lee, E.Y.; Li, X.; Chung, S.R.; Sohn, E.H.; Hong, S. NOAA/AVHRR sea surface temperature accuracy in the East/Japan Sea. *Int. J. Digit. Earth* **2015**, *8*, 784–804. [CrossRef]
4. Petrenko, B.; Ignatov, A.; Stroup, J.; Dash, P. Evaluation and Selection of SST Regression Algorithms for JPSS VIIRS. *J. Geophys. Res. Atmos.* **2014**, *119*, 4580–4599. [CrossRef]
5. Tu, Q.; Pan, D.; Hao, Z. Validation of S-NPP VIIRS sea surface temperature retrieved from NAVO. *J. Remote Sens.* **2015**, *7*, 17234–17245. [CrossRef]
6. Tandeo, P.; Autret, E.; Chapron, B.; Fablet, R.; Garello, R. SST spatial anisotropic covariances from METOP-AVHRR data. *J. Remote Sens. Environ.* **2014**, *141*, 144–148. [CrossRef]
7. Donlon, C.; Rayner, N.; Robinson, I.; Poulter, D.J.S.; Casey, K.S.; Vazquez-Cuervo, J.; Armstrong, E.; Bingham, A.; Arino, O.; Gentemann, C.; et al. The Global Ocean Data Assimilation Experiment High-resolution Sea Surface Temperature Pilot Project. *Bull. Am. Meteorol. Soc.* **2007**, *88*, 1197–1213. [CrossRef]
8. "Level-2" Refers to the Processing Level of the Data, a Nomenclature Used Extensively for Satellite-Derived Datasets, Although the Precise Meaning of the Level of Processing Varies by Organization. The Definition Promulgated by the Group on High Resolution Sea Surface Temperature (GHRSST) Is Used Here. Available online: http://science.nasa.gov/earth-science/earth-science-data/data-processing-levels-for-eosdis-data-products/ (accessed on 21 August 2017).
9. Cornillon, P.; Castro, S.; Gentemann, C.; Jessup, A.; Kaplan, A.; Lindstrom, E.; Maturi, E.; Minnett, P.J.; Reynolds, R. SST Error Budget—White Paper. 2010. Available online: https://works.bepress.com/peter-cornillon/1/ (accessed on 21 August 2017).
10. Schloesser, F.; Cornillon, P.C.; Donohue, K.; Boussidi, B.; Iskin, E. Evaluation of thermosalinograph and VIIRS data for the characterization of near-surface temperature fields. *J. Atmos. Ocean. Technol.* **2016**, *33*, 1843–1858. [CrossRef]
11. Seaman, C.; Hillger, D.; Kopp, T.; Williams, R.; Miller, S.; Lindsey, D. Visible Infrared Imaging Radiometer Suite (VIIRS) Imagery Environmental Data Record (EDR) User'S Guide. Version 1.1. Available online: http://rammb.cira.colostate.edu/projects/npp/VIIRS_Imagery_EDR_Users_Guide.pdf (accessed on 21 August 2017).
12. Schueler, C.F.; Clement, J.E.; Ardanuy, P.E.; Welsch, C.; DeLuccia, F.; Swenson, H. NPOESS VIIRS sensor design overview. *Proc. SPIE* **2002**, *4483*, 11–23.
13. Kilpatrick, K.A.; Podestá, G.P.; Evans, R. Overview of the NOAA/NASA advanced very high resolution radiometer Pathfinder algorithm for sea surface temperature and associated matchup database. *J. Geophys. Res. Oceans* **2001**, *106*, 9179–9198. [CrossRef]
14. Stramma, L.; Cornillon, P.C.; Weller, R.A.; Price, J.F.; Briscoe, M.G. Large Diurnal sea surface temperature variability: Satellite and in situ measurements. *J. Phys. Oceanogr.* **1986**, *16*, 827–837. [CrossRef]
15. Wang, D.P.; Flagg, C.N.; Donohue, K.; Rossby, H.T. Wavenumber spectrum in the Gulf Stream from shipboard ADCP observations and comparison with altimetry measurements. *J. Phys. Oceanogr.* **2010**, *40*, 840–844. [CrossRef]
16. Cayula, J.F.P.; Cornillon, P.C. Edge detection algorithm for SST images. *J. Atmos. Ocean. Technol.* **1992**, *9*, 67–80. [CrossRef]
17. Bouali, M.; Ignatov, A. Adaptive reduction of striping for improved sea surface temperature imagery from Suomi National Polar-Orbiting Partnership (S-NPP) Visible Infrared Imaging Radiometer Suite (VIIRS). *J. Atmos. Ocean. Technol.* **2013**, *31*, 150–163. [CrossRef]
18. Cornillon, P.C.; Stramma, L. The distribution of diurnal sea surface warming events in the western Sargasso Sea. *J. Geophys. Res. Oceans* **1985**, *90*, 11811–11815. [CrossRef]
19. Barnes, S.L. A technique for maximizing details in numerical weather map analysis. *J. Appl. Meteorol.* **1964**, *3*, 396–409. [CrossRef]
20. Wackernagel, H. *Multivariate Geostatistics: An Introduction with Applications*; Springer Science & Business Media, Inc.: New York, NY, USA, 2013.

21. Cressie, N.A.C. *Statistics for Spatial Data*; (Revised Edition); John Wiley and Sons, Inc.: New York, NY, USA, 1993.
22. O&SI SAF Project Team. Low Earth Orbiter Sea Surface Temperature Product User Manual. Available online: http://www.osi-saf.org (accessed on 21 August 2017).
23. Keogh, S.J.; Robinson, I.S.; Donlon, C.J.; Nightingale, T.J. The accuracy of AVHRR SST determined using shipborne radiometers. *J. Remote Sens.* **1999**, *20*, 2871–2876. [CrossRef]

remote sensing

MDPI

Article

Evaluation of the Multi-Scale Ultra-High Resolution (MUR) Analysis of Lake Surface Temperature

Erik Crosman [1,*], Jorge Vazquez-Cuervo [2] and Toshio Michael Chin [2]

[1] Department of Atmospheric Sciences, University of Utah, 135 South 1460 East, Rm 819, Salt Lake City, UT 84112, USA
[2] National Aeronautics and Space Administration Jet Propulsion Laboratory, California Institute of Technology, M/S 300/323 4800 Oak Grove Dr., Pasadena, CA 91109, USA; Jorge.vazquez@jpl.nasa.gov (J.V.-C.); toshio.m.chin@jpl.nasa.gov (T.M.C.)
* Correspondence: erik.crosman@utah.edu; Tel.: +1-801-581-6137

Academic Editor: Xiaofeng Li
Received: 25 May 2017; Accepted: 7 July 2017; Published: 13 July 2017

Abstract: Obtaining accurate and timely lake surface water temperature (LSWT) analyses from satellite remains difficult. Data gaps, cloud contamination, variations in atmospheric profiles of temperature and moisture, and a lack of in situ observations provide challenges for satellite-derived LSWT for climatological analysis or input into geophysical models. In this study, the Multi-scale Ultra-high Resolution (MUR) analysis of LSWT is evaluated between 2007 and 2015 over a small (Lake Oneida), medium (Lake Okeechobee), and large (Lake Michigan) lake. The advantages of the MUR LSWT analyses include daily consistency, high-resolution (~1 km), near-real time production, and multi-platform data synthesis. The MUR LSWT versus in situ measurements for Lake Michigan (Lake Okeechobee) have an overall bias (MUR LSWT-in situ) of −0.20 °C (0.31 °C) and a *RMSE* of 0.86 °C (0.91 °C). The MUR LSWT versus in situ measurements for Lake Oneida have overall large biases (−1.74 °C) and *RMSE* (3.42°C) due to a lack of available satellite imagery over the lake, but performs better during the less cloudy 15 July–30 September period. The results of this study highlight the importance of calculating validation statistics on a seasonal and annual basis for evaluating satellite-derived LSWT.

Keywords: lake surface temperature; sea surface temperature (SST); surface state; lake modeling; numerical weather prediction; surface analysis

---

1. Introduction

Lake surface water temperature (LSWT) is an important environmental parameter for understanding lake ecology, biology, and climate change [1–7]. LSWT is also a critical input variable for numerical weather, climate, and hydrological models [8–10]. While extensive climatological data sets and analyses of satellite-derived sea surface temperature (SST) are available and used in a wide range of applications, satellite-derived near real-time LSWT analyses are largely unavailable due to a number of challenges and limited resources [11–13].

Several key factors contribute to the difficulty to provide reliable and consistently accurate near real-time analyses of satellite-derived LSWT: Lake-specific spatially and temporally variable error sources and uncertainties, cloud contamination of thermal retrievals, gaps in coverage due to clouds, and a lack of in situ lake temperature observations. We elaborate on each of these factors below.

Satellite LSWT estimates are generally less accurate and have more sources of error than oceanic SST retrievals due to the typically larger uncertainties to correct for continental atmospheric air masses, dust and smoke, cloud contamination, water emissivity, and shoreline effects [11,14,15]. Most satellite LSWT retrieval algorithms were designed for ocean surfaces and validated and tuned to oceanic in situ buoy observations [11,13]. Consequently, the effects of variations in lake elevation, atmospheric profiles of temperature and water vapor, dust and smoke sources, and near-shore pixel contamination by adjacent land surfaces are not typically incorporated in the algorithms when they are applied over inland water bodies. Developing lake-specific algorithms for satellite-derived LSWT is an active area of research and several studies have developed improved techniques for LSWT retrievals [11,15–18]. However, these methodologies have not yet been implemented to our knowledge in a near real-time LSWT analysis.

Representativeness errors may also be introduced when attempting to apply satellite skin surface temperature data to estimate bulk lake temperature [10]. The upper layers of a water column in lakes and oceans are associated with complex variability in temperature. The relationship between bulk temperature measured below the surface by buoys and skin temperature measured by surface radiometers and satellites has been found to be a complex function of atmospheric conditions over lakes [19]. As defined by the Group for High Resolution Sea Surface Temperature (GHRSST, see http://ghrsst-pp.metoffice.com/pages/sst_definitions/), the skin temperature that is measured by satellite radiometers is underlain by "bulk" temperature and "foundation" temperature. Bulk temperature is associated with a transition zone located between the water surface and ~1 m below the surface, where some diurnal cycle in temperature is observed. At a deeper (and variable) depth in the water column, "foundation" temperature, or the temperature where diurnal signals are absent is found. Satellite-derived skin water temperatures can be converted to a "foundation" temperature most readily by utilizing nighttime-only satellite retrievals, where the diurnal variations in the surface water layer due to solar heating are absent and skin effects are more predictable [20].

Extensive gaps in the availability of clear-sky satellite thermal retrievals due to persistent and highly variable seasonal cloud cover over many of the mid-latitude regions of the earth where lakes are abundant makes it difficult to obtain representative LSWT analyses on a daily basis [13,21,22]. Consequently, it is not surprising that the majority of satellite-derived LSWT climatological trend studies have focused on the less-cloudy summer season [22,23]. Over the oceans, gap problems resulting from clouds can be somewhat overcome by the combined use of microwave thermal imagery and in situ buoy data, although sampling uncertainty and errors are noted over some ocean regions [24–26]. The large footprint of satellite microwave temperature retrievals and the impacts of sidelobe contamination within 75 km of shorelines make the data difficult to use over lakes [27]. In addition, most global lakes also lack in situ LSWT observations to supplement satellite analyses.

Cloud masking algorithms have also been shown to struggle over lakes. The algorithms are either too stringent (removing good data in regions of large LSWT gradients), further contributing to the aforementioned gap issues, or insufficient at flagging clouds such as thin cirrus, resulting in cloud contamination in the satellite LSWT retrievals. This is another area of active research, and improved cloud detection techniques have been proposed [28–31].

For the aforementioned and other reasons, it is apparent why it is difficult to provide calibrated long-term spatially- and temporally-consistent LSWT analyses, particularly in near-real time, as lake-specific algorithms are to our knowledge not yet incorporated into operational lake temperature processing schemes [13]. In addition, the only global real-time analyses that currently incorporate the use of higher quality climatological data sets (e.g., the ARC-Lake [15,17,30] and Pathfinder SST climatological datasets for lakes exist for the temporal periods of 1991–2011 and 1985–2014, respectively) for prescribing climatological lake temperature when no actual data coverage exists is the Operational Sea Surface Temperature and Ice Analysis (OSTIA) system, which relaxes to the ARC-Lake climatology in the absence of available data [13].

Despite the difficulties in obtaining accurate LSWT analyses, a number of studies have analyzed historical (and in some cases re-processed) multi-year records of satellite-derived LSWT. This research has shown the importance of lake temperature as one of the key indicators of climate change in lakes, which are known as "sentinels" of climate change [1,32]. A number of studies have analyzed both in situ and long-term available clear-sky satellite LSWT retrievals over the past few decades to better understand lacustrine response to climate change [18,22,33–37]. These studies have shown that the lake response to climate change is highly variable, even across global sub-regions. Most of these studies used single sensors and single platform types to avoid the issues associated with blending multiple data sets, and therefore mostly limited their analyses to the less-cloudy (and hence less frequent gap periods) warm season.

Many research studies have shown the sensitivity of numerical weather prediction to variations in LSWT, and the value in obtaining reasonably accurate LSWT values for input into these models. LSWT impact many aspects of simulated weather and climate, including global surface temperatures and precipitation [8,9,38–40], and even small lakes typically unresolved in global models impact regional weather [10].

Over the oceans, a number of studies have illustrated the value of combining satellite retrievals from multiple satellite platforms to increase data availability and decrease coverage gaps [41–43]. However, this topic has remained largely unexplored with lakes, with only a few studies suggesting multi-platform data synthesis [13,21,44]. A key need identified by Fiedler et al. [13], is blended near-real time operational analyses of LSWT for input into numerical weather prediction models. Despite the importance of accurate LSWT for input into climate and numerical weather prediction, relatively few studies have been dedicated to improving near-real time LSWT analyses. The Met Office Operational Sea Surface Temperature and Ice Analysis at ~6 km resolution recently included 248 lakes globally [13]. The blended Real Time Global (RTG) dataset at ~8 km resolution is currently used for lake temperature by the National Center for Environmental Prediction [45], but this analysis has been shown to suffer large biases [12]. The resolution (6–8 km) of these analyses is too coarse to resolve thousands of smaller lakes worldwide.

In this paper, we evaluate the current version of the near-real time (~1 day latency) MUR Analysis (https://mur.jpl.nasa.gov/) of Lake Surface Temperature from 2007–2015 in a small, medium, and large sized lake. While MUR was designed specifically as a sophisticated high-resolution (~1 km) analysis and blending product for global SST, the MUR global analysis product is also processed daily over thousands of the world's lakes captured by the 1-km nominal MUR resolution. The goal of this study is to determine the feasibility of using the near-real time MUR LSWT analysis for input into numerical weather prediction models, as well as the longer period of record of MUR LSWT (2000-present) for climatological studies. Specifically, this study seeks to determine the quality, strengths, and weaknesses of the current MUR analysis processing applied over inland water bodies, and to provide recommendations for targeted improvements to MUR for future releases with respect to LSWT.

The MUR analysis data and in situ buoy validation sets for the three lakes in this study are described in Section 2. In Section 3, the results of multi-year validations between MUR LSWT analyses and in situ buoy observations for the three lakes are shown, and the ability of the MUR LSWT to capture LSWT variability on daily, weekly, seasonal, and interannual time scales, as well as the strengths and weaknesses of MUR LSWT analyses. In Section 4, conclusions and recommended future improvements to the MUR LSWT are presented.

2. Materials and Methods

2.1. Lakes Analyzed

Satellite-derived MUR LSWT analyses and in situ lake temperature data were analyzed at three lakes (Figure 1). These lakes were chosen to represent variations in lake size. Lake Michigan, bordering Wisconsin, Illinois, Indiana, and Michigan, USA is a large lake 190 km wide and 2600 km long. Lake Okeechobee, Florida, USA is a medium-sized lake 48 km wide and 56 km long. Lake Oneida, New York, USA is a relatively small lake 8.0 km wide and 32 km long. The National Oceanic and Atmospheric Administration (NOAA) Great Lakes Environmental Research Laboratory (GLERL) through the CoastWatch program produces a daily near-real time analysis of Lake Michigan temperature named the Great Lakes Surface Environmental Analysis (GLSEA) [46] (https://coastwatch.glerl.noaa.gov/). However, no such operational analyses are produced for Lake Oneida or Lake Okeechobee. These lakes also represent a range of average depth (Lake Michigan, 85.0 m, Lake Okeechobee: 2.7 m; Lake Oneida, 6.7 m) and latitude (Lake Michigan, 44.00°N; Lake Okeechobee: 26.93°N; Lake Oneida, 43.20°N). Lake Michigan partially ices over each winter, and occasionally almost completely freezes over, most recently in 2014. Lake Oneida ice cover is also highly variable, and typically freezes over for at least a month each winter. All three lakes have relatively symmetric shapes and few islands to contaminate the infrared remote sensing retrievals. Lake Michigan and Lake Oneida observe large fractional cloud cover (25–45% according to NCAR regional reanalysis data at http://www.esrl.noaa.gov/psd/), and synoptic-scale storms that frequently traverse these regions, which increase the opportunities for LSWT analyses to suffer from data availability loss due to cloud cover reducing the frequency of clear-sky retrievals. The shallow Lake Okeechobee is far enough south that day-to-day air temperature variability is decreased, with fewer synoptic-scale weather systems and less prolonged cloudy periods, both positive factors for obtaining more frequent clear sky retrievals. The seasonal cycles in air temperature and solar forcing are also smaller at Lake Okeechobee than at Lake Michigan and Lake Oneida, resulting in a smaller annual cycle and less interannual variability in lake temperature at Lake Okeechobee than the other two northern lakes.

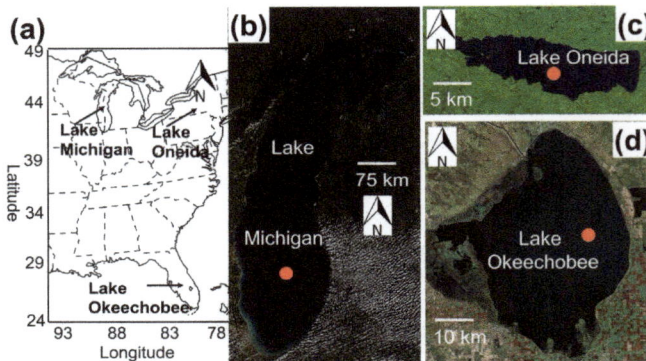

Figure 1. Locations of three USA lakes studied in this paper. (**a**) overview map; (**b–d**) Visible satellite images of the lakes; (**b**) Lake Michigan; (**c**) Lake Oneida; (**d**) Lake Okeechobee. In situ buoy location indicated by red dots.

2.2. MUR LSWT Analyses

The National Aeronautics and Space Administration (NASA) Making Earth System Records for Use in Research Environments (MEaSUREs) Multi-scale Ultra-high Resolution (MUR) analysis, hereafter referred to simply as "MUR", is a global SST analysis field produced daily on a $0.01° \times 0.01°$ grid with an equatorial resolution of 1.1132 km. In addition to coverage over all of the global ocean regions, MUR SST is produced daily over thousands of global inland water surfaces included within the land mask. The MUR SST product has not been previously evaluated or validated for lakes. For the remainder of this document we will refer to MUR SST analysis over inland water bodies as the MUR LSWT. The MUR analysis is described in detail in Chin et al. [43], and only a basic overview is presented here.

MUR LSWT incorporates multiple satellite data sets at different resolutions over multiple time scales within a 5-day window to generate a near real-time analysis (the latency is slightly over a day) that reconstructs small-scale spatial structures of recent and highest resolution satellite-derived LSWT data available while providing the temporal consistency of the data provided over longer time windows by coarser satellite or in situ data. The multi-resolution variational analysis (MRVA) method

(Figure 2) allows data fusion and interpolation using a range of length scales, from 1 km to over 1000 km, specified by a wavelet transform [43].

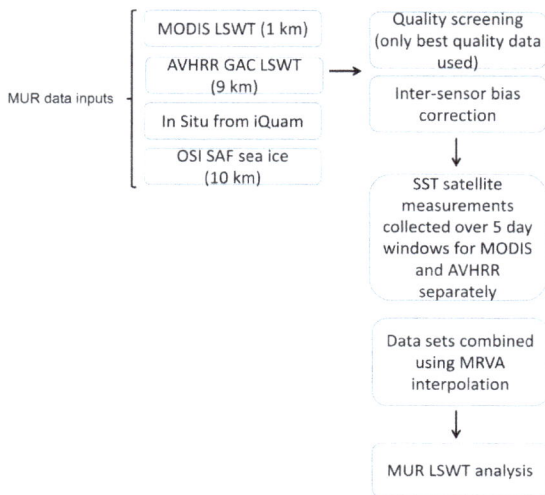

Figure 2. MUR LSWT analysis processing chain.

Over the global oceans, MUR combines over a dozen satellite SST retrievals and buoy near-surface temperature data at different spatial and temporal resolutions over the previous 5 days into a single daily high-resolution SST analysis. Over lakes, only a subset of these data sets is included (Figure 2). MUR provides an estimate of the foundation temperature by utilizing only nighttime satellite data and correcting all retrievals using single sensor error statistics (SSES). MUR utilizes the SSES biases provided with the input data (typically 0.10–0.30 °C) and then these biases are subtracted from the nighttime satellite imagery in MUR to provide a foundation temperature in the final MUR analysis. Inter-sensor bias correction is performed for every data set by MRVA [43]. MUR does not use a background analysis. Therefore, MUR extrapolates SST values from one region to another using the MRVA method when no satellite or in situ observations are available over a given region.

There are several reasons why utilizing nighttime satellite retrievals is preferable. First, nighttime satellite data are typically more easily converted to foundation temperatures, as daytime skin surface and warm layer diurnal heating effects can be up to several °C and highly variable from lake to lake and from day to day, dependent on the lake vertical temperature profile and meteorological conditions such as wind speed [19]. Finally, satellite drift over time could impacts the time of day a satellite crosses a lake. Because a lake surface potentially changes more rapidly from one hour of the daytime to the next, the temporal drift in the satellite could potentially introduce sampling errors [22]. By choosing only night data where changes in LSWT vary less over time, these complications are avoided.

Over lakes, MUR combines satellite-derived SST data from thermal infrared sensors on the two primary polar-orbiting satellites that have been historically used for remote sensing of lakes—the Moderate Resolution Thermal Imaging Spectroradiometer (MODIS) and the Advanced Very High Resolution Radiometer (AVHRR) sensors. Available SST samples from microwave sensors tend to be excluded over lakes by MUR due to a stringent threshold on quality flags associated with such samples. The resolution of the MODIS data is approximately 1 km, while the AVHRR data is ~9 km resolution. Thus, the analysis over Lake Oneida incorporates exclusively MODIS data, given the minimum width of the lake is ~8 km (Figure 1). However, both Lake Michigan and Lake Okeechobee are wide enough for both AVHRR and MODIS data to be included in the LSWT analyses. When available, MUR also incorporates buoy data from the iQuam data network [47]. Over lakes, the only routinely available buoy data in iQuam data network is in the USA Great Lakes, which includes Lake Michigan in this study. No buoy data from Lake Okeechobee or Lake Oneida are included in MUR. In addition to the SST data, MUR also uses an ice fraction ancillary data set at 10 km resolution from the Ocean and Sea Ice Satellite Application Facility (OSI SAF) to help ascertain ice-covered water regions, and parameterizes the temperature as a function of the ice cover [43]. The MUR LSWT analysis for a given day is available a little over a day later. Cloud masking and quality control are applied to the AVHRR and MODIS data sets incorporated into MUR, and only the highest quality possible SST data are used in the MUR LSWT (Figure 2).

For this study, daily subsections of MUR LSWT over each of the three lakes for the 9 year period 2007–2015 were processed and downloaded from the MUR website using OpenDAP https://mur.jpl.nasa.gov/DownloadDataText.php.

2.3. In Situ Buoy Data

Buoy and water analyzer data were obtained to validate MUR LSWT over the three lakes in this study for the period 2007–2015. The in situ data were all collected at a depth between 0.05 m and 0.60 m in order to be representative of near-surface temperature. On Lake Michigan, water temperature at 0.6 m depth was measured by a National Oceanic and Atmospheric Administration National Data Buoy Center (NDBC) platform (http://www.ndbc.noaa.gov/rsa.shtml). All NDBC buoys go through quality control and calibration procedures as outlined by [48]. On Lake Okeechobee, Water temperature sampling was performed at depth of 0.5 m in the morning with a multi-parameter in situ water analyzer. Water temperature sensors used for Lake Okeechobee were calibrated monthly against the National Bureau of Standards thermometers [49]. On Lake Oneida, near-surface water temperature measurements just below the surface (0.05–0.10 m) were recorded in the morning using Hydrolab Datasonde profilers [50]. For both Lake Okeechobee and Lake Oneida, the measurements were conducted such that minimal disturbance was generated in the water column. However, some mixing of the near-surface lake water is to be expected with such approaches, making it difficult to categorize the representative depth of near-surface LSWT being measured. Additional specifications on these three measurement platforms and sensors are given in Sharma et al. [49].

The buoys or sampling locations were located in deep water and at least several km from the shoreline to avoid land pixel contamination of the satellite retrievals (Figure 1). Only nighttime satellite imagery was used to obtain foundation LSWT and to limit the impacts of solar heating on representativeness errors between the in situ and satellite data. However, some of the diurnal surface heating effects likely impacted the daytime Lake Oneida and Lake Okeechobee in situ measurements (this is a known limitation of these two data sets but no nighttime data was available). Under most conditions, the biases introduced by using daytime bulk lake versus nighttime satellite retrievals are expected to be less than 0.5 °C on average [19], although on calm summer days the differences can be much larger (e.g., see discussion for Lake Michigan in Section 3.5).

The surface water samples from Lake Oneida and Lake Okeechobee in situ data from 2007–2014 were collected weekly for Lake Oneida and bi-weekly to monthly for Lake Okeechobee by boat (data was not collected on Lake Oneida when ice covered the lake). A total of 281 daily matchups between daily MUR LSWT and in situ measurements were conducted for Oneida, and 172 daily matchups between daily MUR LSWT and in situ measurements on Lake Okeechobee. The nearest MUR satellite pixel to the in situ observation location was used for the match-up. On Lake Michigan, nighttime hourly buoy data between 0200 and 0400 Local Standard Time (LST) were compared with the MUR nighttime analyses. A total of 1950 daily matchups between daily MUR LSWT and in situ measurements were conducted on Lake Michigan.

3. Results

3.1. Evaluation Metrics

The evaluation of satellite-derived LSWT retrievals against in situ lake temperature measurements has historically been conducted using two widely used metrics: Root-mean squared error (*RMSE*) and bias. In this study, we also evaluate the ability of the MUR LSWT analysis to capture day-to-day variations in LSWT, as well as to provide cycles of climatological lake temperature. Lakes, being shallower than oceans and surrounded by continental landmasses are typically subject to larger temporal variations in surface temperature. Another important consideration for LSWT analyses is the climatological variability of the LSWT, which varies as a function of latitude, lake depth, and other geophysical forcing mechanisms [17]. For the most part, only the *RMSE* and mean bias of LSWT retrievals has been evaluated in the literature. In this study, we also evaluate the seasonal variations in the satellite-derived LSWT bias and *RMSE*.

Following an overview of the sources of error in the MUR LSWT analyses (Section 3.2), an evaluation of the MUR LSWT for Lake Michigan, Lake Okeechobee, and Lake Oneida is presented using three different criteria. For the first criteria, the standard metrics used in evaluation of SST retrievals of *RMSE* and bias of MUR LSWT analyses versus in situ measurement are evaluated for seasonal and annual time scales (Section 3.3). Second, the ability of the MUR analysis to capture short-term (~few weeks in time, ~10 km in space) spatial and temporal variations in LSWT is evaluated (Sections 3.4 and 3.5). Third, we evaluate the seasonal, interannual, and climatological data from 9 years of daily MUR LSWT analyses (Section 3.6).

3.2. MUR-Specific Sources of Error

LSWT satellite-retrievals are subject to a wide array of potential sources of errors. These include errors associated with the atmospheric correction algorithm as well as a number of other factors (e.g., cloud contamination, shoreline effects, etc.). We refer the reader to Hulley and Hook [11] for an overview of LSWT algorithms, and to Crosman and Horel [14] or Fiedler et al. [13] for an overview of other general error sources in satellite-derived LSWT. In this section, we refer only to MUR LSWT-specific sources of error stemming from the MUR LSWT analysis processing of MODIS and AVHRR satellite imagery. As discussed in Section 2.2, the advantages of the MUR analyses include temporal consistency (available every day), multi-sensor (MODIS and AVHRR) platform data synthesis and bias correction, both high-resolution (MODIS) and medium-resolution (AVHRR) thermal imagery, and sophisticated spatial interpolation and gap filling techniques. However, the current MUR analysis processing techniques can also introduce sources of error and have limitations in addition to the various sources of error typically noted (e.g., cloud contamination, atmospheric correction) in LSWT retrievals. Careful analysis of the MUR LSWT input variables, processing techniques, and final analyses have identified the following potential sources of error in MUR LSWT resulting from the processing methodology:

- Errors introduced by MRVA spatial scale used for interpolating the data as the MRVA system is designed for the open ocean. Analysis values from unrepresentative distant lakes or ocean surfaces may be "spread" to other lake surfaces during periods when no clear-sky retrievals are available over a given lake (Section 3.5).
- Errors resulting from spurious or inaccurate ice cover estimates (i.e., incorrectly specifying open water as ice or vice versa).
- Sampling "gap" errors introduced by only utilizing nighttime satellite imagery, which decreases the frequncy of available clear-sky imagery compared to analysis techniques that utilize both daytime and nighttime data.
- Representativeness errors by only utilizing nighttime data. This is not an issue if a daily foundation temperature is deemed to be sufficient for the analysis, which is the current goal of MUR. However,

complications arise on prescribing an appropriate analysis for shallow lakes with climatologically large diurnal temperature ranges through a relatively deep water column.

- Potential under-sampling errors due to the restrictive use of only the data flagged as the highest possible quality in MUR. Some studies have found that the highest quality control consistently throws out large amounts of good data over some lakes [14].

Spurious ice cover in the OSI SAF ice fraction ancillary datasets impacted ~one MUR LSWT analysis time per year on average in Lake Michigan and Oneida, and obviously was not a factor for Lake Okeechobee. An example of effects of the spurious ice cover on a MUR LSWT analysis is shown in Figure 3 for 22 August 2007 (Figure 3).

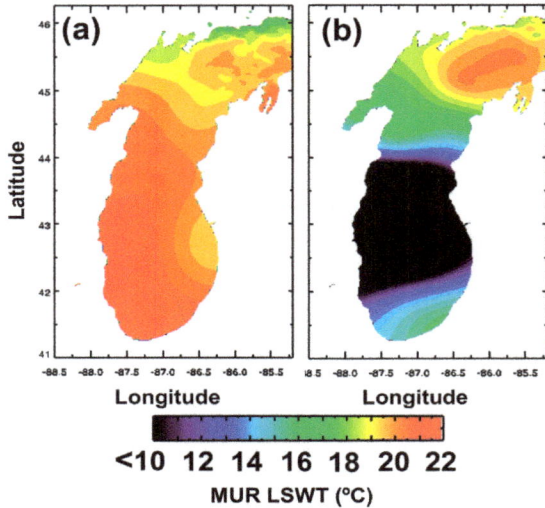

Figure 3. MUR LSWT (°C) analyses on (**a**) 21 August 2007 and (**b**) 22 August 2007. Regions colored in black in (**b**) correspond to temperatures ranging between −4 °C and 10 °C.

Despite the fact that ice cover is not observed over Lake Michigan in August, it is a known problem that spurious ice cover occurs in satellite-derived products over inland lakes due to coastal contamination of microwave imagery used in deriving the mask as the emissivity of the coastline is similar to that of sea ice emissivity [51]. On 21 August 2007, the MUR LSWT analysis agreed well with the in situ buoy observations and provided a realistic spatial map of the temperature variability across Lake Michigan (Figure 3a). There were no ice inputs from the OSI SAF ice fraction data set. However, On 22 August 22, there were 26 false ice inputs that tried to bring the analyzed lake temperature below freezing. These inputs occurred in a region of the lake where no satellite or buoy data was available, thus allowing the false ice to impact the LSWT analysis (Figure 3b). The ice impacts on the MUR LSWT were dramatic and sudden (sudden temperature drops of over 10 °C from one day to the next). These anomalous effects from incorrect ice specification are relatively easy to flag in the data (due to the noted MUR ice input flags and associated unphysical drops in temperature from one day to the next) and are therefore straightforward to quality control and remove. In addition, QC checks to remove spurious ice coverage could be constructed based on climatological checks for spurious ice during historically ice-free months on any given lake. The biases resulting from interpolation of LSWT analysis values from neighboring water bodies were not readily identified, but likely influenced the seasonal variations in MUR versus in situ validation statistics at Lake Oneida, as discussed in Section 3.5.

3.3. In Situ Temperature versus MUR LSWT

In situ buoy temperature measurements taken during the late night (0200–0400 LST) were compared against MUR LSWT (which is an estimate of foundation temperature) at Lake Michigan. The available in situ near-surface water temperatures for Lake Okeechobee and Lake Oneida were only available during the morning hours, and therefore an additional source of uncertainty between the nighttime MUR data and morning in situ observations is introduced. The in situ data was compared to the nearest MUR satellite pixel. MUR defines nighttime as sunset to sunrise, thus any uncertainty introduced in the temporal co-location should be minimized.

The scatter plot of Lake Michigan MUR LSWT versus in situ measurements (Figure 4a) for the period 2007–2015 has an overall cool bias (MUR LSWT-in situ) of −0.20 °C and a *RMSE* of 0.86 °C. Variations are noted in the bias and *RMSE* as a function of season and year (Table 1). On an annual scale, the biases range from −0.41 to 0.47 °C. The bias ranged from −0.02 °C to −0.41 °C during all years but 2014, when surface warming is hypothesized to have warmed the lake surface significantly higher than the buoy water temperature at 0.6 m depth over the summer period (Section 3.5). The annual average *RMSE* also ranged from 0.61 to 1.08 °C on all years except 2014 when it was 1.62 °C (Table 1). Evaluating seasonal averages, the *RMSE* was lowest (0.67 °C) in the fall. This is hypothesized to be due to the weaker stratification of the lake water column during the fall cooling cycle of the lake (warmer water is continually being mixed to the surface as it cools), resulting in less opportunities for representativeness errors between the buoy and satellite observations. Similarly, the higher average *RMSE* of 1.03 during the summer is hypothesized to be due to the high solar insolation effects this time of year and a greater potential for diurnal warming and stratification of the near-surface water column. Evaluation of the time series of plots of annual LSWT also illuminated "lag" errors in the MUR analysis on years with higher *RMSE*. These are errors resulting from the 5-day MUR analysis being unable to respond to rapid atmospheric forcing of surface water temperature. For example, in 2009 the springtime lake temperature warming and fall cooling cycles were accelerated by about 2 weeks each compared to average. In addition, the summertime lake temperatures were more variable on sub-weekly time scales in 2009 than in 2012, which observed the lowest annual *RMSE* during the 2007–2015 period. These factors resulted in increased differences between the MUR analysis and observations during these times when lake temperatures were warming or cooling rapidly, and contributed to the 68% increase in the *RMSE* observed between 2009 versus 2012 for Lake Michigan (Table 1).

Table 1. Comparison of MUR versus in situ LWST data.

Lake Michigan	Bias (MUR LSWT–In Situ, °C)				Root Mean Squared Error (RMSE, °C)			
Year	Spring (MAM)	Summer (JJA)	Fall (SON)	All Months	Spring (MAM)	Summer (JJA)	Fall (SON)	All Months
2007	−0.07	−0.2	−0.39	−0.24	0.61	1	0.78	0.84
2008	0.43	0.1	−0.41	−0.02	0.59	0.72	0.71	0.69
2009	0.14	−0.43	−0.38	−0.29	0.67	1.4	0.7	1.08
2010	−0.13	−0.56	−0.44	−0.41	0.56	0.76	0.82	0.73
2011	−0.03	−0.47	−0.23	−0.28	0.77	0.87	0.75	0.8
2012	−0.09	−0.29	−0.5	−0.32	0.59	0.65	0.68	0.64
2013	−0.15	−0.27	−0.06	−0.16	0.86	0.68	0.42	0.61
2014	1.92	0.89	−0.07	0.47	2.52	2.19	0.55	1.62
2015	NA	−0.79	−0.33	−0.32	NA	1.03	0.7	0.7
2007–2015	0.25	−0.22	−0.31	−0.2	0.9	1.03	0.67	0.86
Lake Okeechobee (* Statistics Only Calculated for Sample Size n > 6)								
2007	NA*	NA*	NA*	0.27	NA*	NA*	NA*	0.73
2008	NA*	NA*	NA*	0.22	NA*	NA*	NA*	0.99
2009	NA*	NA*	NA*	0.27	NA*	NA*	NA*	0.9
2010	NA*	NA*	NA*	0.1	NA*	NA*	NA*	1.11
2011	NA*	NA*	NA*	0.15	NA*	NA*	NA*	0.9
2012	NA*	NA*	NA*	0.28	NA*	NA*	NA*	1.01
2013	NA*	NA*	NA*	0.25	NA*	NA*	NA*	0.81
2014	NA*	NA*	NA*	0.24	NA*	NA*	NA*	0.8
2007–2014	0.13	−0.13	0.46	0.31	0.69	0.66	1.11	0.91

Table 1. *Cont.*

Lake Michigan	Bias (MUR LSWT-In Situ, °C)				Root Mean Squared Error (RMSE, °C)			
Year	Spring (MAM)	Summer (JJA)	Fall (SON)	All Months	Spring (MAM)	Summer (JJA)	Fall (SON)	All Months
	Lake Oneida							
2007	−4.25	−2.05	0.02	−1.57	5.79	3.04	1.68	3.46
2008	−4.03	−2.58	0.23	−2.09	4.62	3.43	1.32	3.3
2009	−5.87	−2.59	0.86	−2.8	6.45	3.02	1.29	4.25
2010	−3.55	−2.43	0.75	−1.73	4.13	3.26	1.31	3.12
2011	−3.79	−1.85	0.88	−1.66	5.35	2.55	1.41	3.58
2012	−2.06	−1.27	0.96	−0.86	2.7	1.69	1.51	2.05
2013	−2.86	−1.75	2.05	−0.81	3.63	2.51	2.71	3.07
2014	−5.09	−3.5	0.9	−2.41	5.87	4.61	1.52	4.15
2007–2014	−3.78	−2.65	0.98	−1.74	4.83	3.51	1.78	3.42
2007–2014	March–15 July	15 July–30 September	1 October–30 November		March–15 July	15 July–30 September	1 October–30 November	
	−3.88	−0.7	1.67		4.71	1.13	2.23	

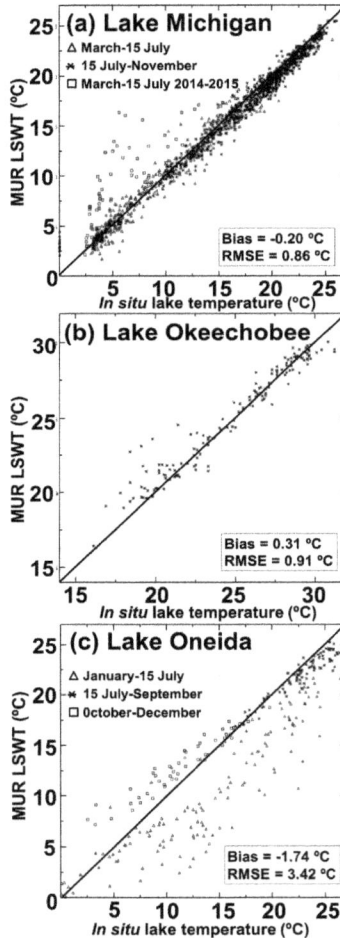

Figure 4. Scatter plots of MUR LSWT analysis versus in situ bulk lake temperature measurements (°C) for period 2007–2015 for (**a**) Lake Michigan and 2007–2014 for (**b**) Lake Okeechobee and (**c**) Lake Oneida. Number of matchups n = 1950 for Lake Michigan, n = 172 for Lake Okeechobee and n = 281 for Lake Oneida.

The scatter plot of Lake Okeechobee MUR LSWT versus in situ measurements for the period 2007–2014 has an overall warm bias of 0.31 °C and a *RMSE* of 0.91 °C (Figure 4b, Table 1). On an annual scale, the biases (*RMSE*) range from 0.10 to 0.28 °C (0.73 to 1.11 °C). The biases and *RMSE* are both decreased on average in the Spring (bias: 0.13, *RMSE* 0.69) and Summer (bias: −0.13, *RMSE* 0.66) compared to the Fall (bias: 0.46, *RMSE* 1.11). The higher bias and *RMSE* in the fall were determined to be impacted by both strong cold fronts impacting the late fall period as well as the more rapid fall cool-down period on Lake Okeechobee compared to a slower spring warm-up (resulting in occasional "lag" temperature errors in the 5-day MUR analysis as discussed earlier).

The scatter plot of Lake Oneida MUR LSWT versus in situ measurements for the period 2007–2014 has an overall cool bias of −1.74 °C and a *RMSE* of 3.42 °C (Figure 4c, Table 1). However, these comparisons show strong seasonality, with a very large cool bias of −3.78 °C and a RMSE of 4.83 °C during the spring. During the summer, the bias (*RMSE*) decreases to −2.65 °C (3.51 °C). By fall, the bias (*RMSE*) decreases further to 0.98 °C (1.78 °C). The hypothesized reasons for these strong seasonal discrepancies will be discussed in Section 3.5. Evaluating MUR after the end of periods with large gap errors due to clouds restricting coverage and also potential ice contamination (15 July through September period), decreased the bias further to −0.70 °C and a *RMSE* of 1.13 °C.

3.4. Evaluation of Spatial MUR LSWT

The MUR MRVA interpolation techniques results in LSWT analyses with both temporal consistency, i.e., daily analyses that do not change too abruptly from day to day despite data availability gaps on some days, and realistic spatial structures by combining satellite thermal infrared retrievals of various resolution and frequency (Chin et al. [43]). Despite the lack of microwave imagery, MUR LSWT was able to effectively retain many of the spatial variations in LSWT typically observed in high-resolution analyses of lake temperature such as those observed in the NOAA GLERL GLSEA over Lake Michigan.

Several examples of MUR analyses for 1 July of different years for Lake Michigan and Lake Okeechobee are shown in Figure 5. Realistic spatial variations in LSWT were observed on most days on these lakes (Figure 5a–e). The structures noted in the MUR analyses were evaluated against several NOAA GLERL GLSEA images during an annual cycle. The MUR LSWT patterns in surface temperature generally agreed with the GLSEA (not shown). For example, the 1 July 2010, 2012, and 2016 GLSEA analyses showed similar spatial patterns in LSWT as the MUR analyses shown in Figure 5a–c. Typically, the spatial structures were linked to geophysical forcing, such as variations in the depth of the lake, upwelling, distance to shore, and vertical mixing of the water column. During periods with limited available satellite imagery (generally due to cloud cover), the analyses would revert to a symmetric spatial temperature pattern resulting from the MRVA technique. An example of such a pattern is shown in Figure 5f, with a symmetric gradient in temperature across the water body. This is the result of MRVA "spreading out" the impacts of both coarser resolution thermal imagery and adjacent water bodies during periods when satellite imagery is not available over a lake during the 5-day MUR MRVA analysis window. Further analysis is needed to better quantify both the frequency of occurrence and overall impact of the MRVA interpolation of temperature values between different distinct water bodies separated by land masses in MUR.

On Lake Oneida, the spatial variations in the MUR analyses were small, likely due to the relatively uniform depth of the main body of the lake and the small size of the lake. However, the MRVA technique discussed previously resulted in contamination of LSWT at Lake Oneida by adjacent water bodies (likely Lake Ontario) and introduced large errors in the analyses during Spring and Fall cloudy periods (Section 3.5).

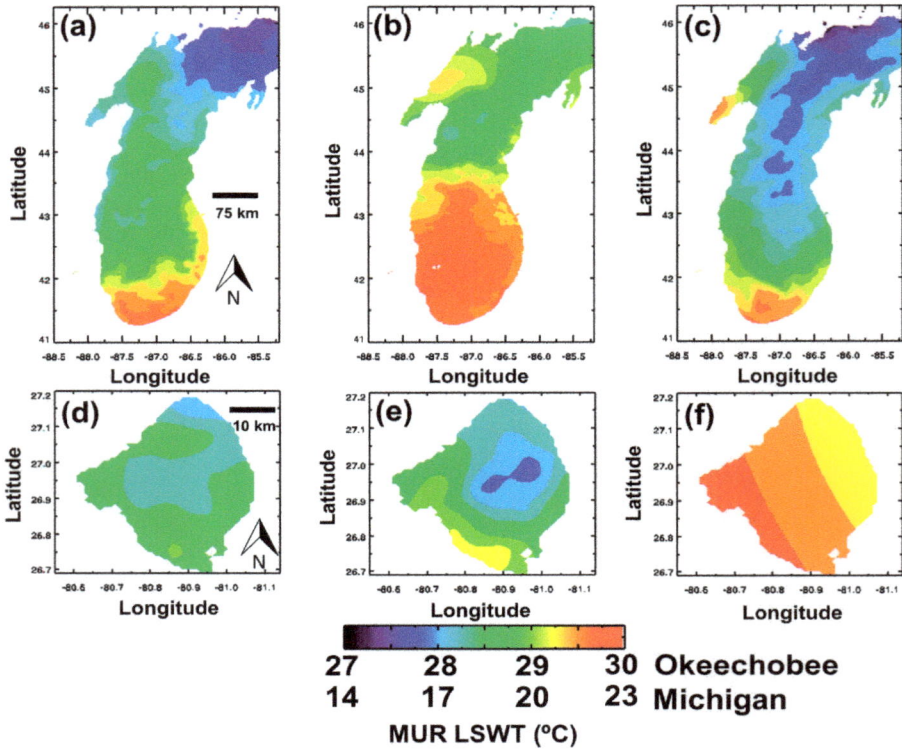

Figure 5. MUR LSWT analyses on 1 July of selected years over (a–c) Lake Michigan and (d–f) Lake Okeechobee. (**a**) 2010; (**b**) 2012; (**c**) 2016; (**d**) 2007; (**e**) 2012; (**f**) 2016.

3.5. Evaluation of Temporal Variability of MUR LSWT

While the validation of in situ versus satellite-derived LSWT "match-ups" of individual thermal images have been conducted for a number of lakes worldwide, none of these studies have analyzed in depth the ability of analyses (such as MUR) that incorporate multi-day satellite retrievals to capture rapid changes in LSWT forced by the overlying meteorology (e.g., summertime surface heating or strong cold fronts). For Lake Michigan, MUR LSWT represents both the rapid spring and summer increase in LSWT and slower fall decreases observed by in situ measurements adequately during all years analyzed with the exception of the Spring 2014 period (Figure 6).

The rapid cool-down associated with cold fronts in September and October 2011 and 2014 were also reflected in the MUR LSWT (Figure 6a,b). In addition, the period of rapid warm-up during June and July 2011 were typically represented by the MUR LSWT analyses for Lake Michigan, with minor lags noted between the in situ and MUR LSWT time series.

The comparisons between MUR LSWT and in situ measurements for Lake Michigan resulted in low biases and *RMSE* on an annual basis (Table 1) except for the May–July 2014 period. During this period, the MUR LSWT versus in situ measurements observed a warm bias of 1.47 °C and a *RMSE* of 2.48 °C (Figure 6b). The large discrepancies between MUR LSWT and in situ buoy data during this period can be ascribed to the following complex factors described below.

An extremely cold winter earlier in 2014 set the stage for unusually high ice coverage on Lake Michigan in early 2014. Consequently, a late ice met resulted in springtime lake temperatures from March through June that remained unseasonably cool (between 2 °C and 5 °C) (Figure 6b). Consequently, when hot summertime high pressure set up over the Great Lakes region in June and July 2014, large air and water temperature contrasts were observed during this time period, with buoy air temperature 6–18 °C warmer than water temperature at 0.6 m depth (See the arrow corresponding to marker A in Figure 7a,b). Between mid-June and mid-July 2014, Lake Michigan surface temperatures warmed over 15 °C (Figures 6b and 7). This set the stage for multi-day shallow thermoclines or surface warm layers that were modulated by the strength of surface wind speeds, leading to unusually large diurnal swings in lake temperature between 2–6 °C amplitude near the surface of Lake Michigan on days with light winds (Figure 7a,c). The very shallow thermoclines resulted in large temperature differences between the warm skin surface (satellite observation) and cooler sub-surface waters at 0.5–1.0 m depth (buoy observations). For example, between 21 and 27 June 2014, the MUR analysis underestimated LSWT by 2–4 °C (Figure 7a). This period was associated with warm air temperatures and light winds below 3 m s^{-1} (Figure 7b,c). The LSWT from MUR, even utilizing only nighttime satellite observations, likely were several °C warmer than subsurface buoy temperatures during calm periods. During the period from 29–30 June 2014, the MUR warm bias was decreased to ~1 °C as wind speeds between 5 m s^{-1} and 10 m s^{-1} likely mixed out the shallow lake thermocline despite large air-water temperature differences (See the arrow corresponding to marker B in Figure 7a,b). During the period between 8 and 15 July, the buoy lake temperature warmed by 6–7 °C, while the MUR analysis LSWT remained relatively constant (See the arrow corresponding to marker C in Figure 7a,b). During the 9–21 July period, the air-water temperature difference began to decrease, likely as the surface warming of the water column began to penetrate deeper below the surface, and by the 15 July, the MUR LSWT analyses were no longer exhibiting the large warm biases observed previously (Figure 7a).

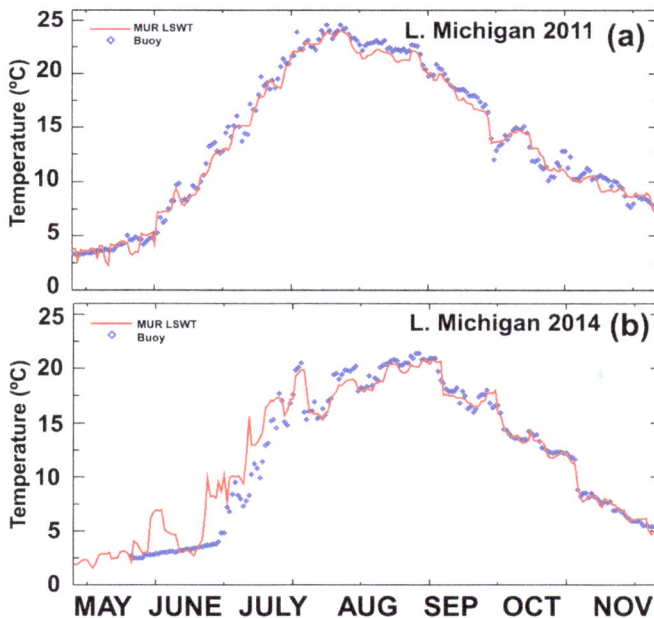

Figure 6. Time series of MUR LSWT analysis compared to in situ measurements for (**a**) Lake Michigan for 2011 and (**b**) 2014.

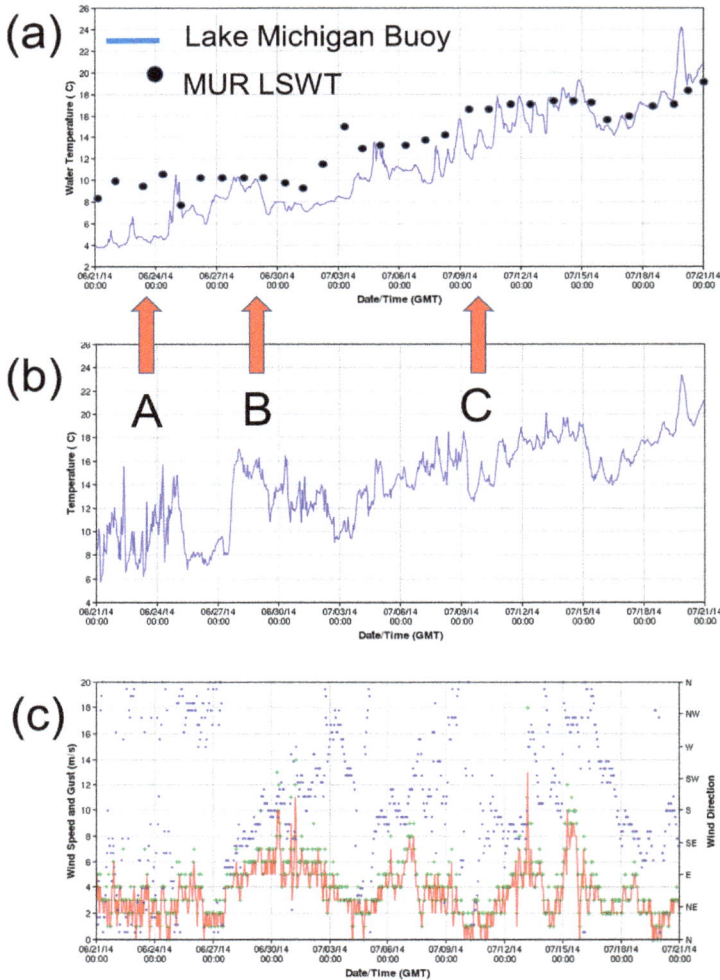

Figure 7. Time series from 21 June 2014 to 21 July 2014 of Lake Michigan buoy (**a**) water temperature (°C, blue line) (**b**) air temperature (°C, blue line); and (**c**) wind speed and direction. Daily MUR LSWT analyses overlaid are indicated by black dots in (**a**).

At Lake Oneida (Lake Okeechobee), once-per week (1–2 times per month) availability of in situ temperature measurements precludes being able to analyze the ability of MUR LSWT to reproduce day-to-day surface temperature variations. However, the MUR LSWT can be evaluated for Lake Oneida (Lake Okeechobee) on weekly (monthly) to interannual scales (Figure 8).

The Lake Okeechobee MUR LSWT analysis is able to reproduce the annual cycle in LSWT (Figure 8a). While the amplitude of the LSWT cycle is lower at Lake Okeechobee, the highly shallow nature of the lake (2.7 m on average) results in periodic variations in the LSWT analysis throughout the year that are largest in the winter (Figure 8a) when cold fronts have the largest impact on air temperature and the corresponding surface temperature of the shallow lake. In the fall season, the 5-day MUR analysis window also is hypothesized to result in a "lag" effect of the MUR LSWT analysis compared to in situ observations, where cold fronts cool the shallow lake relatively rapidly in a period of several days. Consequently, in situ observations are typically biased 0.46 °C cooler than MUR LSWT (Figure 8a and Table 1).

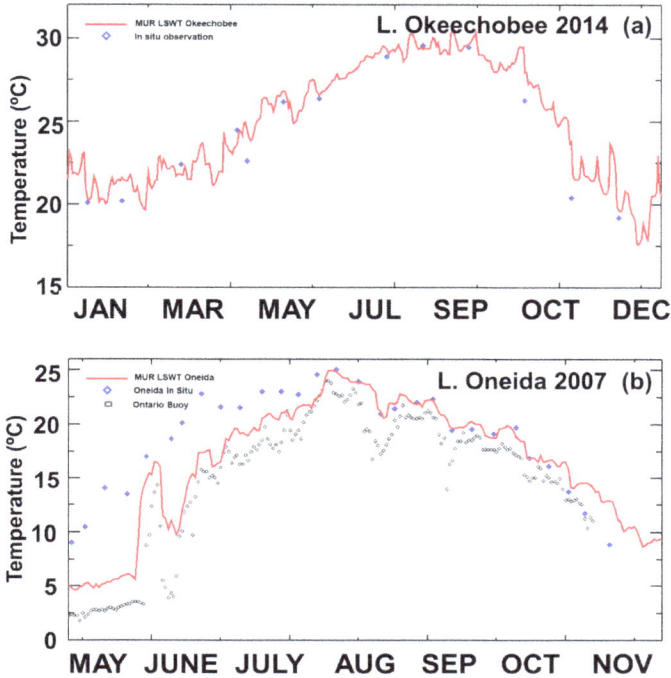

Figure 8. Time series of MUR LSWT analysis compared to in situ measurements for (**a**) Lake Okeechobee for 2014 and (**b**) Lake Oneida for 2007. In situ measurements from a Lake Ontario buoy are also included in (**b**).

The Lake Oneida MUR LSWT analyses are adequate in the mid-July through mid-September period each year, and capture the observed week-to-week variability (Figures 4c and 8b). However, during the spring and fall months, large discrepancies develop between the MUR LSWT and in situ observations. In the spring months, the large errors in the MUR LSWT are believed to be largely due to a lack of available MODIS imagery over the lake due to frequent cloud cover and storms, as well as potential ice contamination. Because of the small dimensions of Lake Oneida (Figure 1), only MODIS imagery is available to the MUR analysis over the lake. During these periods, the thermal infrared retrievals from distant unrepresentative locations (likely Lake Ontario, see how the time series of MUR LSWT for Lake Oneida follows closely in situ buoy measurement from Lake Ontario in both the cloudy spring and fall seasons in Figure 8b) are used by the MUR MRVA technique to "fill in" the data void over Lake Oneida. Because the larger Lake Ontario warms up more slowly than Lake Oneida in the spring, extending these values to Lake Oneida results in the large cool bias of the MUR LSWT versus in situ observations of −3.88 °C and a *RMSE* of 4.71 °C during the period March through 15 July. During the fall the opposite effect is seen with the larger Lake Ontario cooling off more slowly than Lake Oneida, resulting in the warm bias of the MUR LSWT versus in situ measurements of 0.98 °C (Figure 8b).

3.6. Evaluation of MUR LSWT Interannual Climatology and Trends

The mean annual cycle in MUR LSWT (black dashed lines in Figure 9) varies substantially between Lake Michigan, Lake Oneida, and Lake Okeechobee (note that during the winter months on Lake Michigan and Lake Oneida, the MUR LSWT analyses below 0 °C represent ice surface temperature, not lake temperature). The MUR LSWT ranges from near 0 °C in February and March (as partial ice cover melts) to over 20 °C in the summer on Lake Michigan and Lake Oneida. On Lake Okeechobee, a smaller annual cycle of ~10 °C in MUR LSWT is noted due to the more southern latitude and subtropical climate.

A number of interesting characteristics of the seasonal and interannual variability of LSWT of the three lakes can be seen by analyzing the 9-year climatology (Figure 9).

In Lake Michigan, the large thermal inertia of the large lake results in year-to-year temperature variability. For example, in 2012, an anomalously warm winter with no ice cover, Lake Michigan LSWT was ~5 °C warmer from April through July than in 2014, which was a very cold winter with extensive ice formation on the Lake (Figure 9a). During the cooling phase of the lake from September through December, less interannual variations in LSWT are noted. The warming cycle of LSWT occurs from May through July (~3 months), while the cooling cycle is about 4-month duration (September–December).

On Lake Oneida, the impact of regional weather patterns is evident, with spring to summer 2012 also anomalously warm, and 2014 unusually cold (Figure 9c). Overall, Lake Oneida observes less interannual variations in LSWT compared to Lake Michigan, as the smaller thermal inertia of the lake limits the lag and memory of significant cold and warm spells and winter ice cover on the Lake state (Figure 9c). The cooling cycle during the fall in particular, observed typically small interannual variations in Lake Oneida LSWT.

At Lake Okeechobee, the late fall through early spring months (November–March) observe the greatest interannual variability in LSWT. This is largely the result of more frequent synoptic-scale weather systems penetrating south into Florida during the time period, with cooler surface air temperatures rapidly impacting the shallow lake surface temperature. During the summer months, typically small inter-annual variations in LSWT are noted.

The spatial patterns in LSWT observed by MUR also show interannual variability. For example, the LSWT spatial patterns of Lake Michigan LSWT observed on 1 July 2010, 2012, and 2014 vary as a result of atmospheric forcing and lake state (e.g., ice amount, air temperature). The northern portions of Lake Michigan are deeper and hence typically observe cooler water temperatures through the column, but the impacts of the deep reservoir of cold water are modulated by the amount of vertical mixing of this cooler deep water to the surface.

The impact of the lake bathymetry is seen in the 1 July thermal analyses on 1 July 2010 and 2016 (Figure 5a,c). Interannual variations in the extent and frequency of shallow warming over the Lake during periods of light wind speeds and heat waves, as shown in summer 2012 are also noted (Figure 5b).

Evaluation of the entire period of record of the MUR LSWT (2003–2016) illustrates that given the large inter-annual variability in regional climate, 13 years of data are likely insufficient to describe LSWT trends on these lakes with great confidence, although the climate signals will likely grow as additional years of data are added to the period of record (Figure 10). Analysis of lake spatially varying temperature trends across all three lakes in this study resulted in statistically significant trends in LSWT for each of the lakes between 2003 and 2016. On Lake Michigan, a warming trend of 0.03 °C per year was observed, which is lower than other studies, but likely impacted by the very cold winter in 2014. For Lake Okeechobee, a cooling trend of similar magnitude to the warming at Lake Michigan (−0.03 °C per year) was observed. Lake Oneida observed a weak warming trend of 0.02 °C per year was noted. The observed warming at Lake Oneida and Lake Michigan and observed cooling in the MUR LSWT analyses at Lake Okeechobee agrees broadly with the ARC-Lake climatology 1993–2011 [4].

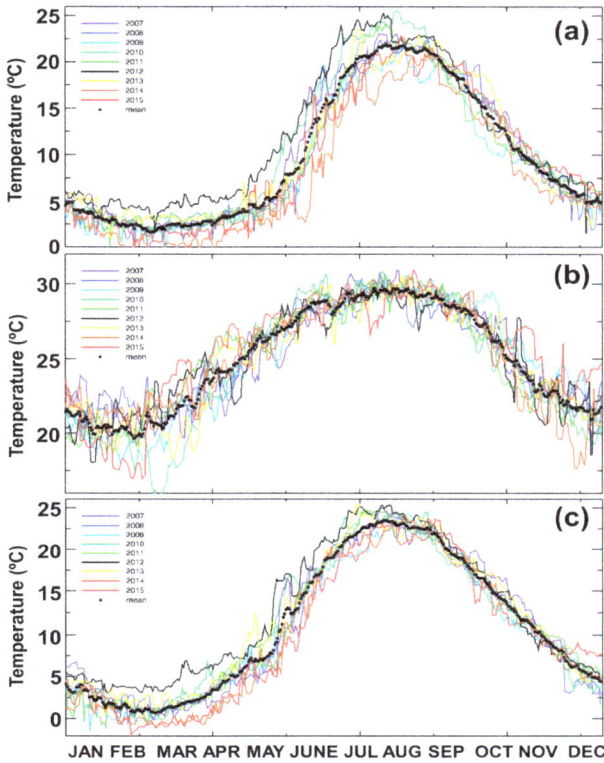

Figure 9. Time series of MUR LSWT analyses (at buoy locations, see Figure 1) for 2007–2015 for (**a**) Lake Michigan; (**b**) Lake Okeechobee; (**c**) Lake Oneida. The dashed thick black line represents the 2007–2015 MUR LSWT mean lake temperature, while individual years are represented by colored lines as indicted in the upper left legend.

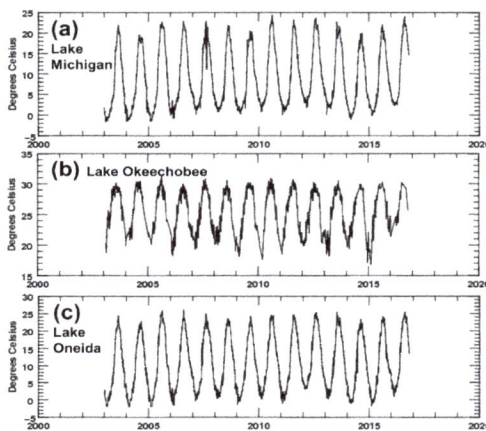

Figure 10. Time series of MUR LSWT 2003–2016 for (**a**) Lake Michigan; (**b**) Lake Okeechobee; (**c**) Lake Oneida.

4. Discussion

The MUR LSWT analyses were compared to buoy temperatures for a small, medium and large lake. The results indicate that the MUR LSWT versus in situ measurements for Lake Michigan (Lake Okeechobee) have an overall bias (MUR LSWT–in situ) of −0.20 °C (0.31 °C) and a *RMSE* of 0.86 °C (0.91 °C). The MUR LSWT versus in situ measurements for Lake Oneida versus in situ measurements have overall large biases (−1.74 °C) and *RMSE* (3.42 °C). For Lake Oneida, problems with spatial interpolation of data from adjacent lakes likely results in large errors in MUR LSWT analyses at this lake for the spring and fall seasons, with much improved analysis during the 15 July–30 September period. The use of daytime in situ validation data at Lake Okeechobee and Lake Oneida compared with nighttime MUR LSWT imagery is a noted limitation of this study, as nighttime bulk lake temperatures at the depth of the buoy observations (less than 1.0 m) are subject to diurnal warming.

The major advantages of the MUR LSWT analyses include daily consistency (analyses generated every day), high resolution (~1 km, which provides analyses over many small lakes that are not resolved in coarser analyses), near-real time production (latency of ~1 day), and multi-platform data synthesis (MODIS and AVHRR), which increases data availability over the typical single-sensor approach. The high resolution of MUR LSWT allows thousands of lakes to be captured that are not included in other currently available LSWT analyses. The synthesis of multiple years of MUR LSWT to produce a "climatological" LSWT for small lakes not resolved by other products (as well as a higher spatial resolution LSWT data set for larger lakes) is promising for future studies utilizing the MUR LSWT dataset for lakes across the world for a number of potential geophysical applications.

Improved analyses of lake temperature are needed for modeling applications [13]. The results of this study are a promising first step to integrate the MUR LSWT analyses worldwide for providing near-real time lake temperature data for input into geophysical modeling systems. The current MUR LSWT would likely be an improvement over what is currently being used for initializing LSWT in a number of modeling systems, as errors in prescribed LSWT from 3 °C to 10 °C have been noted in currently-used LSWT model analyses [12,52,53].

The results of this study have illuminated a number of error sources that lead us to recommend modifications and future improvements to MUR LSWT. These include:

- Utilize additional high-resolution satellite thermal infrared retrievals over lakes to enhance temporal coverage (e.g., Visible Infrared Imaging Radiometer Suite (VIIRS) onboard the Suomi National Polar-Orbiting Partnership spacecraft [54], as well as the (Sea and Land Surface Temperature Radiometer (SLSTR) onboard the European Space Agency Sentinel-3 fleet of satellites [55], and the GOES-16 Advanced Baseline Imager (ABI) currently onboard the most recently launched National Oceanic and Atmospheric Administration Geostationary Operational Environmental Satellite (GOES-16) [56]. As discussed by Chin et al. [43] plans are already underway to incorporate VIIRS in MUR.

- Potentially incorporate daytime imagery over lakes to increase the amount of available data. MUR already has plans to investigate utilizing daytime imagery when wind speeds are high enough to minimize diurnal warming SST effects [19,20].

- Incorporate lake-specific cloud masking and other quality-control procedures to reduce both contamination and removal of clear-sky imagery.

- Improve ancillary ice cover analyses to reduce spurious unphysical ice coverage impacting LSWT analyses, and develop higher-resolution ice cover analyses to improve coverage for small lakes.

- Reduce the spatial footprint of the MRVA technique to preclude non-representative analysis values from adjacent lakes or oceans to be spread to another lake. Plans to flag input footprints as well as to reduce quality threshold barriers to include microwave LSWT samples are both underway for next version of MUR.

- Determine if a shorter or longer analysis window than the 5 days currently used in MUR would improve the LSWT analyses. Over some lakes where cloud cover is low, a shorter window could allow for a more responsive analysis; over cloudy lakes, a longer window may improve analysis coverage.
- Allow climatological LSWT values derived from the MUR period of record (such as shown in Figure 8) or from external data sets such as ARC-Lake (for lakes >5 km in diameter) to become the default field for MUR LSWT during prolonged cloudy gap periods in lakes with no available in situ observations.

5. Conclusions

This study evaluated MUR LSWT analyses, which are available globally at 1 km resolution, for a small, medium and large lake. The results indicate that the daily near real-time MUR LSWT analyses on annual and seasonal time scales have biases below ±0.50 °C and *RMSE* below 1.11 °C for Lake Michigan and Lake Okeechobee, with the exception of 2014 on Lake Michigan (*RMSE* 1.62 °C) where diurnal thermoclines impacted the representativeness of the buoy measurements. Over small lakes where MODIS is the only current source of data, large errors in the MUR LSWT analyses were noted during periods when cloud cover limits data coverage. However, at this time the MUR LSWT is the only high-resolution near real-time global daily analysis available that resolves thousands of lakes with diameters less than 10 km. We conclude that overall, the MUR LSWT analyses show promise for providing real-time analyses of LSWT for lakes larger than a few km in diameter, but a number of improvements and modifications to MUR are recommended.

Acknowledgments: This research and preparation of this paper was partially supported by Eric Lindstrom, NASA, Multi-sensor Improved Sea Surface Temperature (MISST) for (IOOS), NASA Contract: NNH13CH09C, and also partially funded under a contract with the National Aeronautics and Space Administration (NASA) through the Making Earth System Data Records for Use in Research Environments (MEaSUREs) program. We thank Chelle Gentemann for helping to make this research possible. We thank Lars Rudstam and the Cornell Biological Field Station for providing the temperature data for Lake Oneida, and the National Data Buoy Center for hourly temperature data from Lake Michigan. The South Florida Water Management District provided the water temperature data from Lake Okeechobee. We appreciate the expertise of Robert Grumbine on operational user needs for lake temperature analyses. Nate Larson, Will Howard, Alex Jacques also contributed to this research. Thanks also goes to John Horel for his guidance and advice. We gratefully acknowledge the input of three anonymous reviewers, which significantly improved the paper.

Author Contributions: Erik Crosman, Jorge Vazquez-Cuervo, and Toshio Michael Chin conceived and designed the experiments. Jorge Vazquez-Cuervo and Erik Crosman conducted the validation experiments. Erik Crosman, Jorge Vazquez-Cuervo and Toshio Michael Chin analyzed the data. Erik Crosman wrote the paper.

Conflicts of Interest: The authors declare no conflict of interest.

References

1. Adrian, R.A.; O'Reilly, C.M.; Zagarese, H.; Baines, S.B.; Hessen, D.O.; Wendel, K.; Livingstone, D.M.; Sommaruga, R.; Dietmar, S.; Van Donk, E.; et al. Lakes as sentinels of climate change. *Limnol. Oceanogr.* **2009**, *54*, 2283–2297. [CrossRef] [PubMed]
2. MacKay, M.D.; Neale, P.J.; Arp, C.D.; De Senerpont Domis, L.N.; Fang, X.; Gal, G.; Jöhnk, K.D.; Kirillin, G.; Lenters, J.D.; Litchman, E.; et al. Modeling lakes and reservoirs in the climate system. *Limnol. Oceanogr.* **2009**, *54*, 2315–2329. [CrossRef]
3. Bresciani, M.; Giardino, C.; Boschetti, L. Multi-temporal assessment of bio-physical parameters in lakes Garda and Trasimeno from MODIS and MERIS. European. *Ital. J. Remote Sens.* **2011**, *43*, 49–62.
4. Hook, S.J.; Wilson, R.C.; MacCallum, S.; Merchant, C.J. Lake Surface Temperature [in "State Absolute of the Climate in 2011"]. Available online: ftp://ftp.ncdc.noaa.gov/pub/data/cmb/bams-sotc/climate-assessment-2011.pdf (accessed on 10 July 2017).
5. Lenters, J.D.; Hook, S.J.; McIntyre, P.B. Workshop examines warming of lakes worldwide. *Eos Trans. Am. Geophys. Union* **2012**, *93*, 427. [CrossRef]
6. Lenters, J. The Global Lake Temperature Collaboration (GLTC). *LakeLine* **2015**, *35*, 9–12.

7. Woolway, R.I.; Cinque, K.; de Eyto, E.; DeGasperi, C.; Dokulil, M.; Korhonen, J.; Maberly, S.; Marszelewski, W.; May, L.; Merchant, C.J.; et al. Lake surface temperature [in "State of the climate in 2015"]. *Bull. Am. Meteorol. Soc.* **2016**, *97*, S17–S18.

8. Dutra, E.; Stepanenko, V.M.; Balsamo, G.; Viterbo, P.; Miranda, P.M.A.; Mironov, D.; Schär, C. An offline study of the impact of lakes on the performance of the ECMWF surface scheme. *Boreal Environ. Res.* **2010**, *15*, 100–112.

9. Balsamo, G.; Salgado, R.; Dutra, E.; Boussetta, S.; Stockdale, T.; Potes, M. On the contribution of lakes in predicting near-surface temperature in a global weather forecasting model. *Tellus A* **2012**, *64*, 15829. [CrossRef]

10. Javaheri, A.; Babbar-Sebens, M.; Miller, R.N. From skin to bulk: An adjustment technique for assimilation of satellite-derived temperature observations in numerical models of small inland water bodies. *Adv. Water Resour.* **2016**, *92*, 284–298. [CrossRef]

11. Hulley, G.C.; Hook, S.J.; Schneider, P. Optimized split-window coefficients for deriving surface temperatures from inland water bodies. *Remote Sens. Environ.* **2011**, *115*, 3758–3769. [CrossRef]

12. Grim, J.A.; Knievel, J.C.; Crosman, E.T. Techniques for using MODIS data to remotely sense lake water surface temperatures. *J. Atmos. Ocean. Technol.* **2013**, *30*, 2434–2451. [CrossRef]

13. Fiedler, E.; Martin, M.; Roberts-Jones, J. An operational analysis of lake surface water temperature. *Tellus A* **2014**, *66*, 21247. [CrossRef]

14. Crosman, E.T.; Horel, J.D. MODIS-derived surface temperature of the Great Salt Lake. *Remote Sens. Environ.* **2009**, *113*, 73–81. [CrossRef]

15. MacCallum, S.N.; Merchant, C.J. Surface Water Temperature Observations of large lakes by optimal estimation. *Can. J. Remote Sens.* **2012**, *38*, 25–45. [CrossRef]

16. Hook, S.J.; Vaughnan, R.G.; Tonooka, H.; Schladow, S.G. Absolute radiometric in-flight validation of mid infrared and thermal infrared data from ASTER and MODIS on the Terra Spacecraft using the Lake Tahoe, CA/NV, USA, automated validation site. *IEEE Trans. Geosci. Remote Sens.* **2007**, *45*, 1798–1807. [CrossRef]

17. Layden, A.; Merchant, C.; MacCallum, S. Global climatology of surface water temperatures of large lakes by remote sensing. *Int. J. Climatol.* **2015**, *35*, 4464–4479. [CrossRef]

18. Riffler, M.; Lieberherr, G.; Wunderle, S. Lake surface water temperatures of European Alpine lakes (1989–2013) based on the Advanced Very High Resolution Radiometer (AVHRR) 1 km data set. *Earth Syst. Sci. Data* **2015**, *7*, 1–17. [CrossRef]

19. Wilson, R.C.; Hook, S.J.; Schneider, P.; Schladow, S.G. Skin and bulk temperature difference at Lake Tahoe: A case study on lake skin effect. *J. Geophys. Res. Atmos.* **2012**, *118*, 10332–10346. [CrossRef]

20. Donlon, C.; Rayner, N.; Robinson, N.; Poulter, D.J.; Casey, K.S.; Vazquez-Cuervo, J.; Armstrong, E.; Bingham, A.; Arino, O.; Gentemann, C.; et al. The Global Ocean Data Assimilation Experiment High0resolution Sea Surface Temperature Pilot Project. *Bull. Am. Meteorol. Soc.* **2007**, *88*, 1197–1213. [CrossRef]

21. Oesch, D.C.; Jaquet, J.M.; Klaus, R.; Schenker, P. Multi-scale thermal pattern monitoring of a large lake (Lake Geneva) using a multi-sensor approach. *Int. J. Remote Sens.* **2008**, *29*, 5785–5808.

22. O'Reilly, C.M.; Sharma, S.; Gray, D.K.; Hampton, S.E.; Read, J.S.; Rowley, R.J.; Schneider, P.; Lenters, J.D.; McIntyre, P.B.; Kraemer, B.M.; et al. Rapid and highly variable warming of lake surface waters around the globe. *Geophys. Res. Lett.* **2015**, *42*, 10773–10781. [CrossRef]

23. Torbick, N.; Ziniti, B.; Wu, S.; Linder, E. Spatiotemporal lake skin summer temperature trends in the Northeastern United States. *Earth Interact.* **2016**, *20*, 1–21. [CrossRef]

24. Bulgin, C.E.; Embury, O.; Merchant, C.J. Sampling uncertainty in gridded sea surface temperature products and Advanced Very High Resolution Radiometer (AVHRR) Global Area Coverage (GAC) data. *Remote Sens. Environ.* **2016**, *117*, 287–294. [CrossRef]

25. Castro, S.L.; Wick, G.A.; Steele, M. Validation of satellite sea surface temperature analyses in the Beaufort Sea using UpTempO buoys. *Remote Sens. Environ.* **2016**, *187*, 458–475. [CrossRef]

26. Liu, Y.; Minnett, P.J. Sampling errors in satellite-derived infrared sea-surface temperatures. Part I: Global and regional MODIS fields. *Remote Sens. Environ.* **2016**, *177*, 48–64. [CrossRef]

27. Hao, Y.; Cui, T.; Singh, V.P.; Zhang, J.; Yu, R.; Zhang, Z. Validation of MODIS Sea Surface Temperature Product in the Coastal Waters of the Yellow Sea. *IEEE J. Sel. Top. Appl. Earth Obs. Remote Sens.* **2017**, *10*, 1667–1680. [CrossRef]

28. Merchant, C.J.; Harris, A.R.; Maturi, E.; MacCallum, S. Probabilistic physically-based cloud screening of satellite infra-red imagery for operational sea surface temperature retrieval. *Q. J. R. Meteorol. Soc.* **2005**, *131*, 2735–2755. [CrossRef]

29. Hulley, G.C. *MODIS Cloud Detection over Large Inland Water Bodies: Algorithm Theoretical Basis Document*; Jet Propulsion Laboratory-California Institute of Technology: Pasadena, CA, USA, 2009.

30. MacCallum, S.N.; Merchant, C.J. ARC-Lake Algorithm Theoretical Basis Document–ARC-Lake. v1.1, 1995–2009 [Dataset]. The University of Edinburgh, School of GeoSciences/European Space Agency. Available online: http://hdl.handle.net/10283/88 (accessed on 10 July 2017).

31. Fan, X.; Tang, B.; Wu, H.; Yan, G.; Li, Z. Daytime land surface temperature extraction from MODIS thermal infrared data under cirrus clouds. *Sensors* **2015**, *15*, 9942–9961. [CrossRef] [PubMed]

32. Castendyk, D.N.; Obryk, M.K.; Leidman, S.Z.; Gooseff, M.; Hawes, I. Lake Vanda: A sentinel for climate change in the McMurdo Sound Region of Antarctica. *Glob. Planet. Chang.* **2016**, *144*, 213–227. [CrossRef]

33. Schneider, P.; Hook, S.J. Space observations of inland water bodies show rapid surface warming since 1985. *Geophys. Res. Lett.* **2010**, *37*. [CrossRef]

34. Politi, E.; Cutler, M.J.; Rowan, J.S. Using the NOAA Advanced Very High Resolution Radiometer to Characterize temporal and spatial trends in water temperature of large European lakes. *Remote Sens. Environ.* **2012**, *126*, 1–11. [CrossRef]

35. Politi, E.; MacCallum, S.; Cutler, M.E.J.; Merchant, C.J.; Rowan, J.S.; Dawson, T.P. Selection of a network of large lakes and reservoirs suitable for global environmental change analysis using Earth Observation. *Int. J. Remote Sens.* **2016**, *37*, 3042–3060. [CrossRef]

36. Mason, L.A.; Riseng, C.M.; Gronewold, A.D.; Rutherford, E.S.; Wang, J.; Clites, A.; Smith, S.D.; McIntyre, P.B. Fine-scale spatial variation in ice cover and surface temperature trends across the surface of the Laurentian Great Lakes. *Clim. Chang.* **2016**, *138*, 71–83. [CrossRef]

37. Moukomla, S.; Blanken, P.D. Remote Sensing of the North American Laurentian Great Lakes' Surface Temperature. *Remote Sens.* **2016**, *8*, 286. [CrossRef]

38. Zhao, L.; Jin, J.; Wang, S.Y.; Ek, M.B. Integration of remote-sensing data with WRF to improve lake-effect precipitation simulations over the Great Lakes region. *J. Geophys. Res.* **2012**, *117*, D09102. [CrossRef]

39. Kourzeneva, E. Assimilation of lake water surface temperature observations using an extended Kalman filter. *Tellus A.* **2014**, *66*, 21510. [CrossRef]

40. Strong, C.; Kochanski, A.K.; Crosman, E.T. A slab model of the Great Salt Lake for regional climate simulation. *J. Adv. Model. Earth Syst.* **2014**, *6*, 602–615. [CrossRef]

41. Chao, Y.; Li, Z.; Farrara, J.; Hung, P. Blending sea surface temperatures for multiple satellites and in situ observations for coastal oceans. *J. Atmos. Ocean. Technol.* **2009**, *26*, 1415–1426. [CrossRef]

42. Nardelli, B.B.; Tronconi, C.; Pisano, A.; Santoleri, R. High and ultra-high resolution processing of satellite sea surface temperature data over Southern European Seas in the framework of MyOcean project. *Remote Sens. Environ.* **2013**, *129*, 1–16. [CrossRef]

43. Chin, T.M.; Vazquez, J.; Armstrong, E.M. A multi-scale high-resolution analysis of global sea surface temperature. *Remote Sens. Environ.* **2017**. in review.

44. Pareeth, S.; Salmaso, N.; Adrian, R.; Neteler, M. Homogenised daily lake surface water temperature data generated from multiple satellite sensors: A long-term case study of a large sub-Alpine lake. *Sci. Rep.* **2016**, *6*, 31251. [CrossRef] [PubMed]

45. Thiebaux, J.; Rogers, E.; Wang, W.; Katz, B. A new high resolution blended real-time global sea surface temperature analysis. *Bull. Am. Meteorol. Soc.* **2004**, *84*, 645–656. [CrossRef]

46. Schwab, D.J.; Leshkevich, G.A.; Muhr, G.C. Automated mapping of surface water temperature in the Great Lakes. *J. Great Lakes Res.* **1999**, *25*, 468–481. [CrossRef]

47. Xu, F.; Ignatov, A. Error characterization in iQuam SSTs using triple collocations with satellite measurements. *Geophys. Res. Lett.* **2016**, *43*, 10826–10834. [CrossRef]

48. National Data Buoy Center (NDBC), NDBC Technical Document 09-02: Handbook of Automated Data Quality Control Checks and Procedures. 2009. Available online: http://www.ndbc.noaa.gov/NDBCHandbookofAutomatedDataQualityControl2009.pdf (accessed on 27 June 2017).

49. Sharma, S.; Gray, D.K.; Read, J.S.; O'Reilly, C.M.; Schneider, P.; Qudrat, A.; Gries, C.; Stefanoff, S.; Hampton, S.E.; Hook, S.; et al. A global database of lake surface temperatures collected by in situ and satellite methods from 1985–2009. *Sci. Data* **2015**, *2*, 150008. [CrossRef] [PubMed]

50. Rudstam, L.G. Limnological Data and Depth Profile from Oneida Lake, New York, 1975-Present. Web Data on Knowledge Network for Biocomplexity, 2015. Available online: http://knb.ecoinformatics.org/#view/kgordon.35.70 (accessed on 26 June 2017).

51. Lavergne, T.; Tonboe, R.; Lavelle, J.; Eastwood, S. Algorithm Theoretical Basis Document for the OSI SAF Global Sea Ice Concentration Climate Data Record. 2016. Available online: https://www.researchgate.net/profile/Thomas_Lavergne3/publication/306365213_Algorithm_Theoretical_Basis_Document_ATBD_for_the_OSI_SAF_Global_Sea_Ice_Concentration_Climate_Data_Record_v11/links/57baff8108ae202e6a579100.pdf (accessed on 20 June 2017).

52. Blaylock, B.K.; Horel, J.D.; Crosman, E.T. Impact of a lake breeze on summer ozone concentrations in the Salt Lake Valley. *J. Appl. Meteorol. Climatol.* **2017**, *56*, 353–370. [CrossRef]

53. Spero, T.L.; Nolte, C.G.; Bowden, J.H.; Mallard, M.S.; Herwehe, J.A. The impact of incongruous lake temperature on regional climate extremes downscales from the CMIP5 archive using the WRF model. *J. Clim.* **2016**, *29*, 839–853. [CrossRef]

54. Hillger, D.; Kopp, T.; Lee, T.; Lindsey, D.; Seaman, C.; Miller, S.; Solbrig, J.; Kidder, S.; Bachmeier, S.; Jasmin, T.; Rink, T. First-light imagery from Suomi NPP VIIRS. *Bull. Am. Meteorol. Soc.* **2013**, *94*, 1019–1029. [CrossRef]

55. Donlon, C.; Berruti, B.; Buongiorno, A.; Ferreira, M.H.; Féménias, P.; Frerick, J.; Goryl, P.; Klein, U.; Laur, H.; Mavrocordatos, C.; et al. The global monitoring for environment and security (GMES) sentinel-3 mission. *Remote Sens. Environ.* **2012**, *120*, 37–57. [CrossRef]

56. Schmit, T.J.; Griffith, P.; Gunshor, M.M.; Daniels, J.M.; Goodman, S.J.; Lebair, W.J. A Closer Look at the ABI on the GOES-R Series. *Bull. Am. Meteorol. Soc.* **2017**, *98*, 681–698. [CrossRef]

MDPI

St. Alban-Anlage 66

4052 Basel

Switzerland

Tel. +41 61 683 77 34

Fax +41 61 302 89 18

www.mdpi.com

Remote Sensing Editorial Office

E-mail: remotesensing@mdpi.com

www.mdpi.com/journal/remotesensing

www.ingramcontent.com/pod-product-compliance
Lightning Source LLC
Chambersburg PA
CBHW051712210326
41597CB00032B/5458